HUMAN FACTORS IMPACTS IN
AIR TRAFFIC MANAGEMENT

Human Factors Impacts in
Air Traffic Management

Edited by
BARRY KIRWAN, MARK RODGERS AND DIRK SCHÄFER

Routledge
Taylor & Francis Group

LONDON AND NEW YORK

First published 2005 by Ashgate Publishing

2 Park Square, Milton Park, Abingdon, Oxon OX14 4RN
711 Third Avenue, New York, NY 10017, USA

Routledge is an imprint of the Taylor & Francis Group, an informa business

First issued in paperback 2016

British Library Cataloguing in Publication Data
Human factors impacts in air traffic management
    1.Air traffic control 2.Aeronautics - Human factors 3.Air
    traffic controllers - Training of
    I.Kirwan, B. (Barry) , 1960- II.Rodgers, Mark III.Schäfer,
    Dirk
    387.7'40426

Library of Congress Control Number: 2005925518

ISBN 978-0-7546-3502-4 (hbk)
ISBN 978-1-138-26431-1 (pbk)

Contents

PART VI: DISCUSSION

Preface

About this Book

Human Performance and Human Factors have long been considered to be important to Air Traffic Systems performance. This has led to many air traffic service provision companies and organizations either taking on Human Factors professionals or at least taking active interest in the field and acting on generic research and good practices made available in the public domain. Organisations such as the US Federal Aviation Administration (FAA) and EUROCONTROL in Europe have invested heavily in this area, and have qualified staff and programs to continue to develop insight, guidance and solutions related to Human Factors issues in contemporary and future air traffic management (ATM) activities and organizations.

However, in the late 1990s, at several conferences in Europe, a serious question was raised by members of the air traffic community. Essentially it was a challenge, and an important one. Whilst it was not disputed that Human Factors was important and even critical, the ability of Human Factors to deliver realistic solutions that would translate into adopted operational practice was questioned. Where were the published case studies showing Human Factors impacts? Where was the evidence that Human Factors could do more than carry out applied research and write academic papers? Where were the results that would mean something both to controllers at the operational level, and to managers and chief executives in air navigation service provision organizations? These questions made certain amongst the ATM Human Factors community realize two things. Firstly there was a potential 'image problem' for ATM Human Factors. Secondly, and more importantly, it was indeed difficult to find this evidence in the published literature, in readily accessible locations. Although there are many books and articles on Human Factors aspects in ATM, many of these are either written with a peer (Human Factors) audience in mind, or else are dispersed in diverse locations. Often Human Factors case studies do not deal with the ultimate result of how or even whether a solution was adopted and had sustained impact in an operational system.

Therefore, after some discussions, the editors of this book, and more importantly the authors and their respective supporting organizations, decided to rise to this challenge and try to produce a book which would bring together real cases where Human Factors had real impact. A number of people were contacted, and within a relatively short time-scale, a little less than twenty volunteers agreed to provide chapters with an emphasis on operational impacts. This was in fact the rather hard criterion for including chapters in this book – the work had to have had impact in real terms. This excluded work in progress, no matter how important or how promising. Three exceptions were made in the area of methodology as it was felt desirable to show some of the leading edge work in methodology which could yield better impacts of Human Factors work in the future. The result is a set of chapters focusing on four main areas:

- direct operational impacts (e.g. changes to position hand-over protocols),
- impacts on Human Resources functions that directly support operational performance (e.g. improvements in training success rates),
- key methodological 'enablers' for current and future HF work, and
- approaches to running large integrated HF programs

Added to this central core structure, are some introductory points setting the historical scene and contemporary 'drivers' or pressures on Human Factors, and a discussion section at the end of the book considering primarily where Human Factors appears to be working and where further development and/or implementation is required.

In total therefore, this book has aimed to rise to the challenge leveled at ATM Human Factors, so that if anyone asks 'Does ATM HF actually deliver?', then at the least people can cite these case studies and, given the range of issues addressed in this book, answer with an unqualified 'Yes'.

A Guide for the Reader

This book has been written with four audiences in mind. Firstly there is the controller, who needs Human Factors support to be able to carry out his or her critical work to the high standards and levels of performance that have become associated with ATM. Secondly, it is aimed at the managers of ATM centers, so that they are aware of the various Human Factors issues and can see examples of how they were addressed, including where possible an indication of the effort required to develop such solutions. The third type of reader this book is aimed at is of course the Human Factors professional, who may be working in the field of ATM or aviation, and may be tackling similar or related problems to those embedded in the case studies in this book. Some of the case studies are quite candid – showing what did not work as well as what did work. In particular this is true for the section on 'Human Factors Programs', which should be of interest to anyone leading a Human Factors team in an organization. The last type of reader the book is aimed at is those working in related fields (e.g. interface design, procedures development, training, etc.) who wish to understand more about Human Factors, and who want to see what it looks like in practice. Human Factors people rarely work on their own and are continually interfacing with controllers, management, engineers and designers and other disciplines. This is the nature of Human Factors as a 'Systems' discipline, and this also means that usually to achieve any real impact, many others as well as Human Factors people have worked on the project.

The structure of the book has already been alluded to above, but essentially the book is in six sections:

1. Introduction
2. Human Factors in Operations
3. Human Factors and Human Resources
4. Human Factors Methodologies
5. Human Factors Integration Programs
6. Discussion and Conclusions.

However, because many readers may not want to read the whole book, the table on the following pages aims to highlight where to find particular information. For example, if the reader is primarily interested in training, then a good start is chapter eight by Voller and Fowler. If the reader wants to read something about airports in particular, then chapter three by Cardosi is a good start. Therefore Table 1 acts as a 'roadmap' so the reader can quickly find the information of most relevance to his or her needs. To help the reader find relevant case studies, the following information has been inserted into the table:

- why the study was carried out (the 'driver'),
- the life cycle stage of the study (e.g. design, operations),
- the ATM 'function' that was the focus of the study (e.g. airport, military, en route, etc.),
- the major Human Factors approach(es) used (e.g. workload measurement; interviews, human reliability assessment, etc.),
- the type of impact achieved.

For the reader who wishes a more general read and exploration of the book, the introduction chapter gives a brief historical outline of HF in ATM, and discusses the 'drivers' or context for HF at the present time, since such drivers or concerns in ATM-related HF act as positive pressures to achieve results in particular areas, and it is in such high profile areas that HF's utility will inevitably be judged. This section also raises some issues for HF in ATM which, after the four main sections of the book, are returned to in order to evaluate the 'performance' of Human Factors itself, to see if it is delivering what is needed, and how to improve its own performance.

Table 1   Quick-Look Table for Case Study Chapters

Chapter	ATM Function	Life Cycle Stage	Human Factors Function	Impact Area
2	TMA	Operations	Data collection and analysis	Procedures
3	Airports	Operations	Incident analysis and HF analysis	Airport markings
4	all	Operations	Incident analysis	Incident Analysis Method
5	all	Operations	Incident analysis	Incident learning system
6	Military OFIR	Operations	Human Reliability	Procedures and Separation Minima
7	Traffic Flow Management	Operations/ Design	Performance Analysis	Collaborative Decision Making
8	Tower, Approach En route	Training	Training Analysis	Training process and media
9	all	Operations	Selection	Effective controller workforce
10	all	Operations	Critical Incident Stress Management	CISM system
11	all	Operations	Team Resource Management	Training process and media
12	ACCs	Operations	Shiftwork	Shiftwork practices
13	all	Design	Human Performance Measures	Human performance evaluation
14	all	Design	Simulation	Airspace design
15	Approach En Route Military	Design	Human Error Prediction	HMI; safety assurance

Table 1 Quick-Look Table for Case Study Chapters (ctd.)

Chapter	ATM Function	Life Cycle Stage	Human Factors Function	Impact Area
16	all	Design through Acquisition	Technology Readiness Levels/ Integration	Processes for Integration of HF
17	TRACON and Towers	Design	Usability and prototyping	Interface Design
18	all	Design	Safety climate; HMI design	HMI; culture
19	all	All	Integration	Training; HMI; Procedures
20	all	All	Integration	Processes for Integration of HF

About the Editors

Barry Kirwan (PhD) was head of human factors in NATS and led a large program of work from 1996 - 2000 which did achieve impact in the ATM industry, and he and his team won the Guild of Air Traffic Controllers Award for contributions to ATM in 1998. He was also formerly leading human factors initiatives in BNFL in the nuclear industry for five years, and lectured at University to Masters level Ergonomics students for a further five years. He has editorial experience via the successful 'Guide to Task Analysis' which also included a set of 10 detailed studies showing the value of task analysis, and also via a book on Safety Regulation. He has also written his own book on Human Reliability Assessment, and has published more than a hundred and fifty papers, two of which won best paper awards in two separate journals. He is currently working for EUROCONTROL in Brétigny, South of Paris in the field of Safety and Human Factors.

Mark D. Rodgers is currently the Program Director for Human Factors Research and Engineering at the Federal Aviation Administration. Over the past 12 years Dr. Rodgers has held several positions at the agency, most recently Chief Scientific and Technical Advisor for Human Factors in the Office of Aviation Research. Previously, he held positions as an Engineering Research Psychologist at the FAAs' Civil Aerospace Medical Institute (CAMI) and Office of System Architecture. While at CAMI Dr. Rodgers was responsible for directing the development and initiating the national deployment of the Systematic Air Traffic Operations Research Initiative (SATORI) for use as an air traffic training and incident investigation tool. Dr. Rodgers has received numerous awards and honors such as the Air Traffic Control

Association Special Medallion Award, Aerospace Medical Association R&D Innovation Award, FAA Technology Transfer Award, and the FAA Vision of Tomorrow in Safety Award, to name a few. He is the author of over 50 technical reports, journal articles, book chapters, and presentations in the area of human performance and human factors. Dr. Rodgers received his Ph.D. in Experimental Psychology from the University of Louisville in 1991.

Dr. Dirk Schäfer is presently working in the area of Human Factors at the EUROCONTROL Experimental Centre in Brétigny. He is involved in HF coordination and training and has recently led efforts to install a Human Factors Laboratory. He also advises a number of research projects in the field of Human Factors. Previously he held a research position at the German Aerospace Center where he led the ATM simulation group. Dirk Schäfer is the author of numerous publications, predominantly in the field of ATM and Aviation Human Factors, air traffic complexity, free flights, and speech recognition. He received his doctoral degree in Aerospace Human Factors Engineering from the German Armed Forces University in Munich.

List of Abbreviations

ACC	Area Control Center
ABSR	Abstract Reasoning Test
ADS-B	Automatic Dependent Surveillance Broadcast
AERA	Automated En-route Air Traffic Control
AF	Airway Facilities
Air MIDAS	Man-machine Integrated Design and Analysis System
ANS	Air Navigation Service
ANSP	Air Navigation Service Provider
AOC	Airline Operations Center
API	Aircraft Proximity Index
ARS	(FAA) AT Requirements Office
ARSR	Air Route Surveillance Radar
ARTCC	Air Route Traffic Control Center
ARTS	Automated Radar Terminal System
ASAS	Airborne Separation Assistance System
ASDE	Airport Surface Detection Equipment
ASMGCS	Advanced Surface Movement Ground Control System
ASR	Airport Surveillance Radar
ASRS	Aviation Safety Reporting System
ATA	(FAA) AT Acquisitions Office
ATC	Air Traffic Control
ATCC	Air Traffic Control Centre
ATCO	Air Traffic Control Officer
ATCS	Air Traffic Control Specialist
ATCSCC	Air Traffic Control Systems Command Center
ATL	Atlanta (GA) Airport
ATM	Air Traffic Management
ATMDC	Air Traffic Management Development Centre
ATP	(FAA) AT Procedures Office
ATWIT	Air Traffic Workload Input Technique
BOS	Boston Logan airport
CAA	Civil Aviation Authority
CAMI	(FAA) Civil Aerospace Medical Institute
CATC	College of Air Traffic Control
CATS	Crew Activity Tracking System
CBPM	Computer Based Performance Measure
CDM	Collaborative Decision-Making
CDTI	Cockpit Display of Traffic Information
CHI	Computer Human Interface

CISD	Critical Incident Stress Debriefing
CISM	Critical Incident Stress Management
CJI	Controller-Jurisdiction Indicators
CMB	Crisis Management Briefings
COCOM	Contextual Control Model
CORA	Conflict Resolution Assistant
COTS	Commercial Off-The-Shelf
CPDLC	Controller-Pilot Data Link Communication
CPM	Cognitive Perceptual Motor
CPM-GOMS	Cognitive Perceptual Motor – Goals, Operators, Methods and Selection Rules
CRJ	Canadair Regional Jets
CRM	Crew or Cockpit Resource Management
CTAS	Center TRACON Automation System
DCB	Display Control Bar
DCPN	Dynamically Colored Petri-Net
DFS	Deutsche Flugsicherung GmbH (German Air Navigation Services)
DFW	Dallas Fort Worth (TX) Airport
DNV	Det Norske Veritas
DoD	Department of Defense
DOT	Department of Transportation
DSR	Display System Replacement
EATMP	European Air Traffic Management Programme
ECAC	European Civil Aviation Conference
EPA	Education and Public Affairs Inc.
ESL	Emergency Service Level
EWR	Newark (NJ) Airport
FAA	Federal Aviation Administration
FMS	Flight Management System
FPL	Full Performance Level
FPS	Flight Progress Strips
FSL	Full Service Level
GEFA	Göran Ekvall FA Council
GEMS	Generic Error Modeling System
GOMS	Goals, Operators, Methods, and Selection Rules
GPWS	Ground Proximity Warning System
HAZOP	Hazard and Operability Study
HEA	Human Error Analysis
HEIST	Human Error Identification in Systems Technique
HEP	Human Error Probability
HERA	Human Error in ATM
HF	Human Factors
HFACS	Human Factors Analysis Classification System

HFE	Human Factors Engineering
HFU	Human Factors Unit
HLA	High-Level Architecture
HMI	Human-Machine Interface or Human-Machine Interaction
HRA	Human Reliability Assessments
HTA	Hierarchical Task Analysis
HUFA	(LFV) Human Factors in Air Navigation Service Project
IAIPT	(NASA) Inter-agency ATM Integrated Product Team
IANS	EUROCONTROL Institute of Air Navigation Services
ICAO	International Civil Aviation Organization
ICISF	International Critical Incident Stress Foundation
IDA	Institute for Defense Analysis
IFF	Identification Friend or Foe
IFR	Instrument Flight Rules
IMAS	Influence Modeling and Assessment System
IPT	Integrated Product Team
JCS	Joint Human-Machine Cognitive System
JFK	John FG Kennedy (NY) Airport
KSA	Knowledge, Skills, and Attitudes
LATCC	London Area and Terminal Control Centre
LCD	Liquid Crystal Display
LFV	Luftfartsverket (Sweden)
LGA	Laguardia (NY) Airport
LOA	Letters of Agreement
LSM	Lateral Separation Minimum
LTCC	London Terminal Control Centre
MCAT	Multiplex Controller Aptitude Test
MCW	Monitor and Control Workstation
MCW	Monitor and Control Workstation
MHP	Mental Health Professional
MIDAS	Man-Machine Design and Analysis System
MKE	Milwaukee (WI) Airport
MSAW	Minimum Safe Altitude Warning System
MSP	Minneapolis St. Paul (MN) Airport
MTCD	Medium Term Conflict Detection
NAS	National Airspace System
NASA	National Aeronautics and Space Administration
NATCA	National Air Traffic Controllers Association
NATS	National Air Traffic Services
NERC	New En-Route Centre
NEXCOM	Next Generation Air-Ground Communication System
NM	Nautical Miles
NTS	Non-Technical Skill
NTSB	National Transportation Safety Board
NUAC	Nordic Upper Area Control

OASIS	Operational and Supportability Implementation System
ODID	Operational Display and Input Development
OE	Operational Error
OFIR	Open Flight Information Region
OJT	On-the-Job Training
OJTI	On-the-Job Training Instructor/Instruction
OKT	Occupational Knowledge Test
OPM	Office of Personnel Management
ORD	Chicago O'Hare (IL) Airport
PARR	Problem Analysis, Resolution, and Ranking System
PASS	Professional Airway System Specialists
PHEA	Predictive Human Error Analysis
PIF	Performance Influencing Factors
POET	Post-Operations Evaluation Tool
POWER	Performance and Objective Workload Evaluation Research
PRA	Probability Risk Assessment
PRAWNS	Pressure, Runways, Airports/Adjacent sector, Weather, Non-standard information, Strips
PSF	Performance Shaping Factor
PTSD	Post-Traumatic Stress Disorder
PVD	Plan View Display
R/T	Radio/Telephony
RAS	Radar Advisory Service
RDHFL	(FAA) Research Development and Human Factors Laboratory
RGAT	(CAA) Review Group of ATC Training
RIS	Radar Information Service
RVSM	Reduced Vertical Separation Minimum
RWSL	Runway Status Lights
SA	Situation Awareness
SAN	System After Next
SATORI	Systematic Air Traffic Operations Research Initiative
SCC	System Control Centre
SEM	System Effectiveness Measures
SHERPA	Systematic Human Error Reduction and Prediction Approach
SME	Subject Matter Expert
SMS	Safety Management Systems
SOP	Standard Operating Procedure
SSS	Subjective Sleepiness Scale
STARS	Standard Terminal Automation Replacement System
STCA	Short Term Conflict Alert
SUA	Special Use Airspace
SUPCOM	(FAA) Supervisor Committee
TBM	Time-Based-Metering
TBS	Time-Based Separation

TCAS	Traffic Alert and Collision Avoidance System
TCW	Terminal Controller Workstation
TDW	Tower Display Workstation
TFM	Traffic Flow Management
THERP	Technique for Human Error Rate Prediction
TIPH	Taxi into Position and Hold
TLX	NASA Task Load Index
TMA	Terminal Maneuvering Area
TOAST	TRM-Oriented ATC Simulator Training
TOPAZ	Traffic Organization and Perturbation AnalyZer
TRACER	Technique for the Retrospective Analysis of Cognitive Errors in ATM
TRACON	Terminal Radar Approach Control
TRL	Technology Readiness Level
TRM	Team Resource Management
URET	User Requirement Evaluation Tool
VDL	VHF Digital Link
VFR	Visual Flight Rules
VSM	Vertical Separation Minimum
ZAU	Chicago Air Route Traffic Control Center (ARTCC)
ZID	Indianapolis Center

PART I
INTRODUCTION

Chapter 1

Introduction

Applying Human Factors to an industrial domain can be seen as an extended process of negotiation. First there is the question of what Human Factors has to offer. This defines Human Factors from an academic or general perspective. It offers to a domain, such as Air Traffic Management (ATM), what it believes it can do best. Then there is the second question of what the client, in this case ATM, actually thinks it needs. These are the 'drivers' for change in ATM, whether issues of safety, demands for capacity, a temporary or anticipated shortage of controllers, etc. Once applications start occurring, a third question is reached, namely what can Human Factors actually deliver, leading to the inevitable fourth question, concerning what it can not (yet) resolve.

Both sides benefit from this negotiation process. ATM will have some problems resolved and will know the practical limitations of Human Factors. ATM will also re-shape its own perceptions of what is needed – a deeper appreciation perhaps of the necessary sophistication of the approach required to resolve problems relating to human performance in complex systems such as ATM. This appreciation may persuade some ATM organizations themselves to commit to having their own sustained efforts and resources to dealing with Human Factors issues in their organizations, leading to dedicated Human Factors groups working purely in ATM/aviation.

Human Factors in return gains experience on the useful 'range' of its techniques and approaches, and identifies where to carry out further development of methods. Human Factors practitioners find new ways to combine and refine techniques to answer real and pressing industrial questions. They also find new ways to express their own approaches and what Human Factors can deliver, and to justify the resources for studies and access to operational staff that are so often needed. Such experience of 'real world' applications is fed back to those carrying out research and method development, and is discussed at academic and professional conferences, much as doctors discuss new techniques of surgery at their own medical symposia. This process of going from theory to practice, and back to theory, is a never-ending story for applied disciplines such as Human Factors. The important point is that practice is essential, and it is essential to review such practice occasionally and sit back and reflect on where things are going well, and where refinements are needed.

There are numerous books on what Human Factors is, and a number of books on what Human Factors can offer to industries such as ATM. However, there are few books that show what has actually been delivered and has 'worked' in industrial settings – the third question alluded to above. The main objective of this book is therefore to illustrate what Human Factors can deliver to ATM, by presenting a set of case studies that show exactly what *was* delivered. This naturally leads on to a secondary objective, namely the fourth question of where Human Factors needs to improve. Since this book contains only a limited sample of all the Human Factors work ongoing in the domain of ATM, such an objective cannot be called scientific. Nevertheless, in the Discussion chapter at the end of this book, some observations are made by the editors on where it appears that Human Factors needs to either enlarge its scope of application, or develop new methods, to improve its utility to ATM.

However, before such observations, the primary objective must be realized, and so a set of nineteen chapters are presented concerning a diverse range of applications, from detailed studies of human error, to the selection of controllers following the US strike which led to dismissal of thousands of controllers. Each case study outlines how its specific application of Human Factors evolved, what it entailed, where possible how it was resourced, and how the results contributed to operational system performance.

Before presenting such chapters, it is first necessary to briefly define what Human Factors is, and to consider how it has evolved in the domain of ATM over the past forty years or so. This leads to a framework within which to organize the chapters themselves, and also from which to discuss where Human Factors in ATM needs to go in the future.

What is Human Factors?

Human Factors (or Ergonomics[1]) has been defined briefly as 'fitting the task to the man' (Grandjean, 1981), and 'designing for human use' (Sanders and McCormick, 1992), and more lengthily as 'aiming to design appliances, technical systems, and tasks in such a way as to improve human safety, health, comfort and performance' (Dul and Weerdmaster, 1993). An implicit fourth, and operationally interesting, definition (one with which controllers might concur) is 'give us the tools and we will finish the job' (Osborne, 1982). Clearly Human Factors is about giving the human operator an efficient working environment and tools which take account of human strengths and limitations, but it is also about selecting the most suitable operators and giving them the required skills. In this way Human Factors seeks to optimize human performance and thus system performance, but not to the detriment of the health (physical and psychological) of the humans in the system. Human Factors can therefore be said to be 'work-focused', though it also demands 'healthy' work.

Human Factors has its roots in applied psychology, but with substantial inputs over the years from fields as diverse as medicine (e.g. to understand physiological effects on humans of work systems), physics (e.g. to understand perception),

---

[1] In this book no distinction is made between these two terms.

engineering and design. In fact people who are working in Human Factors themselves come from a range of backgrounds such as psychology and engineering, and it is considered a hybrid discipline.

Having briefly defined what Human Factors is, the following sub-section gives a brief historical overview of the evolution of Human Factors in the context of ATM, but also with reference to major events or developments in other industries which have shaped Human Factors approaches generally. This is followed by a summary of some of the contemporary issues in ATM Human Factors which are driving both research and applications today. This then leads on to a framework of Human Factors application areas within which to organize the chapters and make observations about the status of applied Human Factors in ATM.

Human Factors in Air Traffic Management – the Beginnings

Ever since devising tools for human use, since the Stone Ages in fact, mankind has attempted to improve the usability of his (and her) artifacts. Even though early tools may appear primitive from today's perspective, these attempts correspond to the scope of a discipline that modern times have come to term Human Factors, i.e. the systematic elicitation and application of knowledge about human operators and their performance characteristics in order to make man-machine systems perform more efficiently, safely, and reliably whilst maximizing user satisfaction and minimizing detrimental effects on both the user and the environment. In simple words: making humans and machines cooperate effectively.

Man's first attempts to enhance his tools were mostly of implicit and empirical nature: lessons learned during the successful – or unsuccessful – use of a tool influenced the shape and size of the next artifact via the craftsman's experience. It was only comparatively recently that humanity started to take a more systematic and explicit approach when designing tools, and this indeed might be considered as the cradle of early Human Factors.

The roots are difficult to trace, but suffice it to say that around the early 20th century the first attempts were made to systematically study humans at work, typically focusing on manual activities. Time and motion studies were carried out, in order to standardize and improve work cycles. A pioneer in this field, Frederick W. Taylor published his theory on 'Scientific Management' in the early 20th century (Taylor 1911). These principles are often referred to as 'Taylorism' though modern times neglect the fact that apart from analyzing and improving work processes Taylor recommended employers to carefully select and train the work force and to provide positive incentives for improvements.

World War I saw an increasing number of people obliged to interact with hostile or at least stressful environments, both in combat operations and in the military supply chain, and it became paramount to ensure their reliability under these circumstances. As a consequence the military recognized the need to study the properties of human operators as well as their interactions with technical systems. The selection of army recruits, flying aptness tests, and training were among the major fields addressed at that period. Fatigue studies were initiated in ammunition factories leading to a redesign of work-rest cycles (Oborne 1982).

Until the end of World War II the emphasis of what was about to become Human Factors was primarily on testing and selection of operators. However, equipment and machinery became increasingly sophisticated and this placed a threefold increase in demand on the human operator. Firstly, the physical characteristics of the task itself became more demanding, owing for example to higher flying altitudes and speeds, which required greater physical fitness and reactions. Secondly, the interactions with the machinery itself became more complex: an increasing number of displays, levers, and controls in the cockpit for example required a solid understanding of the functioning and control of the system. Thirdly, the interaction with increasingly destructive arms and weapons, and more so the exposure to equally equipped foes aggravated the stress placed upon the operator. It soon became apparent that selection and training alone would not be sufficient to redress the situation and that the system itself would need to be designed so as to better accommodate human performance characteristics and limitations. Anthropometrics (the study of body dimensions) was amongst the earliest attempts to adapt machines to the performance characteristics of their operators (so that machines would 'fit' most sizes of people) – the focus began to shift from 'fitting the human to the machine' to 'fitting the machine to the human'; which is echoed in the claim to overcome 'Procrustean design'[2] (Oborne 1982).

An interesting event illustrates this development: during World War II a number of cases were observed in which pilots in the American Air Force damaged aircraft by retracting the gear after landing. The causes for these mishaps were attributed to the similar design and close physical location of gear and flaps levers. A remedy, quickly introduced at that time, has found its way into the design of most modern aircraft: a wheel-shaped handle on the gear lever and a spoiler-shaped handle on the flap lever have ever since prevented pilots from mistaking one for the other.

The precursors of what would later become air traffic control services were installed as early as the 1930s in the United States, initially on the initiative and under the responsibility of the major airlines. At that time most airlines began to equip their aircraft and airport-based centers with radio communication facilities and the first airport control tower went into operation in 1932. The radio facilities permitted aircraft to report their estimated arrival time and maintenance requirements to the airlines' centers at the airport. The advantages of coordinating

---

[2] The ancient Greek Procrustes lived in Attica and offered delicious meals and free accommodation to anybody requesting his hospitality. His bed, he claimed, would fit each traveler's body size. Regrettably for his unsuspecting guests Procrustes would obtain such an exact match by either chopping off parts of the traveler's legs or stretching his body until he matched the size of the iron bed. This mythical example is used by Human Factors people to warn designers from assuming that an 'average' system design fits most people – e.g. a door frame given an 'average' person's height would cause 50 per cent of people either to bang their head or duck as they went through it.

all aircraft approaching an airport soon became apparent and the three major airlines at that time signed an 'Interline Agreement' to that purpose in 1934. The control centers established in consequence were soon taken over by the US government and provided control services initially only in the vicinity of airports and based on position reporting, a technique generally referred to as procedural control.[3] In 1956, a mid-air collision occurred over the Grand Canyon in the US. This single accident had a profound effect on air traffic control, leading to all air carriers operating under instrument flight rules and 'positive' air traffic control.

During these early years air travel grew significantly as jet aircraft came into operation. The introduction of radar systems assisted the controller in increasingly demanding tasks. The term radar has entered common usage as a definitive word yet it is actually an acronym which stands for radio detection and ranging. The British are principally credited with the development of radar on the eve of World War II. The earliest radars relied entirely upon reflected radio energy referred to as primary radar or skin paint to compute the distance and bearing of aircraft targets. This information was subsequently displayed on a cathode ray tube which showed the aircraft position relative to the radar antenna. By 1952, the Federal Aviation Administration (FAA), which previously relied upon pilot reports, and time and distance to separate aircraft, began to use short range radars called Airport Surveillance Radars (ASRs) for approach and departure control in the vicinity of airports. Four years later, the first long range radars known as Air Route Surveillance Radars or ARSRs were extending radar coverage capability to Air Route Traffic Control Centers (ARTCCs). The FAA continued to extend radar surveillance along major air routes during the 1960s.

A fundamental advance in air traffic control technology was the introduction of transponders to civil aircraft. Developed during World War II for military use as Identification Friend or Foe (IFF), transponders are active devices that, when triggered by a ground signal, send a signal back to the radar site where it is superimposed upon the primary radar return. This, for understandable reasons, is referred to as secondary or beacon radar. In 1960 the FAA began successful testing of a system under which flights in certain 'positive control' areas were required to carry a radar beacon The earliest transponders merely supplemented the primary radar return with an additional mark called the 'beacon slash' because it was depicted as a thin line on the far side of the primary return. However, as secondary radar technology became more sophisticated, it was possible to decode multiple transponder returns and differentiate between aircraft based upon the different codes set in their transponders.

The most significant development came in the 1960s when computer technology was fused with ATC radar to create an integrated system. Since early radars were not capable of tracking targets, controllers moved small markers called 'shrimp boats' around the surface of the screen to maintain the identity of aircraft

---

[3] Initially blackboards were used to note the aircraft identifiers and reported positions. These have evolved into the flight progress strips which are still in use in many control centers and bear a striking resemblance to the format of the original position reports on the blackboards.

targets. With the synthesis of radar and computers it became possible to track individual targets and generate discrete, electronic data tags for each identified aircraft. Depending upon equipment, data tags display such information as aircraft identity, type, speed, and altitude. Networking computers among ARTCCs and Terminal Radar Approach Controls (TRACONs) allowed for the seamless exchange of information and the automation of manual tasks such as handoffs.

In the ensuing years ATC radar has continued to evolve. Air Traffic Control towers were equipped with an increasingly sophisticated series of radars designed to survey the airport surface. These became known as Airport Surface Detection Equipment or ASDEs. The old analog radar in TRACONS was replaced with new equipment employing digital data. New procurements added color displays at terminal and En-route facilities. In the relatively near future, technologies such as Automatic Dependent Surveillance Broadcast (ADS-B) which actively broadcasts aircraft status information will extend the range of ATC surveillance and supplement the radar systems in use today.

Technological progress was rapid, yet the steady growth in air traffic, the fact that aircraft were traveling increasingly fast, and the reduction of safety separation standards left little reason to believe the situation would become less demanding and stressful for the controller. The nature of the controllers' work, however, constantly evolved, and this together with the development of supporting technology triggered the interest of Human Factors research in this domain.

In the late 1940s and early 1950s, in the developing new discipline of Human Factors or Ergonomics, Paul M. Fitts conducted classic experiments about human-machine collaboration in air traffic control. His work lead to a number of publications including the so-called 'Fitts' List' for the allocation of functions (Fitts 1951). Considering human strengths and limitations and the capabilities of technical systems, he argued which functions would be better performed by humans and which functions should be allocated to machines. 'Fitts' list' became a standard reference in many areas, even though Fitts himself advocated a more cautious use and saw its applicability limited to air traffic control. Interestingly, Fitts himself argued that the increasing sophistication of technology would render the list obsolete at some point in the future. It is, however, still frequently referenced today.

The 1960s and 1970s saw a continued interest of Human Factors in selection, and training, anthropometrics (optimization of workstation design to accommodate differing body sizes) and function allocation. The subsequent advancements of technology, especially computer science, raised the researchers' additional interest in mental processes. The term 'Cognitive Psychology' was not yet born, but it dawned upon researchers that understanding mental processes would enable them to build better technology, and that replicating functions of the human brain could greatly enhance what was about to be baptized 'Artificial Intelligence'. In air traffic control Human Factors researchers started to study the mental processes involved in the controllers' work. The term 'mental models' started to evolve referring to the operators' understanding of the environment they were interacting with. This concept wasn't new: controllers would often speak of keeping or losing 'the picture' (Whitfield and Jackson 1982).

The 1970s and 1980s saw a number of dramatic accidents in which human error was a contributory or even dominant cause. The 1977 air disaster in Tenerife claimed 573 fatalities: a Boeing 747 initiated its takeoff though not cleared to do so and collided with another 747 backtrack taxiing on the same runway. The causes of the accident, largely miscommunications between tower control and the aircraft taking off, was exacerbated, at least partially, by the phraseology standards then in place and the lack of visibility due to fog. Two years later a midair collision near San Diego heightened the need for an onboard airborne collision avoidance system, leading to the Traffic Alert and Collision Avoidance System (TCAS) some years later. A ground-based conflict alert system had already been developed in the US. Several other air crashes in this period (see Wickens et al., 1997) led to a realization that problems in the chain of command, team coordination and management style in the cockpit could lead to accidents, and Crew Resource Management (CRM) was developed to counteract these problems.

Other industries were also suffering a period of serious accidents. In 1979 a meltdown in the nuclear power plant at Three Mile Island occurred. The apparent cause of the accident was the operators' failure to detect the correct position of various valves in the cooling water system after a feed water pump malfunction had caused an emergency shutdown of the reactor. This failure was superficially attributable to human error but it appeared that the accident was at least partially caused by incorrect state indicators in the control room (Hopkin, 1982; Kirwan 1994). The reactor was recovered before a full meltdown occurred so that no lives were lost. However, large quantities of radioactive water were released into the environment and the worldwide trust of the public in the safety of nuclear power plants was irreparably damaged.

In 1984 great quantities of toxic chemicals leaked from the Union Carbide Pesticide plant in Bhopal, India, killing some 4000 people and injuring 200,000. Amongst the contributory factors were the loose handling of safety features and maintenance procedures. The accident also demonstrated the advantages of automatic safeguards over a total reliance on humans to activate emergency systems.

Two years later, a meltdown and fire at the nuclear power plant in Chernobyl, Russia claimed at least 30 lives and caused an unprecedented release of radioactively contaminated material into the atmosphere. The accident occurred during a test at reactor shutdown that aimed at testing whether a coasting turbine would still generate sufficient power for the cooling water pumps to operate for a transition period, i.e. before the diesel generators designed for this purpose would be activated. The accident involved a lack of training and clear procedures as well the negligence of safety procedures and standards.

Apart from the more obvious effect of eroding public trust in the safety of air transport and plants dealing with toxic and dangerous substances, notably nuclear power plants, these disasters had a second important consequence. Human error was reported as the primary cause of failure in all of these accidents and started to become the most prominent concern for system safety. Partly as a consequence of these disasters, human error and human reliability were studied in great detail, leading to the application of fault and event trees and other engineering models to Human Factors. The message became clear – in increasingly complex systems it is not enough to rely on training, selection and motivation of the operator to do the right thing – all Human Factors aspects of the work system need to be right, and it is also advantageous to try to predict human errors and their consequences at the design stage, to make system designs more robust. This lesson was learned by the process industries (nuclear power, chemical and petro-chemical) in the 1980s, but is only now becoming accepted in fields such as transportation (including aviation and ATM).

It was at this juncture that certain industries (process and aviation) embarked on a different course to ATM, namely one of enhanced automation. The fact that human error was increasingly cited as the cause of accidents suggested that the overall system performance, reliability, and safety could only be enhanced if the human operator's role would be limited to that of supervising automated systems, and taking over operations in abnormal situations. But it became obvious that in many cases the system and particularly the design of the interface had invited incorrect diagnosis and erroneous inputs. Apart from the more obvious human errors where personnel failed to correctly operate technical systems, a second more indirect type of human error started to be noticed: the designers' failure to develop usable systems.

In a widely regarded publication Bainbridge (1987) reported a number of cases from different industries in which the design of systems and procedures was contradictory to basic Human Factors principles – leaving the operator little chance but to fail. A point of specific concern reported was the task distribution between human and machine which was often guided by the philosophy of automating functions that could easily be automated whilst assigning difficult-to-automate functions to the human operator. This resulted in a situation in which the automation would be active in standard situations whilst the human, now passively monitoring the system, would have to diagnose any malfunction and quickly resolve the problem. Performance problems associated with vigilance and the loss of skills due to an increasingly passive role were amongst the consequences; the slogan 'ninety-nine per cent boredom and one per cent panic' seemed not inappropriate to characterize the resulting work situation.

A prominent example of 'clumsy automation' was the 'mode awareness' problem in early glass cockpit aircraft in the 1980s. Flight Management Systems (FMS) were introduced in these aircraft that assisted the pilot in the planning and execution of the flight, concerning tasks such as fuel, weights, and balance planning and entering the complete flight route into the system. Pilots interacted with the FMS using an input device that allowed switching between different 'pages'. Depending on the flight state and the aircraft configuration, the information displayed on these pages, the data the flight crew was permitted to enter, and the system behavior would vary significantly, often leaving the flight crew wondering what the system would do next. This led to the famous apocryphal pilot phrase

'Why'd it do that?', which signified the fundamental discrepancy between the pilot's mental model or situation awareness of what was happening, and that of the computerized system. This lack of harmony between human and machine led to a number of fatal accidents, called 'automation-assisted', since it was argued that without the automation the accident would not have happened (Wiener and Nagel, 1997; Billings, 1997).

The term 'user-centered automation' (Billings, 1997) became fashionable, referring to a system design that placed the operators' needs and performance characteristics in the center and matched the system design to these. User-centered automation was intended to contrast with what researchers called the then dominant 'technology-driven design'.

In parallel, models of human cognition emerged, the most influential of which was probably Rasmussen's model of skill-, rule-, and knowledge-based behavior (Rasmussen, 1983 and 1986). Other models were more psychological in nature but common to all was the attempt to explain and predict aspects of human cognition, such as attention, perception, and decision-making. Apart from the obvious purpose of explaining human cognition, these models were often used as an inspiration for the design of user-support systems. By now computer technology had advanced to such a degree that automation assistance to the human operator was no longer limited to the execution of simple tasks. Automation could be conceived that, autonomously, under supervision, or in concert with the human, would perform certain cognitive functions, including monitoring and detection, situation analysis, but also decision-making. A pertinent and good example of automation is automated landings, that are in fact more reliable and hence safer than manual landings, although in certain conditions manual methods must still be employed, and so the pilots must still retain their skills.

In contrast to the embracing of automation by a range of other industries, ATM in practice at the time remained very human focused with relatively little automation support. Nevertheless, with the evolution of computer-based systems the Human-Machine Interface (HMI) became an item of central interest to the ATM community, as it was seen as desirable to replace older radar screens with systems that could super-impose more information for the controller, to enable more efficient performance. Legibility and contrast, font size and design were subsequently the subject of research for quite some time. In parallel, the controllers' workload evolved as a central issue for successful system design, arguing that appropriate design of the human-machine interface could help to reduce the operator's workload, thus contributing to overall safety and efficiency. The aim was to increase 'capacity' (volume of traffic) in response to more public demand and accessibility to flight-based travel. It became obvious to many that Human Factors could be a key 'enabler' to increase capacity and hence growth of the industry as a whole. The human element in the ATM system, still the key element, should therefore receive support in order to improve performance.

The philosophy behind this line of thinking was often referred to as 'user-centered design', meaning that the performance characteristics and needs of human operators rather than the technical feasibility should drive the system design. In recent years this philosophy has matured to a line of thinking which considers human operators and automated systems 'partners' who support one another in the execution of tasks. In various domains the computer-based 'assistance systems' of the 1980s and early 1990s have developed into 'agents' that act with a certain degree of independence; yet this development has yet to take place in the domain of air traffic management.

Today, in various Air Traffic Control Centers around the world, most conventional radar-based display systems have been (or are in the process of being) replaced by computer-generated HMIs, allowing much more data to be presented, albeit at the cost of not always necessarily having access to the 'primary' radar information should the HMI fail for any reason. This trend represents the first real step towards automation, since like the process control operator and the pilot beforehand, the controller's perceptions and actions are now mediated through a computerized vision and model of the world. Such basic automation represents the fundamental building block for more serious automation to occur, as discussed next.

Human Factors in ATM – Contemporary Issues and 'Drivers' in ATM

ATM today is an intensive industry. Unlike its nuclear power and process control 'complex system' counterparts, however, it is a relatively open system, controlling aircraft as it does across national boundaries, from one ATM service provider's system to another's. It is thus more interactive than many other industries, and compared to other transportation systems, it is more complex, dealing with three dimensions instead of two for surface transport (rail and automobile), yet having to thread all its traffic ultimately through stopping points (airports) with the certain knowledge that the system cannot be allowed to stop or 'pause' except for limited periods of time, as aircraft will run out of fuel. ATM is therefore a highly interactive and complex system, and one that is not inherently safe. It is of course also intimately linked to aviation and hence the highly cost-sensitive and competitive pressures of the commercial air transport industry. Route complexity and traffic density have increased significantly, and many charter flights fly in the formerly 'quiet hours' – although the air transport network is not yet 'saturated', it is more consistently busy for the controller in certain locations of airspace.

The problem is that this 'trend' is not leveling off, even after such tragic events as the terrorist attacks of 9/11 in the US. The demand for capacity increases, and is still predicted to double in 10-15 years' time. This has led many to consider that the time has come for significant changes to ATM, and for many this means increased controller support in the form of automated tools. Such tools may be oriented towards increasing throughput, such as enhancing the efficiency of arriving traffic and departures, or safety, such as giving controller tools to automatically detect and avoid conflicts well before they register via current short-term conflict alarms. A number of such systems are currently being explored either as research projects or are being trialed in operational centers.[4] Other perhaps more fundamental changes consider exploring the partnership between controllers and pilots, so that control may be 'delegated' temporarily to pilots who will have their own displays of the surrounding traffic (called Cockpit Display of Traffic Information or CDTIs). A further envisaged change, in Europe at least, is to break down the traditional and historical national boundary airspace structure in favor of a more open and flexible airspace system known as 'Single Sky'. Moreover, it is presently being investigated whether abandoning, partly or altogether, the existing fixed route network and permitting aircraft to navigate without reference to a ground-based route structure would have positive effects on the efficiency and safety of the air traffic system.

Such changes are full of Human Factors implications. The general insertion of more automation raises questions not only of workload, but also of situation awareness, and trust in the automation, as well as skill retention and the ability to recover should things go wrong, such as the automation failing or giving erroneous information. The introduction of automation is also known to be capable of affecting not only the workload (often reducing workload on the 'primary' task but increasing it for previously non-existent 'secondary' tasks), but also the way members of the 'team' may work together (e.g. 'R' and 'D' side controllers in the US; tactical and planner positions in Europe; and between controllers on different sectors; and pilots). This can lead to improvements (e.g. more efficient operations and better sharing of workload) or decrements (e.g. a reduction in team situation awareness and detection and recovery of colleagues' errors). It is therefore a Human Factors concern to explore the impact of new automation on controller and pilot working patterns. Social aspects such as the impact of ageing on dealing with these new automation concepts, and the long-term effects of professional stress also need consideration.

There is also the question of safety. Recently in Europe there have been two tragic fatal accidents attributed to ATM, namely the Milan runway collision in 2001 and the Überlingen mid-air collision in 2002. These major accidents have shaken confidence in safety, and have led in Europe to major safety programs including some Human Factors considerations. ATM's 'high reliability' status is now not so secure, and there is a desire to be more proactive about safety and human error.

More generally, it is notable that error forms changed with the advent of automation in other industries (for example the glass cockpits in aviation), and it can be expected that new automation in ATM will bring with it new errors. The aim

---

[4] Some examples of current automation projects are URET (User Requirement Evaluation Tool) & Problem Analysis, Resolution, and Ranking System (PARR) in the US, and their equivalents Medium Term Conflict Detection (MTCD) and Conflict Resolution Assistant (CORA) in Europe, for conflict detection and resolution support respectively, as well as Descent Advisor in the US and Arrival Manager in Europe, both aimed at enhancing airport throughput and efficiency, and Advanced Surface Movement Ground Control Systems for airports, which should enhance efficiency and also prevent runway accidents such as at Tenerife and Milan-Linate.

should therefore be to anticipate and prevent such errors contributing to accidents as has happened in other introductions of automation. Also, errors do occur frequently in ATM, but these errors have a very high detection and recovery rate – it is important to ensure that this propensity for error management is not lost in future systems. These aspects require more focus on human error, and Human Factors working together with Safety and Risk Assessment disciplines.

With a wave of automation and other changes on the horizon and even closer, there is a need to get things right at the design and development stage, or at least prior to operations – learning after the event via accidents will constitute unnecessary loss if life and will seriously damage confidence in ATM and air travel generally. Even if there are no accidents, producing systems that controllers (and/or pilots) do not trust or cannot use efficiently will be very costly to the industry. There is therefore a need to be smarter in the design phase and anticipate problems particularly for the human, determining then how to avoid such problems or enhance their recoverability. This means that much of Human Factors for the future must be proactive rather than reactive, and must occur earlier in the system design lifecycle. Although this area of future automation in ATM and its implications for Human Factors is a fascinating area, one occupying a number of recent books and numerous professional and academic articles, it is not the focus of this book, since a book on achieving operational impacts is necessarily rooted in the present more than the future. Nevertheless, some of the chapters in the book are oriented to studies of recently developed systems, with some automation features. Such chapters foreshadow a way forward for more proactive Human Factors analysis of future ATM systems and tools.

As well as the prospect of automation, there are still general issues associated with current ATM that require support from Human Factors. The training and selection of controllers, the organization of their shifts, and other general Human Resources issues remain challenging objectives for a growing and constantly changing industry such as ATM. There are also new areas in ATM such as Team Resource Management (TRM) and Critical Incident Stress Management (CISM), which aim respectively to share awareness of safety and team performance issues amongst controllers, giving them good understanding of what can go wrong and how to avoid it, and avoiding the psychological impacts of experiencing a near miss incident. TRM mirrors to an extent the Crew Resource Management (CRM) program in aviation more generally, and CISM is related to treatment for post traumatic stress disorder in a number of industries and situations. Both these trends reflect the continuing focus on the human element in ATM, with the latter in particular reflecting a concern for the psychological health of the controller, as was argued for by the definitions of Human Factors cited earlier in this chapter.

The relatively rapid evolution of ATM, the changing nature of the controllers' tasks and equipment, the current drive towards more automation and more flexible airspace, and the more recent focus on psychological and health aspects of ATM, are all aspects of the current context of Human Factors in ATM. Yet this context is complex, dealing with all manner of human facets from controller workload to cognitive models, to the way Human Factors itself is managed and perceived in the industry. It is therefore necessary to reduce this complexity to make the picture clearer, so the next section outlines a simplified framework within which a variety of perspectives and case studies can be viewed and evaluated. This then enables the question of whether Human Factors is achieving true impact in ATM to be more easily answered.

A Framework for Applied Human Factors in ATM

From the above description of the origins of Human Factors and its experience in ATM, it is clearly not so easy to define the full scope of Human Factors, because of the diversity of its subject matter (human behavior and cognition), particularly in the context of complex systems such as ATM. Nevertheless, in an applied context Human Factors can always be elaborated by the areas it focuses on, i.e. showing what it does. Human Factors may therefore be de-constructed into a number of inter-related areas, which are general and relevant to most industrial applications. These areas are as follows (see Figure 1.1):

1. Operational Issues;
2. Human Resources Issues;
3. Methodologies for current and future systems;
4. Human Factors Integration issues.

Operations This area deals with all aspects of day-to-day air traffic management, and focuses especially on what the controllers have to do, for example (Wickens et al., 1997):

● monitor traffic;

● resolve conflicts;

● control aircraft movements;

● manage air traffic sequences;

● plan or route flights and manage sequences;

● assess weather impact;

● respond to emergencies;

● communicate with pilots and other controllers;

● handover position to oncoming controllers;

● deal with and learn from losses of separation (incident reporting);

● use artifacts directly associated with air traffic control duties (HMI, i.e. displays, input devices and communication equipment; electronic flight progress strips; controller tools; etc.).

Operational issues may relate to any of the main functions of ATM, from tower operations, to approach and terminal maneuvering area operations (called TRACON in the US), En-route and Oceanic operations. This area deals with the lifecycle of aircraft operations from 'gate to gate' as controlled by air traffic services.

Human Resources The second area concerns the support of these operational activities, and is termed 'Human Resources'. This area deals with the lifecycle of the controller, including the following:

- selection and recruitment;
- training to become a competent controller;
- maintaining awareness of safety issues and a positive safety culture;
- organizing the operational command structure of air traffic controller operations in a Center;
- organizing the shift and work break system to maintain adequate concentration and avoid fatigue and boredom;
- maintaining job satisfaction and motivation;
- occupational health support, including dealing with stress and stressful events,
- maintaining effective team working;
- managing management/controller relations;
- career progression;
- retirement planning.

Human Factors Methodologies The third area concerns methodologies for measuring and predicting performance. These may include measures for measuring direct performance variables (losses of separation, errors (individual and team), aircraft throughput, etc.) and indirect or 'intervening' variables such as workload, situation awareness, vigilance, as well as more subjective factors such as trust. Such measures usually occur in situations with actual controllers, either measured in live traffic situations, or more usually, in real-time 'human-in-the-loop' simulations. However, for design purposes, non real-time simulations (either fast-time or prototyping simulations) can be used to predict performance. Such approaches vary in the degree to which they represent or model actual controller cognitive behavior (i.e. how controllers think and react to situations), and indeed the area of cognitive modeling, as used in other industries, is of significant potential benefit to the design of future ATM systems. Lastly in this category, are new approaches to predicting human reliability or errors and error recovery in future systems.

Human Factors Integration The fourth area is concerned with 'Human Factors Integration'. This refers to the ways in which Human Factors is organized within an organization to achieve Human Factors goals. This is very much a practical consideration, since having good Human Factors tools and data and ideas is of limited practical value if the organization does not know about it, or if Human Factors is not properly resourced or placed to enable it to make a sustainable impact. This aspect of organizing Human Factors itself, is rarely addressed in Human Factors in applied domains, yet it is critical to realizing solutions to the Human Factors challenges posed by ATM's continuing rapid development.

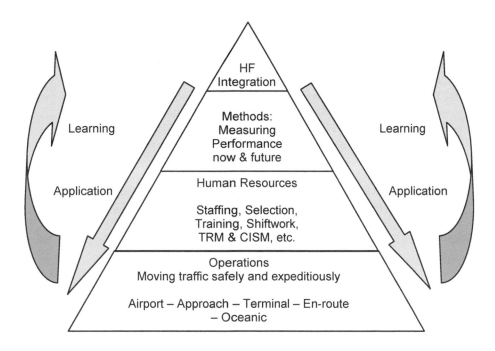

Figure 1.1   Framework of applied Human Factors in ATM

The chapters in this book are therefore placed into these four categories. Not every aspect of each category is fulfilled, partly because the size of the book is limited, but also because there are not always success stories available for each category. This does not mean that if an area is not addressed it is not served well by Human Factors, since many success stories remain undisclosed publicly for a host of reasons. Nevertheless, the Discussion chapter will consider where Human Factors appears to be working well, and where more ingenuity or development is needed. The Discussion chapter also gives a forward vision of future Human Factors needs for the next decade or so, from both a European and a US perspective. The remainder of this book, however, in between this Introduction and the final Discussion chapter, focuses on case studies of actual Human Factors studies making a difference for ATM and controllers, and is therefore available for the reader to peruse what Human Factors has actually achieved and delivered in practice. The editors hope that the reader, via these chapters and their respective authors' efforts, will become more convinced of the importance and practical value of Human Factors in ATM.

References

Bainbridge, L. (1987) 'Ironies of Automation', in J. Rasmussen, K. Duncan, and J. Leplat, (eds.), *New Technology and Human Error*, pp 273-283, John Wiley and Sons, Chichester.

Billings, C.E. (1997), *Aviation Automation: The search for a Human-Centered Approach*, Lawrence Erlbaum Associates, New Jersey.

Fitts, P. M. (1951), *Human Engineering for an effective air-navigation and traffic-control system*, Ohio State University Research Foundation, Columbus Ohio.

Hopkin, D. (1982), *Human Factors in Air Traffic Control*, Taylor and Francis, Bristol.

Kirwan, B. (1994), *A Guide to Practical Human Reliability Assessment*, Taylor and Francis, Bristol.

Oborne, D. (1982), *Ergonomics at Work*, John Wiley and Sons, Chichester.

Rasmussen, J. (1983), 'Skills, Rules, and Knowledge; Signals, Signs, and Symbols, and Other Distinctions in Human Performance Models', *IEEE Transactions on Systems, Man, and Cybernetics*, Vol. SMC-13, No. 3257, May/June 1983, pp. 257-266.

Rasmussen, J. (1986), *Information Processing and Human-Machine Interaction*, North Holland.

Sanders, M. and McCormick, E. (1992), *Human Factors in Engineering and Design*, 7th Edition, McGraw-Hill, New York.

Taylor, F. (1911), *The Principles of Scientific Management*, published 1967 by W. Norton & Company: New York.

US Centennial of Flight Commission (2003), *Essay on Air Traffic Control*.

Whitfield, D. and Jackson, A. (1982): 'The air traffic controllers picture as a example of a mental model', in G. Johannsen and J.E. Rijnsdorp, (eds.), *Proceedings of the IFAC Conference on Analysis, Design, and Evaluation of Man-Machine Systems*, Pergamon Press, London.

Wickens, C.D., Mavor, A.S., and McGee, J.P. (1997) *Flight to the Future: Human Factors in Air Traffic Control*, National Academy Press, Washington DC.

PART II
HUMAN FACTORS IN OPERATIONS

This section contains six chapters, each focusing on changes to operational air traffic facilities and systems. It provides several examples that demonstrate how the application of Human Factors positively affects the operational environment. This is true whether at an En-route centre, a terminal facility, airport, or even the interaction between the Central Flow Control or System Command Centre and the various Airline Operating Centres.

The first chapter by Voller et al. addresses a problem that was encountered at an operational terminal facility in the UK, where incidents tended to occur shortly after position handover between the outgoing and incoming controller. A detailed Human Factors study was undertaken at the facility, using both objective and subjective approaches. These findings were then used to develop a new procedure that focused on memory-aiding to increase position hand-over consistency. The resulting checklist was developed for two different types of controllers, approach controllers and controllers working in the Terminal Manoeuvring Area (TMA). Incidents associated with position handover were reduced after the implementation of the checklist, and the procedure was subsequently formally integrated into the air traffic procedures for the facility.

The second chapter by Cardosi focuses on airport safety, particularly the problem of runway incursions. The chapter commences with a detailed analysis of the occurrence of runway incursions and contributing factors. A number of interventions are discussed, some already implemented, some under development. They range from the enhancement of painted surface markings and the introduction of sophisticated surface-radar systems in the tower cab environment to pilot and controller education and team training techniques. The chapter also discusses regulatory aspects, e.g. in the context of intersection takeoffs. Based on extensive data mining of incident reports, Cardosi discusses many detailed potential incident mechanisms and the solutions most likely to address the occurrence of runway incursions.

The third chapter by Isaac et al. continues in the analysis of human error, but is the first of two focusing on a major effort to develop and apply a model of human performance to operational errors (the second chapter focusing on human error being the chapter by Pounds and Ferrante discussed below). The first half of the chapter describes the model and its adaptation to ATM operational incident analysis to assist in the identification of human performance considerations associated with air traffic control incidents. The second half of the chapter then focuses on a specific incident investigation that used the approach. This case study is unusual in that both the original analysis and the subsequent Human Error in ATM (HERA) based analysis are given, allowing the reader to see the differences in recommendations that result from the original less formal treatment of error causes compared to the more formal HERA approach. The HERA approach was found to provide the investigator with a better understanding of the context surrounding the incident. Additionally, controllers involved in the incident provided investigators with responses that allowed for improved insight into the incident.

The fourth chapter by Pounds and Ferrante focuses on strategies to reduce Operational Errors (incidents) in US operational Air Traffic Control facilities. Several methods and measures that have been developed are described. The first is the Systematic Air Traffic Operations Research Initiative (SATORI), a system used to recreate the incident conditions so that investigators and the involved controllers can understand the relationship between the aircraft movements and any air traffic control instructions that were issued. The second is a new measure of operational error severity. This measure takes into account not only the flight profiles of the involved aircraft, but also the extent to which the controller was aware of the developing situation. Next a data-driven approach for making decisions about safety management is described, particularly so that the causes of incidents can be more reliably identified. The FAA, in a joint effort with EUROCONTROL, integrated the Human Factors Analysis Classification System (HFACS) approach with the HERA approach described in the third chapter in this section, to generate a combined approach called JANUS. Once validated, JANUS should allow for better analysis of Human Factors associated with incidents, and should provide a means to better understand common incident mechanisms in both the US and European States. The final measure in the strategy to reduce operational errors concerns an innovative training approach to help controllers maintain attention. These methods and measures, when taken together, should allow for the development of operational error mitigation strategies that provide for the most effective use of limited resources.

The fifth chapter by Kirwan et al. in this section addresses the decision to reduce separation standards between aircraft operating in the Open Flight Information Region (OFIR), outside normal civil-controlled airspace in the United Kingdom, for aircraft receiving a (usually military) Radar Advisory Service. A safety case was undertaken, of which the major part concerned Human Factors. Through a range of methods including two simulations, predictive error analysis, retrospective error analysis of incidents, and a literature review on related experimental work, it was concluded that lateral reduction (from 5 to 3 nautical miles) should not occur, but vertical reduction (from 5,000 to 3,000 feet) could occur. The Human Factors study continued through the implementation by assessing safety assurance for a six month trial of the new separation minima, and monitoring of the trial itself. Roughly one year later, after no associated hazardous events had occurred, the new minima were endorsed across the UK OFIR. This chapter is an example of the application of Human Factors methods to develop a safety rationale for an important new procedure. It also provides an example where a Human Factors analysis justifies 'no' for a response to a safety critical change.

The sixth and last chapter in this section by Smith et al. is concerned with operations and efficiency of the US airspace system. It starts by outlining the US National Airspace System, and highlighting the different 'agents' in the system, from controllers to personnel at Airline Operations Centers. The airspace system and its associated structure have evolved to ensure that cognitive complexity is not exceeded for any one 'agent' in the system, be it a controller or other operational 'agent' in the system. However, this strategy has the potential of making it difficult for one agent to gain information relating to another agent's domain. This chapter describes a tool that was developed to allow various people involved to access information about recent and planned events in a variety of formats. The tool itself was developed not only by employing key Human Factors principles for software tools, but also by ensuring that the tool meets the needs of the various users. The result is a sophisticated and powerful tool that can be used for such diverse tasks as assisting sector designers in determining how sector performance can be improved by specific changes, or by airline companies to evaluate their flight planning strategies according to fuel-burn data. This chapter therefore in particular shows how good Human Factors can be applied to help operational efficiency improve across the different agents in what is considered to be the largest and most complex single airspace system in the world.

These chapters demonstrate that Human Factors can and indeed has had a positive impact on operational facilties in Europe and the US. They also demonstrate that Human Factors is often called upon as a problem-solver, and indeed can solve problems by detailed analysis and re-structuring of the problem from an integrated human-system perspective. Lastly, they highlight that the best way to improve system performance and safety is often to take these insights back to the controllers themselves and others at the 'front-line' of operations.

Chapter 2

Development and Implementation of a Position Hand-Over Checklist and Best Practice Process for Air Traffic Controllers

Laura Voller, Lucy Glasgow, Nicky Heath,
Richard Kennedy and Richard Mason

Introduction

In 1999 the Human Factors Unit (HFU) of the National Air Traffic Services Limited (NATS Ltd.) in the UK carried out analyses of reported incidents and found that a disproportionate number of incidents appeared to occur within 10 minutes of position hand-over.

Position hand-over requires an incoming air traffic controller to take over control of a sector from an outgoing controller in terms of all the flight management and communication carried out from that position. This can be between two controllers on the same watch when a break is taken or between two controllers on different watches as shifts change. It is essential that the incoming controller receives all of the necessary information from the outgoing controller to safely control the sector without any 'unexpected surprises' that could become a safety risk.

In response to this incident trend, a study was requested by Area Control Services within NATS to be carried out by the HFU at the London Area and Terminal Control Centre (LATCC[5]) in 1999. The purpose of this investigation was to review the hand-over process, establish if there were potential causes or contributing factors of the incidents occurring around hand-over time and suggest improvements to reduce the likelihood of incidents occurring.

This investigation identified several Human Factors recommendations that were used as the basis for a second phase of the project. Members of the HFU, working together with the operational staff at LATCC, developed a position hand-over checklist to help guide the transfer of information between the outgoing and

[5] LATCC is now London Terminal Control Centre (LTCC).

incoming controller during the hand-over process. This checklist was supported by the development of an associated best practice hand-over process that was then operationally trialled and evaluated over a six-month period in 2000/01.

This chapter describes the approach taken in the two studies, the research methods used, the findings and the operational impact the work has had.

Initial Hand-over Study

Data Collection

The initial hand-over study was conducted to identify key areas that may be contributing to the higher occurrence of incidents during and shortly after the hand-over process. To examine current working practices and identify if there were any common features that may be contributing to sub-optimal hand-overs being conducted, several data collection techniques were used. These included:

- video analysis;
- shadowing of hand-overs (combination of observation and controller debriefs);
- analysis of incident reports and Mandatory Occurrence Reports (MORs); and
- review of the Manual for Air Traffic Services (MATS) Part I and II (for current procedures relevant to the hand-over process).

Video Analysis Over ten hours of video recordings were reviewed and 28 hand-overs were observed and analysed. These included hand-over of position both within and between watches. Since this video footage was taken for another reason and re-used for this study, the controllers' attention was not specifically on hand-overs and thus can be considered typical of their natural behaviour.

Shadowing of Operational Hand-overs Two Human Factors specialists visited the Terminal Control (TC) operations room on several occasions to observe hand-overs during light and heavy traffic periods. During these visits the specialists:

- debriefed the incoming controller about any expectations of the shift and recent experience of the sector they were about to control (for example, the volume of traffic, weather effects, particular problems they anticipated);
- observed the hand-over for 10 minutes before and after its occurrence and noted any priorities and tasks that were highlighted as being imminent;
- debriefed the outgoing controller to find out their views on what the incoming controller would be doing next and what the priorities were;
- debriefed the outgoing controller about their experiences when they took over at the beginning of their session and which tasks had been their first priority;
- compared the data of the outgoing controller with the data of the incoming controller and that of the observer for consistency.

Incident Analysis Incident reports were reviewed to identify any issues that may have contributed to the determined primary cause of the incident. Factors which were considered included: time of day, type of error made, whether a sector had been band-boxed (combined with another sector) or split within 10 minutes of hand-over and if there was any non-standard information transferred such as weather, pressure, slow aircraft etc.

Review of MATS I and II MATS I and II for LATCC provide details on the operating instructions to operational staff. These were reviewed to ascertain what current guidance is provided for controllers on information transfer and hand-over and what structure this information is recommended to take.

Discussion of Initial Hand-over Study Results

Analysis of the incidents found there were no common trends in hand-over incidents that related to specific watches or time of day. However, it did reveal that incidents tend to happen in the first six minutes of hand-over and that distractions and imparting of non-standard information increased the potential for a judgement error for someone who has only recently taken over the position and not fully established 'the picture'.

These findings were confirmed by the data collected via the videos and shadowing. Hand-overs were found to generally last an average of 20 to 30 seconds. Frequent distractions were observed during hand-overs (most predominately from the co-ordinator). Although the manual (MATS II) provided a recommended list of information to transfer at hand-over, this was not always communicated at every hand-over within a shift and there was no standard sequence or order of transferring information during hand-overs. The data highlighted that existing working practices placed the onus of understanding on the incoming controller and there were no formal procedures to check that their understanding was accurate. This meant there was a risk that the outgoing controller could be unaware if their colleague had missed important details or that aspects of the hand-over had been misinterpreted.

Controllers reported that they had no formal training regarding when and how to give information on hand-over. However, they were aware that this was an important issue and many had adopted their own procedure for handing over information. The methods adopted tended to vary according to individual styles. For example, some individuals preferred to be given a lot of detailed information while others preferred to get the picture themselves with limited communication. The different styles were particularly noticeable between the way information was provided in watch-to-watch hand-overs as compared to those carried out between members of the same watch.

The results from this initial study identified several areas of the hand-over procedure that could be improved. These included standardizing both the approach adopted during hand-over by the outgoing and incoming controller and what key information was verbally transferred, in what order. Recommendations included:

- provide a structured process for the whole hand-over task;
- develop a standard format for imparting hand-over information (e.g. in a checklist format);
- advice that hand-over of co-ordinators and sector controllers should not be taken simultaneously or in very quick succession; and
- advice that sectors should not be split during or within the first 10 minutes of a hand-over.

Checklist and Best Practice Hand-over Process Development and Trial

Checklist and Best Practice Hand-over Process Development

Based on the findings from the initial study described above, together with a review of existing MATS requirements and relevant safety regulator recommendations, Human Factors specialists and operational staff at LATCC worked together to outline an ideal or 'best practice' hand-over process.

One recommendation made in the initial study was to develop a standard format and process for imparting appropriate hand-over information. Through discussions with relevant operational staff, combined with some basic Human Factors principles and rules of communication, the information necessary to convey during hand-over and the sequence in which it would be best conveyed was determined. Consideration was given to issues such as the order in which priority information should be given, the logical sequence for giving information to aid situation awareness development and the inclusion of feedback to check understanding between the two controllers. The outcome from this work was checked against the MATS manual to ensure it did not contradict the limited guidance that was already provided.

Once all of the necessary information was identified and placed in a logical and appropriate order, a mnemonic (PRAWNS – *P*ressure, *R*unways, *A*irports/*A*djacent sector, *W*eather, *N*on-standard information, *S*trips-to-radar traffic point out) was developed. This then formed a checklist or aide memoire to assist controllers with the transfer and assimilation of key information during hand-over.

In order to construct an appropriate physical checklist and determine an appropriate position for it on each Terminal Control (TC) sector workstation, several visits to the operational unit were made and discussions held with operational control staff.

Consideration was given to whether there needed to be any sector specific differences in the information conveyed at hand-over and whether these differences were significant enough to require a different checklist or mnemonic. By interviewing and working with the operational staff at LATCC it was initially concluded that any sector differences were not significant enough to warrant having several versions of the checklist. However, after trialling the checklist and the best practice hand-over process (which will be discussed in the next section) some work was necessary to tailor the checklist for Approach control sectors' use.

To determine the differences required for Approach control sectors several operational Approach staff were interviewed from each of the airfields (Heathrow, Gatwick, Stansted, Luton and Essex) for their views. On the basis of their feedback a short additional survey was developed and distributed to all Approach controllers at the centre offering them the original checklist (PRAWNS) or several alternative checklists to determine their preferences.

Checklist Operational Trial – Data Collection

Once an ideal hand-over process had been established and a checklist for the transfer of hand-over information been developed, it was agreed that this should be operationally trialled for 6 months to evaluate its impact on hand-overs. Based on the outcome of this trial it would be decided whether to permanently implement the checklist as a new operational procedure.

Initially it had been thought that it would be best to trial the checklist on just one watch first and then roll it out to the other watches if the evaluation of its impact proved positive. However, finding a suitable position for the checklist on the workstation of each sector was not easily achieved so it was much easier to leave them there and trial the checklist with all watches in parallel. This also solved the problem of 'spinners' (controllers working with a watch that they do not permanently belong to) being confused by the checklist if it were to be trialled on just one watch initially. To keep the evaluation process practical however, it was agreed that data would be collected only from one watch unless any controversial issues were raised.

The trial began in October 2000 and continued until April 2001. Data collection took place for 4 months of the 6-month trial.

The same methodology as the initial study was adopted for the trial and, where possible, the same data collection methods were used. This allowed comparisons to be made between the initial hand-over study and the trial, and enabled the effective evaluation of the impact that the new checklist and best practice process had on the safety and effectiveness of hand-overs. Techniques used were:

- shadowing of operational hand-overs;
- analysis of incident reports;
- questionnaires; and
- review of any procedure updates relevant to the hand-over process.

Shadowing The shadowing process used during the first study was repeated. During the operational trial three Human Factors specialists were involved in the shadowing process. Notes were made of the duration of the hand-over, any distractions that took place during the hand-over process (for example, any conversations, work-related or otherwise, that took place within the immediate vicinity of the hand-over). The notes of all the observers were compared following each observed hand-over. The perceptions of both controllers were used as an objective benchmark. Where outgoing controller perceptions of priority tasks were considerably different from the incoming controller's actual actions the difference was noted.

Incident Report Data Approximately three months of MOR data were analysed for the months following the introduction of the PRAWNS checklist into the operational environment (mid October 2000 to end of January 2001). This was compared with the MOR data for the same time period the year prior to the introduction of the PRAWNS checklist. Although these data are limited in amount due to the short time period in which they were collected, they were gathered to provide an initial indication of whether the checklist was able to reduce the occurrence of errors associated with hand-over practices. It is recognized that for a more comprehensive assessment of the impact of the checklist on errors associated with hand-over practices, analysis of incident data over a longer time period would need to be assessed. This is an ongoing activity within NATS.

Questionnaires A questionnaire was issued one month into the operational trial to collect controller opinions regarding the PRAWNS checklist and to ask whether specific aspects of the best practice hand-over process were being implemented and having any effect.

Review of any Procedure Updates Relevant to the Hand-over Process During the trial any changes to operational procedures were carefully monitored to ensure that the trial team were aware of them and any consequences these changes they may have on the hand-over process. This was particularly important as the data were being compared with the initial study data from the previous year.

Discussion of Checklist Trial Results

The data collected from the different methods used (shadowing, incident report analysis and questionnaires) were analysed to determine whether the PRAWNS checklist and defined best practice process had an influence on the following aspects of the hand-over process:

- quality and order of information transfer;
- consistency of information content;
- potential effects of different controller styles;
- effect of distractions during hand-over on completeness of information transfer;

- omissions of standard and critical/non-standard details;
- framework available for the incoming controller to build his picture; and
- discrepancies between outgoing plans and incoming tasks.

The following describes the data analysed and identifies the main issues found.

Shadowing

Eleven full hand-over cycles (where the incoming controller was interviewed both on starting and completing a period of work on the sector) were observed over four separate days. A total of 15 different sector controllers were involved in the observed hand-overs (interviewed as either incoming/outgoing or both). These included regular shift members, spinners, and trainees. All hand-overs occurred at a 'moderate' workload level.

It was found that approximately 90 per cent of controllers used at least some part, if not all, of the PRAWNS checklist in their hand-over. Although it is likely that the presence of an observer increased usage of the checklist, during interviews several controllers said they do regularly use part, or all, of the checklist. They said that they tended to use 'PRAWNS' as a short cut for explaining that nothing had changed, for example controllers often started the hand-over with 'PRAWNS as they were plus....' before going through strips to radar information.

Table 2.1 compares the findings from the present study with the data collected during the 1999 hand-over study. An indication of whether the particular aspect of the hand-over has improved since the previous study is provided in the right-hand column (a tick indicates there has been an improvement during the operational trial; a cross indicates no improvement and a question mark indicates no conclusive outcome).

One difference was that the incoming controller tended to ask questions and prompt the outgoing controller to provide information using 'PRAWNS' more often. This was particularly the case for trainees who seemed to use 'PRAWNS' as a prompt to get more information prior to taking over. This, as seen in Table 2.1, was a change from the initial study where the incoming controller was more inclined to quietly observe and not ask any questions. This change would suggest that use of the PRAWNS checklist made the incoming controller more proactive in the hand-over process.

Increased awareness of the hand-over process seems to have reduced the number of incomplete traffic plans handed over from one controller to another. (It seems the outgoing controllers have made an effort to complete as many plans as possible before hand-over, thus reducing the number of situations the incoming controller needs to actively monitor).

The mean duration of the hand-overs observed during the operational trial was 41 seconds (range 20 – 90 seconds). This increased the average time that hand-overs took prior to the introduction of the checklist by approximately 16 seconds. This finding was confirmed by just under half of the controllers who thought the time taken to do a hand-over had increased; the remaining 51 per cent felt there had been no change. Interestingly, this is balanced by a reduction in the time taken for the incoming controller to appear to 'settle into' the task. It could be that controllers require less time to settle due to longer interaction and the proactive exchange of information.

Table 2.1   Comparison with previous study

Item	PRAWNS Trial	Hand-over Study	Improve-ment
Formal Hand-over observed	91%	79%	✓
Differences in Style	9%	50%	✓
Incoming Quiet Observation only	37%	42%	✓
Incoming asking Questions/prompts	54%	21%	✓
Completed plans prior to hand-over	37%	14%	✓
Discrepancies observed	37%	50%	✓
Display/Chair adjustments	27%	40%	✓
Time of hand-over	Ave. 41sec.	Ave. 25 sec.	✓
Time for Incoming to 'settle into' task	40sec. – 4min.	1 min. – 10 min.	✓
Housekeeping done by Incoming	37%	21%	?
Housekeeping done by Outgoing	18%	Not Recorded	?
Distractions	15 in 11 hand-overs 46% Co-ordinator 54% Other	22 in 28 hand-overs* 35% Co-ordinator 65% Other	X

*data taken from both shadowing and video analysis of 1999 study.

The number of trivial initial tasks carried out by the incoming controller after hand-over, such as chair adjustment, also seem to have declined since the initial study. In addition, there seemed to be fewer discrepancies observed between incoming and outgoing controllers' views of priority tasks since the introduction of the PRAWNS checklist. The differences in hand-over style, particularly between familiar and unfamiliar controllers (e.g. spinners) were also much less noticeable following the introduction of the PRAWNS checklist. This was confirmed by several controllers who commented that the checklist had made the hand-over more standardized. It was also commented that the process had been made more structured, and reduced the chance of information being omitted.

The task of housekeeping (e.g. the addition of pending strips, removal of old strips etc.) was still done more by the incoming controller than by the outgoing controller. This was despite the best practice hand-over process recommending that the outgoing controller conduct any housekeeping activities before the incoming controller takes control of the sector. The recommended quality check during housekeeping by the outgoing controller was, therefore, not necessarily being done to identify any strip/radar discrepancies before hand-over.

Distractions were still found to occur during the hand-over, coming frequently from the co-ordinator, and also from On-the-Job Training Instructors (OJTIs) debriefing their trainee in the vicinity of the sector just after the incoming controller had taken control of the sector. Apart from these distractions, and possibly the housekeeping tasks, all other aspects of the hand-over observed during the trial were improved when compared with the initial hand-over study.

Incident Report Analyses

As mentioned previously, data were only collected for a three-month period so can only provide an initial indication of whether the checklist helped to reduce the occurrence of errors associated with hand-over practices. However, from the analyses it would appear that the number of incidents has been reduced. A summary of the key features present in the incidents analysed from both periods, i.e. 1999 – 2000 and 2000 – 2001, is shown in Table 2.2.

Table 2.2   Incident comparison of before and after introduction of PRAWNS checklist

Feature	Present in incidents Oct 99- Jan 00 (Pre-checklist)	Present in incidents Oct 00- Jan 01 (Post-checklist)
Shift hand-over	2	2
Band-boxed	5	1
Information Transfer related	4	1
Hand-over with different watch Controllers involved	3	1
Involvement of Support Controller (man /boy)	2	1
Housekeeping	1	1
Read-back	4	2
Mentor/trainee	6	1

This shows that the number of incidents that involved band-boxed sectors had reduced and those that could be attributed to information transfer also appeared to have decreased. In addition, there appears to have been a notable reduction in the number of read-back errors made between the two periods recorded. Shift hand-over was still a factor in the occurrence of errors, and the role of housekeeping, particularly between controllers from different watches, was still a possible issue.

Distractions from other controllers, such as co-ordinators, are not reported in incident reports. However, it may be that distractions can be inferred from some aspects of the incident reports analysed such as the presence of a mentor, returning support controller, trying to attract a supervisor's attention, shift hand-over noise etc.

Questionnaire Findings

In order to evaluate the use of the PRAWNS checklist, a questionnaire was produced to elicit feedback from controllers on their experiences and opinions of the checklist. The questionnaire was issued one month after the operational trial began. The response rate from the watch used for data collection during the trial was 100 per cent and responses ranged between 20 per cent and 40 per cent for the other four watches. A cut-off point of 70 per cent agreement or higher was used in the analyses to determine whether each response was significant. The questionnaire was divided into three main sections:

- presentation and usability of the PRAWNS checklist;
- best practice hand-over process;
- longer-term issues associated with the PRAWNS checklist implementation.

The findings associated with each of these sections are discussed next.

Presentation and Usability of the PRAWNS Checklist A summary of all significant responses (that is when 70 per cent or more controllers agreed) is outlined below:

- The content of the checklist was relevant and useful for hand-over.
- The level of information detail on the checklist was considered sufficient.
- The checklist was well laid out and was easy to read and understand.
- The checklist could be adequately seen when seated at the workstation.
- The checklist was positioned in the most appropriate place on the workstation.
- The size of the checklist was considered appropriate.
- No additional information needed to be added or removed from the checklist.
- Controllers said they did *not* always use the checklist. However, the majority of the controllers (between 58 and 75 per cent) used individual parts of it either initially or when they felt the hand-over required it.

- All watches, except one, did not need or want to be able to write on the checklist.

Best Practice Hand–over Process The best practice hand-over process guidelines were posted on the control room door and in the rest room, but no formal introduction to the process was provided. It is believed that in some instances the questionnaire respondents answered the questions as though they related to the PRAWNS checklist rather than the best practice hand-over process. The results of the observation data indicate that controllers were not very familiar with the 'best practice' process (despite 87 per cent indicating in the questionnaire they had read it!).

Longer-term Issues Associated with PRAWNS Implementation In the third section of the questionnaire controllers were asked about the role of co-ordinators during the hand-over process and about the positioning and implementation of both the checklist and best practice process for standard operations. With respect to the co-ordinator:

- Controllers felt that co-ordinators should not be formally incorporated into the hand-over process at all.
- The views on the best time for the co-ordinator to provide input to the hand-over/take-over process were very mixed but with a slight preference to input right at the end.

The main findings relating to the checklist implementation were:

- All watches, except one, were strongly in favour of the PRAWNS checklist being implemented permanently. The watch that differed was not opposed to the checklist but was less in favour of it than the other watches.
- Over 70 per cent of respondents thought that displaying the checklist on the workstation (as it was during the trial) was acceptable in the long-term.

Summary of Findings from Questionnaire A number of controllers made additional comments about the use of the PRAWNS checklist and about the hand-over process and these are summarized below. Positive comments about the checklist were that:

- It was considered a good intervention, and one which helped in the standardization of the hand-over process.
- Trainee controllers found the use of the checklist particularly helpful.
- Many detailed comments referred to the contribution of the checklist as a safety improvement, stating that it was a useful back-up to ensure a complete and structured hand-over.
- Controllers felt that it had 'ironed out' some of the differences in shift hand-over styles.

- Controllers said that although they did not always use the PRAWNS checklist in full, just thinking of the mnemonic itself provided a rapid prompt of what information should be considered for transfer.

Some negative points were also raised and are summarized as follows:

- A few controllers felt the checklist was superfluous, and stressed that experienced controllers should be able to carry out hand-overs and pass on information at the correct level given the situation. This view had also been elicited during the preliminary interview phase of the initial hand-over study. However, the video analysis and observation phases in that study found that although controllers know what they *should* do, they do not always practise their own advice, and in particular tend to drop procedures when under pressure.
- Around 40 per cent of the controllers felt that the checklist had increased the length of the hand-over.
- Some controllers already had their own set procedure that they preferred to use (further investigation indicated that this was particularly the case for Approach controllers).

Differences Between Approach and Terminal Manoeuvring Area (TMA) Sectors
Preliminary analysis of the questionnaire indicated that the order of the checklist might not be as favoured by Approach sector controllers, as for TMA controllers. For example, several Approach controllers reported that the most relevant information is which runway is in operation and this is the information that they normally give first during a hand-over. A series of brief interviews with Approach controllers (see method) revealed that the most important items of hand-over information to the Approach controllers were runway in use, Instrument Landing System (ILS) and gaps. An issue Approach controllers did not agree on was how important pressure was both at the start of a shift and throughout the shift. From these interviews a possible alternative order of information was identified for Approach controllers. These options were then presented to controllers via a questionnaire.

The results from this second questionnaire were inconclusive, with opinions being mixed between PRAWNS and the alternative mnemonic. However, as some controllers work on both Approach and TMA it was decided that the PRAWNS version (with a slight variation to the use of the letter 'A') should be adopted throughout out the Operations room (see Figures 2.1 and 2.2). This enables standardization of the hand-over process operationally thus preventing cross-validated controllers using two different processes. Having only one mnemonic was also considered an advantage for training new NATS controllers in the hand-over process.

P	**Pressure** High - Low - Min Stack
R	**Runway(s) In Use**
A	**Airports** ILS - Gaps - Freqs
W	**Wx** Vis – Avoidance - Winds
N	**Non-Standard/ Priority Info** NSFs – EATs & Holding NavAids – Danger Areas NODE-L Setup - Other
S	**Strips to Display**

P	**Pressure** High - Low - Min Stack
R	**Runway(s) In Use**
A	**Adjacent Sectors** Bandboxed - Split - Freqs
W	**Wx** Vis – Avoidance - Winds
N	**Non-Standard/ Priority Info** NSFs – EATs & Holding NavAids – Danger Areas NODE-L Setup - Other
S	**Strips to Display**

Figure 2.1   PRAWNS for Approach   Figure 2.2   PRAWNS for TMA

Best Practice Process One of the main issues identified during the trial regarding the best practice hand-over process was that insufficient awareness of it was raised amongst the controller population. Questionnaire results and interviews with controllers during the trial indicated that the majority had read the best practice process but could not remember what was contained within it. Observations confirmed that some aspects outlined in the process were regularly not being followed. The following sections describe aspects of the best practice hand-over process, which did not appear to have been fully adopted during the trial.

The results of the initial hand-over study suggested the development of a best practice process that included guidance on:

- distractions;
- splitting sectors around hand-over;
- controllers on the same sector having hand-overs in close succession; and
- the role of the co-ordinator in hand-overs.

This information was not carried forward to the version of the best practice hand-over process issued at the start of the PRAWNS checklist operational trial. This was because the best practice process was being considered for inclusion in the MATS manual and this guidance (although considered best practice) was not considered at the time to be practical enough to enforce. The findings from the trial indicated that these guidance items that had been removed from the best practice process were not resolved and therefore needed to be re-added to the process or addressed in a different way.

Distractions remained an issue during hand-overs, and in particular shift hand-overs. The Human Factors team involved in the trial observed one sector split within the first five minutes of a hand-over, followed by a second hand-over on the same sector some five to ten minutes later. This means that the final controller of the split sector received information third hand via two other controllers who had been working the position for less than ten minutes each. Discussions with group supervisors indicated that it was sometimes necessary for sector splits to occur so soon after hand-overs had occurred. There is a chance that this practice may increase the risk of information getting left out or being misinterpreted due to controllers not having a complete understanding of the situation. The best practice process stressed that the splitting of a sector soon after hand-over should be avoided where possible and instead pre-empting the need to split the sector before the hand-over.

Hand-overs between co-ordinators and between controllers working on the same sector occurring within minutes of each other were also seen during the trial. Whilst it may not be easy to prevent this, the issue may still be worked on to further improve the hand-over process. Perhaps, due to the poor awareness controllers had of the best practice process content, some aspects of it were not well adopted during the trial and these are discussed next.

Housekeeping In the trial, housekeeping did not appear to be considered an integral part of the hand-over process. Different individual working styles meant that housekeeping was done in a variety of ways and at different stages during hand-over. For example, some outgoing controllers tidied strips immediately prior to handing over, others tidied the strips during the hand-over whilst others left the housekeeping tasks to the new controller.

It was recommended that the outgoing controller carry out the housekeeping aspect of the hand-over process to help reduce the likelihood of cognitive errors (i.e. where an aircraft height and its recorded height are different) being perpetuated by the incoming controller. Therefore, prior to hand-over, the outgoing controller should use the housekeeping as a quality assurance check to ensure that any discrepancies are brought to light.

Staying in the Area This was rarely observed during the operational trial, although when questioned, the majority of controllers said they would stay if they thought it was necessary or if the sector had been particularly busy or complex at the time of hand-over.

Revision of the Best Practice Hand-over Process In light of the fact that controllers were not fully aware of the best practice process content, there was still an opportunity to revise the process in line with the trial findings and this was agreed with the operational management involved. An example of the revised best practice process is shown in Figure 2.3. Following this, LATCC controllers' awareness of this best practice process was raised to encourage its use.

HAND-OVERS & TAKE-OVERS OF OPERATIONAL POSITIONS

Pre-Brief before duty and before taking over

- A hand-over produces a workload of its own. *Give careful consideration to the timing of a hand-over.*
- If it is likely that the sector will split within 10 minutes, split the position before a hand-over
- Avoid a new co-ordinator & radar controller taking-over at the same time

During hand-over:

- Avoid distracting controllers involved in a hand-over. (E.g. OJTI briefings should be held away from the hand-over in progress, Co-ordinator inputs should be saved until after the hand-over where possible.)
- The oncoming controller must ensure that they have been able to assimilate all information relevant to a safe hand-over, *eliciting sufficient details from the handing-over controller as necessary.*

Controller Handing-Over	Controller Taking-Over
	Plug in to signal the start of a take-over
Tidy strip display	Evaluate the situation whilst the outgoing controller performs 'house-keeping'
Follow PRAWNS checklist	Question where necessary to check common understanding
Hand-over of control	Take control when satisfied a complete hand-over has been given
Remain to answer any queries and verify hand-over as appropriate	

After hand-over:

- Good practice dictates that the handing-over controller should remain available for a few minutes following the hand-over, particularly in dynamic traffic situations, to provide clarification/assistance regarding any points that may subsequently arise.
- Other controllers on the sector should only impart additional information after a hand-over is complete unless of operational necessity.

Figure 2.3   Revised hand-over/take-over process

Conclusions

In line with the recommendations made in the initial position hand-over study, a checklist (PRAWNS) was developed to assist controllers in the transfer and assimilation of appropriate information during a hand-over. A 'best practice' document was also produced to support the process of handing over and taking over an operational position. It was initially decided that there was no need to tailor the checklist for each Terminal Control (TC) sector and one checklist format (PRAWNS) was trialled on all TC watches over a period of six months.

The conclusions of this trial were as follows:

The operational trial of the PRAWNS checklist and best practice hand-over process demonstrated that significant improvements were made to the hand-over process.

PRAWNS Checklist

- The majority of controllers across all five watches supported the idea of the PRAWNS checklist. It was felt to be a safety improvement that provided a useful fallback to ensure that hand-overs are complete and structured. It also helped to standardize hand-overs and provide a hand-over structure for trainee controllers to follow.
- Controllers considered that the checklist was well laid out, easy to read, located in an appropriate position on the workstation, was an appropriate size, and adequate in detail.

Approach and TMA Differences

- Preliminary analysis of questionnaire findings indicated that the format of the PRAWNS checklist might not be as favoured by Approach controllers as for TMA controllers.
- Interviews and a second questionnaire used to clarify the Approach issues were inconclusive with only just over half of controllers preferring PRAWNS to an alternative checklist solution. On balance the PRAWNS checklist option was recommended for consistency with a minor amendment for Approach controllers.

Best Practice Process

- Controller awareness of the best practice process content was not well achieved during the trial. As a result some information contained in the best practice process appeared not to have been fully integrated into the hand-over process.

- Additional issues raised in the initial study but not carried through to the trial raised concerns relating to distractions, role of the co-ordinator and simultaneous hand-overs that still needed to be addressed.

Incident Analysis Reports

- The comparison of the incident reports during the trial period with incident reports from the same period the previous year indicated that there had been a significant reduction in the number of incidents since the PRAWNS checklist had been introduced.

Observed Hand-Over Improvements

- The time taken to hand-over a position seemed to have increased on average by around 16 seconds. However, the overall time that it appeared to take for controllers to 'get the picture' seemed to be less.
- The increased level of questioning by incoming controllers seemed to indicate that the structured process made it easier for them to ask questions and check that they had understood the situation and information being transferred.
- An increase in the use of a formal structure during the hand-over process was observed during the trial.
- Standardization of the hand-over process, and a reduction in the differences both between shift hand-over and individual hand-over styles was observed.
- The trial showed an increase in the amount of prompting of the outgoing controller by the incoming controller, particularly by trainee controllers during the hand-over process. This finding would suggest that the use of the PRAWNS checklist has made the incoming controller more proactive in the hand-over process and given less experienced controllers support in getting the relevant information from the outgoing controller.
- A reduction in the number of incomplete traffic plans handed over from one controller to another was seen during the trial.
- There was a reduction in the number of trivial tasks carried out by the incoming controller after the hand-over (for example chair adjustments).
- Fewer discrepancies were observed between incoming and outgoing controllers' view of the priority tasks.

Areas for further attention were highlighted, such as the need to address distractions during the hand-over period from co-ordinators who interrupted inappropriately and from OJTIs who conducted debriefs with trainees immediately behind the workstation.

Operational Impact

As a result of the study it was recommended that the PRAWNS checklist be implemented to provide more comprehensive guidance to controllers on the hand-over process. The position hand-over checklist and best practice process (as revised at the end of the trial) were implemented formally as a new procedure within the LATCC Terminal Control (TC) Manual of Air Traffic Services Part II (MATS II) in Spring 2001. The considerable controller involvement in the development and trial of the checklist significantly aided the acceptance of the checklist into operational use.

Since its implementation at LATCC the use of similar position hand-over checklists has spread within NATS to the training college, operational training sections at the units, and has been incorporated in new OJTI refresher training materials. It is also now used in several other NATS Area Control Centres and Airports. In addition to this, its use has spread externally to other Air Traffic Control (ATC) units in Europe and Canada.

Chapter 3

Runway Safety

Kim M. Cardosi

Introduction

The worst collision in aviation history was the result of a runway incursion. It occurred in 1977 at Tenerife and resulted in 583 fatalities. It is rare for a runway incursion to result in a fatal accident. Since 1st January 1990 there have been four fatal accidents resulting from runway incursions at airports with an operating air traffic control tower in the US. Nonetheless, runway incursions clearly pose significant threats to aviation safety and efficiency.

Part One: The Problem

What the Numbers Tell Us

Table 3.1 Classification of runway incursions by year

Fiscal Year	Operational Errors	Pilot Deviations	Vehicle/Pedestrian Deviations	Total Deviations
2000	83	247	75	405
2001	91	233	83	407
2002	75	190	73	338
2003	90	174	60	324

Source: Federal Aviation Administration, Office of Runway Safety (ARI)

Runway incursions are classified as to whether they are attributed to an error on the part of the air traffic controller (i.e., an operational error), pilot (i.e., a pilot deviation), a vehicle driver, or a pedestrian. The categories are not mutually exclusive; rarely, a single incursion is classified as more than one type of error. As can be seen in Table 3.1, while the numbers were stable for Fiscal Year 2000 and 2001, the numbers of reported runway incursions – both those attributed to pilot deviations and those attributed to controller (operational) errors were down by almost 20 per cent in 2002 and 2003. Also, the number of runway incursions attributed to pilot deviations is consistently more than twice the number of reported runway incursions attributed to controller errors.

Within pilot deviations, we find that the smaller general aviation aircraft are more likely to be involved in a runway incursion than commercial aircraft. A study of runway incursions from 1998 through 2001 revealed a rate of five runway incursions per million operations, i.e. takeoffs and landings (FAA, June 2002). While general aviation aircraft comprise approximately 58 per cent of the total operations, they are involved in approximately 65 per cent of the reported runway incursions (FAA, June 2002). Still, there is an average of one serious (A or B[6]) runway incursion involving two jets somewhere in the National Airspace System (NAS) each month.

Several studies have been conducted to examine the causes of these incidents and to identify solutions to the underlying problems. These studies provide a strong foundation for the identification of pilot, controller, and environmental factors that contribute to runway incursions. They also identify potential remedies that include recommended changes to procedures, training, and technological enhancements. This chapter will examine the major findings and recommendations that resulted from these studies and discuss the role that Human Factors has played in helping to: refine our understanding of the causal factors of runway incursions, assess the various recommendations, and decide between competing strategies proposed to mitigate runway incursions.

It is no surprise that nearly all runway incursions are caused by human error. While the opportunities for equipment malfunctions to cause such problems are relatively rare, the opportunities for human error are abundant. The proximity and number of aircraft in the terminal environment, combined with the complexity of operations and the requirement for split-second timing, make the airport surface and proximal airspace extremely unforgiving of (even the inevitable) errors of pilots, controllers, and airport vehicle drivers.

Pilot Errors That Result In Runway Incursions

Official reports of pilot errors that result in runway incursions that are filed by the Federal Aviation Administration (FAA) offer a succinct account as to what the aircraft did to create the incursion (e.g., Aircraft A crossed Runway 25 without a clearance while Aircraft B was on takeoff roll on Runway 25). Until late 2002, these reports rarely contained any information as to why the incursion happened. Because of this, studies of the types of pilot errors that resulted in runway incursions in the United States have focused exclusively on voluntary reports submitted by pilots (such as those submitted to the Aviation Safety Reporting System or ASRS) and information obtained from opinion surveys. While these studies cannot speak as to the relative frequency of occurrence of specific

---

[6] The definitions for the classes of the severity of runway incursions can be found on pages 67ff.

categories of pilot errors that result in runway incursions, they do offer insights as to types of errors that occur, and the factors that contribute to these errors. Most of these studies were conducted by independent researchers. The notable exception was the work performed by the Runway Incursion Joint Safety Analysis Team (JSAT). This team, sponsored by the Commercial Aviation Safety Team (CAST) and the General Aviation Joint Steering Committee (GAJSC), performed a detailed examination of information from several sources, including accident reports published by the National Transportation Safety Board and proprietary reports of pilot deviations from airline databases.

The most commonly cited cause of pilot errors resulting in runway incursions is 'loss of situation awareness' (Kelly and Steinbacher, 1993; Adam and Kelly, 1996; JSAT, 2000). This usually refers to cases in which pilots think that they are at one location on the airport (such as a specific taxiway or intersection) when they are actually at another (such as another taxiway, intersection, or a runway). The other commonly cited causal factor is communication errors between pilots and controllers (Kelly and Steinbacher, 1993; Adam and Kelly, 1996; JSAT, 2000; Cardosi and Yost, 2001). This includes readback errors (i.e., errors in repeating the controller's instruction back to the controller), misunderstanding ATC instructions, and accepting a clearance intended for another aircraft. Factors that have been cited in these studies as contributing to these errors are: poor signage and markings (particularly the inability to see the hold short lines), controllers issuing too much information in a single transmission, controllers issuing instructions as the aircraft is rolling out after landing (when pilot workload and cockpit noise are both very high), and pilots focusing on some tasks at the expense of more important tasks.

These studies also identified the need for cockpit standard operating procedures (SOPs) for ground operations to help ensure that non-essential tasks are completed during relatively low workload and non-critical phases of operation, and (both) pilots are aware of the location of their aircraft on the airport surface, the location of all critical elements in the airport environment (e.g., hold short points, intersecting runways, aircraft on approach, etc.) and their ATC clearance.

Controller Errors That Result in Runway Incursions

Several studies have focused on the types of controller errors that result in runway incursions (Bales, Gilligan, and King; Steinbacher, 1991; Cardosi and Yost, 2001). These studies have analyzed both official government reports of the results of the investigation of these incidents and voluntary reports submitted by controllers to ASRS. The findings of these studies converge on several key points. The most common controller-related factors identified in these studies were:

- forgetting about an aircraft, the closure of a runway, a vehicle on the runway, and/or a clearance that he/she had issued;
- failure to anticipate the required separation or miscalculation of the impending separation;

- lack of, or incomplete, coordination between controllers;
- misidentifying an aircraft or its location; and
- communication errors – readback/hearback errors, or issuing an instruction other than the one the controller intended to issue.

Memory Failures The most commonly identified contributing factor, occurring in 27 per cent of the controller errors examined in one study (Cardosi and Yost, 2001), was the controller 'forgetting' something such as an aircraft (that had been cleared to land, one holding on a runway, etc.), a vehicle on the runway, the closure of a runway or a clearance that he/she had issued. This was a more common type of error than mistakes in estimating or projecting the separation. Another causal factor found in analyses of controller errors is inadequate coordination between controllers. Coordination among controllers in the tower is necessary for efficient and safe operations. Lack of, or incomplete, coordination was noted as a contributing factor in approximately 20 per cent of the controller errors in the tower (Cardosi and Yost, 2001). The most common examples of inadequate coordination between controllers were either a failure of one controller to relay information to another controller or a failure to obtain approval for a specific operation (such as a failure to coordinate a runway crossing).

Communication Errors Communication errors between pilots and controllers are another common factor in runway incursions, occurring in almost 20 per cent of the controller operational errors (Cardosi and Yost, 2001). Catching readback errors is an extremely difficult task for two reasons. First, while the readback is occurring, the controller is naturally thinking about the next clearances to be issued. This fact, combined with the human tendency to hear what we expect to hear, conspire to make catching readback errors a very difficult task. While controller-pilot communications are surprisingly accurate (with only one per cent of the controller transmissions resulting in readback errors), the opportunities for problems are numerous due, in part, to the sheer number of communications. In the tower environment (local control position) there was an average of one readback error every 2.5 hours per frequency and one uncorrected readback error every 7 hours (Cardosi, 1994). Other types of controller errors that result in runway incursions are mistaking one aircraft for another and assuming that an aircraft is somewhere other than its actual location (such as taxiing onto the runway from an intersection as opposed to taxiing onto the runway at the end).

Presence/Absence of a Supervisor A final factor that these studies point to is the absence of a supervisor (who was not working a control position), as a possible contributing factor. In many cases, it was the supervisor that prevented a bad situation from getting much worse. Many voluntary reports submitted by controllers to the Aviation Safety Reporting System lament the need for 'another set of eyes' in the tower cab. 'Another set of eyes' can be electronic as well as human. While the flexibility and speed of human judgment are not able to be duplicated by artificial intelligence, tools that display traffic and alert the controllers to potential problems can be quite useful in preventing accidents and incidents. This is a difficult task for automation to perform because too many false alarms or too little response time each makes the system ineffective. If the system 'cries wolf' too often, then the controller will eventually ignore it. On the other hand, for the system to be useful, it must be able to warn the controller that an action is required with enough time to formulate and execute corrective action. Warnings that serve as little more than an indication as to where to look to witness an accident are as useless as warnings that are more often false alarms than legitimate alerts.

All of the studies discussed cited the need for increased redundancy in the tower. Clearly an 'extra set of eyes,' whether they be human or electronic, (preferably with some analytic ability attached) can go a long way toward providing the necessary oversight to prevent operational errors in the tower. Memory (and to a lesser extent, judgment) will fail at times, no matter how experienced, well-trained, motivated, or conscientious an individual is; it is this fallibility that well-designed automated tools can, and should, be implemented to offset.

Experience The experience level of the controllers involved was not identified as a causal factor; controllers of all experience levels were involved in the incursions studied. However, the controllers with the most experience were acting in a supervisory (supervisor or controller-in-charge) role at the time of the incident and were usually working control positions simultaneously.

Visibility Visibility is not a significant causal factor in runway incursions. Most runway incursions occur during the day in good visibility – the same conditions under which most operate most frequently. Visibility is, however, a factor in accidents. In one study, only 17 per cent of the accidents that occurred on the surface of a towered airport occurred in good visibility during the day (Cardosi and Yost, 2001).

Frequency Congestion Kelly and Steinbacher (1993) were the first to identify radio frequency congestion as a contributing factor to runway incursions by resulting in blocked transmissions, incomplete messages, repeated communications, and misunderstood instructions. This finding was strongly confirmed by controller and pilot opinion in extensive surveys of tower controllers (Kelly and Jacobs, 1998) and of airline pilots on airport surface operations (Adam, Kelly, and Steinbacher, 1994; Adam and Kelly, 1996).

> The survey findings show this interface to be one of the weakest parts of the airport surface system. Surface operations have changed markedly in the recent years as ATC has accommodated more and more traffic. The voice communication that worked effectively with less traffic is now strained to the breaking point during peak traffic periods. At these times, the controllers cannot communicate with the pilots in the way ATC-pilot communication was designed to work. The original design intentionally

included safety measures such as proper timing and readbacks, which are now being dropped so that more ATC instructions can be crowded onto the frequencies at busy times. Yet these are the very times when the consequences of errors may be more critical, and safety measures are needed the most. The complexity of some current operation means that any breakdown of the ATC-pilot interface can be critical to safety. The potential for such breakdown is now greater than ever (p. 7-10).

The same sentiment was later echoed by the Runway Incursion Joint Safety Analysis Team in their 2000 report that stated:

One of the weakest areas of the modern aviation is the industry's continued reliance on a relatively archaic method of communicating information, specifically via one-at-a-time radio transmissions. These transmissions are rather frequently garbled, 'stepped-on', blocked, and otherwise difficult, if not impossible, to understand (p. 36).

While voice communications are surprisingly accurate, the volume and complexity of the traffic at busy airports make voice communications a tenuous safety link. Human Factors studies of controller-pilot voice communications found that radio frequencies in the terminal environment are far more congested than those in the En-route environment. While En-route controllers averaged less than two controller-to-pilot communications per minute (Cardosi, 1993) ground controllers averaged eight controller-pilot communications per minute (Burki-Cohen, 1995). The average for local controllers was three (Cardosi, 1994) and TRACON was 4.5 per minute (Cardosi, Brett, and Han, 1996; note that these numbers are now several years old and so probably represent conservative estimates of today's communication activity; however, they are the only data available on the subject). Perhaps because of the frequency congestion, 27 per cent of the local controllers' transmissions and 33 per cent of the ground controllers' transmissions are responded to by pilots with only an acknowledgement (e.g., 'roger'); an additional 7 per cent of the local control transmissions are responded to with only a microphone click. Twenty-eight per cent of the controller messages on the local control frequencies and 32 per cent on ground control are responded to with a full readback (Cardosi, 1994; Burki-Cohen, 1995). This is dramatically lower than the 71 per cent of the controllers' transmissions En-route and 60 per cent of the TRACON controllers' transmissions that are fully read back. This same series of studies also revealed that the factor most consistent with controller-pilot miscommunications (as defined as readback errors and pilot requests for repeats) is similar call signs on the same frequency. This was a coincident factor in 12 per cent of the miscommunications on the local frequencies.

The results of surveys of controllers (Kelly and Jacobs, 1998) mirror those of the pilots (Adam, Kelly, and Steinbacher, 1994) in their perception of the magnitude of the frequency congestion problem. Nineteen per cent of the controllers from the busiest towers (then 'Level 5') surveyed said that frequency congestion was a 'significant' risk factor for surface incidents. An additional 37 per cent said that it was a 'moderate' risk factor. Twenty-nine per cent of the controllers from the busiest towers said that an 'inability to access the frequency when needed' was experienced 'often'.

Interestingly, when controllers were asked to describe pilot phraseology that is associated with a risk of surface incidents, they cited 'abbreviated or incomplete readbacks', and 'using microphone clicks as a substitute for verbal readbacks.' In fact, 71 per cent of the comments offered were critical of incomplete readbacks. (this figure was derived from the raw data in the appendices and is not contained in the actual report.) Analysis of ATC tapes revealed that, on local control frequencies, seven per cent of the controllers' transmissions were responded to with only a microphone click (Cardosi, 1994). While this lets the controller know that 'somebody got something,' it does not afford any opportunity to catch a communication error. It is certainly the case that full readbacks are the safest pilot response as they at least provide the opportunity for a communication error to be caught before it results in a surface incident. If pilots tried to respond to each transmission with a complete readback, the resulting increase in frequency congestion would be intolerable at busy facilities (p. 7-32).

Microphones Stuck in the Transmit Mode Twenty-six per cent of the respondents and 32 per cent of the controllers from Level 5 towers said that there was a 'significant' risk of surface incidents associated with a stuck microphone. An additional 38 per cent of the controllers said that there was a 'moderate' risk associated with a stuck microphone. Roughly half of the respondents and 62 per cent of the controllers from Level 5 facilities said that transmissions were 'stepped-on' 'often.' Forty per cent of the respondents and 56 per cent of the controllers from Level 5 facilities said that 'blocked readbacks' were experienced 'often.' The report recommends that the FAA implement the use of anti-stuck microphone and antiblocking radio technology for all ATC ground radios and the radios of all aircraft operating in the ATC system (p. 7-30).

Blocked Transmissions Blocked and partially-blocked (also known as 'stepped-on') communications present a serious threat to aviation safety. Nowhere was this fact more dramatically demonstrated than in the 1977 collision on the runway at Tenerife that resulted in 583 fatalities. In this tragic event:

> The Pam Am crew was alarmed by the way in which the Air Traffic Clearance was issued. The captain... feared that...the KLM could possibly take the ATC clearance as a take off clearance and, immediately after the tower controller had said 'Okay', and paused for almost two seconds, he and his first officer jumped in to inform the KLM crew that they were still taxiing on the runway. The message of the Pan Am crew coincided with the message of the tower controller who, at that moment, told the KLM aircraft to wait for takeoff clearance. The coinciding transmission on the same frequency resulted, in the KLM cockpit only, in a strong squeal. Because of this, both vital messages were lost to the KLM crew resulting in the worst collision in the history of

aviation. Every day there are incidents of blocked communication that are less dramatic, but have the same potential for disaster (Extracted from the conclusions presented to the Netherlands Board of Inquiry by the Director General of Civil Aviation).

As the amount of air traffic and radio frequency congestion increases, blocked and partially-blocked transmission present an increasing risk to aviation safety. When a pilot or controller is not able to access a frequency due to a 'stuck mike,' the most fundamental safety net – that provided by voice communications between pilots and controllers – is gone. Partially-blocked transmissions ('step-ons') are far more common than microphones stuck in the transmit position. While these events are typically less dramatic than that of a stuck microphone, they too, have the potential for disaster – as in the Tenerife accident cited above.

Proffered Solutions to the Frequency Congestion Problem

Wider use of Controller-Pilot Data Link Communications (CPDLC) has been proposed to significantly reduce both frequency congestion and the number of communication errors. Currently in the US, CPDLC allows for some routine transmissions in an extremely limited implementation (i.e., only at Miami Center). For example, frequency changes can be automatically uplinked to an equipped cockpit when the controller transfers the aircraft from one sector to another. Expanded use of CPDLC would afford the controller the capability to uplink additional information (such as instructions) to the cockpit. It would also afford pilots the capability to downlink information (requests, acknowledgements, etc.) to the controller. While CPDLC might provide a sorely needed alternative to voice communication between pilots and controllers, it is not expected to reduce the rate of communications errors. Any alternative to voice communication is likely to change the types of errors observed, but is not likely to reduce the error rate. With CPDLC, for example, it would still be possible to send the wrong message to an aircraft or for the pilot to misread the instruction.

NEXCOM (Next Generation Air-Ground Communication System) proposes to increase the efficiency and capacity of air-ground voice communication, in part, by providing digital modulation. The first implementation of NEXCOM is expected to use the VHF Digital Link (VDL) Mode 3 protocol and has the potential for multiplying the number of available voice channels up to four (Kabaservice, 1998). NEXCOM will operate in parallel with the present analog voice system and is expected to be implemented in high and ultra-high altitude sectors by 2008. Selected high density terminal sectors are scheduled to transition to digital NEXCOM by 2015.

One important requirement that NEXCOM may not be able to satisfy in an operationally acceptable way is the need to eliminate blocked and partially-blocked voice transmissions. As traffic continues to increase, the amount of frequency congestion and problems associated with blocked communications will continue to escalate. The incidence of blocked and partially-blocked communications in today's ATC environment has not been systematically studied. However, in an analysis of four hours of TRACON voice tapes (two hours each from Atlanta and Dallas/Ft.Worth), an average of 1.5 per cent of the controller transmissions needed to be repeated after being partially or totally blocked (DiFiore, 2001). Simulation studies have shown that the number of blocked transmissions increases both as the number of communications increases and with the amount of delay (between the onset of the speaker's voice and the beginning of the transmission as heard by the listener) inherent in the system (Nadler, et al., 1990). A high-fidelity En-route simulation study was conducted to assess the level of communication delay that would be acceptable to controllers in the NEXCOM system. This study found that 10 per cent of the communications were blocked (by pilots and controllers combined) with even the lowest communication delays imposed (Sollenberger, McAnulty, and Kerns, 2003).

In a study of communication errors reported to the Aviation Safety Reporting System (ASRS), blocked communications was identified as a factor that contributed to runway transgressions, altitude deviations, loss of standard separation, and pilots accepting a clearance intended for another aircraft (Cardosi, Falzarano and Han, 1998). While similar-sounding call signs are the number one contributing factor to a pilot accepting a clearance intended for another aircraft and other critical communication errors, the risk of a blocked or partially-blocked transmission can compound the problem. When the 'wrong' aircraft accepts a clearance, the pilot's readback can alert the controller (and other pilot) of the misunderstanding – as long as the readback contains a call sign and is not blocked. If two pilots respond simultaneously – as one would expect in the situation in which two pilots think the clearance is for them – at least one readback is likely to be blocked.

While NEXCOM is proposed to incorporate anti-blocking capability, the precise technology that will be used to afford this capability has not yet been defined. Furthermore, the implementation schedule does not project this capability to be available at airports before the year 2015. Projections in the rate of air traffic and the concomitant increase in frequency congestion and blocked transmissions make such a schedule problematic in addressing the future needs of the NAS. Meanwhile, anti-blocking technology has been in use at an ATC facility in the UK for several years. It has also been installed in aircraft by Austrian Air and Britania. Installation of such technology has been advocated by the Allied Pilots Association and the Air Line Pilots Association for over five years.

Part Two: The Role of Human Factors in Suggested Strategies to Reduce Runway Incursions

The strategies proposed to reduce the number of runway incursions and mitigate their effects can be grouped into the areas of changes in procedures, training and education, and technological enhancements.

Technology

The technological enhancements suggested by studies on runway incursions (by the NTSB, 1986; Kelly and Jacobs, 1998; and Cardosi and Yost 2001; among others) range from simple changes to the dimensions of airport markings to a sophisticated surface conflict detection system. At one end are the relatively low-cost improvements in signs and painted markings on the airport surface. The need for reliable and more conspicuous airport markings (particularly runway crossings and hold short lines) that can be seen under all weather conditions has received considerable attention and has resulted in improved signage and markings. Research on further improvements to the current airport markings is in progress. The same studies also identified the need for implementation of surface radar at facilities that are not so equipped, and improvements to the present system, such as the ability to tag radar targets with the aircraft callsign and the addition of conflict detection capability. This recommendation is being implemented with the installation of low-cost surface radar scheduled for busy airports that do not have Airport Surface Detection Equipment Model 3 (ASDE-3) and planned enhancements to surface radar. Ground induction loops that sense aircraft and vehicles passing over the surface (taxiway or runway) is another technology that is being pursued to help prevent runway collisions. The most sophisticated conflict detection tool that is planned for implementation is the Airport Movement Area Safety System (AMASS). AMASS will be implemented at the 33 busiest airports (that already have ASDE-3).

Other technological enhancements include indicators to pilots as to whether or not a runway is occupied. Such enhancements have the advantage of providing the information directly to the pilots who can affect a resolution more quickly than (all things being equal) if the conflict information is provided to the controller who then needs to formulate a resolution and relay it to the pilot. 'Flashing PAPIs' (Precision Approach Path Indicator) lights would signal a pilot on approach if the runway were occupied and Runway Status Lights (RWSL) placed at runway intersections would automatically turn red if an aircraft were on the runway or on final approach to that runway. While the concept of automated runway status lights has received anecdotal approval from both pilots and controllers, it is unclear whether the lights need to be capable of displaying more than 'red' or 'off'; another color (such as yellow) may be required to indicate to the pilot that the system is working when red is not displayed.

Any technological enhancement suggested reducing the incidence and severity of runway incursions can only be successful if it satisfies basic Human Factors requirements. Human Factors requirements help to ensure that the system minimizes the probability of human error and helps to prevent inevitable errors from propagating through the system. The information presented to the pilot or controller must be valid, reliable, and timely. This means that the information must be accurate, trustworthy, and within a timeframe that allows for the information to be processed and acted upon to achieve the desired result. If the system is not designed well from a Human Factors standpoint, then the expected benefits of the system will not be realized. An evaluation of the effectiveness of a specific technology usually requires testing in a well-designed simulation study using a representative set of operational conditions and users (pilots and/or controllers).

Education and Training

There have been several efforts over the last several years to increase pilot and controller awareness of runway safety. There has been an increased emphasis on runway safety in regularly scheduled seminars and a wide distribution of educational materials. Specific training programs are being developed for tower controllers that focus on key areas found to be causal factors in runway incursions.

Since a major contributing factor to controller errors that resulted in runway incursions has been the controller forgetting something (such as a clearance that he/she had issued or the presence of an aircraft), controllers will receive training on the limitations of human memory, the detrimental effects of workload and distraction on working memory, and what can be done to help prevent errors due to memory failures. Also, the controllers' workstation, and tasks (procedures) can be designed to support the controller having the critical information readily available.

Other training for tower controllers geared toward reducing runway incursions will focus on teamwork. Effective teamwork in the tower can catch and correct controller errors before they result in a collision. It is very difficult to determine the effectiveness of any given training program with any more than participants' subjective assessments (the case in point is the plethora of articles on attempts to evaluate the effectiveness of crew resource management (CRM) in the cockpit). Nonetheless, such programs are at least effective in raising the awareness of key issues and usually impart useful information in terms of strategies to help reduce errors.

Changes to Air Traffic Procedures

Several changes to air traffic and flight deck procedures have been suggested to increase runway safety. Two of the most notable recommendations that have been implemented in the last few years concern instructions to a pilot to hold short (i.e., not enter) a runway and holding in position on the runway. In an effort to increase compliance with hold short instructions, pilots are now required to readback all instructions to hold short of a runway, and controllers are required to obtain the pilot's readback of hold short instructions. To reduce the risk inherent in holding on active runways at an intersection, the practice was prohibited at night or whenever the intersection is not visible from the tower (either from out of the window or on surface radar).

There are three other changes to air traffic management procedures that have been recommended by the National Transportation Safety Board (NTSB) and are presently being considered by the FAA. These recommendations concern the issuance of landing clearances to multiple aircraft, holding aircraft on the runway in anticipation of issuance of the takeoff clearance, and requiring a specific clearance to cross a runway. In order to evaluate the relative risk associated with each of these issues, the FAA's Office of Runway Safety (ARI) convened a runway safety procedures working group to gather data respective to the relative risk of the taxi into position and hold procedure. The working group was comprised of representatives from FAA Air Traffic and NATCA (National Air Traffic Controllers Association), and support (data analysis) personnel. All runway incursions that occurred between September 1998 and September 2001 were examined to determine the number of runway incursions that were caused by the types of situations that elicited the recommendations.

Multiple Landing Clearances In 1991, the National Transportation Safety Board (NTSB) recommended that the FAA 'preclude the issuance of multiple landing clearances to aircraft outside the final approach fix' and 'establish a numerical limit so that no more than two landing clearances may be issued to successive arrivals'. The Board recommended that the FAA adopt the International Civil Aviation Organization (ICAO) procedure and not issue a landing clearance to a following aircraft until the preceding aircraft has crossed the runway threshold (NTSB Recommendation A-91-28).

The FAA initially decided not to implement the change, stating that such a change would increase controller workload and could compromise safety. More recently, the working group convened by the Runway Safety Office to revisit the NTSB recommendations to changes in ATC procedure reviewed runway incursions to assess the incidence of incursions that could be attributed to the issuance of multiple landing clearances. Based on this analysis, there is no evidence that issuing multiple landing clearances contributes to the probability of a runway incursion. An analysis of the runway incursions from September 1998 to September 2001 found that only one of the 956 incidents involved the issuance of multiple landing clearances. In support of the recommendation, the NTSB cites the following incident:

> On February 9, 1998, American Airlines flight 1340 (AAL1340), a B-727, landed short of the runway 14R threshold while attempting a Category II ILS approach at ORD and then slid off the side of the runway. (Weather 100 overcast, Visibility ½ mile in freezing

fog). Subsequently, one aircraft landed with debris on the runway and a second aircraft touched down while executing an ATC directed go-around.

However, it is not clear how issuing of multiple landing clearances may have contributed to this accident; nor is it the case that if the landing clearances had been issued differently (i.e., in accordance with the procedure recommended by ICAO Document 4444-RAC/501, 'Procedures for Air Navigation Services – Rules of the Air and Air Traffic Services,' Part V, 'Aerodrome Control Service,' paragraph 15.2), the outcome would have been different. It is most likely that the controllers would still have issued landing clearance to the number 2 aircraft because they were not aware of the fate of the first aircraft. Therefore, the outcome for the second and third landing aircraft would have been the same. Also, if the visibility had been sufficient for the controller to see what had happened to AAL 1340, the controller would have issued a 'go-around' to the second aircraft, irrespective of whether a landing clearance had previously been issued.

Changing the procedure to prohibit the issuing of multiple landing clearances could contribute to inefficiencies and errors, in a number of ways. First, such a prohibition would increase the number of controller transmissions which would, in turn, increase frequency congestion and the probability of a blocked or 'stepped-on' transmission. Second, requiring controllers to refrain from issuing a landing clearance to an aircraft on approach until the previous aircraft has crossed the threshold breaks up the controller's task into subtasks that must be performed under artificial time constraints. Such constraints increase the probability that a person is likely to forget a latter component of the task, in this case, issuing the clearance to land. Also, in terms of task management, controllers are accustomed to being able to issue the landing clearance at the first logical opportunity, thus completing the logical sequence of actions so they can mentally move on to the next required sequence of actions. This sequence of actions has a critical timeframe within which they must be performed to maintain efficient and safe operations. Interrupting this sequence with artificial constraints (i.e., constraints that are not an intuitive or logical component of the task) can cause distraction and lead to errors. Also, controllers are accustomed to using the timing of landing clearance as a cue to (remembering) what other actions need to be taken. For example, a controller could withhold a landing clearance if there is some uncertainty that the runway would be clear in time (e.g., due to an aircraft landing on an intersecting runway or a vehicle on the runway). This would serve as a reminder to the controller to re-verify that the other aircraft or vehicle had completely cleared the runway before issuing the clearance to land. On the other hand, if a series of aircraft is well-sequenced with ample separation, the controller is able to issue the multiple clearances to land and thus complete the task of issuing that particular set of clearances, freeing their attention to scan the airport and monitor the progress of the aircraft.

Third, it is just as critical to consider the effect of prohibiting multiple landing clearances on the pilot. It is clearly beneficial to the pilots to receive the landing clearance early. Many factors, such as blocked radio transmissions and talking aircraft systems (e.g., enhanced GPWS that issue altitude call outs – particularly the call outs at 2500 and 1000 feet) can interfere with successful transmission of the clearance. The earlier in the sequence the pilots receive the clearance, the sooner the pilots are certain of the landing runway and can focus their attention on the other tasks associated with landing and taxiing (such as anticipating the taxi routing). Imposing an artificial time constraint and order to controller subtasks increases the probability that one of the latter actions in the sequence, in this case issuing the landing clearance, would be forgotten. If this were to happen, either the pilot would have to request the landing clearance (thereby contributing to frequency congestion), or (if the frequency were occupied) the pilot would either go around, or land without a clearance.

Presently, controllers can issue multiple landing clearances before ensuring that the runway is clear. However, it is important to remember that the issuing of multiple landing clearances does not relieve the controller of the responsibilities of ensuring that the runway is clear before an aircraft lands and of monitoring the aircraft to ensure that they comply with the controller's instructions (i.e., do what the controller intended). Therefore, there is no advantage to requiring the controller to refrain from issuing a landing clearance to an aircraft on approach until the previous aircraft has crossed the threshold.

Implicit Runway Crossings [14 CFR 91.129(i)] The NTSB also recommended that the FAA Amend Federal Aviation Administration Order 7110.65, 'Air Traffic Control,' to require that, when aircraft need to cross multiple runways, air traffic controllers issue an explicit crossing instruction for each runway after the previous runway has been crossed. (NTSB Recommendation A-00-68). There are two components to this recommendation to amend 14 CFR 91.129(i). The first is the requirement that all runway crossings be authorized only by specific ATC clearance. This could be implemented in different ways. The most sound, from a Human Factors standpoint, would be to issue the clearance to a runway and include the instruction to cross the identified intervening runways. (This is the procedure commonly used in Europe.) An example of this phraseology would be, 'Taxi to Runway two-seven, cross runways one-eight and two-four.'

The second component of the NTSB recommendation, however, states that if aircraft need to be cleared to cross multiple runways, 'controllers should issue an explicit crossing instruction for each runway after the previous runway has been crossed'. The objections, on Human Factors grounds, to the logic of this component of the proposed change is similar to the objection to eliminating multiple landing clearances. Requiring controllers to issue a runway crossing after the previous runway has been crossed, places undue constraints on the controllers and would contribute to inefficiencies and controller errors. First, such a restriction would increase the number of controller transmissions which would, in turn, increase frequency congestion, the probability of a blocked transmission, and the opportunity for errors based upon misunderstanding. Second, requiring the controller to refrain from issuing crossing instructions until after the previous runway has been crossed breaks up a controller's sequence of actions into subtasks that must be performed under artificial time constraints. This sequence of actions has a critical timeframe within which they must be performed to maintain efficient and safe operations. Interrupting this sequence with artificial constraints (i.e., constraints that are not an intuitive or logical component of the task) can cause distraction and lead to errors. In this case, it is likely that such a constraint would increase the probability that the controller would forget about an aircraft waiting to cross a runway and, thus, forget to issue the crossing clearance.

At present, efforts are underway to further assess the use of 91.129(i) and project any changes to the use of the procedure on the effects on capacity, workload, and pilot and controller error.

Position and Hold In the United States, until 31st January 2003, the controller's instruction to the pilot to take the runway and await takeoff clearance – the ICAO equivalent of 'line up and wait' – was 'taxi into position and hold'. This difference in air traffic terminology was complicated by a relatively recent (1st November 2001) change to ICAO terminology changing 'taxi to holding point' – roughly the US equivalent of holding short of a runway – to 'taxi to holding position'. The rationale for the change was that a holding point is a point in space (i.e., a 'fix'). Nonetheless, it was predictably problematic for some of the same pilots to hear 'taxi to holding position' – meaning do not go on the runway – and 'taxi into position and hold' – meaning go onto the runway. Because of this, on 31st January, 2003, the US terminology was changed from 'taxi into position and hold' to 'position and hold'.

The issues surrounding aircraft holding on a runway go far beyond terminology. FAA Order 7110.65, paragraph 3-9-4 'Takeoff Position and Hold,' authorizes air traffic controllers to allow aircraft to taxi into position and hold on a runway pending resolution of traffic conflicts or other issues that preclude immediate issuance of a takeoff clearance. Controllers are required to advise the holding aircraft of the closest arrival traffic that is approaching the same runway and are prohibited from allowing aircraft to hold on the runway at an intersection between sunset and sunrise or at anytime that an intersection is not visible from the tower. The NTSB has recommended that the FAA 'discontinue the practice of allowing departing aircraft to hold on active runways at nighttime or at any time when ceiling and visibility conditions preclude arriving aircraft from seeing traffic on the runway in time to initiate a safe go-around maneuver' (Recommendation A-00-69). Note: The practice of holding on the runway at an intersection at night has already been discontinued in the United States.

Further analysis of incursions in which an aircraft holding in position on the runway was a contributing factor shows that holding at an intersection and holding at the end ('full-length') are two very different operations. Between January 1998 and December 2001, there were 25 runway incursions in which an aircraft holding in position was a contributing factor. Analysis of these incursions showed that incidents involving aircraft holding at the end of the runway ('full-length') and those involving an aircraft holding at an intersection differ both in the factors that contributed to them and in the severity of the outcomes. Each of the 25 incidents examined were attributed to controller error ('operational errors'). During this time period, there were also four incidents in which pilots landed on the wrong (parallel) runway with an aircraft holding in position; these incidents were not included in the analysis as they were attributable to factors other than an aircraft holding in position. Of the 25 incidents, ten involved aircraft holding at an intersection and 15 involved aircraft holding full-length. Since it is not the case that two-thirds of all takeoffs are intersection takeoffs, aircraft holding at an intersection face greater risk of operational errors than if holding at full-length.

Intersection Departures The relative risk of holding in position on the runway at an intersection (for an intersection departure) is higher than the risk inherent in holding at the end of the runway. In fact, in all fatal accidents in the US that involved an aircraft holding in position, the aircraft was holding for an intersection – not a full-length – departure. In half of the runway incursions that involved an aircraft holding for an intersection departure, the controller thought that the aircraft was full length. Therefore, when told to 'taxi into position and hold', the controller thought that the aircraft would be holding at the end of the runway, when in fact, it taxied onto the runway mid-field. This scenario is particularly dangerous since controllers routinely clear an aircraft for takeoff and then, seconds later, clear another aircraft into position and hold on the same runway. Thus, in addition to the risk of an aircraft landing on the same runway on which the aircraft is holding, i.e., a 'landover', an aircraft holding on the runway for an intersection departure is also vulnerable to an aircraft taking off from behind it (a 'takeoff over'). In the case of a runway that is being used solely for departures, the risk of a landover (for an aircraft holding full-length or at an intersection) is negligible (since it is dependent upon an aircraft landing on the wrong runway). However, the risk of a takeoff over (or into) an aircraft holding in position is still an important consideration.

While the data are not available (because they are not collected), it is reasonable to assume that aircraft are more likely to depart from the end of the runway than from an intersection. Since aircraft are more commonly held in position on the runway at the end (i.e., full-length) than at an intersection, and runway incursions with aircraft holding on a runway as a causal factor are as likely to involve an aircraft holding at an intersection as an aircraft full-length, then the risk of a runway incursion associated with holding at an intersection is greater than the risk of a runway incursion given an aircraft holding in position at the end of the runway.

The data set of 25 incidents contained two accidents (Van Nuys, CA, 2001 and Sarasota, FL, 2000). While this could be interpreted that the probability of an accident, given a runway incursion that involves an aircraft holding in position is 2/25 or 8 per cent, it is more accurate to associate the risk with an aircraft holding at intersection. Both of these accidents – and all recorded fatal accidents involving an aircraft holding in position – involved an aircraft holding at an intersection. The probability of an accident, given a controller (operational) error involving an aircraft holding at an intersection, is statistically estimated at 2/10 (two accidents, ten operational errors involving aircraft holding at an intersection) or 20 per cent.

Full-length Departures The relative risk of an accident for an aircraft holding in position at the end of the runway is much lower than that for an aircraft holding at a runway intersection. Nonetheless, the potential for an accident exists with aircraft landing on a runway that is occupied by an aircraft holding at the end of the runway for takeoff. An analysis of these incidents revealed several Human Factors associated with them: time that the aircraft is left holding on the runway, controller working combined positions at the time of the incident, and operations conducted at night.

Elapsed Time Several reports on these incidents indicated that the controller forgot about the aircraft holding in position. Logically, the longer the aircraft is holding in position, the more likely it is that the aircraft will be forgotten. (This is due both to the natural decay of the memory trace and the potential for distractions that interfere with the retrieval of the information.) For this reason, the time that the aircraft was left in position was examined. In each case in which such time is able to be identified, over two minutes elapsed between the time the controller issued the instruction and the resulting event (e.g., landover, go-around).

Controller Working Combined Positions In 40 per cent of the 15 incidents that involved aircraft holding at the approach end of the runway, the controller involved was working combined positions (e.g., working local control and ground control or working two local control positions simultaneously). In an additional 20 per cent of the incidents, the controller involved was also acting as the supervisor. Therefore, in 60 per cent of the operational errors involving aircraft holding in position at the end of the runway, the controller was working more than one position – either by working combined positions or by functioning as the supervisor while working a position. A previous study of 256 tower operational errors found that controllers were working combined positions in 10 per cent of the tower operational errors (Cardosi and Yost, 2001). Therefore, it is reasonable to expect that operational errors involving aircraft holding in position at the end of the runway are more likely – perhaps by a factor of four – to involve a controller working combined positions. This suggests that controllers working combined positions could be a contributing factor in these incidents and that prohibiting the practice of aircraft holding in position while controllers are working combined positions could help to mitigate the risk.

Visibility The reports of these incidents did not contain sufficient information to support an analysis of the effects of visibility.

Day/Night Of the 15 incidents, eight occurred at night. Since there are more operations during daytime hours (and hence, more opportunity for error), yet a higher incidence of these incidents at night, then it is reasonable to conclude that nighttime increases the risk of a runway incursion involving an aircraft holding in position at the end of the runway. These data support the NTSB recommendation that the practice of allowing aircraft to hold at the end of active runways at night be discontinued. (The practice of holding on the runway at an intersection at night has already been discontinued.)

Comparison of Factors Associated with Incidents involving an Aircraft Holding at an Intersection vs. Full-length Of the 15 operational errors involving an aircraft holding in position at the end of the runway:

• 40 per cent involved a controller working combined positions;
• 20 per cent involved the controller also functioning in a supervisory capacity;
• over half occurred at night (Note: Use of Taxi into Position and Hold (TIPH) at an intersection at night was restricted at the time of the incidents examined).

Of the ten operational errors involving an aircraft holding in position for an intersection takeoff:

• none involved a controller working combined positions;
• one involved the controller functioning in a supervisory capacity;
• two occurred at night (Note: Use of TIPH at an intersection at night was severely restricted within the timeframe of the incidents examined);
• one (the Van Nuys, CA 2001 accident) involved marginal visibility (4 miles).

Because there are no identifiable risk factors associated with these incidents, no risk mitigation strategies can be identified from this analysis. The analysis does support prohibiting aircraft holding at an intersection for departure. Note that it is the practice of *holding* or *waiting* on the runway for an intersection departure, not intersection departures per se, that appears problematic. That is, clearing an aircraft for departure from the hold short lines, as opposed to having the aircraft wait on the runway prior to takeoff, appears to be the safer procedure. Efforts are currently underway to determine whether discontinuing the practice of holding aircraft in position at runway intersections would affect airport capacity.

Suggested Changes in Cockpit Standard Operating Procedures (SOPs)

Until recently, there were no cockpit standard operating procedures (SOPs) for pilots for surface operations. Even the 'sterile cockpit rule' which limits non-necessary communications below 10,000 feet, was rarely applied to surface operations, in practice. Several lines of evidence suggested that runway incursions could be reduced if cockpit operations on the surface were standardized to encourage practices that would help to reduce runway incursions. For example, both pilots should be focused on obtaining and ensuring compliance with instructions on the airport surface until the aircraft is clear of all runways. FAA Advisory Circular (AC) 120-74, 'Flight Crew Procedures During Taxi Operations' was the first step toward achieving this goal. This advisory circular has been refined as the result of specific analyses of the data and was issued (as AC 120-74A) in September 2003. The most notable changes that resulted from the data analyses are presented here.

Strategic Use of the Aircraft Landing Light(s)

Pilots have different practices concerning the use of the landing lights while holding in position and cleared for takeoff. Some pilots illuminate all of the landing lights when they taxi into position and hold with the expectation that this makes the aircraft more visible to aircraft on approach and it will help prevent 'landovers'. Other pilots illuminate the landing lights (along with the strobe light, if not automatically controlled) once the takeoff clearance has been received. Thus, they use the landing light as a signal that the aircraft is rolling. Anecdotally, Robert Bragg, the first officer of the Pan Am flight in the Tenerife accident, has been quoted as saying that the crew could see the landing light of the other aircraft (KLM – as they rolled for takeoff head-on toward the back-taxiing Pan Am flight) but they did not realize the plane was rolling for takeoff until he saw that the landing light was vibrating.

Following is an excerpt from the FAA report of an aircraft crossing in front of a takeoff (at a major US airport in 2000) as the result of a controller error and an excerpt from the report on the same incident that was submitted to the Aviation Safety Reporting System (ASRS) by the flight crew of the crossing aircraft. Both excerpts have been de-identified.

> Air Carrier 1 was cleared for takeoff on Runway 25R. Approximately 30 seconds later, the controller told Air Carrier 2, who had just landed on Runway 25L, to cross Runway 25R. Air Carrier 2 was not clear of the runway as Air Carrier 1 was rolling on departure. Air Carrier 1 rotated approximately 4150 feet prior to the taxiway where Air Carrier 2 was crossing. The controller did not remember crossing the aircraft and realized that it was crossing the runway when (the controller) saw Air Carrier 1 rotating and Air Carrier 2 crossing the runway. It was then that the controller realized that she had crossed Air Carrier 2 and not instructed him to hold short.

> We flew a visual approach to Runway 25L…as we were exiting at intersection A6, tower cleared us to cross Runway 25R at intersection A6, I (F/O) cleared the runway. I did see an aircraft in the distance in what appeared to be in takeoff position, but saw no movement of the aircraft. At about half-way across the runway, I noticed the nose rotate

and I told the captain to hurry across. The takeoff aircraft crossed behind us and about 200-300 feet in the air...The motion of an aircraft nose is hard to pick up at these distances. ATC tower controllers frequently taxi aircraft across runways with other aircraft holding in position (ASRS report number 466269).

Several lines of evidence were examined in order to determine whether landing lights – in aircraft with more than one light – should be illuminated while holding in position for takeoff or reserved for when takeoff clearance has been received. First, the relative probability of a landover vs. crossing in front of a takeoff was examined. Second, the effectiveness of preventing a landover vs. preventing a crossing in front of a takeoff with the use of landing lights was examined.

In order to determine the relative frequency of landovers vs. crossing in front of a takeoff, an analysis of all of the runway incursions that occurred in 2001 was conducted. For the purposes of this analysis, all runway incursions that involved an air carrier, air taxi, regional jet, or business jet were included. Runway incursion conflicts between two (smaller) general aviation aircraft were specifically excluded, since the interventions under consideration (and the Advisory Circular in which they would be published) are intended for Part 121, Part 125, and Part 135 operators.

There were 155 incursions in 2001 that met the above criteria. There were two landovers of aircraft holding in position and one accident (involving an aircraft holding at an intersection) in 2001. In addition to this, there were 23 'potential landovers', that is, there was an aircraft cleared into position and hold and another aircraft cleared to land – however, in these cases, the actions of one of the pilots or a controller prevented the landover. This was done either by preventing the aircraft from crossing onto the runway (e.g., the aircraft went past the hold short point, but not onto the runway) or the landing pilot went around. During the same time period, there were 57 instances of an aircraft crossing in front of a takeoff or almost crossing in front of a takeoff.

The FAA runway safety office rates all incursions as to their level of severity, with A the most severe and D the least severe. Of the 57 crossings (and potential crossings) in front of a takeoff, 20 of them were categorized as an A or B. Since there was a total of 32 A- and B-level incidents in these data, crossings and potential crossings in front of a takeoff constituted a total of 62 per cent of the total A- and B-level incursions involving an air carrier aircraft.

A more extensive analysis of runway incursions that occurred over a three year period (Fiscal Years 2000, 2001 and 2002) that were rated as the most severe (A or B) due to controller operational errors or pilot deviations revealed the following:

• 47 per cent of these most severe incursions were scenarios involving crossing in front of a takeoff;

• 15 per cent involved a landover or potential landover;

• 11 per cent involved crossing in front of a landing aircraft.

Focusing on incidents involving larger aircraft (i.e., excluding those involving only general aviation aircraft), crossings, and potential crossings, in front of an aircraft taking off constituted a total of 66 per cent of the total A- and B-level incursions involving these aircraft. Landovers and potential landovers constituted nine per cent.

When an aircraft is on final approach, pilots cannot reasonably be expected to 'see' an aircraft holding in position, especially at night. Whether the landing lights are on or off would be expected to have little, if any, effect on the pilot's ability to 'see' the holding aircraft – even at night – for two reasons. First, when landing, the pilot's attention is not focused on the end of the runway. Pilots are trained to look at the far end of the runway and the horizon when landing, so that they do not 'spot' the runway and land short. While pilots should be accustomed to scanning the runway before landing, a stationary object (as it would be seen during the day) or set of lights (as it would be seen at night, or in haze) would have a tendency to blend into the background; it would not attract attention like a moving object would – especially in the periphery of the visual field.

On the other hand, if the use of the landing light was reserved to indicate that the aircraft had received a takeoff clearance, the landing light could be a very useful cue to an aircraft on an intersecting taxiway expecting to cross the runway. Since it can be very difficult to discern movement of an oncoming aircraft, reserving the use of the landing light when initiating takeoff roll would provide other pilots with an indication of the intent of the aircraft in position, that is, whether it is holding in position or taking off.

Therefore, the strategic use of the landing light as an intervention for runway incursions has a much greater potential for preventing a crossing in front of a takeoff than it does for preventing a landover for two reasons. First, the relative frequency of a landover vs. that of crossing in front of a takeoff indicates that crossing in front of a takeoff is a much more likely event. Second the strategic use of the landing light has the potential for being much more effective for signaling pilots and vehicle drivers downfield that the aircraft is rolling than it does for increasing the conspicuity of the aircraft to pilots on approach. For these reasons, it was recommended that when holding in position for takeoff, the landing lights should be off until takeoff clearance is received. Based on this Human Factors analysis, the following recommendation was added to the air carrier SOPs (AC 120-74A):

When holding in position for takeoff, the landing lights should be off until takeoff clearance is received; in this way, it provides an indication to ATC and other aircraft that the aircraft is rolling.

Cockpit Procedure to Aid Controller Memory

As we have seen, a common contributing factor in surface operational errors was the controller 'forgetting' something. In runway incursions involving an aircraft holding in position, controllers have forgotten about aircraft holding in position and aircraft on approach. Logically, the longer the aircraft is holding in position, the more likely it is that the aircraft will be forgotten (this is due both to the natural decay of the memory trace and the potential for distractions that interfere with the retrieval of the information.)

Between January 1998 and December 2001 there were 12 incidents in which an aircraft was holding in position on the runway and another aircraft was cleared to land on the same runway. These incidents were examined to determine the time between the issuance of the instruction to position and hold on the runway and the resulting event (e.g., landover or landing pilot initiated a go-around.). Of these 12 incidents, this elapsed time was able to be determined in 9 of them. In the other three cases, the timeline of the voice transmissions was not transcribed in the report, and only vague references to the timing of the sequence of events (e.g., 'few moments later' and controller issued the instruction to taxi into position and hold and 'then became busy with an aircraft whose data block did not acquire and another aircraft that did not answer several radio calls'). In all nine of the remaining incidents, more than two minutes elapsed after the clearance was issued. Depending on the type of aircraft and the distance between the hold line and the runway, it can take up to one minute for an aircraft to line up and stop on the hold position. As a result of these findings, it was recommended that the following be added to the FAA's Advisory Circular on Flight Crew Procedures During Taxi Operations (AC 120-74A): 'If you expected an imminent takeoff and you have been holding in position for more than approximately 90 seconds, contact ATC.' While the wording that was eventually incorporated into the revised circular (dated 9/03) is much more convoluted than the wording that was recommended, the intent is the same.

Part Three: Existing Challenges

The challenges that exist to promote safer surface operations on an international scale go beyond developing technologies and strategies to prevent runway incursions and mitigate the effects of human error. A constant challenge is to continue to refine our understanding of the causes of runway incursions. This requires improving the ways in which runway incursion data are collected, recorded, and analyzed. Furthermore, development of an international standard for the definition of a runway incursion would support a comparison of information from different sources. Specific analyses of the data and targeted studies are also needed to address critical issues in several areas. Finally, a standard and objective metric for determining the severity of the outcome of runway incursion and a method of predicting the risk of a runway incursion (given specific factors) remain to be developed. Efforts are underway in all of these areas, however, they are integrally linked. For example, an effective model for predicting the risk associated with runway incursions under certain circumstances can only be developed after sound analyses have been conducted to determine causal and contributing factors.

Measuring the Effectiveness of Specific Interventions

A fundamental, but difficult, task is to measure the effects of specific risk mitigation strategies on the number of runway incursions or on accidents involving a runway incursion. Accidents as a result of runway incursions are mercifully rare; therefore, a measure other than accidents must be used to assess the effectiveness of an intervention. Interventions include educational programs, changes to air traffic or flight deck procedures, changes to signs or markings on the airport, or implementation of any technology to prevent incursions or accidents due to incursions. The problem with evaluating the effect of a specific intervention is that mitigation strategies are continuously being implemented and refined. While this is a critical and laudable activity, it makes the statistical measurement of the effects of any given strategy impossible. When changes in the trend of the incursion data are observed, the effects of all of the changes that have been implemented, along with the possible interactions among these changes, must be considered. This is in addition to the (potentially short-lived) effect of the heightened level of awareness of pilots and/or controllers that also may have occurred as the result of any given change.

Data Collection and Recording

The foundation of an effective program to increase surface safety, or of any accident prevention program, is the collection and analysis of valid and reliable data on critical incidents. Such data are necessary to support operationally useful analyses that can point to causal factors and potential remedies. Toward this end, there are plans to revise the current methods used for investigating and recording controller operational errors and pilot deviations. FAA investigations of operational errors that result in runway incursions describe what happened and attempt to determine the underlying factors that contributed to the errors. The events are investigated by a specialist from the facility who is knowledgeable of the operations at that facility. The reports include information on the operational conditions at the time of the incident and sometimes contain the controller's account of what happened.

While the FAA reports of operational errors that result in runway incursions contain a great deal of useful information, the ways in which the data are collected and recorded raise several concerns. First, there is the perception (mostly among pilots) that not every operational error is reported and included in the database. This is supported by analyses of ASRS reports that show many more reports of incidents that could qualify as runway incursions than are officially recorded as such. It appears likely that minor controller errors that result in an incident that meets the definition of a runway incursion, but has no risk of collision associated with it, and has little or no chance of being discovered, would go unreported. This is supported by the fact that while the majority of runway incursions that are reported are attributed to pilot error, the proportion of the more serious runway incursions (referred to as A- and B-level incursions) is equally distributed between controller and pilot errors. While the issue of underreporting of controller errors that result in runway incursions has yet to be objectively investigated, if true, then the data may be biased towards including some type of errors over others. This could skew the information available of identifiable causes and the remedies that the data point to. A second concern is that the investigation is of a nature that may not lend itself to full disclosure on the part of the controllers involved or generate an interest in preventing such errors. Thus, there may have been factors that contributed to the error that were not identified in the reports.

The FAA reports of pilot deviations that result in runway incursions are extremely succinct. The reports usually lack any detail as to why the incident occurred, thus, their usefulness is extremely limited. Reports submitted by pilots to the Aviation Safety Reporting System (ASRS) on the other hand, are often very helpful in identifying why certain errors happened, since the reporters often include details that they would be reluctant to report to investigative officials. However, they often lack critical objective details and, more importantly, cannot provide incidence data (that is, they cannot provide information on *how often* something happens). Recent programs have encouraged airline pilots to submit confidential reports to safety programs within their own airline in return for the same immunity from prosecution offered by ASRS. Whether these reports are as candid and detailed as the ones submitted directly to ASRS remains to be seen.

Since there is no formal mechanism for pilots to instigate an objective investigation into a runway incursion (to get the voice tapes analyzed, for example), controllers are the de facto gatekeepers of the runway incursions that are reported. Critical factors, such as closest point of approach, are based on the controllers' estimates. This is likely to be one reason why the number of recorded runway incursions due to pilot deviations is consistently more than twice of the incursions due to controller (operational) errors. It could also be argued, however, that there are more opportunities for pilot errors on the airport surface than there are for controller errors. Loss of positional awareness, sometimes related to the pilot being unfamiliar with the airport, is often a contributing factor to runway incursions that are caused by pilots. For the pilot, each airport presents a slightly different task, and in this sense, a different set of opportunities for error.

Specific Analyses to Support Operational Decisions

Several operational questions in the area of runway safety have benefited from ad hoc data analyses. Analysis of runway safety data has pointed to mitigating strategies and has helped to decide between competing strategies. There are many operational questions that would benefit from (yet to be conducted) data analyses. For example, research is needed (and in progress) to determine specific characteristics of the airport layout that may contribute to runway incursions. Also, the relative risks of specific aspects of an operation (e.g., holding aircraft on a runway at night) need to be assessed and, if necessary, compared to the effect on efficiency and capacity of discontinuing the practice. This information then needs to be considered within the concept of acceptable risk. In the context of the failure rate of an individual aircraft component, for example, the accepted failure rate is usually specified. For overall operations, however, or surface operations in particular, a numerical criterion for acceptable risk has yet to be defined.

Definition and Categorization of Runway Incursions

There is no agreed-upon international standard for the definition of a runway incursion. This makes comparisons across data sets difficult. The FAA defines a runway incursion as:

> Any occurrence at an airport (with an operating control tower) involving an aircraft, vehicle, person, or object on the ground that creates a collision hazard or results in a loss of separation with an aircraft taking off, intending to take off, landing, or intending to land (FAA, July 2002, p. B-1).

Another category of incidents that is also recorded is 'surface incidents'. The FAA defines a surface incident as:

> ...any event during which unauthorized or unapproved movement occurs within the movement area or an occurrence in the movement area associated with the operation of an aircraft that affects or could affect the safety of flight (FAA, July 2002, p. B-2).

This does not mean, however, that *all* incidents in which a pilot takes evasive action to avoid an aircraft, vehicle, object, or person on the runway qualifies as a runway incursion or is recorded. The definition would exclude, for example, an incident in which a pilot is instructed to abort a takeoff in order to avoid a collision with a landing aircraft on an intersecting runway, as long as the aborted takeoff was able to stop short of the intersecting runway.

There are also no objective criteria for classifying the severity of a runway incursion. Until relatively recently, the severity of incursions was not even discussed. This meant that statistically, an incident that involved an aircraft that crossed the hold line but not the edge of the runway with another aircraft on takeoff roll was equal to an incident that involved an aircraft that crossed the runway with another aircraft taking off. Because of this, a system was developed to classify the severity of runway incursions into four categories depending on the potential for collision. These categories are defined in the Runway Safety Blueprint as follows (FAA, July 2002, pp. B1-2):

A. Separation decreases and participants take extreme action to narrowly avoid a collision, or the event results in a collision. These runway incursions are typified by: the immediate need for corrective action by ATC and/or evasive action. The collision either occurs or is narrowly avoided by chance. These runway incursions are characterized by events, with at least one aircraft traveling at a high speed or speed sufficient to cause substantial damage with the potential for injury or fatalities.

B. Separation decreases and there is a significant potential for collision. Level B events are characterized by the need for a time-critical corrective/evasive action that was, or could have been, taken by ATC or the flight crew. These situations are typified by critical errors, which under different circumstances or aircraft performance or the timing of ATC clearances or instructions, could have lead to a barely avoided collision.

C. Separation decreases but there is ample time and distance to avoid a potential collision. In cases where no corrective action was taken, the collision risk is reduced by a significant level based on available time, distance margin, and aircraft performance. These situations are typified by critical errors, which under different circumstances or aircraft performance or the timing of ATC clearances or instructions, could have led to a higher potential for a collision. Participants in these incidents usually do not come in close proximity at high speed.

D. Little or no chance of collision but meets the definition of a runway incursion.

The variables that are considered in determining the severity of the runway incursions are:

•   available reaction time – how much time the pilot, controllers, and/or vehicle operators had to react to the situation based on aircraft type, phase of flight, and separation distance;

•   evasive or corrective action – the magnitude of the maneuvers required to avoid a runway collision;

•   environmental conditions – day/night, weather, visibility, surface conditions;

•   speed of aircraft and/or vehicle;

•   proximity of aircraft and/or vehicle – closest point of approach in vertical and horizontal distance.

At present, these variables are qualitatively, but not quantitatively, defined. For example, 'proximity' is defined as:

•   'near miss or collision – aircraft/vehicle traveling at high-speed narrowly missing one another or colliding on the runway';

•   'very close – aircraft/vehicle approached one another at a high rate of speed';

- 'close – aircraft/vehicle approached one another at a low/moderate rate of speed';
- 'close – aircraft/vehicle did not approach one another' (FAA, June 2002, p. 12).

A model is currently under development to reduce the subjectivity, and increase the inter-rater reliability, in the categorization of the severity of runway incursions. Variables, such as proximity, will be quantitatively defined and other variables (such as available response time, severity of the avoidance maneuver executed) will be quantified to the extent possible. These variables will be weighted in the algorithm that defines the levels of severity. In addition to adding objectivity to the way in which incursions are rated as to their severity, such a model will provide the foundation for the development of a predictive model of risk. A predictive model would forecast the effect of changes on an airport surface, such as the addition of a runway or taxiway, on the probability of runway incursions.

Conclusions

The science of Human Factors continues to play a critical role in runway safety in several areas. As we have seen, Human Factors analyses of the data on runway incursions has led to the development of specific mitigation strategies. Some of these are currently being implemented; others are under consideration. The results of such data mining will grow even more productive as the information collected in the recording of pilot and controller errors that result in runway incursions continues to be refined. A richer and more detailed Human Factors database will allow for more sophisticated analyses into the causal factors of pilot and controller errors. Such analyses will point to additional strategies to reduce the number of incursions and mitigate the effects of inevitable incursions before they result in collisions.

References

Adam, G., Kelly, D., and Steinbacher, J. (1994), *Reports by Airline Pilots in Airport Surface Operations: Part 1. Identified Problems and Proposed Solutions for Surface Navigation and Communications*, MITRE Report Number MTR 94W0000060, MITRE, McLean, VA.

Adam, G. and Kelly, D. (1996), *Reports by Airline Pilots in Airport Surface Operations: Part 2. Identified Problems and Proposed Solutions for Surface Operational Procedures and Factors Affecting Pilot Performance*, MITRE Report Number MTR 94W0000060v2, MITRE, McLean, VA.

Bales, R., Gilligan, M. and King, S. (1989), An *Analysis of ATC-Related Runway Incursions with Some Potential Technological Solutions*, MITRE Report Number MTR-89W00021, MITRE, McLean, VA.

Burki-Cohen, J. (1995), *An Analysis of Tower (Ground) Controller-Pilot Voice Communications*, DOT/FAA/AR-96/19, FAA, Washington DC.

Cardosi, K. (1993), *An Analysis of En Route Controller-Pilot Voice Communications*, DOT/FAA/RD-93/11, FAA, Washington DC.

Cardosi, K. (1994). *An Analysis of Tower (Local) Controller-Pilot Voice Communications*, DOT/FAA/RD-94/15, FAA, Washington DC.

Cardosi, K., Brett, B. and Han, S. (1996) *An Analysis of TRACON (Terminal Radar Approach Control) Controller-Pilot Voice Communications*, DOT/FAA/RD-96/66, FAA, Washington DC.

Cardosi, K., Falzarano, P. and Han, S. (1998), *Pilot-Controller Communication Errors: An Analysis of Aviation Safety Reporting System Reports*, DOT/FAA/AR-98/17, FAA, Washington DC.

Cardosi, K. and Yost, A. (2001), *Controller and Pilot Error in Airport Operations: A Review of Previous Research and Analysis of Safety Data*, DOT/FAA/AR-00/51, FAA, Washington DC.

DiFiore, A. (2001), *Time Required for Transmission of Controller Messages in the TRACON Environment: An Analysis of Voice Tapes*, Unpublished manuscript.

Federal Aviation Administration's Office of Runway Safety (June 2002), *FAA Runway Safety Report: Runway Incursions Trends at Towered Airports in the United States*, U.S. Department of Transportation, Federal Aviation Administration, Washington DC.

Federal Aviation Administration's Office of Runway Safety (July 2002), *Runway Safety Blueprint 2002-2004*, U.S. Department of Transportation, Federal Aviation Administration, Washington DC.

Joint Safety Analysis Team (JSAT) (2000), *Results and Analysis*, JSAT.

Kelly, D. and Jacobs, G. (1998), *Reports by Airport Traffic Control Tower Controllers on Airport Surface Operations: The Causes and Prevention of Runway Incursions*, MITRE Report No. MTR 98W0000033, MITRE, McLean, VA.

Kelly, D. and Steinbacher, M. (1993). *Pilot Surface Incident Safety Study*, MITRE Report No. MTR 92W0000116, MITRE, McLean, VA.

Nader, E., DiSario, R., Mengert, P., and Sussman, E.D. (1990), *A Simulation Study of the Effects of Communication Delay on Air Traffic Control*, DOT/FAA/CT-90/6, FAA, Washington DC.

National Transportation Safety Board (1986), *Special Investigation Report – Runway Incursions at Controlled Airports in the United States*, NTSB/SIR-86/01.

Sollenberger, R., McAnulty, M., and Kerns, K. (2003), *The Effect of Voice Communications Latency in High Density, Communications-Intensive, Airspace*, DOT/FAA/CT/TN03/04, FAA, Washington DC.

Chapter 4

Human Error in European Air Traffic Management: from Theory to Practice

Anne Isaac, Paul Engelen and Martin Polman

Introduction

This chapter deals with an approach to Air Traffic Management (ATM) human error analysis that is being developed for the European ATM system. This new approach aims to determine not only how and why human errors are contributing to incidents but also how to predict the likelihood of these errors occurring and thus how to improve human reliability within the system. This approach is called HERA – Human Error in ATM.

The chapter covers three major issues. Firstly the theoretical aspects of an *error model* based on human performance within the ATM environment are detailed. The *technique*, which is based on the model, is then discussed, including details of the classification systems for determining, in some detail, the nature and causes of the errors within incidents. Secondly the technique itself is examined through an example in terms of its efficacy when used in real incident investigation activities. Lastly the impact of the use of such a method within incident investigation is discussed with regard to the ongoing safety management of the air traffic control system.

Background

The Air Traffic Management (ATM) system in Europe is very safe, but the demands from the aviation industry will inevitably require changes in the way ATM conducts its business. These changes are already impacting on the operations environment in which procedures are changing to increase traffic. In harmony with these procedural changes are the increases in the technology to 'assist' the controllers. These advances in technology and procedures are not new in aviation, and much has been learned from the introduction of new technologies on the flight-deck (Billings, 1997; O'Leary, 1999).

These major shifts in work practices will affect both controller and pilot performance (i.e. datalink), and new opportunities for error could arise, particularly in the 'transition period' during which new systems and practices are introduced. These developments suggest that the ATM system is at the beginning of a long period of significant change and evolution, a period that will possibly see new error types, and potentially increased error rates. These issues point to the need for the development of an approach to better understand errors, to monitor error trends, and to develop effective error reduction and management strategies.

Theoretical Background

Development of the Human Error in ATM (HERA) System

The Human Error in ATM (HERA) project started in 1999 with the development of a new approach to the Human Factors issues associated with ATM incidents within Europe. The development of a unique incident investigation method and technique dedicated to the air traffic control system occurred in five stages (Isaac et al., 2002).

1. *Literature review:* academic and industrial research findings on human performance models and taxonomies of human error were reviewed. This review included models ranging from Reason's (1990) model of slips, lapses and mistakes, to Rasmussen et al.'s (1981) skill, rule and knowledge based behavior framework, and Endsley's (1996) situation awareness approach. In total over fifty models/taxonomies were reviewed dating from 1948 to 1999.
2. *Selection of the chosen conceptual approach:* the literature review identified the most appropriate framework from human performance research. Within this approach the information-processing models from Marteniuk (1976) and Wickens (1992) were chosen.
3. *Review of current and future ATM systems:* a review of the human performance aspects of ATM was conducted to ensure that all significant aspects of these activities were included. Different functional areas of ATC were considered, e.g. tower, approach, en-route, etc. This phase also considered future ATM systems, such as computerized conflict detection support tools, electronic flight strips, and Datalink technology.
4. *Adaptation of the chosen conceptual approach:* the chosen model and approach were adapted in light of the ATM context. In addition, the controller's mental 'picture', or representation, of the ATM scenario was added, as this was seen as essential to the job of an air traffic controller, and the breakdown of 'the picture' was considered a precursor to some incidents in ATM.
5. *Specification for the HERA technique:* also based on the literature review, two techniques were identified which would support the HERA approach. These techniques, were TRACEr – Technique for the Retrospective Analysis of Cognitive Errors in ATM (Shorrock and Kirwan, 2002), which represents a

flow-chart based taxonomy; and the Tryptich Pyramid model (Isaac, 1999), which presented the possibility to integrate ATM contextual issues with the cognitive failures.

The literature review and the analysis of controller tasks led to the need for the inclusion of a number of high level 'cognitive domains', that were concordant with the information-processing model, and allowed analysis of different types of failure at successively detailed levels. This hierarchical approach to error analysis was based partly on Rasmussen's Skills, Rules, and Knowledge (SRK) classification system, and also followed other successful classification system developments (Taylor-Adams and Kirwan, 1995). The basic structure of these taxonomies reflects three types of understanding of the aetiology[7] of an incident: what happened, how it happened (internally in terms of cognition), and why it happened (due to internal and/or external events, states, or pressures). The three perspectives in this error analysis are given the titles of Error Type, Information-Processing Level (or internal mechanism of malfunction), and the Contextual Conditions (or contextual aspects). If all three can be understood, then the reasons the error occurred will also be accurately understood, and hence effective error prevention can be developed. A final grouping of the Error Detail lent the HERA model its overall architecture and turned it from a simple classification list of error forms, into a true taxonomic system.

As has been mentioned, the notion of context was also incorporated into the HERA approach. It was recognized from the outset that a contextual approach was needed, since errors do not occur in an abstract or generic environment (Dougherty, 1993). Therefore HERA identified aspects of the controller task (or the ATM function), the behavior at the time of the error (Fleishman and Quaintance, 1984), and the equipment/information being used (input devices, displays, etc.). These contextual classifications, including the Contextual Conditions themselves, would capture much of the actual context of the incident.

The HERA Technique

The HERA technique firstly identifies, from the original event, the factual background data and also the sequence of events in a shortened format. The technique then guides the analyst through the detailed process of identifying each error and its causal routes. This process starts with the identification of the first erroneous act in the sequence. The analyst is asked to record how, if at all, the error was detected; how, if at all, the error was recovered; and whether it could be considered causal, contributory, compounding or non-contributory to the overall sequence of events. The analyst identifies the task, the information which was associated with the event and then, in sequence, the Error Type itself (ET), the Error Detail (ED), the Error Mechanism (EM), the associated Information-Processing Level (IP) and any Contextual Conditions (CC).

---

[7] i.e. the science of typology and causes of diseases.

For each of the errors and their root causal factors a set of tables and flow-charts were developed to assist the analyst. These were developed from the psychological and safety science literature as explained above. An example of the flow chart relating to the Working Memory Error Mechanisms can be seen below.

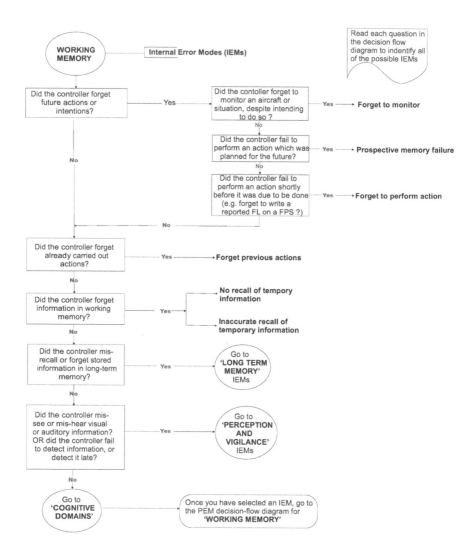

Figure 4.1   Flow chart associated with working memory error mechanisms

The analyst then identifies the associated Contextual Conditions from the reference table which has eleven sections and over 200 options. Further information regarding this technique can be found in (Isaac, Shorrock, and Kirwan, 2002).

The Application of HERA to Previously Analyzed Incident Reports

During the development of the technique, about 80 incidents (from several European and overseas sources) were re-analyzed based on incident reports in the public domain. The results indicated that, most of the time, more than one error was made by one or more individuals and that several contextual conditions contributed to the errors. The following were the most significant results with respect to the cognitive classification of the Error Types observed.

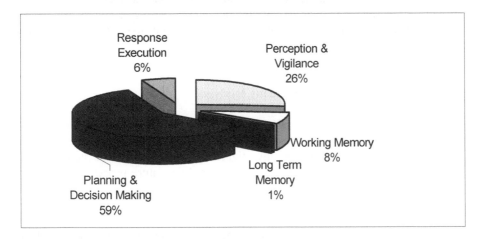

Figure 4.2   Classification of error types

Each cognitive area was further broken down into the actual error details recorded. The results indicated that the highest number of errors was found in Planning and Decision-Making – 59 per cent – more precisely the problems were in the area of incorrect decision or plan. The second area of concern was seen in Perception and Vigilance – 26 per cent – and more precisely in the problems of hearback and mishearing. These results are difficult to compare with previous studies as no previous methodology analyzes errors in such a precise cognitive way. However previous studies have classified decision-making as a problem (Rodgers and Nye, 1993), but with no analyses in depth. Pilot-controller communication has also been cited as a problem area (Cardosi and Murphy, 1995) but again these results make no reference to related factors.

Within the same 80 incidents, the Contextual Conditions were also identified. The results indicate that the most prominent contextual conditions were linked with teamwork related issues (lack of communication, lack of co-ordination between the controllers and/or other personnel, etc.). The second most prominent contextual condition was related to the traffic (sector capacity, after-peak traffic, complexity and mix of traffic) and the airspace/aerodrome structures (airspace design, complexity, etc). The third most problematic issue was associated with workplace design and Human-Machine Interface (HMI) and the fourth with procedures and documentation.

A small validation exercise followed this development to ascertain whether the technique could be used by several different specialist personnel in ATM and whether it was intuitive for several users from different European countries. The results were very positive and now the second phase of the project has been launched, which involves the development of training the incident investigators in European countries with this new technique.

However the most important test for such a method and technique has to be when the practitioners in incident investigation and the safety managers responsible for the overall safety management of the ATM system start to explore the use of the technique.

The Practical Application in Incident Analyses

At the end of the 1990s an air traffic control organization within Europe was in the middle of a major reorganization. At this time it was also confronted by two serious incidents. The first incident was investigated by members of the new incident investigation section in a way that was considered the future process. The second incident was investigated in a similar way to previous methods and the report was unpublished due to a misunderstanding during transfer to the new reorganized incident investigation group. When this report was rediscovered, the new investigators had a feeling there was more behind the incident, but were not able to address this problem, or the issues of the findings. A decision was therefore made to reconsider this report. Because of the problems of staffing, the investigation took almost 2½ years to complete and in this time period a new lead-investigator was, by coincidence, involved with the development of the new HERA technique. As one of the first investigators who had actually worked with the technique in its development, he noticed possibilities which he considered would allow the investigation of the Human Factors aspects of this incident, which had not been addressed before. Finally the second investigation was published in May 2001.

The way the investigation process was finally carried out gave the unique possibility to compare the findings of both investigations.

The Incident – December 1998

A Boeing 747 left a major northern European airport for a flight to the Far East. After 2 minutes the flight-path of the Boeing 747 conflicted with another (light) aircraft, which had left a smaller aerodrome in the same country for a flight to a neighboring European country. The aircraft passed each other in opposite directions with only half a mile horizontal separation and about 370 feet vertically (the prescribed separation minima in Terminal Manoeuvring Areas are 3 nautical miles or 1000 feet.). The pilots of both aircraft were not able to see each other, as there was poor weather, but TCAS[8] warned the Boeing 747 of the other aircraft in its vicinity.

Reconstruction of Events – Fact-Finding

The light aircraft – a Piper 46 Malibu – was departing from a small aerodrome situated below the Terminal Manoeuvring Area (TMA) of the major northern European airport. Since the smaller airport is a non-controlled aerodrome the Malibu departed under Visual Flight Rules and subsequently had to obtain a clearance by radio to proceed under Instrument Flight Rules from the major airport's Approach Control to continue its flight to the neighboring country. The Approach controller identified the aircraft on radar and subsequently issued the required en-route clearance and instructed the Malibu to climb to flight level 060.

The Malibu had planned to join the airway to the south at the navigational beacon 'HAM' (Hamper). This beacon is situated in the center of the major airport's TMA and is used to define several standard departure procedures for traffic departing from the major airport. Therefore the Approach controller decided that he had to instruct the Malibu to turn to the southwest approximately 3 nautical miles before the aircraft reached 'HAM'. In this case the aircraft would stay in the area of his responsibility and would remain clear of the majority of the outbound traffic from the major airport. As a reminder he attempted to add the text 'HAM' in the aircraft label on his radar screen by making a computer input, but this input appeared not to be accepted by the ATC computer system.

At the same time a commuter aircraft – an Embraer Brasilia – departed from the major airport, not on an instrument departure but on a heading to the Northeast. The next aircraft to depart the major airport was a Boeing 747 on a 'PANIK' departure. After 'PANIK' the intended flight-paths of both aircraft were the same, but the Boeing 747 was faster and had to climb to a higher flight level. Therefore the Approach controller called the adjacent sector of the major airport's Area Control Centre (ACC) and requested permission to clear the Brasilia on a direct track (northeast) to its destination. The ACC controller asked whether the Brasilia would climb to its final flight level – 090 – to which the Approach controller confirmed this request. The ACC controller replied that he would descend traffic inbound to the major airport to flight level 100 instead of flight level 070 which was the standard agreement.

At the moment the Approach controller cleared the Brasilia as arranged, the ACC controller instructed an inbound Boeing 737 to proceed to the navigational beacon 'SKY' (Skyway), while descending to flight level 100. Despite the fact that this aircraft was still outside the major airport's TMA, the ACC controller

---

[8] TCAS or Traffic Alert and Collision Avoidance System.

transferred the Boeing 737 over to the Approach controller, so he would be able to solve the separation problem between the Brasilia and the Boeing 737. In the meantime the Malibu was flying 7 miles northeast of 'HAM'.

The Boeing 737 was not displayed on the radar screen at this point as it was still too far away. In order to be able to see the traffic the Approach controller needed to change the range of the radar screen from 36 to 48 miles. At that moment the Boeing 747 called Approach having just departed from the major airport on a 'PANIK-departure'. The controller immediately issued a clearance to climb to flight level 090 and attempted to take this aircraft 'under control'.[9] However, because the system was still presenting the radar range menu, this action resulted in a sudden 'explosion' of the radar picture as the range changed from 48 to 16 miles. As a result the Approach controller lost his oversight over the traffic under his control for a short period.

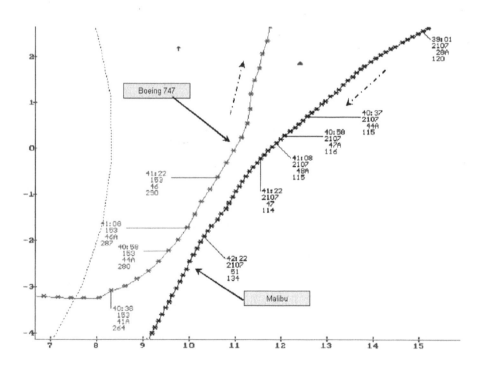

Figure 4.3   Radar track printout

In the meantime the Malibu was approaching the position at three miles northeast of 'HAM', the point at which the controller intended to turn the aircraft on a heading to the south-west, but no instruction was given to the Malibu. The

---

[9] This means that the aircraft is assigned to the Approach controller by a computer input, and he or she is able to make computer inputs for this flight in the ATC-system.

Boeing 747 was turning to the opposite direction and also heading for 'HAM'. Both aircraft were approaching each other rapidly, the distance at that moment was approximately 3½ miles horizontally and 300 feet vertically.

The Approach controller then turned his attention back to the separation problem between the inbound Boeing 737 and the outbound Brasilia. After a couple of questions and instructions to the Boeing 737, he was looking back to the Boeing 747 on the 'PANIK-departure' and almost immediately realized the serious conflict situation between this aircraft and the Malibu. However, the instruction to the Boeing 747 to take evasive action was too late to prevent a serious incident from happening. The aircraft passed each other by only half a mile horizontally. It was only because of the presence of the Traffic Alert and Collision Avoidance System (TCAS) on board the Boeing 747, and the reaction to it by the pilots, that there was a vertical distance between the aircraft of 372 feet.

Conclusions of the First Investigation

The first investigation, which had little formal procedure, focused on the 'technical side' of the incident. The radar data and radio-telephony conversations were compared and the Approach controller was requested to submit a written statement. Therefore the conclusions of this investigation – which are summarized below – were aimed at the actual controller involved, and the technical problems which occurred during the incident. The conclusions were:

• The Approach controller had insufficient knowledge of the ATC system regarding computer inputs. Therefore valuable time was wasted when the controller unsuccessfully attempted to make a specific input.
• The co-ordination conversation between the Approach controller and the ACC controller was carried out with great difficulty. Therefore there were no positive and clear agreements made about the handling of the traffic involved.
• Because the Approach controller erroneously changed the settings of the radar range, he temporarily lost the oversight of the traffic under his control.
• The attention of the Approach controller was negatively influenced by the reasons mentioned above.
• Despite several aircraft on standard instrument departures via the beacon 'HAM', the Approach controller did not issue adequate altitude restrictions to the Piper 46 Malibu from the smaller airport.

Conclusions of the Second Investigation

During the second incident investigation special attention was paid to the Human Factors aspects and therefore the Approach controller and the ACC controller were interviewed on their recollection of the incident. The interviews revealed that the Approach controller did not simply 'forget' the Malibu. Also it became clear that the technical mishaps, detected during the first investigation, were not causal factors as assumed before, but were only contributing to the fact that the incident occurred.

The Approach controller planned to solve the potential conflict situation between the Malibu and other traffic departing from the major airport later in time. Therefore conflicting clearances were issued to the aircraft; clearances which would lead directly to a loss of separation if no additional or amended clearances were issued. This decision – which is not unfamiliar amongst air traffic controllers – contains a certain risk: the risk of distraction at the moment the controller should actually issue the additional clearance(s).

The investigators were convinced that the traffic flow to and from national aerodromes, through the major airport TMA, was difficult to combine with the normal but already complex flow of traffic to and from the major airport. Therefore postponed solutions for similar traffic situations occur frequently.

The main reason that the Approach controller was distracted at the moment he was supposed to turn the Malibu to the Southwest can be found in the telephone conversation between the Approach and the ACC controller. This conversation was carried out unsatisfactorily for both parties involved. It appeared that clear arrangements were made, but neither controller understood the other's objectives, motivation or reactions. The Approach controller did not mention the reason why he wanted to clear the Brasilia on a direct course. The ACC controller did not ask for an explanation and did not pass any information in return about the traffic problems he would have, due to this specific action. The ACC controller reacted on this feeling of discontent by transferring the inbound Boeing 737 almost immediately over to the frequency of the Approach controller. For the Approach controller this action resulted in a completely new separation problem – at a location where he was not used to handle traffic – which he had to solve immediately. Therefore his feeling of discontent changed to feelings of indignation and he started to discuss this matter with the colleague next to him. This particularly distracted the thoughts of the Approach controller, resulting in forgetting to carry out his planned action.

Finally the investigators also discovered other contextual conditions that negatively influenced the attention of the Approach controller. The most important was the fact that just before the incident took place, two visitors entered the Approach control room of the major airport. The Approach controller was informed about this visit, but a colleague of the controller reacted in a very direct and explicit manner telling the visitors that they were not welcome at that moment. The Approach controller was disturbed by the way his colleague handled this situation and in an attempt to make an apology that the planned visit could not take place, he briefly greeted the visitors. This occurrence took place when the Malibu had already passed the point at which the controller had planned to turn it to the southwest.

The Practical Application in Safety Management

Recommendations

This section describes two different sets of conclusions regarding one incident. Both sets of conclusions could have led to one or more recommendations which would, each in their own way, help to prevent incidents like this from happening again in the future.

However, the outcome of the first investigation did not lead to published recommendations, as the investigation process at that time was not the same as it is today. In the former approach of incident investigation within the ATC organization, it was considered to be the responsibility of the line manager to take required actions to prevent re-occurrences. But looking at these conclusions today, it is fair to say that *if* recommendations were written at the time, the following would have been the most likely:

- Retraining should be considered to assure that the controller fully understood the computer input functions in the ATC system.
- A study concerned with the layout of the computer input device and how it could be changed to reduce the chance of making unintended and unwanted changes in the settings of radar range.

Will these recommendations be of great help in preventing a similar incident from happening again? The answer to this question is not easily given; maybe they would have been useful if combined with a little luck, we will never know. The retrained controller is placed back on position, which is now correctly configured for communication. 'End of story...'

However, during the second investigation it was realized there had been many issues which had not been addressed:

- the problem of distractions from visitors;
- the reason the controller omitted immediate action to separate the aircraft;
- the failed co-ordination between the controllers.

But this is something we know now. It is something the investigators learned – by applying HERA – from all the information which was already there, and additional interviews with the controllers involved. This was unknown at the time of the first investigation and it is certain that the above potential recommendations were the only results of the investigation which was carried out then.

As mentioned earlier, the investigation of this incident was picked up by an investigator who had some experience with the HERA principles and technique. The experience he had with the use of the technique in another incident gave him a feeling that there was more behind the incident described in this paper. Maybe better lessons could be learned from it as better recommendations were derived.

The second investigation revealed much more information than the first. This made it possible to perform a much better analysis, leading to more complete conclusions. Finally, the investigation concluded with the following most significant recommendations:

- Introduce Team Resource Management (TRM) training for controllers of the organization to stimulate better inter- and intra-team relations and to create a better understanding of the handling of the air traffic by adjacent centers.
- Study the standard departure and arrival routes to/from smaller aerodromes below the major airport TMA and look for possibilities to 'redesign' these routes so they will be situated outside the major airport TMA to avoid conflicting traffic situations between this traffic and the traffic to/from the major airport.
- Study (internationally) whether it is possible to add a technical support tool, like Short Term Conflict Alert (STCA), to warn Approach controllers adequately when dangerous traffic situations develop.

It is obvious that the two sets of recommendations are totally different. The first set tried to prevent this type of incident from happening again by giving extra training to controllers on the existing system (procedures, equipment handling) without suggesting any changes to the system. This is a rather simple and easy solution. Using these recommendations will, therefore, keep these 'latent' errors in the system or procedures and also suggests that the only thing that did not work was the human operator.

The second set of recommendations attempted to prevent another incident like this by giving help to the controller to deal with complex situations. This help can be divided into three areas: people, procedures and equipment. All these areas are addressed in the recommendations mentioned. This indicates that the total system must 'learn' and become safer.

General Comments

In the modern ATC environment all the elements (human, machine and procedures) are subject to highly defined standards. Equipment such as radar consoles, R/T-sets, information-processing systems, etc. undergoes precisely defined test programs and fault analyses before becoming operational. ATC procedures are becoming increasingly subject to underlying safety case analysis and risk assessment. Controllers are highly trained and constantly undergo retraining and competency checking.

Considering all these variables the ATC environment should be a totally safe environment in which there are few incidents. But incidents – with a causal role in ATC – are still taking place. Some of these incidents are so serious that only the factor of 'luck' keeps them from developing into accidents. It is therefore absolutely necessary that thorough incident investigation be performed to reveal the factors which are causing incidents from the ATC environment.

In comparison with a pilot, a controller does not work with checklists. Controllers are in the middle of a constantly changing environment and decisions made are continuously evaluated and adjusted if needed. As traffic figures rise, the work of the controllers becomes more demanding and the total system (human, machine and procedures) are pushed to their limits more frequently. In such an environment the human is very inventive, not only to 'survive', but also to perform the tasks demanded as well as possible. This inventiveness leads to a way of working which can occasionally differ slightly from the way of working which was planned when equipment or procedures were designed, or when training was given. The way tasks are performed in this situation are not always predictable and also differ between controllers.

The way incident investigation is performed must be able to address these specific areas of the ATC environment. When an investigator is able to build a complete overall picture of all the elements which were present in an incident, he or she is more able to write the most effective recommendations.

By applying the HERA technique during investigations we experienced the following advantages:

- Investigators are able to perform the analyses of the gathered facts and information more thoroughly.
- Investigators are able to have better interviews with involved controllers.
- Controllers have a feeling that the way they are performing is better understood. This leads to a more open interview with more results.
- The investigation is performed on a 'wider' area of possible weaknesses in the organization.
- The final incident reports therefore contain more effective recommendations.
- Management is enabled to get a better overview of the actual way the controllers perform their tasks in the ATC environment and what items are influencing the total environment, both positively and negatively.

Of course there are also disadvantages. The following three are considered to be the most significant:

- Introducing the use of the HERA technique asks for a good communication with controllers as the material is not easy to understand. A lot of explanation and examples have to be given to prevent controllers, who are new to the technique, from turning away from the investigation.
- The organizational culture must be one which is open and respectful; otherwise investigations cannot be performed in a complete way. This is particular the case when a controller is faced with what they believe is a complex and difficult technique such as HERA.

- Not all types of errors can be analyzed with the technique. Sometimes mistakes or mishaps are taking place because someone did something wrong as a result of not knowing things which should be known, i.e. mis-applied procedures.

Conclusions

The HERA technique was developed as a new approach to the Human Factors analysis of ATM incidents. It has taken a considerable amount of time to review, discuss and refine. Although at each stage the ATM community was asked for their input, it is impossible to evaluate such a complex technique without the co-operation of the real ATC investigation and safety groups. This co-operation has been possible with the present ATC organization.

The way this organization performed its investigations in the past did lead to results. But by applying the HERA technique they are convinced that the results have become better than in past years. It is still early to say, but the experiences with the technique so far have been positive. The technique proved to be valuable in this organization's constantly improving investigation process, although it is not possible to determine whether this process leads to fewer incidents and accidents. This can only be considered in time.

References

Billings, C.E. (1997), *Aviation Automation: The search for a Human-Centered Approach,* Lawrence Erlbaum Associates, New Jersey.

Cardosi, K.M. and Murphy, E.D. (1995), *Human Factors in the Design and Evaluation of Air Traffic Control Systems.* U.S. Department of Transportation, FAA, Washington, DC.

Dougherty, E.M. (1993), Context and human reliability analysis, *Reliability Engineering and System Safety,* Vol. 41, pp. 25-47.

Endsley, M.R. (1996), 'Design and evaluation for situation awareness enhancement', in *Proceedings of the Human Factors Society 32nd Annual Meeting,* pp. 97-101, Santa Monica, CA, Human Factors Society.

Fleishman, E.A. and Quaintance, M.K. (1984), *Taxonomies of Human Performance: The Description of Human Tasks,* Academic Press, London.

Isaac, A.R., Shorrock, S.T., and Kirwan, B. (2002), 'Human Error in European Air Traffic Management: the HERA Project', *Reliability Engineering and Systems Safety* Vol. 75(2), pp. 257-272.

Isaac, A.R. and Ruitenberg, B. (1999), *Air Traffic Control: the human performance factors.* Ashgate, Aldershot.

Martiniuk, R.G. (1976), *Information Processing in Motor Skills,* Holt Rhinehart Winston, New York.

O'Leary, M. (1999), 'The British Airways Human Factors Reporting Programme', *Proceedings of the 3rd Human Error, Safety and System Design Conference,* Liege, Belgium.

Rasmussen, J. (1981), *Human Errors. A Taxonomy for Describing Human Malfunction in Industrial Installations.* Risø National Laboratory, DK-4000, Roskilde, Denmark.

Reason, J. (1990). *Human Error,* Cambridge University Press, Cambridge, UK.

Rodgers, M.D. and Nye, L.G. (1993), 'Factors associated with the severity of operational errors at Air Route Traffic Control Centers' in M.D. Rodgers (ed.) *An Examination of the Operational Error Database for the Air Route Traffic Control Centers,* (DOT/FAA/AM-93/22) FAA, Washington, DC.

Shorrock, S.T. and Kirwan, B (2002), 'The Development of a Human Error Identification Technique for ATM', *Applied Ergonomics,* vol. 33, 319-336.

Taylor-Adams, S.E. and Kirwan, B. (1995), 'Human Reliability Data Requirements' *International Journal of Quality and Reliability Management* Vol. 12(1), pp. 24-46.

Wickens, C. (1992), *Engineering Psychology and Human Performance (Second Edition).* Harper-Collins, New York.

Chapter 5

FAA Strategies for Reducing Operational Error Causal Factors

Julia Pounds and Anthony S. Ferrante

Introduction

The US Federal Aviation Administration (FAA) oversees the largest, safest, and most complex aviation system in the world, relying on a workforce of highly trained air traffic control specialists who interact with an environment of radar, computers, and communication facilities to maintain the safety and efficiency of the system. In fiscal year (FY) 2000 alone the US air traffic system handled 166,669,557 operations. Calculated as a percentage of facility activities, the operational error (OE) rate per 100,000 activities increased from .60 in calendar year (CY) 1999 to .69 in CY00 and .74 in CY01, then declined by 11 per cent to .66 in CY02[10] (FAA, 2003a). Although air traffic declined after the events of 11th September 2001, the OE rate reflects the continuing need to identify mitigation strategies.

The FAA has historically tried to understand and prevent the incidence of OEs. To accomplish this, an elaborate and detailed incident reporting system evolved to capture causal factors related to their occurrence. The data were intended to supply information about where to target intervention strategies, with several initiatives focusing on one critical component of the system – the air traffic control specialist (ATCS). Because the controller is the closest person to the air traffic situation and the last point of prevention, much attention was logically focused on controller performance. Indeed, a controller's action or inaction provides a channel through which pre-existing system vulnerabilities can be manifested. As Fisher (2002) noted, human error is the mechanism that translates this potential for making a mistake into an occurrence. Thus, performance of the controller will always be at the sharp end of the operational system (Dekker, 2002) in the complex and multifaceted environment of ATC. Whereas past initiatives focused on remedial training and targeted deficient performance, several recent initiatives have focused

---

[10] Calculations of rates use fifteen decimal places but are rounded to two places for the table on page 6 of the FAA Administrator's Fact Book.

on human performance within, and as it interacts with, the larger ATC system, and viewed the human element as a fundamental part of this complex environment.

Initiatives

In recent years, the FAA Air Traffic Evaluations and Investigations Staff began several programs that focused more attention on skill building and performance maintenance rather than on remedial training. This approach is based on the philosophy of adult education and individual responsibility for maintenance of best performance rather than viewing training from a directive 'schoolhouse' approach, often disparagingly referred to as the 'blame and train' method. Initiatives included fielding automation to re-create traffic situations, developing safety metrics, analyzing incident data to identify performance enhancement opportunities, and sponsoring research to further develop the capability to identify causal factors.

Incident Re-Creation

As computer capabilities increased, the idea that computer processing could be harnessed to re-create operational errors became a reality. During the 1990s, the FAA Civil Aerospace Medical Institute (CAMI) collaborated with the FAA Air Traffic Evaluations and Investigations staff and Atlanta air route traffic control center (ARTCC) to develop automation to graphically re-create radar data that were routinely recorded by En-route air traffic control facilities.

Referred to as the Systematic Air Traffic Operations Research Initiative (SATORI), it was developed, tested, and fielded to all En-route facilities and regional quality assurance offices with the goal of gaining 'a better understanding of the interaction between the various elements of displayed information, verbal interactions, and the control actions taken by air traffic control specialists' (Rodgers and Duke, 1993, p. 1). SATORI is still currently in operational use and enables its users to re-create segments of operational traffic in a format similar to what was displayed to the ATCS, for example, showing relative location and separation, speeds, and headings of aircraft. Among other things, SATORI can display full and limited data blocks, beacon targets, and conflict alerts. Video and audio are synchronized, and the air traffic situation can be displayed in four dimensions.

At En-route facilities, SATORI systems enable the facility quality assurance staff to re-create OE situations for the controllers involved. It is important to note that the system was not intended to be used to 'call' an OE; that is, to identify when an OE occurred. An OE was first determined to have occurred and then SATORI was used to review the situation. Systems located at regional and headquarters offices enable further incident review. FAA Order 7210.56 provides guidance for the use of all replay tools.

As technological advances are made, the ATC system must adjust to these changes and ensure that radar reduction tools are used correctly and consistently throughout the system in order to provide the most accurate re-creation possible (FAA Order 7210.56, pp. 1-4).

After an OE is identified (or 'called'), SATORI re-creations are useful in determining aspects of controller and/or pilot performance involved in the event. Re-creations can also be used to review peak periods of traffic flow ('pushes') and the effects of weather on traffic flow. Viewing the re-creation also helps to target specific skill enhancement programs for those employees involved in the event. In addition to helping with the identification of performance issues, re-creations of randomly selected traffic samples not related to OE situations have also been productively used to assess controllers' technical proficiency in relation to training.[11]

The first of its kind to harness the capability of computers to re-create air traffic situations, SATORI re-creations have also been used to provide assistance to other agencies' investigations of incidents involving aircraft. These have included the National Transportation Safety Board (NTSB), US Department of Justice, and the National Aeronautics and Space Administration (NASA).

In addition to performance management and investigation, re-creations of traffic samples not related to OEs have been used in Human Factors research at CAMI. For example, traffic samples were used to study sector complexity, controller workload, and performance (Manning et al., 2001; Mills, Pfleiderer, and Manning, 2002). TRACON SATORI, a prototype system for the terminal environment, was used to examine how controllers use information about aircraft relative position to maintain 'the picture' of the traffic situation (Pounds, in review).

Used effectively, the capability to re-create traffic situations can bring about beneficial changes in procedures, airspace, and future ATC systems. SATORI re-creations are currently being used in diverse ways to enhance system performance. It is anticipated that next generation re-creation tools currently being developed will continue to provide added value.

Calibrating Incident Severity

The FAA Evaluations and Investigations Staff monitors the frequency of operational errors to determine the system vulnerabilities contributing to each incident so that they can be identified and reduced. Once a relevant separation standard is violated by ATC, an OE is recorded. Every violation of separation standards provides an important opportunity for lessons to be learned and system

---

[11] Technical performance issues consist of areas of knowledge and application that might benefit from training. These issues are not necessarily areas of deficiency. An employee may demonstrate overall acceptable technical proficiency but might benefit from technical training in the application of a particular skill or task (FAA Order 7210.56, p. 3-1).

improvement, although not all operational errors share the same characteristics. Separation standards and procedures differ depending upon, for example, the type of airspace, weather conditions, type of aircraft, and altitude.

A study conducted by Rodgers and Nye (1993) investigated whether the number of aircraft being worked by the controller or the traffic complexity at the time of the OE was related to its severity, defined in terms of vertical and horizontal separation between aircraft. Three categories were created by assigning a maximum of ten possible points each to the horizontal and vertical separation reported.[12] Based on the total point value, OEs were partitioned into categories of major (20 points), moderate (14-19 points), and minor (13 or less points) severity.[13]

Results of the Rodgers and Nye study were counterintuitive and demonstrated that neither number of aircraft nor traffic complexity was significantly related to major, moderate or minor OE severity, although the analyses revealed ways that the reporting process, and thus the resultant data, could be improved. The authors recommended gathering more normative rather than descriptive data, increasing the reliability of the reported data, and using a re-creation capability to permit investigators to review the dynamics of the air traffic situation.

The US Department of Transportation Office of Inspector General (December 2000) recommended that the FAA Air Traffic Investigations Division tackle the problem of modeling and defining severity for OEs in flight to describe the degree that the applicable separation standard was violated. The purpose was to group airborne OEs as low, moderate, or high severity and thus be able to focus resources on the most severe events and to identify factors related to specific categories of events. Data about systemic causes of OEs could then be used to more explicitly direct action towards prevention of future occurrences.

The FAA Air Traffic Evaluations and Investigations Staff developed a classification system to distinguish between OEs on these dimensions. Classification categories were developed to reflect the operational environment. In the first version of the index, an OE was classified on each dimension according to its severity – the extent to which separation distance was reduced – as low (39 points and below), moderate (40-89 points), or high (90 points and above) using a 100-point scale. Objective distances were used to minimize subjective interpretations of the data. Actual radar data from numerous operational errors representing En-route facilities nationwide were used to test the adequacy of the categories and the classification methodology. The components of the model included elements associated with loss of standard separation, such as the relationship of the aircraft in conflict to one another (e.g., converging versus diverging courses), closure rate, and level of ATC involvement – whether the event was a 'controlled' or 'uncontrolled' OE. The point distributions for En-route radar

---

[12] Vertical separation was subdivided depending upon whether the incident occurred below or above 29,000 feet (FL290).

[13] Of the 1053 OEs in their sample, only 15 (.01 per cent) were coded as 'major' severity.

OEs are shown in Table 5.1. A similar table was developed for OEs in terminal and En-route airspace with single-site radar.

Table 5.1   Radar OE severity index En-route chart (FAA Order 7210.56)

VERTICAL SEPARATION 1,000 feet required	POINTS	HORIZONTAL SEPARATION* 5-mile separation requirement	POINTS
Less than 500 feet	25	Less than ½ mile	25
500 feet to 599 feet	20	½ mile to 0.999 mile	25
600 feet to 699 feet	16	1 mile to 1.499 miles	24
700 feet to 799 feet	12	1.5 miles to 2 miles	24
800 feet to 899 feet	6	2 miles to 2.499 miles	23
900 feet to 999 feet	2	2.5 miles to 2.999 miles	22
VERTICAL SEPARATION 2,000 feet required	**POINTS**	3 miles to 3.499 miles	20
Less than 500 feet	25	3.5 miles to 3.999 miles	16
500 feet to 599 feet	25	4 miles to 4.499 miles	10
600 feet to 699 feet	24	4.5 miles to 4.999 miles	5
700 feet to 799 feet	24	**CLOSURE RATE**	**POINTS**
800 feet to 899 feet	23	700 knots and greater	10
900 feet to 999 feet	22	300 knots to 699 knots	8
1,000 feet to 1,099 feet	20	100 knots to 299 knots	6
1,100 feet to 1,199 feet	18	Less than 100 knots	4
1,200 feet to 1,299 feet	16	**FLIGHT PATHS**	**POINTS**
1,300 feet to 1,399 feet	14	Converging – Opposite Courses	20
1,400 feet to 1,499 feet	12	Converging – Crossing Course	18
1,500 feet to 1,599 feet	10	Same Course	10
1,600 feet to 1,699 feet	8	Diverging/Non-Intersecting	0
1,700 feet to 1,799 feet	6	**ATC CONTROL FACTOR**	**POINTS**
1,800 feet to 1,899 feet	4	Uncontrolled	20
1,900 feet to 1,999 feet	2	Controlled with TCAS RA	15
		Controlled with no TCAS RA	4

The initial step in calculating error severity was to determine the lateral and vertical proximity between the involved aircraft. The horizontal and vertical distances were defined as the minimum separation based on the radar data just prior to aircraft divergence. Situations with faster aircraft closure rates, coupled with converging, opposite direction flight paths were assumed to present a greater threat than slower closure rates and diverging flight tracks. The idea was that head-on encounters with high rates of speed, coupled with minimum radar data separation distance prior to divergence should account for the greatest total point value in the model. An assumption was made that aircraft not crossing each other's paths greatly reduces the threat to safety, and consequently, does not receive severity index points (see Table 5.1). In addition, if aircraft are not converging, closure rates become less significant when the other parameters are properly factored.

This model does not imply in any way that any minimum separation less than those required by FAA Order 7110.65 is acceptable from an operational perspective. The main purpose of the ATC system is to preserve safety. To learn from the occasional shortfall in the ATC system, the controller's action prior to the OE was also included in the assessment of the total severity of an event. A 'controlled' OE was defined as an OE where the controller was aware of the impending conflict and took corrective action to increase separation. An 'uncontrolled' OE was defined as an OE where the controller was unaware of the conflict, took no corrective action, and/or became aware of the conflict but did not have enough time to effectively mitigate the loss of separation.

The severity model assigns point values on each dimension with a maximum point value of 100. Component categories are used to allot point values corresponding with their relative significance during the event. For example, vertical separation of less than 500 feet in En-route airspace requiring at least 1000 feet of vertical separation would be assigned a point value of 25; closure rate of 700 knots and greater would be assigned 10 points. These values would be summed with points assigned to the other categories. The remainder of the 100 points left after the assigned points have been subtracted is the 'safety factor.' Table 5.1 shows the distribution of points possible for En-route OEs. Different tables of point values are used for En-route and terminal OEs where 3-mile or 2.5-mile horizontal separation was required and for events when in-trail separation standards were governing. Figure 5.2 shows a hypothetical example of how the points in each category of the severity index can be used graphically to display the various elements of one OE.

This severity index is an algorithmic approach, based on objectively measurable and observable variables in the operational environment. As the method was further developed, the initial three categories were split into four: High Severity (A); Moderate Severity, uncontrolled (B); Moderate Severity, controlled (C); and Low Severity (D). The four category model made it conceptually similar to, and potentially confusable with, less objective methods being used by other groups to classify incident types such as runway incursions.

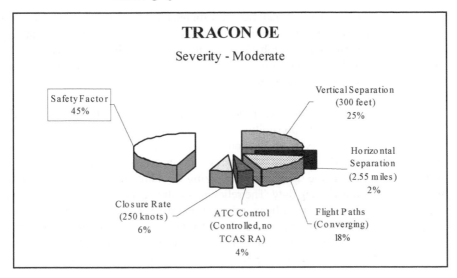

TRACON OE

Severity - Moderate

Figure 5.1   Sample of a moderate severity OE, with the per cent of total points for each element shown

Data-driven Focus on Performance

Analysis of OE data by the FAA Office of Investigations revealed several recurring causal factors, for example, readback/hearback errors and position relief briefings. As a result, interventions were initiated to address them. A series of videos was also produced to communicate several types of complex system vulnerabilities.

Readback/Hearback Errors Readback/hearback errors occur if the pilot incorrectly repeats back to the controller the instruction or information just received from the controller and if the controller fails to catch the pilot's incorrect response. As an example, a readback/hearback error would be noted when the controller instructs 'Piper 123 climb and maintain 1 – 0 thousand,' but the pilot repeats back to the controller 'Roger. Piper 123 climbing to 1 – 2 thousand.' If the controller does not catch the pilot's error in the response, it would be classified as a readback/hearback error. Although these are human errors, they do not necessarily lead to an OE.

Awareness programs targeting readback/hearback errors were developed by headquarters staff as well as by regional and facility groups. Facility programs included 'tape talks' where voice recordings for the ATCS on a position were reviewed by the ATCS and/or facility staff specialists to assess communication performance. The Air Traffic Investigations Division Staff produced a video – 'Preventing Readback Errors' – highlighting how different influences, such as ambient noise and distraction, can contribute to the occurrence of this type of error.

The video was sent to all facilities as a mandatory briefing item. Based on its initial success, this awareness program became an annual event with January inaugurated as Hearback – Readback Awareness Month, which included constant emphasis on good communications skills, random tape monitoring to highlight examples of correct phraseology, and positive coaching.

Position Relief Briefings Position relief briefings take place when one controller assumes (takes over) responsibility for a position from another controller, transferring responsibility from one controller to the other. The position relief briefing is a standard operating process designed to optimize transfer of responsibility while at the same time minimizing the additional workload associated with the task of transferring duty. The relieving controller previews the position and then indicates to the controller being relieved that the verbal briefing may begin. A checklist covers items to be noted prior to the relieving controller assuming responsibility for the position. The relieving controller observes the position and then the controller being relieved points out any abnormal items, traffic situation, and any other issues of concern using the checklist. Thorough coverage of the checklist items is meant to ensure that the relieving controller 'has the picture' of the situation. That is, the controller being relieved ensures that the relieving controller sees all relevant information, understands the situation, and is aware of any potential conflicts or problems. After the relieving controller assumes responsibility for the position, the controller being relieved observes the overall position operation to determine if assistance is needed and provide or summon it as appropriate.

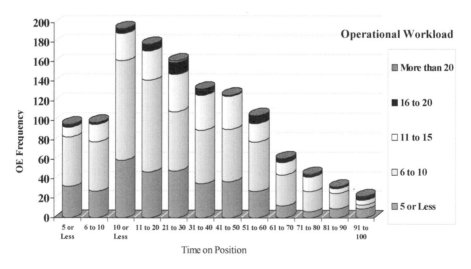

Figure 5.2   Number of 1056 OEs occurring relative to the controller's time on position and workload (FY 1997 through FY 2000)

Trends in the OE data showed that OEs frequently occurred within 10 minutes of a controller taking over a position. Figure 5.2 shows the distribution of 1,056 OEs between 1997 and 2000, and the number of OEs that occurred within 5, 10, and 20 minutes of the controller taking over a position. As a percentage of the total number of OEs (Figure 5.3), approximately 9 per cent occurred within the first 5 minutes, 18 per cent within the first 10 minutes, and 35 per cent within the first 20 minutes on position.

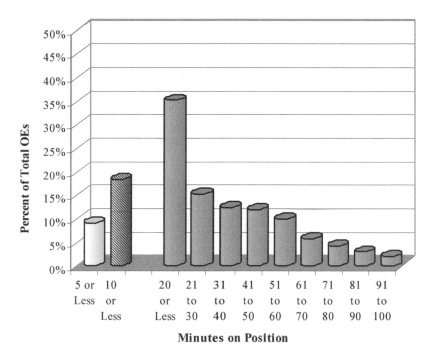

Figure 5.3   **Per cent of 1056 OEs occurring relative to the controller's time on position (FY 1997 through FY 2000 from Figure 5.2)**

An initiative to address this problem required all managers to validate position relief checklists, as well as provide a capability for recording the briefing. Shift supervisors were required to ensure the use of a position relief briefing checklist and, where available, ensure that position relief briefings were recorded. Controllers were trained and encouraged to accept position responsibility only after they were fully aware of the traffic, and the relieving controller and the controller being relieved were to establish an appropriate overlap period to complete the transfer of responsibilities. Some facilities mandate a specific amount of time that the relieved controller shall remain at the position. Other facilities adopted this as a good practice and permit the overlap period to vary depending upon traffic demands. The intent in this latter practice is that basing the overlap period on traffic demand reduces the likelihood of potential distractions associated with multiple personnel remaining on the same operational position unnecessarily.

Video Briefing Materials The Air Traffic Investigations Division Staff produced four videotapes to focus awareness on different types of causal factors found to be related to OEs. Copies of these were sent to all FAA air traffic facilities to be used for briefing materials.

Accidents often occur after a series of inconsequential events that create links in a chain during which individuals have had a number of opportunities to intercede and break the chain. The 'Break the Chain' video illustrates how events, if uninterrupted, can culminate in an accident and how attending to details can help break the chain of events and prevent accidents and incidents. The episode portrayed in the video is fictional but was compiled from actual events. A small business-owned aircraft is carrying a group of business executives to a meeting. The flight departs later than expected due to the passengers' late arrival and delays in air traffic services. The original plan was to depart ahead of developing stormy weather; however, the delay and the quickly developing storm result in moderate icing conditions. An aircraft mechanic, the passengers, both pilots, an airway facility staff specialist, and multiple air traffic control specialists (both terminal and En-route) all either contributed in some way to the chain of events or failed to interrupt the sequence when they had the opportunity.

'Collision Course: What are the odds?' depicts an actual event, demonstrating how rare and improbable events can and do occur. Individually, each event would be fairly benign. However, as part of a chain of events, seemingly inconsequential errors can set in motion a series of events that cannot be undone. Outwardly insignificant errors committed, for the most part, by controllers in two air route traffic control centers, put two large jet aircraft on a converging course and that information could not be relayed from the ground to the aircraft because radio communications were lost. The video demonstrates how the failure to accomplish routine procedures such as switching an aircraft to another frequency, becoming momentarily distracted, or being complacent about an evolving situation can potentially result in large and unexpected adverse outcomes, although a midair collision was avoided in the event depicted.

'Consequences of Simple Omissions' is a compilation of actual events. This video illustrates examples of how small omitted actions, lack of attention, failure to follow operational practices, compounded by poor facility practices, and lack of self-discipline or professionalism resulted in incidents ranging from operational errors to fatal accidents. The video advocates adhering to individual professionalism and remembering the importance of maintaining standards and accuracy for safety.

The video titled 'Preventing Readback Errors,' mentioned earlier, focused on communication strategies to reduce misunderstandings between controllers and pilots. Strategies include reducing the complexity of each communication and chunking information within the communication to facilitate understanding.

Information is presented in a humorous manner to illustrate strategies to overcome common blocks to good communication.

Identifying OE Causal Factors

The need for a formal reporting process was recognized early in the FAA's evolution and by 1965, FAA Order 8020 had established the ATS System Error Reporting Program. Early on, recommendations were developed regarding the conduct of incident investigations and the use of the resulting information, many of which were incorporated into the Order (O'Connor and Pearson, 1965). For instance, O'Connor and Pearson suggested that any reporting system should be based on a system view, including controller performance, the influence of personal capacities and skills, the design and operation of the system, and modifiers of the working environment such as supervision, operating procedures, health, morale, and work schedules. Notably, recommendations made by O'Connor and Pearson remain as relevant today as they were then:

> Air traffic control has become an increasingly complex system involving men and equipment in a continuous and dynamic decision-making function. Future developments point to the rising use of complex equipment, including high-speed computers, as aircraft speeds and the system's load continue to rise. By projecting current trends, it can be anticipated that future system changes involving equipment, personnel, and/or procedures point to the need for longer and longer lead times as complexity grows. The above considerations point to the need for close scrutiny of the ongoing system failures and/or incidents in order to provide the most accurate feedback information for system correction or modification (p. 1).

Additionally, O'Connor and Pearson recommended an approach to system error evaluations, asserting that 'the man-machine system will never be perfectly reliable, efficient, and error free because of the inherent limitations and idiosyncrasies of the human component' (p. 1) and that a deliberate and objective process would be better than 'shooting in the dark for sources of error and possible solutions' (p. 1).

Later, other changes were periodically made to the incident reporting form. For example, in 1997 a version was fielded having three additional causal factors under the category of 'Inappropriate use of Displayed Data'. This category was associated with the controller's use of the radar display and situation awareness associated with use of the radar data. The three added items – 'Failure to Detect Displayed Data', 'Failure to Comprehend Displayed Data', and 'Failure to Project Future Status of Displayed Data' – were included to address the controller's situation awareness. This change proved initially successful, with subsequent OE data showing that use of the category 'Inappropriate use of Displayed Data – Other' decreased and the data that would otherwise have been attributed to the 'Other' category distributed across these three new categories (Rodgers, personal communication). Although this change brought finer detail to the description of the OE, the issue of reporting reliability raised by Rodgers and Nye (1993) was still an issue.

Currently, the FAA Air Traffic Investigations Division oversees and coordinates the OE reporting process governed by the FAA Air Traffic Quality Assurance Order 7210.56 (FAA, 2002). The order 'provides specific guidance on investigation, reporting, and recording types of incidents that impact the quality of air traffic services' (p. i). Specifically, section 5-1-2 (p. 5-2) stipulates:

5-1-2 SUSPECTED EVENT

a. In order to maintain an effective Air Traffic System, it is imperative that we identify all deficiencies within our system and take appropriate corrective actions necessary to fix any associated problems. Operational errors and deviations are reported for just that reason, so those problems (either systemic or individual) can be corrected to enhance system integrity. The identification of operational errors and deviations without fear of reprisal is an absolute requirement and is the responsibility of all of us who work within our system.

b. Accordingly, it remains Air Traffic Policy that any employee who is aware of any occurrence that may be an operational error, deviation, or air traffic incident (as defined in paragraph 4-1-1, Definitions), immediately report the occurrence to any available supervisor, controller-in-charge (CIC) or management official.

In order to develop information so that data-driven decisions about causal factors and intervention strategies could be made, the Office of Investigations determined that a method for identifying causal factors related to human performance was needed – a method that viewed human performance as one among several potential points of system vulnerability. Resulting information about Human Factors could then be proactively used to mitigate the potential for future incidents. This effort was responsive to goals in the FAA's 1999 Strategic Plan, including the goal to 'eliminate accidents and incidents caused by human error' (FAA, 1999b). The FAA's National Aviation Research Plan for 1999 also echoed the intended outcome of developing enhanced measures of human performance and increased understanding of factors that lead to performance decrements.

An initial effort was undertaken in coordination with the FAA's Civil Aerospace Medical Institute (CAMI) to determine whether retrospective analysis of existing OE reports could extract additional useful information by using a Human Factors approach (Pounds and Isaac, 2001; Pounds and Scarborough, 2002). The retrospective analyses relied on data from existing OE reports that were based on standardized procedures specified by FAA Order 7210.56. Outcomes from this work suggested that standard reporting procedures did not require facilities to collect the type of data necessary to perform a thorough Human Factors analysis. That is, although the report forms captured descriptive data about the OE, little information was collected about events and causal factors preceding and during the loss of separation.

As these retrospective analyses were being completed, the FAA Office of Aviation Research entered into a collaborative agreement with EUROCONTROL and signed Action Plan 12 (AP12): Management and Reduction of Human Error in ATM. The initial goal of this project was to examine two existing techniques for identifying human error and to determine whether they could be harmonized into one technique.[14] If so, the FAA and EUROCONTROL member states would be able to use the technique retrospectively to examine existing incident reports for information related to Human Factors trends in the data, and to leverage this information to develop and share mitigation strategies.

As AP12 activities progressed, ATC subject matter experts from both FAA and EUROCONTROL judged that the harmonized technique – JANUS – also showed promise as a supplement to existing reporting processes (Pounds and Isaac, 2001). That is, rather than serving merely as a retrospective data mining tool, the new technique might also have value if integrated directly into existing OE reporting processes. Based on this hypothesis and after successful harmonization (Pounds and Isaac, 2001), the course of AP12 was modified to include further refinement and testing of the technique. Validation activities posed unique challenges to both the FAA and EUROCONTROL. Based on discussions of these differences, it was decided that parallel and complementary approaches be used based on the particular requirements of each to conduct the validation activities. Thus the FAA and EUROCONTROL developed the technique to reflect their individual system needs, resulting in two structurally parallel techniques.

The FAA method is a structured interview process that leads the analyst through a series of questions designed to identify the mental error and the contextual conditions surrounding it. Questions related to the controller's perceptions, memory, and decision-making processes that lead to execution of a plan are included. The method of using the technique considers the dynamic ATC traffic environment and treats each OE as a potential chain of events (or human errors) that result in the final loss of separation.

This line of work expanded on efforts by the Air Traffic Evaluations and Investigations Staff to improve causal factor information. For an inexperienced analyst of an OE, the technique is a potential aid to ask the right questions for

---

[14] The two techniques used for the harmonization were the Human Factors Analysis and Classification System (HFACS; Shappell & Wiegmann, 2000) and the model for Human Error in Air Traffic Management (HERA; EATMP, 2003). The Human Factors Analysis and Classification System (HFACS), a Human Factors taxonomy for analysis of existing aviation accident databases, was originally developed for the US Navy to investigate military aviation accidents. HFACS is currently being used by the FAA to analyze civil aviation accident databases. HERA is a model of human error for air traffic control that was developed as a tool to increase the effectiveness of human error identification.

eliciting causal factors and relevant Human Factors. For an experienced analyst, the technique ensures that he or she considers a broad range of causal factors rather than relying on experience and focusing on 'the usual suspects.' Figure 5.5 illustrates the general categories of contextual conditions included in the technique.

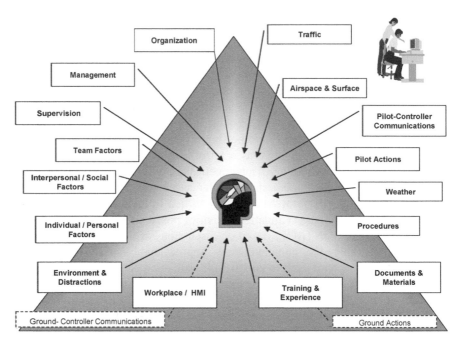

Figure 5.4   FAA JANUS taxonomy showing the categories of contextual conditions included in the technique

A research project to test and validate the JANUS technique was conducted in collaboration with the Air Traffic Evaluations and Investigations Staff and the National Air Traffic Controllers Association. A parallel project was conducted by EUROCONTROL. The research proposed to answer several basic questions related to validity: Does the technique work? How well does it work? Is it better than the current method? Is it ready for operational implementation? Will the results from the technique help to improve safety management?

Validation activities also posed unique challenges to both the FAA and EUROCONTROL. Based on discussions of the organizational differences, it was decided that a harmonized approach could be defined that would allow the other organization to leverage the potentially complementary work, findings, and lessons learned while also meeting the particular requirements of each organization. Methodologies for testing the technique were adapted to accommodate the respective organizations' testing environments, including use of operational resources and test constraints.

Results from the validation activities conducted to date by both organizations suggest that information provided by the JANUS technique will help improve safety management through the more effective identification of Human Factors associated with OEs. For example, Figure 5.5 shows the groups of mental processes that were identified in the beta test data from 79 OEs and the percentage of critical points where the category was identified relative to each facility type. In the beta test, multiple categories could be associated with each critical point. In general, the mental processes associated with planning and decision-making and with perception and vigilance were most frequently reported, although these differed somewhat by facility type. Although no statistical differences were computed between these groups, the relative differences between facility types makes sense. The critical points in the OEs from En-route centers were associated with more factors of Planning and Decision-Making whereas the critical points in OEs from the terminal facilities were associated with more factors related to Perception and Vigilance.

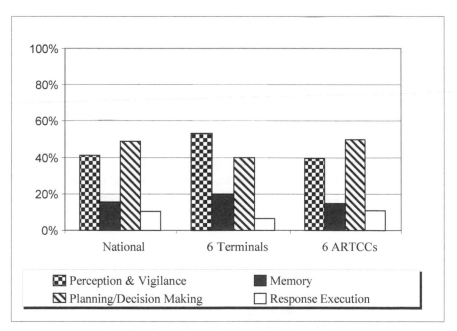

Figure 5.5   Categories of mental processes reported during the JANUS beta test

Mitigation strategies and programs can be developed to address these categories. One of these is described in the next section. However, a definitive answer about the usefulness of this approach can only be scientifically determined by a longitudinal analysis of data gathered using this technique. The information can then be evaluated for its usefulness as strategies are identified with which to mitigate the potential for future operational errors. Indeed, the ancient Roman symbol of JANUS, after which the harmonized technique was named, was attributed with the ability to look back so that an understanding of past events could lead to insight about future events.

Keeping the Mental Edge

The National Air Traffic Professionalism (NATPRO) project is an example of how information identified by OE analysis can be turned into strategy and skill enhancement. NATPRO is a new training approach sponsored by the FAA Air Traffic Investigations Division. Rather than relying solely on knowledge-based training, this approach integrates the concept of 'performance coaching,' using an awareness seminar coupled with a practicum.

Facility personnel first participate in the coaching clinic developed by Breedlove (2003) that covers the knowledge and critical skills of effective coaches. The coaches then conduct the seminar and practicum at their facility. The coaches help participants understand how the seminar concepts relate to performance by pointing out the connection between the seminar material and the practicum experience when participants practice the activities. For example, the coach might try to distract the participant during practice to demonstrate the influence of environmental distractions on performance.

The initial NATPRO program is focused on cognitive skills related primarily to visual attention, such as detection of information, focusing on relevant information and multitasking. The practicum includes activities to exercise the mind and improve concentration through distributed practice. The basic concept is not only limited to concentration. The NATPRO program concept is designed to also include other ATC skills such as auditory attention, decision making, and planning.

The seminar on cognitive skills was developed to increase an individual's understanding of the mental process of concentration, factors that affect mental processes, and how mental skills relate to performance. Although the skills are generic, it is hoped that by improving general skills, individuals will demonstrate a corresponding improvement in controller performance. Once armed with the knowledge provided by the seminar, participants then experience their own strengths and limitations during the Practicum. Interactive web-based computerized skill challenges permit participants to gain insight about their own skills. By testing themselves against the computer and experiencing how performance can vary in relation to factors such as distraction, fatigue, boredom, and so on, participants gain increased understanding of their own performance and identify strategies to improve it. Putting the challenges in a web-based application gives individuals and teams the opportunity to compete for high scores, should they desire to do so. Although the competitive aspect is available to participants, it is not required. It is included to enhance the experience for those participants who want to engage in interpersonal competition.

Admittedly, the successful transfer of skills from this program to actual air traffic control performance will be difficult to measure objectively. There are several measures that may be partially influenced as a result of this training, including a reduction in the number of operational errors; however, because the air traffic system is complex, there are many factors not related to controller preparation and awareness that also affect the number of operational errors. Other less tangible measures may include an increase in efficiency and higher job satisfaction. Success of the program could be captured by evaluating data from several sources. Initially, it is planned to compare participants' skill levels and subjective evaluations of the program before and after participating in the program. However, whether skill enhancement realized in the NATPRO program will translate to working traffic cannot be objectively evaluated at this time. Measurement of performance using high-fidelity air traffic simulations or medium-fidelity standalone simulation tools would supply more substantive evidence for evaluating the effectiveness of the program.

If this type of program is successful, it would demonstrate a more cost-effective, personally rewarding delivery of training. The NATPRO approach may improve air traffic safety and efficiency by increasing the controller's attention and perception skills, sustaining a highly skilled work force. Although targeting ATC performance, the skills themselves are generic. Participants would be expected to also benefit by transfer of skill to other activities beyond their professional ATC work environment.

Conclusions

A philosophy of individual leadership combined with joint accountability guides these efforts and others, all part of the FAA's ongoing efforts to maintain and enhance aviation safety by identifying system vulnerabilities and developing mitigation strategies. Air traffic control is a high-demand, high-consequence activity, and controllers must rely on their mental skills to successfully orchestrate air traffic. The initiatives described here reflect a human-centered approach but situate the human as the crux of a larger, more complex, dynamic system of other people and computers in a fast-paced environment of radar information, communications, and aircraft movement. Although any single initiative may not be a solution in itself – no silver bullet that will immediately eliminate all human error and every operational error – mitigation and reduction is a journey of many single steps that progress in the same direction toward prevention of these events.

Acknowledgements

Ms. Christine Soucy, a safety investigator with AAT-200 wrote and helped direct the videos for the OE awareness program. The NATPRO project was developed by Mr. Randall Breedlove with assistance from Mr. Jimmy Mills (ASO-150) and the Air Traffic Southern Region Learning Council.

References

Breedlove, R. (2003), *NATPRO. Creating the cognitive advantage for high performance ATC professionals,* unpublished.

Dekker, S. (2002), *The field guide to human error investigations*, Ashgate, Aldershot, UK.

EATMP (2003), *Short report on human performance models and taxonomies of human error in ATM (HERA),* HRS/HSP-002-REP-02, EUROCONTROL, Brussels.

FAA (1999), *National Aviation Research Plan,* Federal Aviation Administration, Washington, DC.

FAA (1999b), *Strategic Plan*, Federal Aviation Administration Washington, DC.

FAA (2002), *Air Traffic Quality Assurance*, FAA Order 7210.56, Federal Aviation Administration, Washington, DC.

FAA (2003), *Administrator's Fact Book,* Federal Aviation Administration, Washington, DC.

Fisher, B. (2002), 'Study looking at runway incursions identifies contributing factors and recommends solutions', *ICAO Journal*, Vol. 57(1), pp. 13, 27-28.

Manning, C.A., Mills, S.H., Fox, C., Pfleiderer, E. and Mogilka, H.J. (2001), *Investigating the validity of performance and objective workload evaluation research (POWER)*, DOT/FAA/AM-01/10, Federal Aviation Administration Office of Aviation Medicine, Washington, DC.

Mills, S.H., Pfleiderer, E.M. and Manning, C.A. (2002), *POWER: Objective Activity and Taskload Assessment in En Route Air Traffic Control.* DOT/FAA/AM-02/2. Federal Aviation Administration Office of Aviation Medicine, Washington, DC.

O'Connor, W. F. and Pearson, R. G. (1965), *ATC System Error and Controller Proficiency*, Federal Aviation Agency, Office of Aviation Medicine, Washington, DC.

Pounds, J. (in review), 'Using static and relational information to maintain the "picture" in ATC'.

Pounds, J. and Isaac, A. (2001), 'Preliminary Test of Two Models for Human Factors Analysis of Operational Errors', *Proceedings of the 11th International Symposium on Aviation Psychology*, pp. 1-5, Columbus, Ohio.

Pounds, J. and Isaac, A. (2001), *Development of an FAA-EUROCONTROL approach to the analysis of human error in ATM,* DOT/FAA/AM-02/12, Federal Aviation Administration, Office of Aviation Medicine, Washington, DC.

Pounds, J. and Scarborough, A. (2002), 'Retrospective Human Factors analysis of US runway incursions' *Proceedings of the ICAO NAM/CAR/SAM Runway Safety / Incursion Conference,* Mexico City, Mexico.

Rodgers, M.D. and Duke, D.A. (1993), *SATORI: Situation assessment through the re-creation of incidents,* DOT/FAA/AM-93/12, Federal Aviation Administration, Office of Aviation Medicine. Washington, DC.

Rodgers, M.D., Mogford, R.H. and Strauch, B. (2000), *Post-hoc analysis of situation awareness in ATC incidents and major aircraft accidents,* in M.R. Endsley and D.J. Garland, (eds.), *Measurement and analysis of situation awareness in complex systems,* Erlbaum, Mahwah, NJ.

Rodgers, M.D. and Nye, L.G. (1993), 'Factors associated with the severity of operational errors at air route traffic control centers', in M. D. Rodgers (ed.), *An examination of the operational error database for air route traffic control centers,* DOT/FAA/AM-93/22, Federal Aviation Administration, Office of Aviation Medicine, Washington, DC.

Shappell, S.A. and Wiegmann, D.A. (2000), *The Human Factors Analysis and Classification System – HFACS,* DOT/FAA/AAM-00/07, Federal Aviation Administration, Office of Aviation Medicine, Washington, DC.

Chapter 6

Reducing Separation in the Open Flight Information Region: Insights into a Human Factors Safety Case

Barry Kirwan, Steven Shorrock, Richard Scaife and Paul Fearnside

Introduction

In 1997 a study was launched to determine whether the then-current aircraft separation minima between aircraft in the Open Flight Information Region (OFIR: class F and G airspace, UK) receiving a Radar Advisory Service could be reduced safely, in order to help optimize controller workload. In particular, it was seen as desirable to reduce the lateral and vertical separation minima from 5nm and 5000 feet to 3nm and 3000 feet respectively. The four-year study comprised a number of inter-related elements, but Human Factors played a dominant role throughout. This chapter explains the context and motivation behind the study, the Human Factors approaches that were used, the decision-making rationale, and the arrangements for the safe introduction of the new vertical separation minima in the UK OFIR.

Background

This first section explains relevant aspects of the UK airspace, the air traffic service in the Open Flight Information Region, and the motivation that existed for a reduction in separation minima, which led to the study taking place.

UK Airspace Classifications

In the UK, the airspace within which aircraft operate is classified according to the type of service provided to pilots by Air Traffic Control (ATC) and the ratings required of pilots and avionics systems in order to operate in that airspace. There are six classes of airspace in the UK, namely Classes A, B, D, E, F, and G. The International Civil Aviation Organization (ICAO) also defines Class C airspace, but there is no Class C airspace in the UK.

Classes A, B, D and E are collectively known as Regulated Airspace, within which pilots must follow all instructions given by ATC.[15] Pilots are required to obtain a clearance from ATC before entering Regulated Airspace, with the exception of Class E, where a clearance is not required if the pilot is operating under Visual Flight Rules (VFR).

Class F and Class G airspace are collectively known as Unregulated Airspace (also known as Open Flight Information Region or OFIR). This differs from Regulated Airspace in that pilots are not required to follow ATC instructions: they may elect to maintain their own separation from other traffic and receive advisory information only from ATC, or receive no ATC service whatsoever. Unregulated Airspace is sometimes referred to as being 'outside controlled airspace', which suggests that there is no control over what occurs in Unregulated Airspace. This is not always the case, as traffic is sometimes monitored on radar, and if the Air Traffic Control Officer (ATCO) feels that there is a danger of the prescribed separation minima being breached, action will be taken dependent upon the type of service being provided.

Air Traffic Control within Unregulated Airspace

Within Unregulated Airspace, two types of radar service may be provided by ATC, a Radar Information Service (RIS) and a Radar Advisory Service (RAS). When providing a RIS, the ATCO will pass information to pilots regarding other traffic in their vicinity, including bearing, range and if possible the altitude or flight level of other traffic. Under a RIS it is the pilot's responsibility to see and avoid other traffic. Under a RAS, the ATCO will still pass information on nearby traffic to pilots, but in addition will provide advice on the prevention of a loss of separation. Separation standards define the minimum distance (laterally and vertically) that must be maintained between aircraft receiving a service from ATC. If these minima are breached, then, depending on the severity of the breach, an incident report may be filed by either the ATCO or the pilot. The advice provided by the ATCO under a RAS for the prevention of losses of separation may include climb, descent or turn instructions, or any combination of these. Under a RAS the pilot may elect not to follow the advice provided by the ATCO, but the pilot must declare that he/she will maintain their own separation.

Under a RAS in Unregulated Airspace, the prescribed separation minima were (at the time of the study) 5000 feet vertically (using SSR Mode-C) and five nautical miles (nm) horizontally provided the RAS traffic is separated from unknown traffic. These minima had been in place for over 20 years, when both radar systems and aircraft avionics systems were significantly less advanced than today.

---

[15] An exception being if the pilot receives a TCAS (Traffic Alert and Collision Avoidance System) Resolution Advisory (RA) for an emergency collision avoidance maneuver.

Motivation for a Reduction in Separation Minima

In the mid-1990s, press reports increased public awareness of the growth of air traffic within UK airspace. This growth was seen as placing increasing pressures on Air Traffic Service Units (ATSUs) to increase capacity whilst still maintaining the required levels of safety within their areas of responsibility. Most publicity about traffic growth figures is confined to the major ATC centers, dealing with regulated airspace, but the problem extends to Unregulated Airspace as well. Traffic growth in combination with the current separation minima meant that the limits of airspace capacity would be reached, resulting in reduced quality of service to pilots and increased ATCO workload.

The Civil Aviation Authority (CAA) Directorate of Airspace Policy (DAP) therefore commissioned a report on radar separation standards in Unregulated Airspace in April 1995. One of the recommendations of this report was an action on the National Air Traffic Services (NATS) Radar Separation Standards Working Group (RSSWG) to review existing radar separation standards in Unregulated Airspace.

A sub-group of the RSSWG was formed to progress this recommendation with the aim of reducing the Vertical Separation Minimum (VSM) from 5000 feet Mode-C to 3000 feet Mode-C and the Lateral Separation Minimum (LSM) from 5 to 3 nautical miles in Unregulated Airspace.

The RSSWG OFIR Sub-Group was at the time a multi-disciplinary group comprising representatives from DAP, the Directorate of Operational Research and Analysis (DORA), the CAA Safety Regulation Group (SRG), Military Air Traffic Operations (MATO) and NATS Air Traffic Management Development Centre (ATMDC) Human Factors Unit (HFU). The Sub-Group was tasked with evaluating the proposed reduction in separation minima paying particular attention to safety, airspace capacity, quality of service, and controller workload.

Since the separation standards had been in place for many years, since the early days of air traffic control radar technology, and altimeter and transponder technology, it was considered timely to see if such standards could be reduced, to ease controller workload, and possibly to give a better service. Workload can be a significant issue, since in the OFIR, some aircraft are not Mode-C equipped, hence vertical separation is not possible against such aircraft (as their altitude is unknown). Also, in the OFIR there may be aircraft performing complex and unpredictable maneuvers (e.g. aerial photography, aerobatics) and maintaining 5nm against such aircraft can be unfeasible. It was therefore suggested that 3nm lateral and 3000 feet vertical separation standards could be used on the basis that radar and transponder accuracy had increased significantly since the introduction of the separation minima.

The question was therefore whether such reductions in separation minima were safe. This was seen from the outset as mainly a Human Factors question, since such reductions could affect controller performance. Additionally, however, this was partly a mathematical modeling question, in terms of the impact such minima reductions might have on increasing the number of conflicts or 'encounters'. The following section outlines the Human Factors approach, and the subsequent section describes the overall decision rationale, wherein the mathematical modeling results were found to support the resultant Human Factors arguments. The final section then describes the Human Factors support to the operational implementation of the decisions made.

Human Factors Approach

A hybrid team was assembled, of safety analysts and operational researchers, military personnel, and Human Factors specialists, to evaluate the safety adequacy of the proposed new minima. This section will only focus on the Human Factors elements of the study, but it is noteworthy that Human Factors did in fact dominate the decision-making in this study (Shorrock and Kirwan, 1998; Shorrock et al., 2000).

Overall, the methodology was hybrid in nature, looking for evidence from several sources. First, it was necessary to understand properly the context of handling traffic in such airspace, carried out mainly by military controllers. Given the gravity of changing separation minima, it was also deemed necessary to have some degree of real-time simulation, to see how controllers would actually react to the proposed separation standards. For the purposes of assessing safety implications, it was also necessary to review past safety events that could bear on the decision about separation criteria. Lastly, because the separation criteria had not been experienced, it was also decided to try to predict potential problems in the anticipated conditions of new separation minima. Throughout the study, the researchers were not looking for compelling convergent evidence to prove or disprove the concept. Rather, if, on balance, the weight of evidence for any of the components of the study was negative, this would lead to a negative decision, i.e. to retain the existing standards. This conservative approach was taken because the controllers and the decision-makers alike place safety as the highest priority.

A series of evaluation studies therefore took place over a period of approximately 18 months:

- visits to three military Air Traffic Service Units (ATSUs), involving observation and interviews with military controllers;
- a preliminary real-time simulation to evaluate the impact of reducing separation standards, to see controllers' capability to adapt to the new standards, and to observe errors;
- a second and more controlled experimental study recording more data and variables including errors, losses of separation, situation awareness, workload (NASA-TLX), and statistically analyzable subjective data, plus debriefs;
- an investigation into several years worth of OFIR incident data;

- an identification of potential human errors in the handling of traffic in the OFIR to see how errors could occur, propagate, and be detected/corrected, using a fault and event tree modeling approach;
- a contrast of the error identification with the incident reports to evaluate any actual evidence of these errors in real incidents;
- an evaluation of the literature on controller situation awareness and detection rates for un-requested turns or descents by aircraft.

Types of Data and Analysis Methods A number of types of data and associated analysis methods were used throughout the study in order to gain a balanced and comprehensive perspective on the impacts of separation minima reductions on human performance. These data and methods are shown in Table 6.1, and are discussed in the following sections. Figure 6.1 shows the main aspects of the study schedule.

Table 6.1   Types of data and analysis used in the study

Subjective analysis	Objective analysis	Predictive analysis
Interviews and observations	AIRPROX data	Event tree analysis
Direct simulation debriefs (controllers and instructors)	Real-time simulation measures (erosions; heading and FL changes)	Fault tree analysis
Indirect measures (performance ratings)	Small-scale published studies	Detailed error analysis

Task Name	Duration	1997	1998	1999
1 Terminal Control Simulation	66d			
2 Area Control Simulation	96d			
3 AIRPROX Review	118d			
4 Error Analysis	22d			
5 Evaluation of Small Scale Studies	666d			

Figure 6.1   Study schedule

The main elements of this overall approach are detailed further below.

Visits to Aerodromes

Three one-day visits to military aerodromes were made to discuss with controllers their perspectives on separation minima and potential reductions. There was general interest in a reduction in separation standards. However, some factors were raised which were later explored in the simulations, such as the danger of 'capture' of controller situation awareness while elsewhere on the radar screen a 3nm separation 'erodes', the point being that 3nm erosion had less time to recover before it became serious, compared to 5nm. The visits also gave the principal Human Factors researchers opportunities to experience the ways of working of the military controllers, and the types of traffic they were dealing with.

Simulations

Two small-scale real-time simulations were conducted in the radar training simulator at the Central Air Traffic Control School (CATCS) in RAF Shawbury, to test the application of reduced separation minima.

CATCS Shawbury houses a high-fidelity training simulator. As such, the responses of the regular military controllers in the simulation were likely to be based on a realistic situation, and the controllers were familiar with this training facility. The simulations aimed at gathering a range of opinions, behavioral responses and safety-related data from controllers operating in different types of ATC environments. The first simulation was conducted using Terminal (Aerodrome) controllers and the second simulation using Area controllers. The first simulation was not a controlled study; it was to present the conditions to controllers to gauge their reactions. The second study was more methodological and controlled in nature. In both simulations, the following separation criteria for unknown traffic were assessed:

- 5nm/5000 feet Mode-C;
- 5nm/3000 feet Mode-C;
- 3nm/5000 feet Mode-C; and
- 3nm/3000 feet Mode-C.

The earlier discussions with controllers had suggested that the main areas of interest for study were in workload, safety, and 'frustration', the latter meaning the number of seemingly 'needless' turn or climb/descent instructions that needed to be given to pilots in order to maintain separation. Frustration was therefore to an extent representing the converse of being 'expeditious', and giving a good quality of service. Accordingly the following subjective and objective data were collected for each exercise:

1. subjective ratings of workload using the NASA Task Load Index tool (TLX);
2. questionnaires covering the usage of reduced separation minima;

3.   number and details of erosions of the separation minimum in force during the exercise;
4.   heading and flight level (FL) changes issued to aircraft;
5.   controller de-briefs; and
6.   open discussions with controllers.

The CATCS simulation facility was run by two full-time instructors, who could control and manipulate traffic situations. This enabled realistic simulations, including some non-nominal events and maneuvers by aircraft. Controllers found the simulated traffic realistically challenging. This 'fidelity' aspect of the simulator was important to the study, since unregulated airspace is characterized by unpredictable traffic behavior, and it was important that such a characteristic was simulated.

Another useful feature of the simulator facility was the replay function, so that after a simulation run, any aspect could be replayed. This was useful in one case for example, where situation awareness seemed to have suffered, though the controller was unaware of this. The controller in question could then be shown not only where the simulation finished, but also, by running the simulation forward in time by several minutes, a significant conflict appeared that the controller had not anticipated.

First Simulation (Terminal Control) Three controllers undertook a number of familiarization exercises and measured trials using progressively reduced separation standards over four days. [16] Each day had one familiarization exercise and three measured exercises to test each of the reduced separation conditions. Each exercise lasted for 50 minutes.

The results of the simulation revealed that the number of erosions of separation reduced as the simulation progressed. This coincided with progressive reduction of separation minima, although several other factors may have contributed to this trend, including practice effects and the continuation of standard working practices with the option to separate with reduced distances if necessary. Additionally, a reduction of erosions would be expected, as given the same level of traffic, there is less 'need' for traffic to get closer (i.e. rate of erosions is not independent from separation minima). However, an erosion from, for example 3nm to 2nm, generally carries more risk than an erosion from say 5nm to 4nm, and the controllers were well aware of this.

Seven erosions occurred throughout the simulation. The erosions were most often associated primarily with lateral separation (e.g. misjudging a turn; or getting 'boxed in' by conflicting traffic). Other erosions were associated with visual

---

[16] Note that this introduces 'sequence effects' into the simulation – however the simulation stakeholders believed it was important to gain acceptability that controllers were not immediately introduced to radical reductions in separation (e.g. 3nm and 3000 feet on day one). Ideally there would have been more training time before the measured simulation, but resources were not available for this to happen.

monitoring and distraction. At least three of the erosions could still potentially have occurred under the 3nm/3000 feet Mode-C separation criteria.

Controllers initially felt that judging 3nm on a radar display which could only display 5nm range rings and 10nm-wide airways was difficult, but overcame initial difficulties with practice.

Controllers stated that reduced separation reduced pressure and allowed for more thinking time, and a reduction in co-ordination and workload. Whilst workload changes could not be tested statistically, subjective ratings suggested lower workload during reduced separation exercises.

In summary of the first simulation, the subjective evidence was in favor of the reductions, but some safety concerns remained since erosions were still occurring with the reduced separation minima, particularly in the lateral dimension.

Second Simulation (Area Control) Four military controllers participated in the second simulation. The controllers performed four medium intensity exercises per day, so that the controllers performed each condition for each area.

There were no statistically significant differences in the number of heading changes and FL changes between the conditions.

Seven erosions of separation were observed and recorded. Five of the erosions involved the late detection of a conflicting aircraft (e.g. the controller was aware of an aircraft, but the aircraft turned into confliction). The remaining erosions were due to either misjudgement of a turn or misidentification of an aircraft.

The subjective ratings from the questionnaires suggested a gradual but modest increase in the controllers' median ratings of quality of service, usage of the reduction, timeliness of actions, and efficacy of avoidance strategies as the simulation progressed, and a small decrease in perceived risk over the three reduced separation conditions.

There were no trends in the perceived workload over the four conditions, therefore subjective workload did not vary as a function of separation criteria. This appeared to be because the controllers adapted and used any extra airspace capacity gained from reduced separation to give better service to aircraft receiving a RAS.

The controllers made several comments in open discussions and on debrief questionnaires regarding the separation standards throughout the four days. These are summarized below against each reduced separation condition.

5nm/3000 feet Mode-C: Some controllers stated that this condition allowed aircraft to maintain their FL with 3000 feet Mode-C separation. The use of this condition differed according to the controller and the exercise. One controller noted that this condition was useful for slow-moving aircraft against unverified conflictors.

3nm/5000 feet Mode-C: The most notable comments for this exercise were that 3nm separation took the pressure off controllers, especially when they spotted conflictors late or were against slow-moving or climbing conflicting traffic. Some controllers stated that this condition reduced stress. Again, there were differences between controllers in the usage of reduced separation minima.

In the open discussion, the controllers agreed that the 3nm separation made controlling easier. However, 3nm can sometimes be more difficult to gauge visually on the radar display. This is because the airways, being 10nm wide, provide a baseline against which to estimate 5nm. The controllers stated that the 3nm criteria reduced the frequency of 'avoiding action' turns, and would do so in an operational setting. In particular, the reduction made it easier to avoid unverified Mode-C aircraft and non-Mode-C aircraft.

3nm/3000 feet Mode-C: This condition received the most favorable comments, combining the positive comments of the two previous conditions. The controllers generally agreed that this condition was used to the greatest extent. They believed that it improved expedition, helped to prevent 'knock-on' conflictions, and helped with planning.

Controllers' order of preference for the simulated conditions was as follows:

1. 3nm / 3000 feet;
2. 5nm / 3000 feet;
3. 3nm / 5000 feet; then
4. 5nm / 5000 feet.

The results of this simulation suggested that controllers found the 3000 feet VSM the most beneficial separation criterion, but that the 3nm criterion was helpful for resolution of difficult situations. In fact, some controllers preferred to have 3nm available, but it was noticed in the simulation that this occasionally resulted in riskier conflict situations. For this reason, and because it had not been possible to run extended controlled simulations, other sources of evidence and types of analysis were considered.

AIRPROX Review

A review of AIRPROX (aircraft proximity, or air miss) reports from 1991 to 1995 was conducted to determine the possible effects of separation standards on incidents involving aircraft receiving a RAS, as well as the types of errors that controllers were currently making. A total of 31 reports were analyzed. Of these reports, 25 involved aircraft in receipt of a RAS from civil controllers and six involved aircraft in receipt of a military service.

Each report reviewed entailed a loss of separation and occurred in Unregulated Airspace (Class F and G airspace) between 2000 feet and FL240. Twenty-four of the 31 reports were classified by the relevant panel or working group as 'No risk of collision', whilst seven of the pilot-reported AIRPROX reports were deemed to compromise safety (three military and four civil). None of the incidents was classed as having 'actual risk of collision'.

Several types of information were extracted from the review:

1. summary description of the AIRPROX;
2. risk classification as decided by the Joint AIRPROX Working Group (JAWG) or Joint AIRPROX Assessment Panel (JAAP);
3. classification of controller error and Performance Shaping Factor relating to controllers and pilots involved, according to the human error types contained within the 'Technique for the Retrospective and predictive Analysis of Cognitive Errors in ATC' (TRACEr: Shorrock and Kirwan 2002);
4. possible effects of the current separation standards and reduced separation.

The 'causal factors' were translated into error modes, mechanisms and Performance Shaping Factors (PSFs) using TRACEr. Some of these error types and factors are described below. Note that some of the reports were allocated more than one causal factor.

Several of the error types could have a greater effect with reduced separation. Four errors involved *misjudgments of heading*. This observation has significant implications for reduced lateral separation, because the effects of misjudgments are likely to increase where the controller is aiming for little more than separation minima. Four errors occurred where the controller was *late to notice* a conflicting aircraft. This error could also be affected by reduced separation, where the controller has first seen an unknown aircraft and then turns to deal with other traffic for a period of time before returning to check the unknown aircraft.

Four errors occurred where *avoiding action instructions or traffic information* were delayed, even though the controller had seen that the conflicting aircraft were present.

The reports noted some other factors that appeared to influence the performance of the controller. The PSFs were *Mode-C/Secondary Surveillance Radar (SSR) problems* (e.g. where no Mode-C, SSR label and symbol reflections, label overlap, edge of range) and *high workload* (due to traffic mix, staff shortages, traffic load). These factors are indirect contributors to risk, and could influence performance in reduced separation conditions. Workload could reduce generally, but increase when the reduced separation minima are eroded. Secondly, problems of label overlap could increase. The other factors are unlikely to interact significantly with reduced separation standards.

The separation standard was not thought to be a significant factor in 24 of the 31 reports (e.g. the controller failed to notice the presence of conflicting aircraft, or separation was negated by a fast jet). In seven reports, it was unclear how reduced separation would have affected the incident, particularly with respect to the timings involved in controllers calling traffic.

In conclusion, the AIRPROX review suggested that the current separation criteria did not have a large effect on the incidents. However, in a small number of cases, different separation minima could potentially have affected the incident.

Predictive Human Error Analysis

Human Error Analysis (HEA) is a way of predicting what errors could occur, based on an analysis of how the operations should be done, together with an understanding of human error and factors that can affect error and performance. The analysis is supported by errors known to have contributed to AIRPROXs, and errors that were observed during the Shawbury simulations. This is a comprehensive approach that can identify errors that might not have been recorded in incident reports, or that have not been observed in simulations.

HEA can be directed forwards to predict what errors would be likely as a situation unfolds. Alternatively, it can be directed backwards to consider an undesirable event (a so-called 'top event') and then determine what errors could lead to such an event. Both of these approaches were used in this study. The first approach gives a general picture, and uses 'Event Tree Analysis'. The second approach gives a more detailed picture, and uses 'Fault Tree Analysis'. In addition to these approaches, 'Human Error Identification' was employed to predict more independent errors that relate to the sub-tasks involved in separating aircraft. This section summarizes the combined findings.

Findings of the Error Analysis

The HEA revealed a number of possible controller errors. Most of these errors could occur with the current separation standards, but their effects or their frequency could be different with reduced separation minima. In many cases, time pressure is a major factor that hinders error detection and recovery. However, some errors might be prevented with reduced separation. The identified errors and some factors that could help prevent errors are discussed below.

Planning (positive) Reduced lateral and vertical separation should allow more time for planning ahead, as there are fewer conflicts to resolve and thus lower R/T (radio-telephony) load. Controllers could allow aircraft to stay level, and climbs and descents to maintain separation would be easier to achieve and possibly more economical. Reduced lateral separation would also allow more time to plan ahead, due to the frequency of heading changes outside controlled airspace. This should help to prevent further problems in tactical control.

Judgment (negative) Misjudgments are a particular area of concern for reduced lateral separation, because OFIR traffic generally tends to maneuver more laterally than vertically. It is more important, therefore, to make a correct first judgment of the required heading. Lateral judgments that appear to be safe at first can be 'foiled' at short notice, and so a controller could have to reverse or tighten the turn. This situation is only relevant to the reduction of lateral separation.

When trying to achieve vertical separation the controller could misjudge required climbs or descents. Alternatively, the controller might suggest climbs or descents that the aircraft cannot make, or that are negated by weather conditions.

Reduced vertical separation could be less forgiving of such errors. Importantly, however, if the controller were to misjudge a required climb or descent, he or she could still opt to stop the climb or descent and turn the aircraft to achieve lateral separation.

To balance this argument, reduced separation would provide more space for tactical maneuvers, and so possibly more time to make judgments before an erosion occurs. Also, reduced lateral separation could reduce needless turns that can create secondary conflicts. It could be possible for the controller to direct RAS-supplied aircraft between two oncoming aircraft rather than make a large diversion. This is, however, subject to the effects of maneuvers by the oncoming aircraft.

Memory (negative): Controllers could forget to issue a planned instruction after a distraction. These instructions include changes to cleared FLs or headings, or traffic information. Controllers could forget other information that has been received recently, such as cleared flight levels or speed restrictions. Whilst these errors are possible regardless of separation standards, reduced separation allows less time to address situations that might follow, and places more demand on the pilot to sight traffic or request or implement avoiding action.

Perception (positive and negative): Late detection of conflicting aircraft is a relatively common controller error in AIRPROX reports. If the controller did not notice the presence of traffic in the first place, then reduced separation would have no effect on this error. In other circumstances, the controller could detect traffic at long range, then deal with other issues before intending to check the position of the traffic later. With reduced separation minima, it is possible that controllers might fail to check on traffic for longer than at present, knowing that traffic will take more time to cover the distance to separation minima. However, controllers should spend less time co-ordinating aircraft and issuing instructions. Thus, it is unclear how much time controllers are likely to be distracted.

Reduced lateral separation to 3nm would mean that the ATCO would be required to judge 3nm on the radar screen. This would be more difficult to do than 5nm, since 10nm-wide airways provide a reference which would not apply if 3nm were adopted. Reduced vertical separation, however, increases the visual demand in detecting gradual changes in digital Mode-C FLs, both for unknown aircraft and in ensuring that RAS-supplied aircraft do not bust their cleared FLs.

The vigilance demand on controllers is likely to increase if aircraft are separated at the proposed separation minima, because the controller will need to be more vigilant to notice any deviations by traffic or the RAS-supplied aircraft, since erosions are likely to be more serious with reduced separation. This could result in the controller becoming fixated on these potential conflicts, particularly with 3nm lateral separation. Reduced separation also offers less time to see whether avoiding action has been successful.

Communications (positive and negative) Controllers could make a number of communication errors, for instance where the controller makes a slip of the tongue or omits information from an instruction. Examples include saying the wrong heading or FL, saying 'right' instead of 'left', omitting the phrase 'avoiding action', or using non-standard phraseology. The pilot might 'step on' a message by trying to use the R/T at the same time as the controller, or could confuse headings and FLs. Again, whilst these errors are possible regardless of separation standards, reduced separation, particularly 3nm, allows less time for detection and correction. However, if the controller is engaged in fewer communications, there should be fewer opportunities for error and possibly more time to detect and correct communication errors.

Human-Machine Interface (negative) A key PSF identified during the HEA related to the Human-Machine Interface (HMI). Two HMI issues could be affected by reduced separation. The first is SSR garbling, where transponder information of two nearby tracks is corrupted. In this case, the controller would have to try to achieve lateral separation. Reduced vertical separation would create more pressure on the controller to achieve this, and 3nm would be easier to achieve than 5nm, but with a reduced margin for error. A second issue is label overlap. This could affect reduced lateral separation where aircraft separated by 3nm are 'hidden' beneath labels (particular non-Mode-C traffic). It is possible that the controller might fail to detect the movements of an unknown aircraft if it is hidden beneath a label. Label overlap could affect reduced vertical separation if the Mode-C were obscured, and the unknown aircraft climbed or descended.

In summary, a number of potential concerns were raised by the HEA, focusing more on the lateral dimension. These issues could be summed up by stating that the reduction in separation minima should lead to the occurrence of fewer problems and hence less workload, but when problems did occur, there would be less time to detect and resolve them. Since this was a qualitative study only, it was not possible to predict whether the potential gains outweighed the potential losses, but it was conceivable that at least for the lateral dimension, there could be a net degradation in safety terms.

Decision-Making Rationale

General Considerations

Although the current separation standards were seen as unexpeditious, they were familiar, and few serious AIRPROXs had occurred, even given the unpredictable nature of unknown traffic outside controlled airspace. The main problem was that the current standards, combined with traffic growth, might increase controller workload. Also, whilst dealing with a 'needless' de-confliction task, a more important problem could occur elsewhere and escape the controller's situation awareness.

The reduced separation options appeared to have several advantages, such as reducing overall workload, improving the use of the airspace, allowing a better quality of service, and potentially allowing more efficient conflict avoidance and better planning. Reduced separation (particularly vertical) was supported by all of the controllers who were involved in this study. Additionally, the simulations showed that controllers could work with these reduced separation minima, although erosions still occurred but none of these were considered serious by subject matter experts.

Lateral Separation

Controllers used lateral separation more for tactical conflict avoidance, because 5000 feet Mode-C vertically often cannot be obtained. Reduced lateral separation should provide more space for tactical avoidance. It could prevent some detours and sometimes reduce the risk of secondary encounters. Hence, 3nm could be more expeditious than 5nm. Also, 3nm is more useful than 3000 feet Mode-C for slow-climbing RAS-supplied aircraft or when against conflicting non-Mode-C aircraft, where vertical separation cannot be used.

These benefits must be balanced against the risks associated with reduced lateral separation. The major concerns raised in this study were increased time pressure, separation judgement, and error likelihood during tactical collision avoidance situations following the erosion of minimal separation.

Time pressure appeared to be more severe with lateral separation than vertical separation. Generally, lateral maneuvers tend to erode separation more quickly than vertical maneuvers (although high performance military aircraft can be an exception to this). A conflict might therefore develop too quickly for the ATCO safely to intervene.

Another consideration was that climbs/descents tended to be more stable than lateral maneuvers. An aircraft in climb tends not to oscillate between climbs and descents, whereas a maneuvering aircraft might alternate left and right turns in an unpredictable pattern (e.g. if practising turns or carrying out aerial surveys). This means that aircraft maneuvering laterally need to be monitored more closely. This requires extra vigilance and produces extra workload as aircraft approach minimum separation. This was borne out in the interviews throughout the study.

A recurrent theme during the interviews and the simulations was that if the standards were reduced to 3nm, controllers would aim for more than this. The simulations showed that under the current separation standards controllers aimed for around 10nm separation and frequently obtained at least 5nm, but occasionally eroded 5nm. With a 3nm criterion, controllers aimed for around 5nm, generally obtained 3nm, and occasionally eroded 3nm. This observation was reinforced by controllers' statements of how much separation they would aim for. The problem is that an erosion of 3nm separation is clearly of greater concern than an erosion of 5nm. The evidence accrued for the reduction of lateral separation standards therefore suggests that there are significant safety concerns, from a Human Factors perspective, which might well outweigh the potential benefits of this option.

Vertical Separation

Reduced vertical separation could also present some problems for controller and aircraft performance. Gradual changes in digital height readouts can be difficult to detect when an aircraft is flying straight and level or when the controller is not monitoring the aircraft specifically, but is monitoring the whole display.

Intermittent Mode-C and SSR garbling could present difficulties for vertical separation. Fast climbing or descending jets and 'pop up' traffic also allow less time to react with 3000 feet Mode-C than with 5000 feet Mode-C. Finally, weather conditions could prevent an aircraft from achieving its usual climb rate. Importantly, however, the controller can still try to achieve lateral separation in these situations.

Nevertheless, reduced vertical separation has a number of potential advantages. The controllers in this study believed that reduced vertical separation was needed more than reduced lateral separation, and was useful to terminal control because aircraft are already climbing or descending.

Additionally, 3000 feet Mode-C vertical separation would be easier for the controller to utilize than 5000 feet Mode-C. Thus, reduced vertical separation, whilst still giving a good deal of aircraft separation (three times more vertical separation than in regulated airspace), will provide more FLs and is likely to be used more than the current vertical separation standard, because it will be easier to obtain 3000 feet Mode-C than 5000 feet Mode-C. Furthermore, 3000 feet Mode-C offers the controller more separation options for tactical avoidance if an unknown aircraft climbs or descends towards a RAS-supplied aircraft. In these cases, the controller can try to achieve either lateral separation, or vertical separation, or both.

The Decision

In summary, the weight of evidence in this study suggested that reduced lateral separation to 3nm is of greater safety concern than reduced vertical separation to 3000 feet Mode-C. Of the three reduced separation options evaluated, the new option of least concern, in terms of the impact on the controller's performance, was 5nm/3000 feet Mode-C. Reduced vertical separation should improve planning and expedition, but imported less additional risk than reduced lateral separation. In order to determine if the vertical reduction is acceptable in risk terms, quantitative collision risk modelling needed to occur (i.e. a form of fast-time simulation modelling possible aircraft 'encounter' conditions and situations in this type of airspace). This modelling was conducted in parallel with the simulation work by another group within NATS, and the results supported the Human Factors conclusion and preference for vertical reduction only. Therefore, it was concluded that the lateral reduction in separation minima was inadvisable, and that the vertical reduction would not increase risk, but would yield benefits. The next stage was to see how the reduction might work in practice, during a long adaptation period (to see if safety remained intact, and if benefits did indeed arise). This required a live trial over an extended period. However, before such a study involving real controllers and aircraft, safety needed to be assured via a pre-implementation safety study.

Pre-Implementation Study

The above decision led to Phase 2 of the work, which was to establish a live trial of the reduced vertical separation minima. It was decided to trial the proposed vertical separation minima reduction in six co-operating military and civil ATSUs, which were representative of the rest of the ATSUs in the UK. In fact these ATSUs also included one of the more challenging OFIR areas (the Vale of York). The trial comprised four main phases:

- a pre-trial safety investigation using HAZOP (Hazard and Operability Study) to consider the safety aspects of the trial, what could go wrong both with the participating ATSUs, and also adjacent ATSUs (hence considering potential 'risk export'), and to establish reversionary procedures in case adverse events occurred;
- an initial live trial of six months duration, with six ATSUs, including a range of daily and weekly measures to determine if errors or other problems were occurring, and testing the adaptation process of controllers to the new vertical minima;
- a further six months live trial with the same ATSUs, in parallel with consultation with other ATSUs in preparation for the whole of the UK OFIR region adopting the new standard;
- a safety case preparation phase, and an acceptance phase by the regulators, and then implementation across the UK OFIR regions, with a post-implementation monitoring phase.

These pre-implementation aspects are elaborated upon below.

Live Trial HAZOP

Since the next steps involved carrying out a live trial, it was necessary to assure the safety of the trial itself, i.e. so that no accidents or increase in risk-bearing incidents should occur during the trial period or as a result of the trial. In order to facilitate such assurance, a Hazard and Operability Study (HAZOP) was carried out. HAZOPs (Kletz, 1974; Kirwan, 1994) are used to identify potential hazards and associated control mechanisms. They are particularly useful in novel situations, as they rely on structured brainstorming. They are used in chemical, oil and gas, and nuclear power industries in many countries.

The objectives of the proposed 'live-trial' HAZOP exercise were as follows:

1. to identify all potential hazards, errors, scenarios, failure modes, factors, and other pertinent information relating to the safety of aircraft receiving a RAS in a reduced vertical separation condition;
2. to additionally consider any specific hazards relating to the trial itself and the transition period to UK-wide OFIR implementation of the proposed new separation minimum.

A one-day HAZOP was convened and run by a senior Human Factors investigator. The participants included a number of military controllers and a civil pilot familiar with flying in OFIR airspace, two hazard analysts and another Human Factors practitioner involved in the study.

A number of scenarios were presented at the outset of the HAZOP, and discussed by the participants. The scenarios involved aircraft receiving a RAS during the trial, and then explored events that could happen both within the trial period and afterwards, during the transition to UK-wide implementation

The traditional set of HAZOP keywords (Kletz, 1974) were used (including No/Not, More, Less, Early, Late, etc.), to prompt the experts into identifying hazard scenarios. Additionally, there are a number of components or 'elements' of the scenario that can be focused on to help identify potential hazards (these include: controller; pilot of RAS aircraft; other pilot; radio/telephony (R/T) communications; paper flight strips; radar displays).

The hazards of interest were mainly in connection with other aircraft not receiving a RAS, e.g. a controller in a participating unit might forget to modify separation before handing over the aircraft to a non-participating unit, such that the aircraft was immediately in a loss of separation situation when being handed over to the non-participating unit. The primary event of concern throughout the HAZOP was the potential for a mid-air collision between a RAS aircraft and an unknown aircraft. The latter aircraft might or might not be using Mode-C. Additional and more likely 'hazards' however, were in terms of short-term losses of separation, and also in terms of loss of credibility in the trial, and loss of trust in the proposed reduction. The HAZOP also focused on dangers for adjacent non-participating units, and the potential dangers of the participating units going back to 5000 feet vertical separation minima after having become familiar with 3000 feet VSM. The HAZOP therefore not only considered what could go wrong with the live trial, but also informed the implementation strategy itself.

All identified potential hazards were entered into a hazard log, so that the subsequent trial could look for evidence of these hazards or precursors to them. In practice, no significant hazards were identified, but in any event, reversionary procedures were developed. This meant that any participating unit could, of its own volition and without question, decide to opt out of the trial if it deemed it necessary. Furthermore, special procedures were set up to investigate rapidly any airproxs where the separation reduction may have played a part. Neither of these procedures needed to be invoked during the ensuing year-long trial period.

The hazard log, along with other Human Factors data collected, were used in conjunction with other analyses performed by the Directorate of Airspace Policy

(DAP), Safety Regulation Group (SRG) and the Directorate of Operational Research and Analysis (DORA) to construct a safety case to justify the conduct of a live trial to evaluate the reduction of the vertical separation minimum. The trial then began in October 1999.

Live Trial Support

Following the HAZOP, a live trial was carried out to examine the benefits and drawbacks of reducing the vertical separation minimum using Mode-C from 5000 feet to 3000 feet, when applied to aircraft receiving a RAS and conflicting with unknown traffic in Unregulated Airspace. This regional trial was carried out in the north east of the UK and involved both civil and military ATSUs. A trial instruction was used to define the temporary procedures to be adopted by trial units and contained guidance on how to proceed with the trial.

The trial design comprised the following conditions, initially to be conducted over a six-month period, but then extended to a full year:

- Baseline – for the first month of the proposed trial controllers operated using (then) current procedures to provide baseline data via questionnaires covering key issues following each session, which would then be used to compare the current procedure to the proposed reduced separation procedure. These key issues included workload levels, number of erosions of separation, quality of service, usage of separation procedures, and R/T loading.

- Trial – immediately following the baseline, one month of further data collection using the revised procedures was conducted. This enabled a direct comparison to the baseline phase with the addition of specific questions on the usability of the reduced separation minimum, including advantages and disadvantages of the new separation procedure and success rates of applying the new procedure.

- Monitoring – following the direct comparison between trial and baseline, the final four months of the proposed trial were conducted using a weekly questionnaire to gather information on controller concerns. This provided more general information on the progress of the trial and could alert the project team to any problems with the new procedure not identified during the first months of the trial.

Prior to the trial, a Human Factors representative visited all six ATSUs to explain what was going to happen, and to introduce the personnel to the various daily and weekly questionnaires and recording forms that would be used, as well as the reversionary procedures. The trial therefore generated a large amount of data which was sent back to the Human Factors Unit for analysis on a weekly basis throughout the trial period. These data had to be analyzed relatively quickly, in case any undesirable trends were arising.

Present Situation

A mid-trial review was conducted three months into the trial to review the evidence obtained to date. At the end of the initial six-month trial period, it was decided, based on positive responses from the ATSUs and interest from other ATSUs, to continue the trial, leading up to UK implementation after 12 months. This went ahead, the formal safety case was made and accepted by the regulators, and reduced vertical separation was implemented nationwide in OFIR airspace, and has now been in operation for several years.

Conclusions

This was a challenging study from the outset, with a proposed change to safety related airspace policy. Human Factors was largely given the responsibility of trying to determine the safety adequacy of such proposed separation minima reductions. There was no simple, ready-made method for deriving this answer, and so the approach used a number of intersecting measures to try to see whether it would lead to a net increase or decrease in safety. In such approaches, one hopes that several measures will indeed converge onto a single and unmistakable answer, but in reality it often ends up requiring a careful balancing and weighing up of evidence. Suffice it to say, that there were many long and involved discussions among the assessment team and with the internal client as to the arguments and counter-arguments, the factors, the scenarios, etc., in the process of developing a final risk-based argument. However, throughout all of these discussions, safety was always seen as paramount, and 'production bias' never influenced the decision-making process.

The study was effective in determining the safety adequacy of the proposed changes, deeming one safe, and one unsafe. This was not at any point quantified against target risk or safety criteria. Rather the argument was instead that risk should not be substantively increased against its current levels. Lateral reduction was considered to have the potential for substantive increase in risk, vertical reduction was deemed to have negligible net effect. Human Error Analysis, used in other industries but relatively new in ATM, was pivotal in the decision-making process, as it enabled a step from what we knew then, to how it might work and fail in the future.

One aspect of the study that was commented on by a non-Human Factors ATM specialist, was that the study effectively said 'No' to one of the proposed reductions. This was something of an initial surprise to certain interested parties, i.e. that Human Factors would not simply sit on the fence, but would actually construct decisions based on its analyses, and would not shy away from hard decisions in the face of imperfect data. The decision was not an easy one to make, since evidence was only 'suggestive' of problems – there was no hard evidence either way. Nevertheless, the assessment team needed to make a 'judgment call', and did so. This shift in perception of Human Factors as being capable of delivering answers to difficult, real-world problems was to the benefit of the Human Factors Unit. As for the decision itself, the subsequent trial period, extensive data collection and analysis, and success of the vertical reduction and its implementation have to date ratified that decision.

Acknowledgements

The authors would in particular like to thank Mike Strong, Kevin Dowling and Graham Vernon, and all the controllers and instructors who participated in the study.

References

Kirwan, B. (1994), *A Guide to Practical Human Reliability Assessment*, Taylor & Francis, London.

Kletz, T. (1974), *HAZOP and HAZAN – Notes on the Identification and Assessment of Hazardz*, Institute of Chemical Engineers, Rugby.

Shorrock, S.T., Kirwan, B., Scaife, R. and Fearnside, P., (2000), 'Reduction of separation minima outside controlled airspace' *IBC Aviation Safety Management Conference*, Copthorne Tara Hotel, London.

Shorrock, S.T. and Kirwan, B. (1998), *Evaluation of the Human Factors Impact of Reduced Separation Standards Outside Controlled Airspace: CATCS Area Simulation Study*, NATS ATMDC Memorandum Report 9803, National Air Traffic Services Ltd., London

Shorrock, S.T. and Kirwan, B. (2002), 'Development and Application of a Human Error Identification Tool for Air Traffic Control', *Applied Ergonomics*, Vol. 33, pp. 319-336.

Chapter 7

Distributed Work in the National Airspace System: Providing Feedback Loops using the Post-Operations Evaluation Tool (POET)

Philip J. Smith, Mark Klopfenstein,
Joe Jezerinac and Amy Spencer

Introduction

The focus of this chapter is on the need to provide operational staff within the National Airspace System (NAS) with better feedback about the impacts of their decisions, helping them to understand how their actions interact with those of others to affect the performances of individual flights and the NAS as a whole. The discussion begins with a view of the NAS as a distributed work system that started to reach its limits in the 1990s in terms of its ability to meet demand by users, and that has been rapidly evolving in order to deal with this constraint.

Research is then reviewed indicating that, as rapid changes in the architecture of the NAS began to occur starting in the mid-1990s, Federal Aviation Administration (FAA) traffic managers and airline dispatchers responsible for strategic planning often did not have adequate information and feedback to learn and adapt as effectively as possible. Based on this observation, a new tool, the Post-Operations Evaluation Tool (POET) was implemented for the FAA to provide these individuals with easy access to an integrated source of NAS data. POET provides feedback about the outcomes of plans for traffic flows and airline schedules as well as for individual flights. POET is now an operational system that has been fielded by the FAA, and is currently being used extensively by traffic managers, airline dispatchers and a variety of other system analysts and researchers. The design of this tool is reviewed, and examples are provided of its use by FAA and airline staff to enhance their understanding of current performance in the NAS and to assess alternative approaches for improving performance in the future.

The Distribution of Roles and Responsibilities

In order to deal with cognitive complexity, the NAS in the United States has evolved into a highly distributed system, including business and general aviation flights, as well as commercial airline flights. The military also has an increasingly significant presence in the NAS. One sense in which the NAS is a distributed system is due to the fact that these different users and their associated organizations all work somewhat independently in order to conduct their operations. Even within an organization such as an airline, however, roles and responsibilities are highly distributed. A typical airline management structure includes (among others) a group labeled 'flight operations' (pilots and their management), another labeled 'airline operations control', a third focusing on 'maintenance' and a fourth in charge of 'marketing'.

Distributed Work in FAA Operations

Like the airlines and other NAS users, the FAA is a highly distributed organization. Even if we focus just on those parts of the FAA charged with daily operations, we find that work is distributed among a number of organizations and individuals with very different types of tasks.

Organizationally, Traffic Flow Management (TFM) within the FAA is in charge of making daily and hourly decisions about how to structure air traffic flows to ensure safe, efficient performance. This includes decisions about restricting arrivals to an airport if its capacity has been reduced (due, for instance, to a closed runway) by instituting a ground delay program. It also includes decisions to move flights, for example, that would normally fly through Chicago Center to reroutes further north through Canada or south through Memphis Center to avoid thunderstorms around Chicago.

These decisions concerning traffic flows are made by a hierarchically organized and geographically dispersed set of organizations. This includes the Air Traffic Control Systems Command Center (ATCSCC) which is responsible for the facilitation of traffic flows nationally, and Air Route Traffic Control Centers or ARTCCs (see Figure 7.1) such as New York Center or ZNY, which are responsible for high altitude traffic flows within their own boundaries and for helping to coordinate flows to and from other Centers. It also includes Terminal Radar-Approach Control (TRACON) facilities which are responsible for lower altitude flows surrounding certain high density traffic areas such as those around New York and Southern California, and traffic managers in some airport Air Traffic Control (ATC) Towers.

The traffic managers at these various FAA facilities must coordinate with each other as well as system users to develop strategic plans to deal with weather, traffic congestion and other dynamic constraints that arise in the NAS. They must also work in close coordination with the air traffic controllers at their facilities who are responsible for the tactical separation of flights within individual sectors. In short, within the FAA itself, TFM and ATC tasks are highly distributed both geographically and organizationally.

Distributed Work in Airline Operations

As discussed briefly above, work is highly distributed among and within the organizations representing the various NAS users, who must work directly or indirectly in coordination with FAA operational staff. One such illustration of how work is distributed among the NAS users is the functioning of Airline Operations Centers (AOCs). Each airline has an AOC in which dispatchers and other staff (crew and aircraft scheduling, weight and balance, etc.) deal with daily and hourly strategic planning and tactical operations. Among other responsibilities, AOC dispatchers develop the flight plans for aircraft (routes, altitude profiles, fuelling, etc.) and share joint responsibility with the flight crew for the real-time operation of the flight. Most AOCs today also have Air Traffic Control Coordinators, who are responsible for communication and coordination with FAA TFM staff regarding strategic planning and tactical issues (Kerns et al., 1999; Smith et al., 1997).

The Need for Feedback in this Distributed System

The driving forces for the evolution of the NAS into this distributed system included two primary goals:

- limiting the cognitive complexity of the task for any one individual, thus making each individual job feasible in terms of knowledge, skill and workload requirements;
- providing overlap or redundancies in responsibilities as a safety net to ensure that, even if one individual or component of the system made a slip or a mistake (Norman, 1981) or overlooked some important event, another individual would note the problem before it became serious (Reason, 1990).

As with most classic architectures for distributing work, this system decomposition limits the extent to which interactions occur among the various sub-components (Smith et al., 2000; Smith, Beatty et al., 2003). Thus, the assumption underlying this architecture for the NAS has been that, for most of the time, if each individual or organization completes its tasks satisfactorily, then the need for interaction with other individuals or organizations will be minimized. This also assumes that interactions are expensive and time consuming and should therefore occur sparingly.

However, as demand began to reach capacity limits in a number of areas within the NAS in the 1990s, and as economic pressures caused the airlines to seek ways to further increase efficiency, a number of significant changes were made in the 'rules of the game' (Smith, Billings et al., 2000), moving from management by directive, toward management by permission and then further toward management by exception for routing decisions. These changes altered the locus of control and gave the airlines and other NAS users greater flexibility in terms of filing routes.

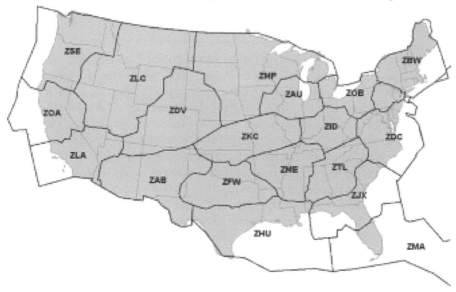

Figure 7.1   En-route centers in the National Airspace System (NAS)

Other changes associated with the use of ground delay programs shifted the parameter of control used by the FAA to a more abstract level, again giving NAS users greater flexibility (Smith et al., 1999). In addition, changes in the nature and location of air traffic (increased use of regional jets, increases in business jets and general aviation traffic, increased demand from smaller airports, new demands for national security resulting in dynamic changes in airspace blocked out for military use, etc.) lead to changes in airspace utilization. These significant changes had two important implications:

- In order to adapt effectively to changes in airspace utilization and to changing roles, responsibilities and procedures, FAA traffic managers and airline dispatchers needed better information about performance in the NAS in general, and feedback about the impacts of their own decisions in particular.
- In many cases, such feedback indicated that there was an opportunity to improve system performance, but the necessary data or knowledge resided with an individual in another organization (such as knowledge a dispatcher needed from a traffic manager about departures out of Newark). Thus, there was also a need to better support either synchronous or asynchronous interactions among AOC and TFM staff in order to disseminate the 'tribal' knowledge necessary to resolve the problem identified as a result of better feedback (McCoy et al., 1999).

Illustrations of this problem were provided by a study conducted by Smith, Billings et al. (2000) in which dispatchers made comments about the introduction of new procedures such as the following:

'Under the expanded NRP [National Route Plan], it's like shooting ducks in the dark.'
'The problem with the expanded NRP is that there's no feedback. Nobody's getting smarter. Someone has to be responsible for identifying and communicating constraints and bottlenecks.'
'It used to be the weather that was the biggest source of uncertainty. Now it's the air traffic system.'

In follow-up studies, FAA traffic managers made similar observations from their perspective in relation to the Severe Weather Avoidance Program (SWAP):

'We'll put flights on SWAP routes during a weather event, but I never know what the real impact is on those flights because I do not get sufficient feedback about actual performance.'

The Design and Implementation of POET

When evaluating software, three critical criteria are whether the product is:

- useful;
- usable; and
- used (effectively and profitably).

The previous section presented data suggesting the need for a feedback tool like POET, which is to say it is potentially useful. This section discusses the design features intended to make it usable. Then the next section discusses its widespread use by FAA and airline staff.

In designing POET, the goal was to provide easy, integrated access to NAS-wide data about what is planned and what actually happens each day. POET is now an operational system that has been fielded by the FAA and is used extensively by FAA TFM and airline AOC staff, as well as by research and development groups and system analysis groups at the FAA, the airlines and numerous universities and aviation-related companies.

At the heart of POET is a database that provides near-real time access to data about all the planned and actual operations in the NAS over the past 60 days. Near-real time means that there is a 2 hour delay between events in the NAS and access to information about them via POET. Below, we run though a sample use of POET to both illustrate its functionality and to highlight the design decisions intended to make it highly usable.

Figure 7.2 shows the Homepage for POET, and emphasizes that there are two conceptually different ways to identify a collection of flights for analysis when using POET. One way is to conduct 'flight-based' searches, which typically focus on collections defined by a city pair or call sign. The second approach is to use an 'airspace-based' search, which defines the collection of flights in terms of some region of airspace (such as flights filed through sector ZOB48 from 1400-1800Z). The Homepage also offers the user the ability to request 'summary reports', which automatically run a series of POET commands to produce canned reports for certain types of analyses which are required routinely.

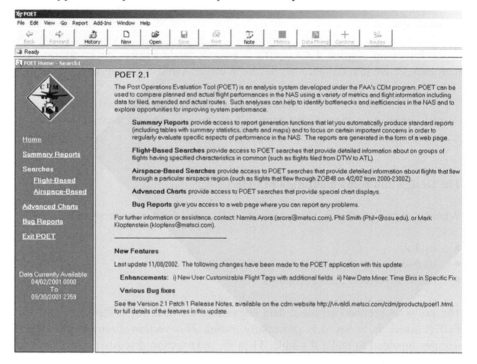

Figure 7.2   POET Homepage

Figure 7.3 illustrates how flight-based searches are completed in POET. The user builds a query by selecting items from menus indicating basic choices such as airline, departure airport, arrival airport, departure center, filed or actual departure date and filed route. Advanced users can construct more complex queries by selecting the Additional Selection Criteria button. In this example, the search that has been defined focuses on flights filed to depart from DFW to EWR, JFK or LGA during the period 4th to 10th September 2001 between 1800-2359Z.

The first two of nine design principles discussed in this chapter with regard to the development of POET apply this window:

Design Principle 1: Provide the user with visible feedback about the current state of the software and the system its displays represent (Norman, 1990; Rasmussen et al., 1994). In Figure 7.3, this is achieved by displaying the currently defined query in the search box at the top of the display. Each time the user makes a selection from one of the Search Builder menus, that selection is added to the search box at the top of the Search Builder Window.

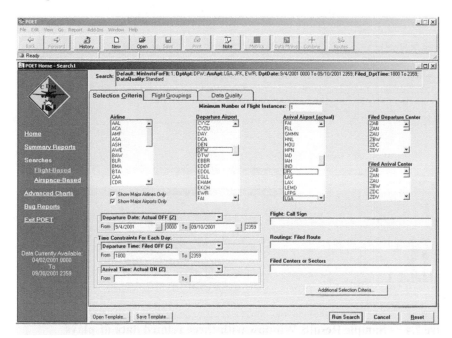

Figure 7.3   Search Builder page in POET

Design Principle 2: Design the system to provide the rich functionality required by advanced users while still allowing less experienced users to easily complete basic searches (Gould, 1995; Weiss, 1993). This goal is achieved through the use of the basic menus presented on the Search Builder Window, which focuses the user's attention on choices available to complete most of the flight-based searches that are of interest. In addition, the easily accessible but less salient Additional Selection Criteria function (button) provides a richer vocabulary for advanced users when defining queries. The Summary Reports option on the Homepage provides a second, even easier way for novice users to complete searches, assuming that the information they want is available in one of the canned summary reports.

Figure 7.4 shows the results of this flight-based search. Note that the top portion of this display again shows the search that has been run, and that the bottom portion is a Results Table that compares planned with actual performance. In this case, the 29 flights from DFW-EWR had an average airtime that was 4.3 per cent greater than planned. These quantitative metrics are available in different graphical displays in the Charts Window in the lower left hand corner. Also, filed and actual routes for the flights included in this search are shown in the Map Window in the lower right hand corner.

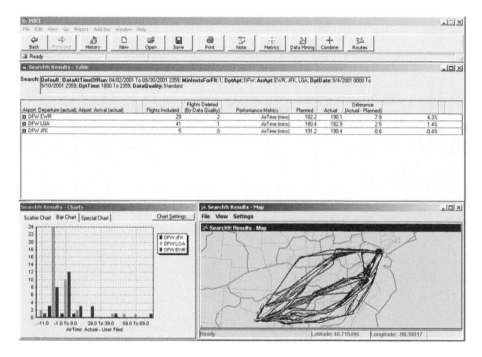

Figure 7.4   Sample Results Window with three related data displays

Note that this display illustrates the application of three additional principles:

Design Principle 3: Reduce the user's short-term memory load by providing a display that helps the user to recall the context of the current information display (Shneiderman, 1998). In this case, the context is defined by the search used to retrieve the flights of interest. The description of the search that was run at the top of the screen is especially important if the user has run a series of related searches and needs to make sure he or she is looking at the right results at some point in the analysis.

Design Principle 4: Reduce information-processing demands by displaying information in a consistent location and format (Marcus, 1995). Not only does the Search Results Window show the user the current context (search), it does so with an identical display (in terms of location and content) to the one that was displayed on the Search Builder Window as the user generated the search.

Design Principle 5: Allow the user to view the results with a variety of different representations (Bennett et al., 1997; Tufte, 1983, 1990, 1997; Watzman, 2003). The Results Window allows the user to view performance metrics in a tabular format in the Results Table, as well as in the form of a bar chart or scatter chart. Depending on the user's needs and the nature of the search results, the user may find one representation to be more perspicuous than another.

Figure 7.5 shows how POET provides a compact, flexible display of a variety of different performance metrics for the collection of flights of interest. In the Results Table shown here, the user has added statistics dealing with average Offtime, as well as the percentage of flights that experienced circular airborne holding or a changed arrival fix. (These additional metrics can be added using the View menu or the Metrics button on the toolbar.) This design feature illustrates another design concept:

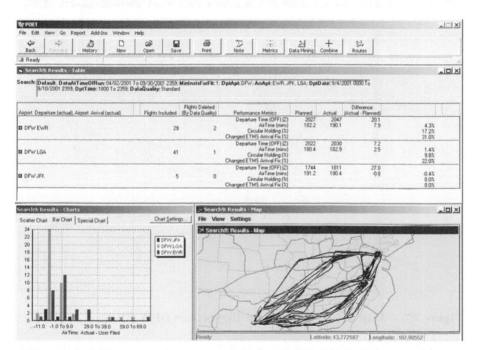

Figure 7.5   Addition of other performance metrics to the Results Table

Design Principle 6: After the search has been completed, allow the user flexibility in determining what performance metrics to display (Preece, 1994; Weiss, 1993). In applications like POET, the user may not know what performance metrics are most interesting until some preliminary results have been seen. In this case, the user might notice that there is evidence of holding on the Map Window and therefore decide to include Circular Holding as a performance metric.

ch9: Results - Table

Default; DataAtTimeOfRun: 04/02/2001 To 09/30/2001 2359; MinInstsForFlt: 1; DptApt: DFW; ArrApt: EWR, JFK, LGA; DptDate: 9/4/2001 0000 To 9/10/2001 2359; DptTime: 1800 To 2359; DataQuality: Standard

Departure (actual), Airport Arrival (actual)	Flights Included	Flights Deleted (By Data Quality)	Performance Metrics	Planned	Actual	Difference (Actual - Planned)	
EWR	29	2	AirTime (mins)	182.2	190.1	7.9	4.3%
bgroup By Airborne Delays: Circular Holding							
Held (17%)	5	1	AirTime (mins)	182.2	218.4	36.2	19.9%
09/04/2001 ABC1162			AirTime (mins)	179.0	255.0	76.0	42.5%
09/04/2001 XYZ1158			AirTime (mins)	198.0	246.0	48.0	24.2%
09/07/2001 XYZ812			AirTime (mins)	172.0	195.0	23.0	13.4%
09/10/2001 XYZ92			AirTime (mins)	176.0	197.0	21.0	11.9%
09/07/2001 ABC1138			AirTime (mins)	186.0	199.0	13.0	7.0%
09/04/2001 ABC92			AirTime (mins)	192.0	null	null	null
Not_Held (83%)	24	1	AirTime (mins)	182.2	184.2	2.0	1.1%

Figure 7.6   Subgrouping based on whether flights experienced holding

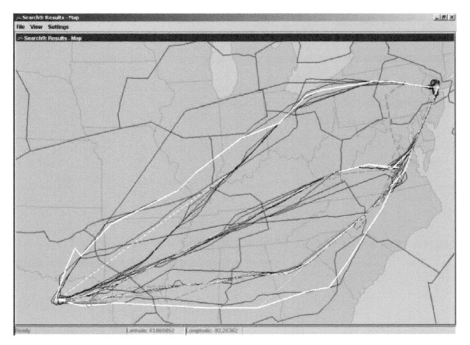

Figure 7.7   Map display showing the flights from DFW-EWR

Another application of this principle in POET is that the user can change the Charts Window to show graphs pertaining to all of the different performance metrics. Because POET accesses a very large database (information on over 50,000 flights per day for the past 60 days), users need data mining tools to support searches for patterns of activity. As an example, if a user wanted to know how many flights were experiencing circular airborne holding, the user could look at the map displays and count these flights. However, since there are a number of such patterns that are of widespread interest, POET has a set of data mining tools that use algorithms to look for certain patterns. The user can request that POET apply such pattern detection algorithms and show the results as subgroups within the Results Window, as illustrated in Figure 7.6. Figure 7.7 shows the map display for flights in a chosen area. Filed routes are in white; routes with circular holding are shown as dashed lines; routes with no circular holding are in black (POET actually displays routes in color). In this case, 17 per cent (five of the 29 flights) were put in circular holding patterns. On average, those flights put in holding had increases in airtime of 19.9 per cent, while those that were not put in holding had airborne delays of 1.1 per cent. Note that in this display, we have also opened up additional details about the flights that were put in circular holding patterns.

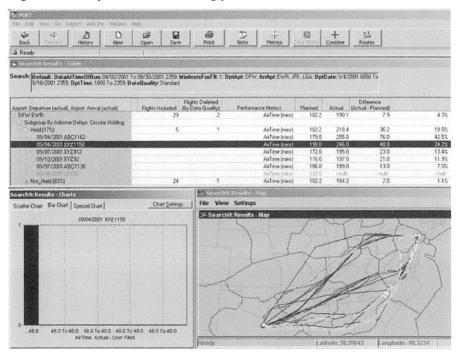

Figure 7.8   Linking different representations of the same 'object'

Two additional design principles are demonstrated in this example:

Design Principle 7: For large data sets which may contain detectable patterns, employ data mining algorithms to find these patterns for the user in order to reduce workload. Along with a table with relevant descriptive statistics, provide displays that help the user to visualize the patterns that have been found (Tufte, 1983, 1990, 1997).

Design Principle 8: Give the user control to decide whether and when to look at certain aspects of the results in greater detail (Weiss, 1993). Figure 7.8 and Figure 7.9 provide two additional examples regarding the design of POET. Figure 7.8

shows how the results in the different Results Windows are linked. The filed and actual routes are shown in white for the flight highlighted in the table. The bar chart shows a single value of 48 minutes as the flight is highlighted in the table. Highlighting any one of the rows in the Results Table causes the filed and actual routes for the included flights to be highlighted on the map, and to be displayed on the Chart (with all three representations visible at the same time).

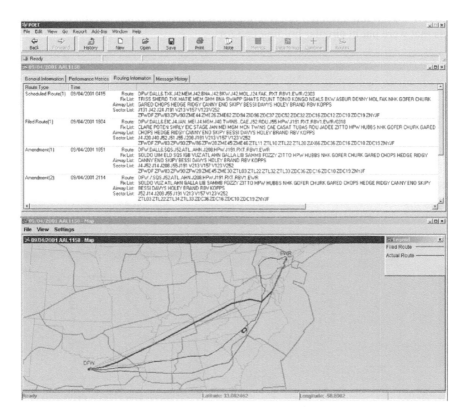

Figure 7.9   Additional details on a selected flight

The applicable general design principle is:

Design Principle 9: When an integrated display cannot be used to show different properties or representations of the same object, provide a highlighting function that allows the user to select that object in one of the representations and see it automatically highlighted in all of the other views or representations (Carswell and Wickens, 1987; Goettl et al., 1991). Finally, Figure 7.9 shows an illustration of more detailed information about individual flights that can be accessed by hyperlinks (accessed by double clicking on any individual flight shown in the Results Table).

The Design and Implementation of POET – Summary

The discussion above serves to introduce the basic functions embedded in POET, while emphasizing the underlying design principles that guided the interaction design. While any one of these principles may seem 'obvious' by itself, it was not so easy to determine how to put them all together to produce an information retrieval and visualization tool that was usable for both novice and expert users. Thus, POET provides a case study to help illustrate how the application of certain basic Human Factors design principles can be effectively integrated.

Actual Uses of POET

It is not enough for a tool to be useful and usable. In order to be of value, it must also be used effectively and profitably. At this writing there are over 140 individuals who actively use POET, many of them on a daily basis. These users belong to the various FAA facilities (including ATCSCC, ARTCCs, TRACONs and Headquarter Offices), airlines, and FAA research organizations. Below, we provide several examples of the actual use of POET by both the FAA and the airlines. These serve to further illustrate its design features, as well as to demonstrate how it has been used over the past 3 years since it became available.

FAA Use of POET

POET is being used regularly in most of the 21 ARTCCs to provide feedback on operations. The following is a list of how traffic managers at Indianapolis Center (ZID) have been using POET:

- establish baseline performance databases on new sectors to compare past to actual performance to determine if the fiscal expenditure for some change was warranted;
- contrast the actual arrival and departure feeds to the advertised acceptance rates. This provides the necessary feedback facilities need to ensure that constant pressure is placed on the airport without overtaxing the Air Traffic System. ZID does this for airports that lie within and outside ZID airspace;
- compare, analyze and establish numerous route structure changes within the NAS to maximize throughput while minimizing customer impact;
- systemically measure sector loading to determine, identify and correct problem areas;
- compare customers' performance expectations to actual events that, in turn, quantify the level of customer service provided by the FAA;

- identify and evaluate flow-constrained areas and weather events that dramatically impact the NAS;
- evaluate NAVAID (navigation Aid) outages for user impact and preplan alternative routes to minimize this impact;
- provide visual playbacks to customers and their FAA counterparts on flights or flows that present concerns. ZID is utilizing video-streaming programs to record POET map displays and export playbacks to those stakeholders. This capability has allowed all the parties to visually observe a common situation and then discuss their concerns and evaluate alternative solutions.

Sector Loading Evaluation A more detailed example of how POET is being used by the ARTCCs is in the evaluation of overloading of ZID97. ZID was experiencing heavy volume through one of its super high sectors, ZID97, and a TFM analyst was assigned the task of understanding this traffic and making recommendations as to how to reduce the load. First, using POET he was able to analyze performance when the sector was busy (see Figure 7.10).

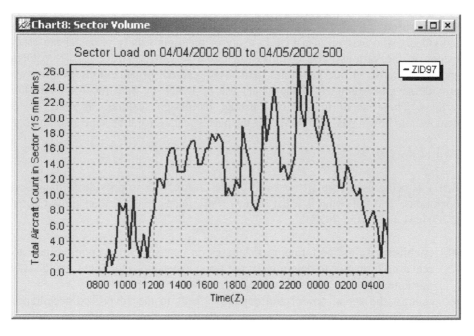

Figure 7.10   Sector load on ZID97 showing 2200 – 0000 as one of its busiest periods

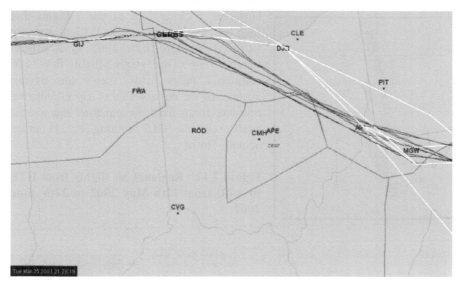

Figure 7.11   Flights flying through ZID97 that were originally not filed to fly through this sector

The analyst next used POET's mapping functions to compare the filed vs. flown routes of the flights going through this sector. This analysis showed that one of the reasons for this sector being busier than normal was because ORD, MSP and MKE departures were receiving direct clearances that placed those flights in ZID97 airspace as shown in Figure 7.11 (The white lines are the filed routes, while the black lines are the flown routes).

Finally, the analyst examined the loading of nearby sectors and found that the high sector, ZID87, just below ZID97 could handle a few more flights. Several recommendations were identified to assist in controlling the ZID97 volume problems, which included capping certain flights (holding them at lower altitudes to fly under the sector) and not allowing ZAU center departures to be cleared 'Direct' through ZID airspace during the busy times.

Route Evaluation of Flights from HTS to ATL Another example of POET use by an ARTCC is in the evaluation of alternative routes. In this example, the usual route for jet aircraft from HTS (Tri-State/Milton J. Ferguson Field) to ATL is via the fix VXV (see Figure 7.12). However, for traffic flow management reasons the FAA wanted the regional jets (CRJs) to file a slightly different routing via position fix ODF, so they would not impede the flow of the faster non-regional jets. In order to make the case to the airlines that the impact of this routing will be minimal, POET was used to assess the performances of flights flying the different routes.

Figure 7.13 below shows that, on initial comparison, the 5 flights that flew the route over ODF flew an average of 372 miles, which is 64 miles farther than planned. Conversely, the 107 flights flying the route over VXV averaged 331.3 miles.

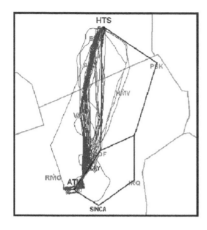

However, a closer look at the 5 flights that filed via ODF reveals that 1 flight was fix balanced on 23rd June 2002 as shown in Figure 7.14. This single flight flew 186 additional miles. This skews the overall average of the other 4 flights. By eliminating this one flight from the equation the overall average of the ODF routing is 341 miles actually flown.

Figure 7.12   Regional jet flights from HTS to ATL from 12th May 2002 to 24th June 2002

Airport: Departure (actual)	Flights Included	Performance Metrics	Planned	Actual	Difference
HTS	115	Distance	316.9	333.3	16.5
Subgroup By Route: Filed Route					
HTS., BULEY. J186. ODF. MACEY2. ATL (4%)	5	Distance	308	372	64
Route: Spatial Comparison of Filed to Flown Routes					
Similar (100%)	5	Distance	308	372	64
HTS., ECB., AZQ., VXV. MACEY2. ATL (94%)	107	Distance	315	331.3	16.3
Route: Spatial Comparison of Filed to Flown Routes					
Similar (93%)	100	Distance	315	330.1	15.1
Dissimilar (7%)	7	Distance	315	349.1	34.1
HTS., HMV., ODF. MACEY2. ATL (1%)	1	Distance	321	322	1
Route: Spatial Comparison of Filed to Flown Routes					
Dissimilar (100%)	1	Distance	321	322	1
HTS., PSK. J53. SPA. ODF. MACEY2. ATL (1%)	1	Distance	407	354	-53
Route: Spatial Comparison of Filed to Flown Routes					
Dissimilar (100%)	1	Distance	407	354	-53
HTS., PSK. J53. IRQ. SINCA3. ATL (1%)	1	Distance	467	347	-120
Route: Spatial Comparison of Filed to Flown Routes					
Dissimilar (100%)	1	Distance	467	347	-120

Figure 7.13   POET search results comparing filed to flown distances for flight from HTS to ATL from 12th May 2002 to 24th June 2002

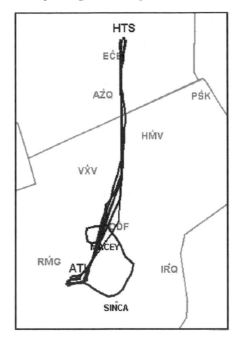

Figure 7.14   Fix-balanced flight from HTS to ATL via ODF

Based on this analysis, showing that the mileage difference is on the order of 10 miles, the recommendation was made that a 30-day test be conducted where the airlines file all CRJ flights via ODF, and that they conduct a follow up evaluation to see if there are any significant performance penalties before implementing this routing on a more permanent basis.

Understanding Changes Pre/Post 11th September 2001 The FAA's ATCSCC was interested in what kinds of changes in demand took place as a result of the national tragedy of 11th September 2001. POET was used to examine how the mix between different user classes (commercial, air-taxi, general aviation, etc.) changed before and after that date.

Figure 7.15 shows a sample of their search results for a particular sector of interest. The figure shows that a decrease in commercial flights was offset by increases in other user classes resulting in approximately the same number of flights through this sector. In a similar manner ATCSCC analysts used POET to examine changes in requested cruise altitudes, aircraft types, and the split between major versus regional carriers.

Search1: Results - Table

Search: Default; DataAtTimeOfRun: 04/02/2001 To 09/30/2001 2359; Flights Flown Through Selected Sector(s): Sector= ZID88; And/Or=OR; Start Date=09/10/2001; Start Time (hhmm)=1100; End Date=09/10/2001; End Time (hhmm)= 2200

	Flights Included	Flights Deleted (By Data Quality)	Performance Metrics	Planned	Actual	Difference (Actual - Planned)	
Flights	543	0	AirTime (mins)	106.6	108.4	1.8	1.7%
Subgroup By Flight: User Class							
O (1%)	4	0	AirTime (mins)	78.3	89.0	10.8	13.7%
C (59%)	320	0	AirTime (mins)	111.2	114.8	3.7	3.3%
T (19%)	104	0	AirTime (mins)	85.3	86.5	1.1	1.3%
G (20%)	110	0	AirTime (mins)	113.7	109.3	-4.4	-3.9%
F (1%)	5	0	AirTime (mins)	124.6	107.8	-16.9	-13.5%

Search1: Results - Table

Search: Default; DataAtTimeOfRun: 10/02/2001 To 12/26/2001 2359; Flights Flown Through Selected Sector(s): Sector= ZID88; And/Or=OR; Start Date=11/01/2001; Start Time (hhmm)=1100; End Date=11/01/2001; End Time (hhmm)= 2200

	Flights Included	Flights Deleted (By Data Quality)	Performance Metrics	Planned	Actual	Difference (Actual - Planned)	
Flights	521	0	AirTime (mins)	105.9	105.8	-0.1	-0.1%
Subgroup By Flight: User Class							
T (20%)	106	0	AirTime (mins)	86.1	89.2	3.1	3.6%
C (52%)	273	0	AirTime (mins)	110.8	111.6	0.7	0.7%
G (17%)	89	0	AirTime (mins)	95.7	94.9	-0.8	-0.8%
O (2%)	10	0	AirTime (mins)	71.1	69.4	-1.7	-2.3%
M (2%)	11	0	AirTime (mins)	211.0	202.6	-8.4	-4.0%
F (6%)	32	0	AirTime (mins)	136.2	121.3	-14.8	-10.9%

Figure 7.15 POET search results for flights through ZID88 broken out by user class (C = commercial, T = air taxi, G = general aviation, M = military, F = freight, O = other). The top figure is for flights on 10th September 2001 and the bottom is for flights departing on 1st November 2001

Search7: Results - Table

Search: Default; DataAtTimeOfRun: 04/02/2002 To 06/30/2002 2359; MinInstsForFlt: 1; Airline: ; DptApt: ORD; DptDate: 06/01/2002 0000 To 06/30/2002 2359; Filed_DptTime: 1800 To 2200; DataQuality: Standard

Airport Departure (actual) Departure Time Bin: Filed OFF	Flights Included	Flights Deleted (By Data Quality)	Performance Metrics	Planned	Actual	Difference (Actual - Planned)
ORD 1920	44	0	Departure Time (OFF) (Z)	1922	1930	8.0
ORD 1925	35	1	Departure Time (OFF) (Z)	1926	1941	14.3
ORD 1930	42	2	Departure Time (OFF) (Z)	1932	1953	21.0
ORD 1935	24	1	Departure Time (OFF) (Z)	1936	1939	2.5
ORD 1940	9	8	Departure Time (OFF) (Z)	1942	1955	13.4
ORD 1945	28	17	Departure Time (OFF) (Z)	1947	2000	13.0
ORD 1950	51	1	Departure Time (OFF) (Z)	1952	2006	14.2
ORD 1955	94	4	Departure Time (OFF) (Z)	1957	2012	14.7
ORD 2000	157	3	Departure Time (OFF) (Z)	2002	2017	15.9

Figure 7.16 Differences in departure delays with changes in departure time

Airline Use of POET

At present over 12 airlines are using POET to benefit their operations. The following are examples of such use.

Overhead Congestion Delays One major airline uses POET to analyze delays caused by overhead congestion (the presence of already airborne flights that limit or impact the departures out of a particular airport). Using POET they examine the trends in the overhead traffic, and determine alternative departure times that would avoid departing into congested airspace. Several of these alternative departure times were implemented, and using POET they were able to verify the reduction in delays to these flights (see Figure 7.16). As shown, flights filed to depart between 19:30-19:34 had on average 21 minutes delay vs. flights filed to depart 5 minutes later which had on average 2.5 minutes of delay. According to the airline, adjustments to three such flights have resulted in annual delay savings of 21,900 minutes equating to $941,481/year.

Another airline reports that they use POET in a similar manner to understand delays that may routinely occur on specific flights by looking to see how they fit in the flow at a specific airport during peak times. This type of feedback tells them if minor adjustments to their schedule can reduce these delays.

Flight Planning Evaluation Two airlines are using POET to evaluate their flight planning systems and strategies. One has indicated that they first look at reports on city pairs that may be over- or under-burning fuel as compared to what was planned, in order to identify flights meriting further analysis. They then use POET to see exactly what was filed versus what was actually flown in order to identify any early descents or delayed climbs that would impact their fuel plans. They can then adjust their routes and climb/decent profiles to match what is really flown, resulting in more accurate fuel burn estimates. This increases confidence in their flight planning system and ultimately reduces the amount of extra fuel a Dispatcher or Captain may feel the need to add to deal with possible inaccuracies in the flight plan.

The second airline uses POET as a means of validating its flight planning strategies for routings between various city pairs in order to overcome operational constraints in the En-route system. Specifically, this airline compares what route was filed with the route ultimately flown, and looks for traffic congestion that may be causing the reroutes.

Flight Safety Issues POET has been a valuable resource to a major airline in providing crew resource management debriefings in operational issues such as loss of separation and altitude deviations. Using the map animation functions (see Figure 7.17) they have created animation files (showing traffic moving over time on the map) for use in crew debriefings and recurrent training, and for discussions with the involved ATC facilities.

Evaluation of Fix Balancing Operations POET was used by a major airline to examine how well the FAA was balancing traffic between Atlanta's four arrival fixes, and the impact on their flights of the route adjustments imposed by these fix balancing operations. This analysis was used to support the need for dual Standardized Arrival Routes (STARs) as part of the FAA's Atlanta Airspace Redesign. As a result of the new STARs, the need for fix balancing over the MACEY fix was significantly reduced, saving the airline over 16,000 airtime minutes annually.

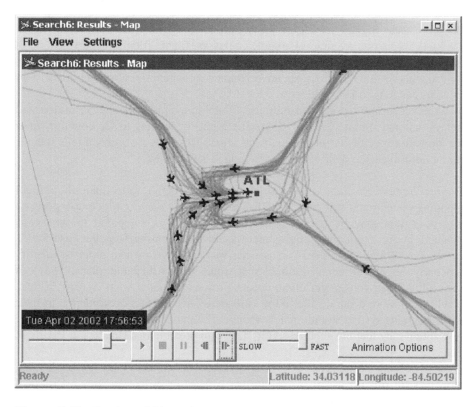

Figure 7.17   Replay of flights into Atlanta

Conclusions

The emphasis of this case study has been twofold. First, in a highly distributed system like the NAS, in order to ensure effective performance, it is critical to provide operational staff with feedback concerning the impacts of their actions. This is especially true when significant changes are being made to the design of the system, such as with the introduction of new sectors, the implementation of the National Route Plan or the use of Coded Departure Routes, Playbook plays and capping as tools for traffic flow management. In addition, such feedback needs to integrate data representing a number of different perspectives, for example letting dispatchers see the system from a traffic flow management perspective and vice versa.

The second emphasis is on the need to design and evaluate software tools in terms of their usefulness, usability and actual use. The ultimate measure of 'success' is whether a new cognitive tool is being used effectively to change performance in the NAS. Many factors influence the level of success, including the extent to which the tool meets real needs (usefulness) and the extent to which it is usable both in terms of the details of its design and in terms of how it fits into the broader work environments where it could be useful. In that sense, this chapter has been meant to provide a case study highlighting these considerations in the design of a tool to enhance performance in an evolving system where control, data, knowledge, processing capacities, goals and priorities are highly distributed.

Acknowledgements

This work has been supported by the FAA Collaborative Decision-Making Program and the FAA Human Factors Research and Engineering Division. Special appreciation is also due to Roger Beatty (American Airlines), Elaine McCoy (University of Illinois), Judith Orasanu (National Aeronautics and Space Administration Ames Research Center), Jon Mintzer, Scott Ayers, the Airline Dispatchers Federation, and all of the dispatchers and traffic managers who participated in the research that helped generate the ideas embedded in POET. Credit is also due to the other staff at The Ohio State University, Metron Aviation, Cognitive Systems Engineering and AMT Systems Engineering that assisted in the implementation and testing of POET.

References

Bennett, K.B., Nagy, A.L., and Flach, J.M. (1997), Visual Displays, in G. Salvendy (ed.), *Handbook of Human Factors and Ergonomics,* (2nd edition) New York, NY: Wiley, pp. 659-696.

Carswell, C. M. and Wickens, C. D. (1987), Information Integration and the Object Display, *Ergonomics*, 30, p. 511-527.

Goettl, B. P., Wickens, C. D., and Kramer, A. F. (1991), Integrated Displays and the Perception of Graphical Data, *Ergonomics*, 34, pp. 1047-1063.

Gould, J. (1995), How to Design Usable Systems, in Ronald Baecker, Jonathan Grundin, William Buxton and Saul Greenberg (eds.), *Readings in Human Computer Interaction: Toward the Year 2000,* 2nd Edition. San Francisco: Morgan Kaufman Publishers, Inc., pp. 93-121.

Kerns, K., Smith, P.J., McCoy, C.E., and Orasanu, J. (1999), Ergonomic Issues in Air Traffic Management, in W. Marras and W. Karwowski (eds.). *Handbook of Industrial Ergonomics,* New York: Marcel Dekker, Inc., pp. 1979-2003.

Marcus, A. (1995), Principles of Effective Visual Communication for Graphical Interface Design, in R. Baecker, J. Grundin, W. Buxton and S. Greenberg (eds.), *Readings in Human Computer Interaction: Toward the Year 2000,* 2nd Edition. San Francisco: Morgan Kaufman Publishers, Inc., pp. 425-443.

McCoy, E., Smith, P.J., Obradovich, J. and Orasanu, J. (1999), The Dissemination of 'Tribal Knowledge' in the Air Traffic Management System, *Proceedings of the Tenth International Symposium on Aviation Psychology,* Columbus OH, pp. 310-315.

Norman, D. A. (1981), Categorization of Action Slips. *Psychological Review*, 88, pp. 1-15.

Norman, D.A. (1990), *The Design of Everyday Things,* New York, NY: Doubleday.

Preece, J. (1994), *Human-Computer Interaction,* Harlow, England: Addison-Wesley.

Rasmussen, J., Pejtersen, A. M. and Goodstein, L. P. (1994), *Cognitive Systems Engineering,* New York, NY: Wiley.

Reason, J. (1990), *Human Error,* Cambridge, England: Cambridge University Press.

Shneiderman, B. (1998). *Designing the User Interface: Strategies for Effective Human-Computer Interaction,* Reading MA: Addison Wesley Longman, Inc.

Smith, P.J., Beatty, R., Spencer, A. and Billings, C. (2003), Dealing with the Challenges of Distributed Planning in a Stochastic environment: Coordinated Contingency Planning. *Proceedings of the 22nd IEEE Digital Avionics Systems Conference,* Indiannapolis, IN.

Smith, P.J., Billings, C., Chapman, R., Obradovich, J., McCoy, E. and Orasanu, J. (2000), Alternative 'Rules of the Game' for the National Airspace System, *Proceedings of the Fourth Conference on Naturalistic Decision Making,* Stockholm, Sweden.

Smith, P.J., Billings, C., Chapman, R.J., Obradovich, J., McCoy, E. and Orasanu, J. (1999), Alternative Architectures for Distributed Work in the National Aviation System, *Institute for Ergonomics Technical Report #1999-8*, Ohio State University, Columbus OH.

Smith, P.J., McCoy, E. and Orasanu, J. (2000), Distributed Cooperative Problem-Solving in the Air Traffic Management System, in G. Klein and E. Salas (eds.), *Naturalistic Decision Making,* Mahwah, NJ: Erlbaum, pp. 369-384.

Smith, P.J., McCoy, E., Orasanu, J., Billings, C., Denning, R., Rodvold, M., Gee, T., and Van Horn, A. (1997), Control by Permission: A Case Study of Cooperative Problem-Solving in the Interactions of Airline Dispatchers and ATCSCC, *Air Traffic Control Quarterly*, 4, pp. 229-247.

Tufte, E.R. (1983), *The Visual Display of Quantitative Information,* Chesire, CT: Graphics Press.

Tufte, E.R. (1990), *Envisioning Information,* Chesire, CT: Graphics Press.

Tufte, E.R. (1997), *Visual Explanations,* Cheshire, CT: Graphics Press.

Watzman, S. (2003), Visual Design Principles for Usable Interfaces, in J. Jacko and A. Sears (eds.), *The Human-Computer Interaction Handbook: Fundamentals, Evolving Technologies and Emerging Applications*, Mahwah, NJ: Lawrence Erlbaum Associates, Publishers, pp. 263-285.

Weiss, E. (1993), *Making Computers People-Literate*, San Francisco: Jossey-Bass Publishers.

PART III
HUMAN FACTORS AND
HUMAN RESOURCES

This section focuses on the broad area of Human Resources, an area that is concerned with the placement of qualified staff, in sufficient numbers and with appropriate skills, attitudes, and cognitive resources (including awareness) to do the job well.

The first chapter by Voller and Fowler is concerned with the area of training. As with several other chapters in this book, it starts with a problem that was difficult to solve, and ends with Human Factors considerations and findings being integrated, in this case into the training development system. In this particular example, too many students were failing their ATC courses at NATS, and it was not understood why. A long and detailed study took place running over a period of three years, providing insights into training scheduling and the student assessment process that significantly improved the success rate. This was achieved with a very large initial increment in training effectiveness, with a general training improvement sustained since that time. This chapter is a classic example of taking an ill-defined problem and by carefully applying Human Factors methods and detailed analysis allowing for the distillation of a solution that provides significant benefit to all parties concerned.

The second chapter in this section by Broach complements the first one and concerns selection of controllers. In 1981 after the major US air traffic controllers' strike during which around 11,000 controllers were dismissed, there was an unprecedented selection and recruitment need. This chapter details the background of the selection tests that had been developed prior to the strike and their subsequent application that was able to deliver a competent workforce with short notice. The various tests and selection processes are explained, and their validity and 'fairness' to adverse impact is explored. This chapter represents an example of Human Factors delivering in a very difficult and serious circumstance, and demonstrates that investment in Human Factors research before an issue develops can have significant pay offs.

The third chapter by Leonhardt focuses on the important and relatively new area of Critical Incident Stress Management (CISM). The theory underlying CISM and how stress develops amongst air traffic controllers, witnessing or in some cases even contributing to a critical incident (such as a significant loss of separation or an accident), and can become a significant and chronic problem, are described in the first part of the paper. This description includes an analysis of the characteristics of professional controllers that may predispose them to critical incident stress reactions as well as the nature of the air traffic control task. The resultant ATM CISM 'peer' approach that has developed in countries such as Germany is then described. This approach is growing in popularity, and is seen as useful and practical. This CISM approach was indeed applied immediately after the mid-air collision over Germany on 1st July 2002. The paper concludes with a discussion of further development needs with CISM and the benefits CISM has brought for the German ANSP to date.

The fourth chapter by Woldring et al. concerns the adaptation of Crew (or Cockpit) Resource Management (CRM) principles and training approaches to controllers, in what has become known as TRM (Team Resource Management). The emphasis is on teamwork and helping controllers understand how teams do and do not work effectively, with a special focus on team roles, communication, situation awareness, decision-making and stress management. During the past decade, TRM has come from being a concept in a study group, to a program adapted and taken up by approximately twenty countries in Europe. TRM has therefore been a Human Factors success story in Europe. The chapter describes the history of the program, its objectives, and the way in which tailoring and adaptation of the original prototype course to a variety of European cultures took place. The chapter ends with a view towards the future of TRM, which considers linking TRM with simulator exercises to ensure behaviors are effectively transferred to actual ATC situations.

The last chapter in this section by Della Rocco and Nesthus is a comprehensive review of the complex area of shiftwork practices in US ATM, based on scientific studies starting in the 1970s. Fundamentally, it addresses the area of shiftwork scheduling regimes and their associated impacts on controller health and performance at work. The first two-thirds of the chapter provides a detailed scientific analysis of the various shift systems and factors that need to be taken into consideration when shiftwork is utilized. The last part of the paper discusses counter-measures to performance decrements associated with shiftwork, such as exercises and optimization of shift scheduling. However, it is here that there is a realization of the work that remains to be done in terms of transitioning the extensive research results and understanding of shiftwork effects, to the practicalities of the field. This process has begun, with an educational program on the effects of shiftwork for all US ATC specialists. This chapter presents the state of the art regarding the development of better shiftwork systems for future ATM service providers and identifies where further research is required.

Traditionally human resources and manpower planning have been concerned with staffing, selection, and training and these areas do indeed continue to play an important role, especially in the light of an air traffic control task that is evolving and increasingly complex in its cognitive demands. In addition to these 'classical' disciplines, new approaches have arisen, such as Team Resource Management (TRM) and Critical Incident Stress Management (CISM). Some of these 'newer' approaches aim at providing operators with a better understanding not only of the task, but also of its nature so as to help them to make best use of their resources whilst avoiding detrimental effects. This is interesting because, to a degree, these approaches involve training the operators themselves in Human Factors principles.

Chapter 8

Human Factors Longitudinal Study to Support the Improvement of Air Traffic Controller Training

Laura Voller and Abigail Fowler

Introduction

The National Air Traffic Services Ltd (NATS) Department of Air Traffic Control Training & Simulation (DAT&S), formerly known as the College of Air Traffic Control (CATC), trains all UK En-Route and Approach controllers for NATS. To achieve this trainees are selected and recruited to undertake either one year of Approach training or one and a half years of En-Route training at the college before moving on to operational training at a unit.

A failure rate of up to 40 per cent during college training was the case in the mid-1990s and this was clearly unacceptable, representing a significant waste of training resources and money within the company. Also unacceptable was the impact this failure had on staff morale. For trainees not making it through the college and for the instructors not imparting their skills with successful outcome this failure was demoralising. This, coupled with increasing air traffic growth in the UK in the 1990s as well as a forecast retirement wave amongst Air Traffic Controllers (ATCOs), placed challenging demands on the ATCO training system within NATS.

The need for more ATCOs to handle more aircraft and the need to train new skills, due to the introduction of computer assistance tools, resulted in demands for increased throughput of 'robust' ab initio ATCO trainees and improved training standards. The Human Factors Unit (HFU) within NATS was requested by the General Manager of the college in 1996 to conduct a study to identify the causes of failure in training and identify ways of improving the quality and pass rate.

This longitudinal study was based on in-depth interviewing of trainees and instructors at the college. It began at the end of 1996 and was completed in March 1999 although some monitoring and follow-up data collection continued into 2000. This study involved gathering feedback from trainees and instructors at various stages throughout the training process to gain an understanding of the problem areas, concerns, and reasons for failure, as well as positive aspects of the college training process. Three different courses of trainees in particular were tracked, from start to finish, through the different stages of training at the college. Feedback at each stage was collected using in-depth (group and individual) interviews. Feedback was also obtained from interviewing:

- college instructors;
- trainees who had completed training at the college and were undergoing training for validation at an operational unit; and
- failed trainees, at various stages of training, who were undergoing a training review (some who were terminated from training and others who were re-coursed as a result of this review).

When a trainee fails a module of the course at the college they undergo a training review where the trainee and the relevant college staff go through the trainee's training record and discuss the difficulties that were experienced. From this review process the college produce a training review report which identifies the reasons that the college attribute to the trainee's failure. Based on the training review the college staff decide whether the trainee will have their training terminated or whether they will be granted a re-course to continue training.

The feedback obtained from interviewing failed trainees, together with training review reports produced by the college, was used to identify common areas of difficulty experienced amongst the college trainees who failed. It was also used to compare the reasons that the trainees and the college attributed to the trainees' failure to see whether they perceived the reasons for failure to be the same or different and to identify possible indicators of trainee failure. Based on the outcome of the review (i.e. whether the trainee was terminated from training or re-coursed), potential indicators of failure and common difficulties amongst trainees were also identified.

Each significant change that was introduced to the training process during the time of the study was noted to ensure that any impact it had was considered when analysing the feedback data.

All of the interview feedback was analysed to summarize and consolidate the findings from each interview conducted. All of the issues raised which related to perceived problems or concerns within the training process were tabulated. In addition, positive comments about the training process or improvements noticed were recorded.

The table of issues was then analysed to:

- compare and contrast the issues raised by the different groups during the project;
- draw out the most commonly raised issues;
- identify the main, persistent issues;
- identify the instructor-related issues and module-specific issues;

- show the human performance related issues; and
- identify any issues addressed by changes made during the study.

All analysed findings were considered in light of Human Factors and training theory and principles such as the systematic approach to training, the requirements of adult learners, cognitive skills training and other learning principles and best practice to identify recommendations for change within the training process.

This Human Factors study successfully identified problems within the training process. The main problem areas requiring improvement and predictors of trainee failure were distilled from a large number of issues raised. In addition to identifying problem areas, there were several noted areas of significant improvement in the training process that were able to be evaluated as a result of changes made during the study.

The findings gained from this Human Factors study for the college improved the training throughput and standards, thereby producing tangible cost benefits for NATS as well as morale benefits for all parties involved in the training process. This chapter describes the study that was undertaken and the outcomes it achieved.

Background

ATCO training at NATS starts with an induction week in which the new trainees get to know each other and take part in several team-building activities. During the next ten weeks the trainees have classroom lessons in the basics, including aviation law, ATC, navigation, meteorology, etc. Additionally, they carry out a few initial simulation exercises as part of the first Aerodrome course and after this do a two-week flying training course in a light type aircraft. This is followed by a week on leave before returning to begin the first part from the next training stage. The next course is the Aerodrome 2 course. It also includes some classroom lessons covering theory and basic ATC topics, but it mainly consists of practical training including briefings for the practical exercises that dominate this course. During their simulator lessons the trainees learn how to handle aerodrome traffic in the terminal area of an aerodrome. Aerodrome 2 ends with the qualification to start on-the-job training (OJT) at an airport or airfield with an Aerodrome Control Tower under the supervision of operational ATCOs. Another two weeks break marks the completion of this stage of the training.

Some trainees complete their college training at this point and go on to train as Aerodrome controllers who will work in a tower rather than on radar at a centre. Those that do return to the college divide into two groups and either do the radar skills course or have a few weeks of on-the-job-training (OJT) at an operational unit (usually an airport). The Radar Skills course is a practical course to teach the basic skills for using the radar facilities later. At the unit, those trainees who carry out OJT have the chance to experience live traffic and speak to the ATCOs there to gain experience of what it is like to work in an operational setting. If traffic permits trainees can practice their R/T (radio-telephony) ability under supervision of an ATCO. [17] After five and a half weeks, trainees swap training roles so that those who have completed the OJT, change to radar skills and vice versa.

After this stage, trainees are divided into streams as either airport trainees in the Approach stream or as Area Controllers in the Area stream. The streaming decision is taken during and at the end of the Aerodrome 2 course. Normally the decision by the college is based on the current demands of the operational units although trainees do have the chance to influence the decision by requesting a desired unit.

The Airport trainees continue training by starting the Approach course. During this time they will learn how to separate arriving and departing aircraft and to establish proper landing sequences for aircraft using radar equipment. Airport trainees passing the final examination are posted to an operational unit which marks the end of their college training. Before arriving at their dedicated units they will undergo an airline familiarisation course at British Airways which gives them information about the structure of an airline and the chance to use the airline flying simulators to gain some understanding of the pilots' perspective.

Area trainees stay at the college a little bit longer than Approach trainees. After the Radar Skill and OJT training their first Area specific course is the Area Control 1 course (ACS 1). This course is known as the procedural course but it teaches trainees more about radar methods than conventional procedural ATC techniques (separation of aircraft based on time and distance calculations). All courses, Approach and Area Control, are purely simulator based. The Area trainees' college training does not end with the ACS 1 course. After this course they take a short break then return to another nine weeks of OJT at a NATS unit (this time typically an Area Control Centre (ACC)). On OJT the trainees get a structured introduction in unit and local procedures. They are tested on what they learn at the unit and are required to present an aviation topic to other ATCOs and observe the work of an ATCO as project activities while they are there.

Area control trainees also have a British Airways airline familiarisation course; which starts immediately after the OJT. Finally, Area trainees return to the college to do a second Area course (ACS 2). Upon successful completion of this course the trainees become qualified to control En-Route air traffic under supervision in their new units. At this point they are posted to an operational unit to continue training there.

Figure 8.1 below illustrates the training process at the college at the time of this study. It does not include operational training at the units where sector-specific training takes place. The numbers in brackets represent the duration of each stage of training in weeks. Since the study some minor changes have been made and some more significant changes are now under development.

---

[17] Although this may change in the future as a result of European Licensing changes.

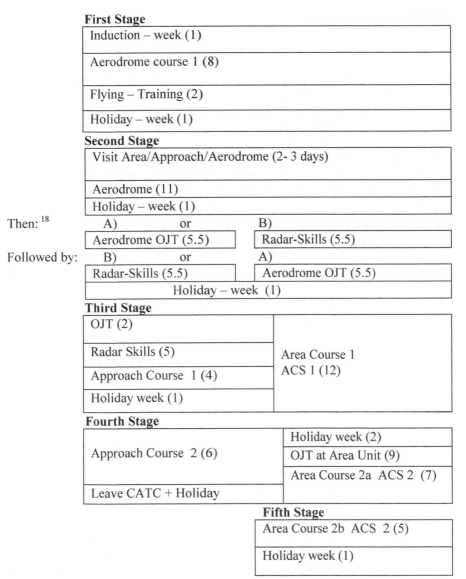

First Stage

Induction – week (1)
Aerodrome course 1 (8)
Flying – Training (2)
Holiday – week (1)

Second Stage

Visit Area/Approach/Aerodrome (2- 3 days)
Aerodrome (11)
Holiday – week (1)

Then: [18]

A) or	B)
Aerodrome OJT (5.5)	Radar-Skills (5.5)

Followed by:

B) or	A)
Radar-Skills (5.5)	Aerodrome OJT (5.5)
Holiday – week (1)	

Third Stage

OJT (2)	Area Course 1
Radar Skills (5)	ACS 1 (12)
Approach Course 1 (4)	
Holiday week (1)	

Fourth Stage

Approach Course 2 (6)	Holiday week (2)
	OJT at Area Unit (9)
	Area Course 2a ACS 2 (7)
Leave CATC + Holiday	

Fifth Stage

Area Course 2b ACS 2 (5)
Holiday week (1)

Figure 8.1   Illustration of the College of Air Traffic Control (CATC) training process and its timescales[19]

---

[18] At this point the trainee group is divided two: half do Aerodrome OJT while the others do the Radar Skills course and then afterwards they swap so that each group does both options before completing Stage 2.

[19] This illustrates the training process as it was at the time of this study rather than the current process as some changes have since taken place.

Method

Data Collection

Interviews Three courses of trainees (course numbers 99, 101 and 103) were tracked through their training at the college and interviewed at regular intervals. Three additional courses of trainees (91, 93 and 95) who had recently completed their training at the college and been posted to London Area and Terminal Control Centre (LATCC[20]) at West Drayton were also interviewed to give their views on college training having successfully passed it and moved on.

College instructors from each of the different stages of the course (Aerodrome, Radar Skills, Approach and Area) were also interviewed.

Throughout the project as many failed trainees as possible were interviewed, irrespective of what course they were from and at what stage of the training they had failed.

To broaden the experiences captured, trainees from three other courses (104, 105 and 106) were also interviewed during the latter part of the study. Several instructor meetings at the college were attended to capture additional instructor views of the monitored training changes that had been made.

Figure 8.2 illustrates the timing of all of the interviews carried out with instructors and trainees (in training at the college, at operational units and those who had failed) during the project. These are shown in relation to the phases of training that trainees had reached, and the main changes that occurred during these trainees' training at the college. A combination of group and individual interviews was used. The individual interviews were mainly with either trainees who had failed a stage of the training or instructors consulted on an informal basis. The individual interviews with failed trainees and with instructors are not individually identified in Figure 8.2, but the time-scale in which these interviews took place is illustrated.

Training Review Report Data In addition to interview data and liaison with the college instructors, fifty training review reports, written by the college when a trainee fails an aspect of the course, were collected for analysis.

Twenty of the fifty training review reports collected were for the 21 failed trainees who were interviewed (one training review report was never collected for a failed trainee who was interviewed at the beginning of the project). The remaining thirty reports were for trainees who had failed during the time that the project was running, but who could not be interviewed.

Conduct of Interviews (Group and Individual) A series of prepared open questions were used to stimulate conversation with the trainees and instructors during the

---

[20] This centre has now been replaced by the new London Area Control Centre (LACC) at Swanwick while the West Drayton centre remains as the London Terminal Control Centre (LTCC).

interviews, but the interviewees were encouraged to bring up any additional points or issues that they thought were relevant to their experience of the training process at the college.

The interviews were conducted in a quiet room and lasted for between thirty and ninety minutes. Most interviews were conducted in groups of three to six people, however all failed trainees and some informal instructor interviews were conducted on an individual basis. All interviewees were told that the interviews were to be kept anonymous. In addition, all trainees were assured that anything they said would not impact their training in any way.

Notes were taken during the individual interviews whereas some group interviews were tape recorded to aid the capture of information. Once the interviews were completed the tapes were transcribed and then erased.

Recording of Changes Each significant change made to the training process during the study was recorded so that, during analysis, the feedback obtained in the study could be used to assess the impact that each change had on the training process as perceived by the trainees and instructors. By comparing the relevant feedback obtained before and after each change was introduced, assumptions could be made as to whether the change was perceived to have resolved problems, introduced new problems or had no noticeable impact.

There were three significant changes considered likely to have a potential impact on the data gathered during this project.

1. The first of these changes was the transition between the former training process and the new Review Group for ATC Training (RGAT) process that was introduced in 1994. Despite the fact that this change took place before the start of this project, the possibility of this change having an impact on the views of the trainees interviewed who had completed their training at the college and were training for validation was thought worth consideration.
2. The second significant change was the introduction of a new assessment process. This new assessment process was first tried on the radar skills module of the course in July 1997 and six months later, in January 1998, introduced on all of the six course modules during the time-scale of the project. Interim findings from this study assisted the college in the introduction of this change. Feedback from the initial trial on the radar skills module was positive, hence the change was rolled out to all of the other college courses. This change in assessment had the potential to affect much of the data gathered during the project.
3. The third significant change was the introduction, in April 1998, of an instant re-course on the first occasion when a trainee failed a module of the training. The instant re-course allowed the trainees to fail two aspects of the training before a training review would be conducted instead of holding a training review when a single aspect of the training was failed.

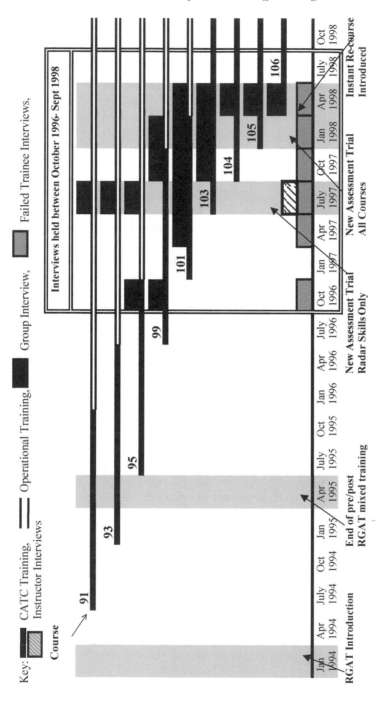

Figure 8.2 **Illustration of the timing of interviews conducted in relation to the stages of training and main changes in the college training process**

Data Analyses

Trainee and Instructor Interview Data The interview transcripts, along with any notes taken from interviews that were not recorded, were each assessed in turn to summarize the main issues raised in each group or individual interview conducted. The summarized results from each interview were then analysed to consolidate the findings from all of the groups interviewed (instructors, trainees under operational training, trainee courses at the college, and failed trainees). All of the issues raised which related to perceived problems or concerns within the training process were combined together in a large table.

Issues were compared across the different groups over time during the project, and across the different stages of training (i.e. different courses within the training process). Finally, issues within Table 8.1 were analysed in relation to the different changes made during the project to assess the impact, if any, that each change was perceived to have.

Failed Trainee Data (From Interviews and the College Training Review Reports) The 21 transcribed interviews from the failed trainees, together with the 50 training review reports collected, were summarized in a large table (See Figure 8.4). This table was analysed to compare the trainees' perception of why they failed with that of the college. The most frequently raised issues associated with trainee failure were identified from the trainee interview data and the twenty associated college reports. The frequency with which each issue was raised from each of the two data sources, and the percentage of cases each issue referred to, were then used to compare similarities and differences in the reasons associated with failure between the opinions of the trainees and the college staff.

The study also examined the relationship between the issues that the college associated with trainee failure (obtained from the 50 college training review reports), and the outcome of the training review process (i.e. whether the trainee was terminated from training or re-coursed to continue training). The issues associated with trainee failure, identified from the 50 college training review reports, were quantified by a frequency count. The frequency count for each issue was used to calculate the percentage of trainee cases the issues applied to. These data were tabulated and compared with the outcome of the training review for each failed trainee to see whether certain issues could be linked to termination from training.

Analysis of Changes made to the College Training Process As mentioned earlier, three significant changes were made to the college training process (introduction of RGAT, new assessment, and instant re-course) which were considered likely to have had an impact on the interview data gathered during this project. A comparison of issues raised before and after each change enabled inferences to be drawn regarding the impact of the changes (those issues that were related to the change in some way that were not raised again after the change were assumed to have been addressed by the change).

In order to assess the effects of the instant re-course introduction, the number of failed trainees over a two-year period was examined. The number of failed trainees in the twelve months leading up to the instant re-course introduction (May 1997 – April 1998) was compared to the same period after the instant re-course introduction (May 1998 – April 1999). The majority of information for this comparison had already been captured during the project, but additional information was collected in order to have a full year of 'post instant re-course' information for comparison. The post-instant re-course trainees were also monitored for a period of time after their college training in an attempt to evaluate this change further. The information collected was compared numerically and broken down into the outcomes of failure (i.e. whether the trainees resigned, were terminated or re-coursed). This breakdown was then used to compare the outcome of failures for trainees both pre and post instant re-course introduction.

Discussion of Results

Findings from the Interview Data

The collection of all sources of feedback data spanned a 28-month period. In total, during the course of the study 133 trainee and instructor interviews were conducted. Fifty training review reports were collected from the college (20 for failed trainees who were interviewed and 30 additional reports for trainees who were not interviewed).

In total 104 different issues were raised which related to perceived problems or concerns within the college training process, although not all of these issues were frequent or persistent throughout the duration of the study. Although all of these issues were considered within the study here, only a sub-set of the main, more persistent issues are discussed further.

There was a considerable amount of overlap between the issues raised by the different groups interviewed. The instructors raised 37 different issues and of these 62 per cent (23 issues) were in common with different groups of trainees interviewed. The instructors had a similar number of issues in common with successful trainees (in training, and who had completed the course at the college) as with failed trainees. There was also considerable overlap between the issues raised by successful and unsuccessful trainees.

In order to gauge whether the issues raised changed over the duration of the project, the issues raised in the first half of the project time-scale were compared with those raised in the second half of the project duration. Although this was a rather crude method to adopt, the findings indicated that in the first half of the project, 69 different issues were raised by the different groups interviewed, compared with 60 issues in the second half of the project. There were 31 issues in common in these two halves of the project, which would suggest that, although the number of issues raised did not change significantly, the nature of the issues raised varied across the duration of the project.

Figure 8.3 illustrates the distillation process that was used to identify the main issues that were considered necessary for the college to address initially. The 38 issues which were no longer raised as the project continued were considered to be either 'historic' by the end of the project or 'addressed' due to changes made during the project (this will be discussed further later in this section). The 29 issues first raised in the latter part of the project were mainly 'module-specific' issues and are also discussed later in this section.

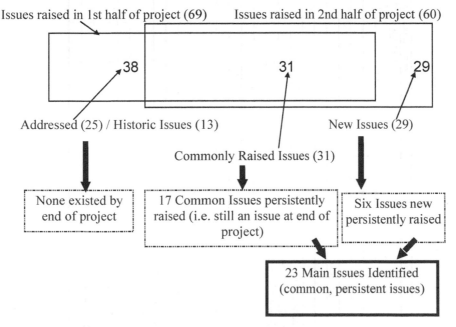

Figure 8.3   The categorisation and distillation of issues

Main Issues (Common, Persistent Issues)

Issues were considered commonly raised and persistent if they were raised by three or more interviewed groups including at least one group interviewed towards the latter part of the course (i.e. the issues were still relevant at the end of the project). Of the 31 issues which were raised in both the first and second half of the project time-scale, 17 of these were considered to be commonly raised, persistent issues. In addition to these, 6 'new' issues were raised in three or more interviews towards the end of the project. Together, these 23 issues made up the main issues considered common and persistent amongst the different groups interviewed, and therefore the main issues for the college to initially address. Table 8.1 below shows the 23 main, persistent issues identified.

Table 8.1　The 23 main, persistent issues identified

NO.	MAIN ISSUES IDENTIFIED	Instructors	LATCC ACC	99 AD 1	95 Area	99, 101 103	Failed (CATC)	101 Area 2	103 Area 1, Approach	104 Radar Skills	105 AD2	106 AD1
1	Assessment is too subjective	*	*	*	*	*	*		*		*	*
2	More one-to-one feedback needed	*	*	*	*	*	*			*	*	*
3	Too much negative feedback		*	*	*	*	*		*	*		
4	Reports contain inadequate feedback	*	*			*	*	*	*			*
5	More consolidation time needed	*	*	*		*	*	*	*			
6	Instructors assessment is inconsistent	*	*		*			*		*	*	*
7	Verbatim learning does not encourage understanding	*	*		*	*		*	*	*		*
8	Instructor training needed (assessing/reporting progress)	*				*		*		*	*	
9	Time available for reports, demos, debriefs is too limited	*			*	*		*	*	*		
10	Mentor/tutor instructors would help				*	*	*			*	*	
11	Training process too rushed generally	*	*		*	*	*	*	*			
12	Area 2 Module needs to be longer		*	*	*	*	*		*			
13	Remedial training facility needed		*		*	*	*		*			
14	Radar Skills AC/Apr teaching differences cause problems	*	*			*	*		*			
15	Severity of problems on reports not clear		*		*	*	*		*			
16	The college too failure focused					*	*		*	*		
17	Some Instructors seem to lack motivation					*	*		*	*	*	
18	More feedback /report details needed in 'formative' training					*	*	*	*	*	*	
19	Formative reports don't need box markings/comments essential							*	*	*	*	
20	Box markings need to be introduced in consolidation phase							*	*		*	
21	Box markings MUST be supported by comments to be of value							*	*			*
22	Only two middle box markings used (all four should be)							*	*	*	*	
23	Still difficult to gauge progress from feedback given							*	*	*	*	

These 23 main issues could be loosely classified into three main 'themes' or problem areas as follows:

- quality and quantity of feedback given to trainees;
- inconsistency and subjectivity of instructor assessment, and;
- problems caused by time limitations within the training process.

This finding provided the college with a manageable number of problem areas to focus on improving, while providing additional detail of the specific concerns of the instructors and trainees within these three areas. Narrowing the problems down to this extent made recommendations possible to identify.

The six issues in italics in Table 8.1 were new issues raised towards the end of the project. These issues related to the feedback and assessment problem areas, and were specific to the new assessment process.

Module-Specific Issues

Of the six different training modules at the college, the Aerodrome 2 and the Area 2 modules were identified as causing the greatest number of problems for the trainees and instructors. Seventy-six per cent (76 per cent) of the failed trainees interviewed failed during these two modules, and 63 per cent of the total module-specific issues raised were related to these two modules.

The most frequently raised issues regarding the Aerodrome 2 module were the amount of 'cross instruction' (inconsistent instruction given on the same topic by different instructors), the way this cross instruction is resolved, and the lack of positive feedback provided during this early stage of learning. Also mentioned was the need to have early exposure to the operational ATC environment, the need for more mid-module consolidation time and the need for more direct instruction on the basic skills contained within this module.

The most frequently raised issues regarding the Area 2 module were the need to lengthen this module and the need for the Area 1 module to be reduced to limit the gaps in time between the completion of the Radar Skills module and the start of the Area 2 module. In addition, the emergency exercises were considered misplaced within the Area 2 module and there was a need for more mid-module consolidation time.

Instructor-Related Issues and Instructional Technique Issues

Instructor-related and instructional technique issues were also identified providing an indication of how the training content and delivery could be improved in future. Trainees are currently taught individual skills which are then combined within consolidation exercises at the end of the training module, just prior to assessment. The ability to multi-task (carry out several of the skills together) appears, to some extent, to be 'assumed' at the college when it is actually a skill in its own right that requires attention.

Other topics that were not considered to be explicitly taught or taught thoroughly enough were the role of the co-ordinator (how to be an effective co-ordinator), the understanding of aircraft performance characteristics and sufficient variety of radar techniques and when they are appropriate in different circumstances. The use of demonstrations and video playbacks of real ATC scenarios may be useful in a group discussion forum where trainees can discuss, amongst themselves and with an instructor, the team communications that were or were not effective, the alternative radar technique options that might have been possible etc. This could be an effective method of developing adaptive expertise in trainees and providing them with a broader knowledge base from a limited number of scenarios in a relatively short period of time.

The instructors were alleged to frequently interrupt the trainees during the practical exercises just before the trainee was able to act on the plan they were constructing. This is due to the fact that trainees typically take longer than instructors to process the necessary information to resolve ATC problems. The frequent disruption of this planning process can make it difficult for trainees to progress and positively reinforce their learned skills.

A concern raised by both trainees and instructors was the lack of exposure to the operational ATC environment. Regular contact with the operational environment would ensure that instructors remain current with operational practice. It would also increase their credibility with the trainees. Earlier exposure to the operational environment (particularly a centre where most trainees are likely to end up working) might be beneficial to trainees to give them a better appreciation of the true ATC context and job.

Indicators of Trainee Failure

Figure 8.4 below summarizes and compares the main reasons identified, from both the trainee and the college perspectives, as being associated with the difficulties experienced by the trainees and their subsequent failure during training. These data were obtained from the twenty failed trainee interview transcripts and the training review reports produced by the college for these trainees.

The trainee feedback identified four issues as the most frequently raised issues associated with failure:

- the need for more time to practice skills;
- the inadequacy of feedback;
- the amount of pressure experienced from fear of failure; and
- the predominantly negative training environment.

These four issues were similar to the four issues most frequently mentioned in the college training reports although they were expressed in a slightly different way. For example, the college reports identified 70 per cent of the trainees as slow learners whereas 70 per cent of the trainees felt they required more time to grasp the skills being taught.

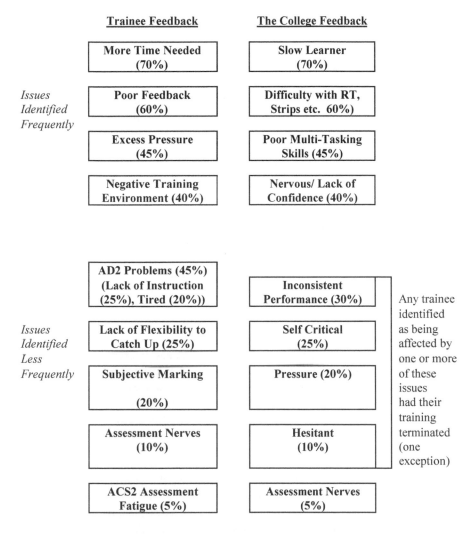

Figure 8.4   Comparison of trainee and the college perceptions of reasons for trainee failure (n=20)

The college identified 60 per cent of the trainees as having difficulties with technical issues like radio-telephony (R/T) skills and strip management, whereas 60 per cent of the trainees felt that the feedback they received during training was not adequate to gauge their weaknesses. Forty-five per cent of the trainees were identified by the college as having difficulties 'multi-tasking' or dealing with several skills simultaneously. The same trainees said they needed more time to consolidate their skills. These trainees are likely to find difficulty combining skills at which they are not yet fully proficient. Finally, the college identified 40 per cent of the trainees as being nervous or lacking in confidence. At least this many of the interviewed trainees said their confidence or morale suffered from excess pressure (45 per cent) and the negative training environment they experienced (40 per cent). None of these most frequently mentioned issues, identified by the college or the trainees, were specific to a certain phase of the training.

In addition to the four issues most frequently identified from the trainee and the college feedback, there were several other less frequently raised issues identified. These issues do not link as directly to each other as the more frequently mentioned issues do. However some potential associations were identified. For example, 30 per cent of the college reports identified trainees as showing inconsistent performance, whereas 20 per cent of the trainees felt that their performance was assessed inconsistently because of the subjectivity in the marking process.

The college described several trainees as being self-critical, showing difficulty when under task pressure, showing inconsistent performance or being hesitant. These traits may appear similar to being nervous or lacking in confidence, however (with only one exception), any trainees identified as being self-critical, suffering when under task pressure, being hesitant or showing inconsistent performance, had their training terminated. Trainees identified by the college staff as suffering from nerves or lack of confidence did not typically get terminated from training. Therefore, it is important to distinguish between these traits to predict a trainee's likely training outcome.

Twenty-five per cent (25 per cent) of the trainees interviewed felt that the college needed to be more flexible in offering remedial training time when a trainee had a setback. This seemed to be related to the trainees who felt that more consolidation time was needed generally. Among the less frequently raised issues, two were related to specific courses or phases of training. One was related to the Aerodrome 2 module and the other to the Area 2 assessment. Twenty-five per cent of the trainees said that on the Aerodrome 2 module there had been a lack of direct instruction during the practical exercises which left them vulnerable to making errors which could go on undetected or corrected by the instructors. Also, 20 per cent of the trainees experienced extreme fatigue by the end of this module that was thought to have lowered their performance at critical assessment times.

The issues identified from the 50 training review reports were similar to those from the first 20 reports used to compare the college feedback with the trainee feedback, suggesting that these findings are fairly consistent and representative. The same nine issues were found to be the most frequently raised issues with only slight differences in the order they were raised, and percentage of cases in which they were identified.

Three additional issues were identified which had not been obvious from the initial 20 reports examined. These were:

- low motivation;
- 'personality related' problems; and
- problems associated with milestones assessments.

Reasons Attributed to Trainee Failure   **Outcome of Training Review**

Difficulty with R/T and Strips (68%)

> *> 3 + other issues = training*
> *terminated*

Slow Learner (56%)

> *< 2 issues = trainee re-coursed*

Poor Multi-Tasking Skills (54%)

> *(2 exceptions, trainees terminated:*
> *trainees had both already had 1*

Nerves/ Lack of Confidence (44%)

> *re-course)*

Inconsistent Performance (30%)

> *> 1 issue = training terminated*

ATC Task Pressure (28%)

> *(3 exceptions, trainees were all re-*
> *coursed:*
> *1 trainee recognized as a marginal*

Too Self Critical (12%)

> *performance on previous modules;*
> *1 trainee self critical but only other*

Poor Reaction Time/ Hesitant (14%)

> *milestones and initial RT difficulties;*
> *1 trainee had inconsistent*

Personality Related Problems (8%)

> *hesitant and had poor multitasking*
> *skills.)*

Low Motivation (6%)

Milestone Only (14%)

> *trainee re-coursed*
> *(1 exception, trainee terminated:*

Assessment Nerves (2%)

> *trainee had already been re-coursed*
> *once.)*

Figure 8.5   Total college feedback on reasons attributed to trainee failure (n=50)

It was still the case that, with only three exceptions, any trainee who was identified by the college as being affected by one or more of the following issues was terminated from training:

- showing inconsistent performance;
- suffering when under ATC task pressure;
- being too self-critical;
- poor reaction time or being hesitant;
- problems related to personality;
- low motivation.

Of the most frequently occurring issues (difficulty with basic skills like strips and R/T; slow learner; poor multi-tasking skills; and nervousness/lack of confidence), it was generally the case that any trainee who was identified as being affected by three or more of these four issues was terminated from training. Any trainee who was affected by two or fewer of these four issues tended to be re-coursed. This suggests that these difficulties, or the combination in which they are experienced, could be used to help predict trainee failure.

Impact of Changes Made to the Training Process

Commonly raised issues were provided to the college as interim findings at intervals during the project to guide the college's implementation of changes to the training process. Three significant changes (the introduction of the RGAT training process, a new assessment process and the instant re-course) were considered in relation to the data collected during the project.

The first of these three changes, the introduction of the RGAT training process, raised a difference in feedback between trainees on the 91 course and those on the 93 and 95 courses. The trainees on the 91 course expressed less positive memories about their training at the college than those on the 93 and 95 courses. This difference was considered to be most likely to be attributable to the transition to the RGAT training process in the time between these courses. Although based on limited evidence, the change to the RGAT training process appeared to be a positive change.

At the end of the study the commonly raised issues were assessed in relation to the other two changes. Those issues which related to these changes, and which ceased to be raised after the change was implemented, were assumed addressed by the change. The new assessment process eliminated 14 issues of all the issues raised within this project. The instant re-course change eliminated at least six issues. Therefore these two changes alone appeared to reduce the negative feedback obtained during the duration of the study by approximately 20 per cent. The feedback and more detailed findings relating to each of these two changes are discussed below in turn.

Changes in the Assessment Process

Before the introduction of the new assessment process, this feedback study had detected some negative feedback from trainees and instructors regarding the regular assessment on each module and the final 'milestone' assessment that came at the end of each module. The regular assessments were said to provide the trainees with no un-assessed learning time and restrict the trainees from asking questions or trying out new techniques. It was also said to create difficulties for the instructors in having to teach and assess simultaneously, as these two requirements often conflicted. The 'milestone' assessment was very unpopular with trainees and instructors alike and was reported not to be a good indicator of progress because it was very dependent upon the trainees' performance on a single day when the stakes were high and the trainees were nervous. It did not assess consistency of performance well either.

The new assessment process was introduced in July 1997 on the radar skills module initially, and then across all modules in January 1998. This change was introduced partially as a result of the interim results of this project, which highlighted the need for un-assessed learning time and more focus on training than assessment in the early stages of learning. It provided a phase of un-assessed leaning called the 'formative phase' which allowed the trainee to learn all of the new skills introduced on that module before any assessment took place. The formative phase of the module was followed by a 'consolidation phase' just prior to the assessment phase. The assessment phase comprised between 4-10 'summative exercises' each of which was assessed and used together by the college to gauge the trainees' progress and decide whether the trainee had reached a standard acceptable to pass that module. The latter summative exercises replaced the former milestone assessment. The feedback reports, written by the instructors for the trainees after each exercise, included only comments on the formative exercises unlike the feedback reports produced for the summative exercises that included comments and a box marking.

The overall feedback as a result of the new assessment process was extremely positive from both the trainees and the instructors. One of the main problems addressed by the change in assessment process was the 'fear culture' or 'failure focus' said to exist at the college. This change enabled the college to separate out the ATC task pressure and the non-ATC pressure caused by constant assessment and daily fear of job loss through failure. The elimination of the latter allowed the trainees to concentrate on learning and the instructors to concentrate on teaching. With this came an improvement in trainee-instructor communication and trust. In addition, the formative phase encouraged the trainees to try out different techniques and ask the instructors more questions. This phase also provided the trainees with the un-assessed learning time that had frequently been reported as necessary before the change in assessment was introduced.

Instant Re-Course

The results presented here are only an initial look at the impact of the instant re-course change. For the full effect of the instant re-course introduction to be appreciated, monitoring would need to continue at least until the final outcome of training is known for all of the trainees within this sample of data and this can take a considerably long time.

Before the introduction of the instant re-course, this feedback study had detected some negative reports from trainees and instructors, regarding training reviews. The trainees felt under significant non-ATC pressure because of the possibility of being terminated at their first training review. It seemed to some trainees that the college was just a drawn out selection process rather than being more focused on training. This feedback was provided to the college as interim results and this helped to influence the decision to introduce instant re-coursing.

With the introduction of the instant re-course, if and when they failed for the first time, trainees would automatically be re-coursed onto the next available course. This took away the need for a training review to take place every time a trainee failed. Instead, a training review only needed to take place if a trainee failed for a second time during the training process.

Initial feedback gained, regarding the introduction of the instant re-course, was positive. By introducing the instant re-course, the college was seen by the trainees to be putting the onus on training, rather than their de-selection. The pressure was lifted significantly for trainees. However, the possibility of being terminated from training after failing twice still remained of course. This means that while trainees were given a second chance if they failed a course, the incentive remained for them to work hard during training in order to succeed.

For the instant re-course to be a full success the re-coursed trainees needed to be monitored throughout their training until their final training outcome was known. Increasing the pass rate at the college would not be the advantage it may seem unless these trainees successfully validated once posted. Otherwise the effect of the instant re-course may be a delay in the time it takes the trainees to fail. If trainees who have been granted a re-course go on to succeed at training once posted, then any increase in passed and posted trainees would be a positive increase in the pass rate of the training process.

To determine the final training outcome of the trainees in the sample, their progress was monitored until their operational training outcome was known. Although one trainee's outcome was not determined, Figure 8.6 below shows the situation as at August 2000.

This shows that some of the trainees did in fact still fail their college training, however, thirty-three trainees (sixteen more than before the instant re-course was introduced) went on to be posted to an operational unit for further training. This means that of the 46 trainees who failed an aspect of training pre instant re-course, 63 per cent were terminated from college training and only 37 per cent were posted for operational training. This is compared with 31 per cent of the 49 post-introduction of the instant re-course being terminated at the college and 67 per cent achieving a posting to continue training at a unit. The trend in the number of trainees passing the college training and being posted to operational units, compared with the number of trainees being terminated from training at the college was reversed. Within the time-scale of the study it was not possible to do further follow-up analysis to assess the outcomes of operational training for these trainees but the initial indication was that this change was a favourable one.

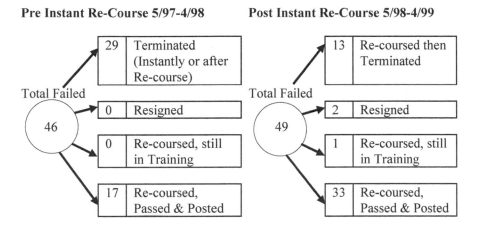

Pre Instant Re-Course 5/97-4/98   Post Instant Re-Course 5/98-4/99

Figure 8.6   Outcome of trainees failing a module in the years pre and post instant re-course introduction (As at August 2000)

The introduction of the instant re-course allowed for different rates of learning and consolidation among the trainees, but at the same time maintained the high standards required. It allowed many slightly slower, but no less able, learners to continue with their training, which would otherwise have been terminated. In addition, trainees and instructors were less inclined to feel the emphasis at the college was on de-selection but rather training which was a significant morale boost. This re-course change helped to reduce the 'fear culture' and 'failure focus' problems that had been identified by the study.

Conclusions

Having identified an historic pass rate of around 60 per cent, it is significant that within one year of the college acting upon the main Human Factors Study findings, the pass rate was increased significantly and this increase has been sustained. Due to the number of variables involved there is no available figure of percentage increase which can be specifically attributed to the Human Factors study. However, the resource and associated cost savings were considerable. Less obvious, but equally important, was the positive effect achieved upon trainee and staff morale of achieving such improvements.

The study successfully identified the problems with the training process, as they were perceived by the trainees and instructors at the college. The main problem areas requiring improvement and the indicators of trainee failure were identified providing the college with an appropriate number of priority issues to focus on and address.

Several of the main problems perceived within the training process were the same for successful and unsuccessful trainees as well as instructors. There were 23 main issues identified, which fell into three main problem areas to be addressed. These were:

- lack of quality feedback;
- lack of consolidation/practice time; and
- inconsistent/subjective assessment.

The areas identified as needing improvement were: specific issues relating to individual modules within the college training course; issues related to instructors and the instructional techniques used; and other Human Factors issues. For example the Aerodrome 2 and Area 2 modules were found to cause the most difficulties for both trainees and instructors alike. Most trainees who failed at the college failed on these two courses.

In addition to identifying problem areas, there were several noted areas of significant improvement in the training process as a result of changes made during the study. These changes (to the assessment process and training review process) reduced or eliminated 20 per cent of the problems raised over the duration of this study. In particular, issues addressed by these changes included the elimination of the 'fear culture' and 'failure focus' that the college was perceived to have prior to the implementation of these changes.

The introduction of the instant re-course resulted in a significant reduction (from 29 to 13) in the number of the terminated trainees at the college. Several issues were identified as predictive of trainee failure. The fact that the issues identified as being related to trainee failure were similar to those issues raised by successful trainees would indicate that these issues were not the *causes* of failure. However, the fact that the outcome of the training reviews could be classified (i.e. whether the trainee was re-coursed or terminated from training) according to the number and combination of issues experienced would suggest that these issues could be used to predict the likelihood of failure before the event. This could be used to help identify weaknesses likely to lead to failure early enough to work on correcting them. Some traits were identified which, if associated with failed trainees by the college, always resulted in the trainees' termination from training. It was recommended that these data be fed back into the selection process to improve the selection of trainees most likely to succeed at the college. These traits were:

- being too self-critical;
- reacting poorly to task pressure;

- slow reaction times/hesitancy;
- inconsistent performance;
- inappropriate personality traits for the job; and
- low motivation.

The main problem areas identified in this study were addressed at least in part by the college. Feedback quality was improved by separating out training and assessment, allowing instructors to provide better quality feedback to trainees during the formative training phases. By changing the assessment to formative and summative phases more time for consolidation and practice was provided. The formative phase allowed the trainees to practice new skills and a discrete consolidation phase was created prior to the summative assessment phase. Further training on the provision of feedback reports was recommended to the college to further improve the quality of feedback the trainees received as well as the consistency of assessment.

The findings specific to certain college training courses (for example the Approach and Area aspects of the Radar Skills module) were not addressed immediately by the college although considerable thought was given to these issues. Since the study, significant changes have been planned to improve these areas of training in line with the study recommendations that were to separate the Approach and Area aspects of this course so that trainees do not learn two ATC domains in parallel.

The study raised a recommendation that more time be spent on the training of some basic skills (e.g. R/T, strip marking and movement etc.) in the Aerodrome 2 module to ensure all trainees receive a good understanding of these skills before progressing further through the training course. In addition, it was recommended that more time be spent on some additional skills that were not being taught very explicitly, for example, co-ordinator skills (how to be an effective co-ordinator); the understanding of aircraft performance characteristics; and a wider variety of radar techniques.

Some instructional techniques and improved delivery methods were also suggested. This includes the use of demonstrations and video playbacks of real ATC scenarios in a group discussion forum where trainees could discuss, amongst themselves and with an instructor, the team communications that were or were not effective, and the alternative radar technique options that might have been possible etc. This could be one possible method of providing the trainees with a broader knowledge base from a limited number of scenarios in a relatively short period of time. Again, the college has since worked on improving most of these areas and has implemented changes to meet most of the recommendations that were raised by this study.

This study provided direct benefit to the problem of overcoming a shortage of ATC staff by helping to prevent the loss of capable trainees within the training system and improving their chances of successful validation. Furthermore, the perceived success of this work helped NATS managers to better appreciate the contribution that Human Factors could make, to the extent that the Human Factors Unit now has an early and significant involvement in the development of ATC training improvements within the company.

Acknowledgements

The authors wish to thank all of the trainees and college instructors who participated in the study for their time and considered views and the college staff who helped to arrange the interviews and supply the training review reports and other information used in this study.

References

Buckley, R. and Caple, J (1992), *The Theory and Practice of Training,* 2nd Edition, Kogan Page Ltd.

Collins, A., Seely Brown, J. and Holum, A. (1991), 'Cognitive Apprenticeship: Making Thinking Visible', *American Educator, The Journal of The American Federation of Teachers.*

Donohoe, L., Lamoureux, T., Atkinson, T., Kirwan, B., Phillips, A. and Brown, L. (1999), Human Factors Suppport to Training for Future Air Traffic Controllers, *Proceedings of the International Aviation Training Symposium,* FAA Academy Oklahoma City, Oklahoma, USA.

Patrick, J. (1992), *Training: Research and Practice,* Academic Press.

Zemke, R. and J. (1984), '30 Things We Know for Sure About Adult Learning', *Innovation Abstracts,* Vol. 6, No 8.

Chapter 9

A Singular Success: Air Traffic Control Specialist Selection 1981-1992

Dana Broach

Introduction

Matching human capabilities with job requirements is a fundamental Human Factors problem. The Federal Aviation Administration (FAA) was confronted with this issue in 1981 on an unprecedented scale when more than 11,000 of about 15,000 air traffic control specialists went on strike and were summarily fired when they failed to return to work. The FAA was faced with an immense organizational challenge – rebuilding its core, technical, and highly trained air traffic control specialist (ATCS) workforce. From late 1981 through mid 1992, the FAA rebuilt this critical workforce through application of Human Factors research findings in a successful, large-scale testing, screening, and training program. The abilities of over 400,000 applicants were assessed using a test battery based on Human Factors research in the preceding decade. Approximately 10 per cent of the top-scoring applicants were hired and underwent a second-stage screening at the FAA Academy between 1981 and 1992. Just over half (56 per cent) of those new controllers successfully completed the Academy screening program and were placed into field training. Between 60 and 90 per cent of Academy graduates went on to successfully complete field training and become Certified Professional Controllers. In this chapter, the Human Factors research roots of the 1981-1992 ATCS selection test battery will be described. The reliability, validity, fairness, and utility of the battery will be examined. The chapter closes with a discussion of workforce demographic trends that will require the FAA to yet again consider, on a large scale, the fundamental problem of air traffic control specialist selection.

Human Factors Roots

The US aviation system expanded in the late 1960s, placing a heavy burden on the Federal Aviation Administration and its employees. The most visible and public employees were the air traffic control specialists. In reaction to their job burdens, the controllers engaged in a series of public actions to draw attention to their complaints of old equipment, staff shortages, and job stress. In reaction, the Department of Transportation appointed a blue-ribbon committee to study the career of the air traffic controller. The effectiveness of the 1964 civil service battery was questioned in the final report of the Air Traffic Controller Career Committee (Corson, Berhard, Catterson, Fleming, Lewis, Mitchell, and Ruttenberg, 1970). Not only were there problems with aptitude testing, the committee also concluded that existing measures of controller for performance, against which to assess the validity and utility of the selection processes, were weak and uninformative. The committee also found, as had previous investigators, that the attrition rate in training was as great as 50 per cent in some years. Based on the committee recommendations, the FAA entered into a research contract to '…analyze the existing procedures for selection in depth, to note important gaps which mitigate against improved selection and high quality of performance, to locate (within the time constraints of one year) assessment techniques to fill those gaps, and to conduct a field validation study to establish their validity' (Education and Public Affairs EPA Inc., 1970).

The first step was to review the existing selection procedure for controllers. Completed in July 1970, this report was quickly reviewed by the scientific community, which concluded that there was little that was new in the EPA report (Dille, 1970). Working scientists with substantial experience in research on controller selection expressed reservations about the centerpiece of EPA's solution – an orientation center and procedure to screen applicants (Coulter, 1970a). EPA next produced a report in which 'attributes of a good air traffic controller' were identified, based on a review of available job materials and visits to air traffic control facilities (Harding, 1970). As with the first report, reservations were expressed by the FAA research community about the job analysis performed by Colmen and his associates. The major reservation was that the report proposed establishing a controller selection program on the basis of what were 'essentially existing (written and verbal) job descriptions rather than on what is traditionally differentiated as job analysis' (Coulter, 1970b; emphases in the original). The comments attached to the covering memorandum noted that the extensive list of worker attributes required in the controller job was not the result of a conventional job analysis. As a consequence, the reviewers had 'limited optimism' regarding outcomes from implementation of an assessment founded on a list based more on assumptions than scientific evidence. Subsequent reports from Colmen and associates fared no better with the scientific community. Two concerns were consistently expressed: (a) the need for a thorough, comprehensive, and detailed job analysis; and (b) the need for reliable, useful, and interpretable criterion measures (see Collins, 1970 for example). Despite these concerns from the technical community within the FAA, the contract with EPA continued to run its course over the next several years. In late December 1970, Phase III of the contract was approved 'with full understanding of the limitations brought about by the lack of adequate criterion measures of performance' (Wormser, 1970). Data collection began in late spring 1971 at the FAA's Mike Monroney Aeronautical Center, and was completed by the end of the year.

The final report under the contract was delivered in 1972. A test battery comprising six paper-and-pencil tests was recommended. Close examination of the study data indicated that test scores did 'not differentiate significantly enough between each of the options and between high and low activity facilities' (Dailey, 1972). Two other findings were noted by Dailey. First, the proposed test battery did appear to differentiate between the controllers assigned to Flight Service Stations and other facilities. Second, the tests appeared to be 'most valid' for controllers assigned to En-route centers and terminals using Instrument Flight Rules (IFR). These tests were also equally valid for controllers assigned to Visual Flight Rule (VFR), although the average scores were lower for VFR than IFR controllers.

The tests recommended by EPA included the Controller Decision Evaluation (CODE) test. The CODE test grew out of Human Factors research on the evaluation of man-machine system performance. It was designed to '... abstract the essential decision-making processes' of controllers (Buckley and Beebe, 1972, p. 1). In the course of research on systems design issues such as display design, researchers noted that (a) there were individual differences in CODE scores between controllers, and (b) that those scores correlated with scores on other criteria such as performance in full-fidelity simulation and supervisory ratings of efficiency. In the adaptation of this technique for personnel selection, films of air traffic situations were presented, and participants asked to identify the aircraft that would violate separation standards (Dailey and Pickrel, 1977). Scores on the CODE test added significantly to a composite for predicting on-the-job success (Milne and Colmen, 1972). Subsequent work by Dailey and Pickrel from the Office of Aviation Medicine focused on translating the filmstrip version of the CODE test into a paper-and-pencil measure that would satisfy the requirements of the Civil Service Commission. The result of their work was the Multiplex Controller Aptitude Test (MCAT).

Predictive Validation

The 1964 Civil Service Commission test battery continued in use through the 1970s while predictive validation studies of the proposed new test battery were conducted. At the same time, the FAA undertook a significant modernization of the air traffic control system. The centerpiece of the modernization was the use of computers to distribute flight plan data, integrate aircraft beacon and ground-based radar data, and display the information directly to controllers on a plan view display. These technologies provided the FAA with the capabilities to handle ever-increasing traffic as the national aviation system continued to expand. This expansion put significant pressure on the ATCS training systems. However, dissatisfaction with the training system grew in parallel with increased traffic. In response, the FAA commissioned yet another study, this one by the Institute for Defense Analysis (IDA; Henry, Ramrass, Orlansky, Rown, String, and

Reichenbach, 1975). This review concluded that improved controller selection methods could have substantial monetary implications for the agency.

Cooperative research on ATCS selection was already underway by scientists from the FAA Office of Aviation Medicine (Dailey and Pickrel, 1984a, b), Civil Aero Medical Institute (CAMI; Collins, Boone, and VanDeventer, 1984), and OPM (Rock, Dailey, Ozur, Boone, and Pickrel, 1982). Beginning in 1976, CAMI was assigned the task of maintaining a longitudinal database on newly hired controllers, including Civil Service Commission aptitude scores, biographical data, and training-related data (Collins, Boone, and Vandeventer, 1980). A series of studies resulted in the validation of a new written ATCS aptitude test battery (Rock, Dailey, Ozur, Boone, and Pickrel, 1982). By early 1979, discussions had begun with the Office of Personnel Management (OPM; successor to the Civil Service Commission) to implement the new test battery. The new battery comprised three tests: the MCAT; an assessment of abstract reasoning from the OPM inventory of tests; and a test of air traffic control occupational knowledge (Lewis, 1978). While plans for implementation of the new battery were being formulated, the tension between the FAA and the Professional Air Traffic Controller Organization (PATCO) continued to mount, culminating in the illegal strike of August 1981. The FAA immediately faced a daunting requirement: rebuild a critical, technical workforce, all the while in the media spotlight. One of the tools immediately available to the agency was the new controller test battery.

Description of the ATCS Written Test Battery

Three tests comprised the new 1981 test battery: the Multiplex Controller Aptitude Test (MCAT), the Abstract Reasoning Test (ABSR), and the Occupational Knowledge Test (OKT). The MCAT was a timed, paper-and-pencil test simulating activities required for control of air traffic. Aircraft locations and direction of flight were indicated with graphic symbols on a simplified radar display (Figure 9.1); an accompanying table provided relevant information required to answer the item, including aircraft altitudes, speeds, and planned routes of flight. MCAT test items required examinees to identify situations resulting in conflicts between aircraft, to solve time, speed, and distance problems, and to interpret the tabular and graphical information. The ABSR was a 50-item civil service examination (OPM-157). To solve an item, examinees determined what relationships existed within sets of symbols or letters. The examinee then identified the next symbol or letter in the progression or the element missing from the set. A sample ABSR item is presented in Figure 9.2. The OKT was an 80-item job knowledge test that contained items related to seven knowledge domains generally relevant to aviation, and specifically relevant to air traffic control phraseology and procedures. The OKT was developed as an alternative to self-reports of aviation and air traffic control experience. The OKT was found to be more predictive of performance in ATCS training than self-reports (Dailey and Pickrel, 1984b; Lewis, 1978).

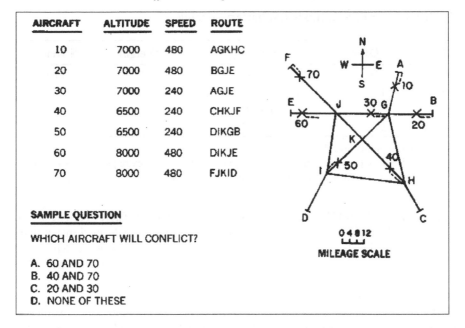

Figure 9.1   Example item from Multiplex Controller Aptitude Test (MCAT)

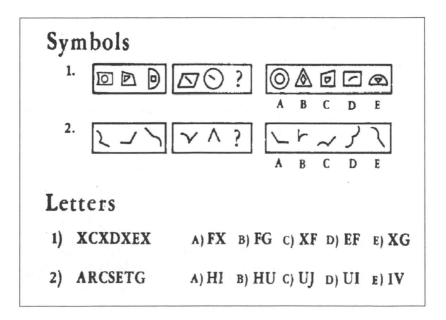

Figure 9.2   Example item from Abstract Reasoning Test (OPM-157; ABSR)

Table 9.1   ATCS Aptitude test battery scoring

Test	OPM #	Scoring	Weight	N Items
MCAT	510	N Right	2	110
ABSR	157	N Right – (0.25*N Wrong)	1	50
OKT	512	N Right	a	80

Note: Extra points awarded as follows for OKT raw scores: 0-51 = 0 extra points; 52-55 = 3 extra points; 56-59 = 5 extra points; 60-63 = 10 extra points; 64-80 = 15 extra points for computation of the civil service rating.

The scoring of the MCAT, ABSR, and OKT is presented in Table 9.1. The sum of the weighted MCAT raw score and the ABSR score was transformed into a linear composite ranging from 19.5 to 100. This 'Transmuted Composite' (TMC) score was used to determine employment eligibility. For applicants without specialized prior experience, education, or superior academic achievement, a minimum TMC of 75.1 was required to qualify at the entry level GS-7 grade (Aul, 1991, 1997, 1998). If an applicant achieved the minimum TMC, extra points were awarded on the basis of scores on the OKT, as noted in Table 9.1. The final civil service rating (RATING) was the sum of TMC, extra points on the basis of OKT scores, and any adjudicated veteran's preference points. Ranking, referral, and selection were based on an applicant's final RATING. In general, only those competitive applicants with a RATING of at least 90 were selected by the FAA for employment as controllers. The distributions of applicant and selectee RATING scores are presented in Figure 9.3, based on over 200,000 applicant records archived at CAMI for research purposes.

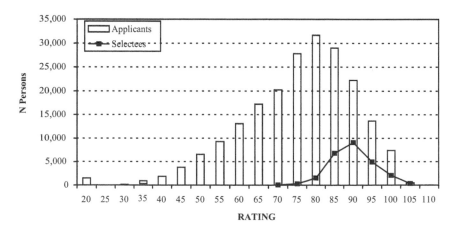

Figure 9.3   Distribution of RATING for applicants and selectees

Psychometric Characteristics

Reliability in testing refers to the degree to which scores on a test are free from errors of measurement (American Educational Research Association, American Psychological Association, & National Council on Measurement in Education, 1999). Test-retest reliability is an estimate of the degree to which a person will obtain approximately the same score when retested after some time interval. The test-retest correlation for the MCAT was estimated at .60 in a sample of 617 newly-hired controllers (Rock, Dailey, Ozur, Boone, and Pickrel, 1981, p. 59). Parallel form reliability estimates the degree to which an applicant will obtain approximately the same score on a different version of the same test. The parallel form reliability, as computed on the same sample, ranged from .42 to .89 for various combinations of items (Rock et al., p. 103). Internal consistency estimates the degree to which the items in a test are homogenous (Ghiselli, Campbell, and Zedeck, 1981). Lilienthal and Pettyjohn (1981) examined internal consistency and item difficulties for ten versions of the MCAT. Cronbach's alpha for the ten versions ranged from .63 to .93; the alphas for 7 of the 10 versions were greater than .80. The available data suggest that the MCAT had acceptable reliability but was vulnerable to practice effects. In contrast, no item analyses, parallel form, test-retest, or internal consistency estimates of the ABSR (OPM-157) test have been reported by the FAA. Therefore, no conclusion can be drawn about the measurement properties of the ABSR. Published data are available for the OKT. Parallel form reliability for the OKT ranged from .88 to .91 (Rock et al., p. 65, 70). The internal consistency estimate of reliability (Kuder-Richardson Formula 20 (KR-20); Kuder and Richardson, 1937) for a 100-item version of the OKT was .95 (Rock et al., p. 51). Lilienthal and Pettyjohn reported Cronbach alphas for ten versions of the OKT ranging from .85 to .94 on a sample of about 2,000 FAA Academy air traffic control students. However, test-retest estimates of reliability have not been published. The available data suggest that the OKT had acceptable reliability.

Validity generally refers to the appropriateness, meaningfulness, and usefulness of specific inferences made on the basis of test scores (American Educational Research Association, American Psychological Association, & National Council on Measurement in Education, 1999). Validity in the specific context of employee selection, as discussed in the federal *Uniform Guidelines on Employee Selection Procedures* (Equal Employment Opportunity Commission, 1978), refers to the degree to which test scores used for making employment decisions are predictive or correlated with important and/or critical work outcomes, elements, or behaviors. The ATCS aptitude test battery has been validated against two classes of work outcomes in previous studies: (a) performance in the FAA Academy initial training programs (SCREEN; Della Rocco, 1998); and (b) outcomes of on-the-job field training (OJT) at the first assigned field facility (STATUS; Manning, 1998).

As the final civil service RATING was the basis of operational personnel selection decisions, this retrospective analysis examined the validity of the final RATING as a predictor of these two criteria for the 15,876 controllers who completed the FAA Academy and were placed into field training between 1981 and 1992. The field training STATUS criterion was coded as follows: 1 = *Reached Full Performance Level*; 2 = *Still in Training (Developmental)*; 3 = *Switched facilities*; 4 = *Switched options*; and 5 = *Failed*. As recommended by Manning, persons who left the agency for reasons other than performance were excluded from the validity analysis. Because the samples had been truncated by selection first, on RATING, and second, by selection on FAA Academy score (SCREEN), corrections for restriction in range were made as shown in Table 9.2, using the formulae presented by Ghiselli, Zedeck, and Campbell (1984).

Table 9.2   Correlation matrix structure for regression analyses

	RATING	SCREEN	STATUS
RATING		$r_{e_{(RATING)}}$	$R_{i_{(SCREEN)}}$
SCREEN	r_s		$R_{e_{(SCREEN)}}$
STATUS	r_s	r_s	

Note: Sample correlation matrix structure shown below the diagonal, corrected matrix structure above the diagonal. r_s = sample correlation; $r_{e_{(RATING)}}$ = correlation corrected for explicit selection on RATING; $r_{i_{(SCREEN)}}$ = correlation corrected for incidental restriction in range due to selection on SCREEN; and $r_{e_{(SCREEN)}}$ = correlation corrected for explicit selection on SCREEN

Correlations between civil service RATING and FAA Academy SCREEN scores were corrected for direct restriction in range due to explicit selection on RATING. The corrections were based on the population standard deviations presented in the Tables 9.3 through 9.6. Correlations between FAA Academy (SCREEN) and field training outcomes (STATUS) were corrected for direct restriction in range due to explicit selection on the FAA Academy score. Finally, correlations between civil service RATING and field training outcomes were corrected for incidental restriction in range due to selection of the samples on FAA Academy SCREEN scores. The uncorrected, zero-order correlations are presented in the lower left corner of each correlation matrix and the corrected correlations in the upper right corner. The corrected correlation matrices for each sample were submitted to regression analysis to estimate the validity of RATING as a predictor of FAA Academy SCREEN and field training STATUS.

In view of the iterations of the FAA Academy initial training programs, described by Della Rocco (1998), and differences in field training as described by Manning (1998), the sample was divided into four groups: (a) 1982-85 FAA Academy Terminal program graduates assigned to training in the terminal option

(Table 9.3); (b) 1982-85 FAA Academy En-route program graduates assigned to field training in the En-route option (Table 9.4); (c) 1986-92 FAA Academy Non-radar Screen assigned to field training in the terminal option (Table 9.5); and (d) 1986-92 FAA Academy Non-radar Screen graduates assigned to field training in the En-route option (Table 9.6).

On one hand, the analysis found that the final civil service RATING had acceptable validity as a predictor of performance in initial training at the FAA Academy for all four groups. The uncorrected correlations between RATING and SCREEN were statistically significant and ranged from .18 to .22 across the four groups. The corrected correlations between civil service RATING and FAA Academy SCREEN ranged from .46 to .50. On the other hand, regression analyses of the corrected correlation matrices found that RATING was a relatively poor predictor of field training STATUS in three of the four FAA Academy and field training variations (Table 9.7). The standardized regression weight (β) for RATING was not statistically significant for graduates of 1982-85 Terminal Screen program assigned to field training in the terminal option ($\beta = -.025$, *ns*) or for 1986-92 Non-radar Screen program graduates assigned to either terminal ($\beta = -.041$, *ns*) or En-route ($\beta = -.010$, *ns*) field training. However, RATING did enter into the regression equation predicting STATUS in En-route field training, with a standardized regression weight of -.107 ($p \leq .001$), for graduates of the 1982-85 En-route Screen program, as shown in Table 9.7.

Table 9.3   Descriptive statistics and correlations for 1982-85 Terminal Screen graduates assigned to Terminal OJT

Variable	Sample[a]			Population[b,c]			Correlations[d]		
	N	*M*	*SD*	*N*	*M*	*SD*	(1)	(2)	(3)
(1) RATING	2,814	92.31	5.72	99,887	74.93	14.24		.49	-.17
(2) SCREEN	3,174	79.45	5.48	4,686	73.82	11.32	**.22**		-.30
(3) STATUS	3,174	88.0%	FPL				-.04	**-.15**	
		12.0%	NOT FPL						

Notes: [a]STATUS outcomes for first assigned field facility only as of March 1997
[b]RATING based on CAMI records for examinees tested 1982-85
[c]SCREEN based on Terminal Screen entrants who did not withdraw, 1982-1985
[d]Uncorrected correlations in lower left half, corrected in upper right half. Bold faced correlations significant at $p \leq .01$

Fairness Air traffic control is a career field in which female and minority workers have been historically under-represented relative to the American population at large. The fairness of the written ATCS aptitude test battery was recently examined in a series of papers as the first step toward assessing to what degree, if any, the battery may have served as an 'engine of exclusion' (Seymour, 1988) of women and minorities from the ATCS occupation (Young, Broach, and Farmer, 1996; Broach, Farmer, and Young, 1997). The term fairness has no single technical meaning (American Educational Research Association, American Psychological Association, & National Council on Measurement in Education, 1999, p. 74). There are four principal ways in which fairness is used in psychological testing: (a) equality of outcomes of testing; (b) opportunity to learn; (c) equitable treatment in the testing process; and (d) lack of bias. The idea that fairness is defined as equality of outcomes for socio-economic or demographic groups is not generally accepted in the professional psychological testing literature (American Educational Research Association, et al., p. 75). Opportunity to learn applies to educational assesment rather than the assessment of aptitude. This leaves fairness as equitable treatment and freedom from bias as concerns for ATCS selection testing. Fairness as equitable treatment encompasses the notion that examinees are given comparable opportunities to demonstrate their standing on the construct(s) assessed by the test. Other aspects of fairness from this perspective include appropriate testing conditions, equal opportunity to become familiar with the test, and equal access to practice materials. Fairness as to the lack of bias focuses on the degree to which test characteristics – deficiences – or the manner of use result in different meanings for scores from different groups. The fairness of the ATCS selection test battery was assessed from both perspectives during its use from 1981 through 1992.

Table 9.4   Descriptive statistics and correlations for 1982-85 En-route screen graduates assigned to En-route OJT

Variable	Sample[a]			Population[b,c]			Correlations[d]		
	N	M	SD	N	M	SD	(1)	(2)	(3)
(1) RATING	4,021	92.56	5.37	99,887	74.93	14.24		.50	-.26
(2) SCREEN	4,276	82.02	7.17	4,686	73.82	11.32	.21		-.35
(3) STATUS	4,276	66.1%	FPL				-.05	**-.23**	
		33.9%	NOT FPL						

Notes: [a]STATUS outcomes for first assigned field facility only as of March 1997
[b]RATING based on CAMI records for examinees tested 1982-85
[c]SCREEN based on Terminal Screen entrants who did not withdraw, 1982-1985
[d]Uncorrected correlations in lower left half, corrected in upper right half. Boldfaced correlations significant at $p \leq .01$

The ATCS test battery was administered in a group setting with proctors for the test session and scored by computer. From that perspective, all examinees had equal oppportunity to demonstrate their aptitude. While biographical data collected by CAMI from newly hired controllers suggested differences in preparation for the

OPM test by applicants, the differences in test score were small (1-2 points) for the MCAT and ABSR components, but larger for the OKT (5-6 points; Broach, 1991). However, these differences were not reflected in the overall composite. These analyses suggested that the test battery was fair in terms of equitable treatment of applicants.

Table 9.5   Descriptive statistics and correlations for 1986-92 non-radar screen graduates assigned to Terminal OJT

Variable	Sample[a]			Population[b,c]			Correlations[d]		
	N	M	SD	N	M	SD	(1)	(2)	(3)
(1) RATING	2,765	93.31	5.15	106,201	76.33	14.67		.46	-.10
(2) SCREEN	3,145	78.28	6.07	12,756	72.26	11.80	**.18**		-.14
(3) STATUS	4,276	95.0%FPL					.03	**-.07**	
		5.0%NOT FPL							

Notes: [a]STATUS outcomes for first assigned field facility only as of March 1997
[b]RATING based on CAMI records for examinees tested 1985-1992
[c]SCREEN based on Non-radar Screen entrants who did not withdraw, 1986-1992
[d]Uncorrected correlations in lower left half, corrected in upper right half. Boldfaced correlations significant at $p \leq .01$

Table 9.6   Descriptive statistics and correlations for 1986-92 non-radar screen graduates assigned to En-route OJT

Variable	Sample[a]			Population[b,c]			Correlations[d]		
	N	M	SD	N	M	SD	(1)	(2)	(3)
(1) RATING	3,969	93.79	5.06	106,201	76.33	14.67		.49	-.13
(2) SCREEN	4,449	80.26	5.89	12,756	72.26	11.80	**.19**		-.25
(3) STATUS	4,732	77.3%FPL					-.01	**-.13**	
		22.7%NOT FPL							

Notes: [a]STATUS outcomes for first assigned field facility only as of March 1997
[b]RATING based on CAMI records for examinees tested 1985-1992
[c]SCREEN based on Non-radar Screen entrants who did not withdraw, 1986-1992
[d]Uncorrected correlations in lower left half, corrected in upper right half. Boldfaced correlations significant at $p \leq .01$

Table 9.7   Civil service RATING validity in predicting field training outcome by sample, based on corrected correlations

	Sample				
Period	FAA Academy	Field OJT	R	β-RATING	β-SCREEN
1982-85	Terminal Screen	Terminal	**.30**	-.03	**-.29**
1982-85	En-route Screen	En-route	**.36**	**-.11**	**-.30**
1986-92	Non-radar Screen	Terminal	**.14**	-.04	**-.12**
1986-92	Non-radar Screen	En-route	**.25**	-.01	**-.24**

Note: Boldfaced correlations significant at $p \leq .001$

This leaves consideration of test bias. Bias, from the perspective of psychological testing, refers to the regression model of test bias for which there is a reasonable professional consensus, as embodied in the 1999 Standards for Educational and Psychological Testing, rather than a socially constructed standard regarding test use (Sackett and Wilk, 1994; Gottfredson, 1994). Fairness as to the lack of bias, and under the *Uniform Guidelines on Employee Selection Procedures* (Equal Employment Opportunity Commission, 1978), encompasses two issues. First, the effect on protected groups arising from use of a particular cut score on the predictor must be evaluated. A selection rate for any protected group that is less than four-fifths (4/5 or 80 per cent) of that of the majority group will '... generally be regarded by the Federal enforcement agencies as evidence of adverse impact' (29 CFR 1607.4.D). Second, where the use of a selection procedure results in adverse impact, the *Uniform Guidelines on Employee Selection Procedures* require that the test user evaluate the degree to which differential predictions of future job performance are made from selection test scores by subgroup (29 CFR 1607.14.B.(8).(b)).

Gender Previous research on ATCS selection tests suggested that mean score differences by gender were insignificant (Rock, Dailey, Ozur, Boone, and Pickrel, 1984a, p. 476) and that, overall, 'the evidence for adverse impact against women, based on this sample, was marginal, at best' (Rock, et al., p. 507). This conclusion was based on results of their 1984 study in which 57 per cent of men (n = 3,835) passed the written test, compared to 45 per cent of women (n = 1,473). The adverse impact ratio in this case was .78, rather than the .80 required under the 'four-fifths rule of thumb.' Young, Broach and Farmer (1996) analyzed archival records for 170,578 applicants. Based on test scores, OPM determined the eligibility of applicants for employment by the FAA. OPM codes indicating that an applicant had either failed the test ('IA') or scored too low for consideration ('IS') were categorized as test failures. All other ineligibility codes were categorized as 'other ineligible,' and excluded from the analysis. OPM codes indicating eligibility were categorized as 'eligible' for employment. The adverse impact analysis compared the proportion of women considered as eligible, compared with that of men. The analysis is presented in Table 9.8.

Table 9.8   Adverse impact analysis by gender based on OPM eligibility codes

	Sex		
OPM Eligibility	Males	Females	Row totals
Eligible	69,056	14,564	83,620
	(58.1%)	(42.0%)	
Failed test	49,902	20,077	69,979
	(41.9%)	(58.0%)	
Column totals	118,985	34,641	153,599

Note: Percents are column percentages

A significantly lower percentage of women (42.0 per cent) than men (58.1 per cent) were coded by OPM as eligible for employment as controllers, based on test scores ($Z = -3.58$, $p < .001$). The ratio of female to male selection rates was .72. Using the 4/5ths rule of thumb of the *Uniform Guidelines on Employee Selection Procedures*, it appeared that the use of scores on the written ATCS aptitude test battery to determine eligibility for employment in the ATCS occupation between 1981 and 1992 resulted in statistically significant adverse impact on female applicants.

Given the finding that there appeared to be adverse impact against women, the *Uniform Guidelines on Employee Selection Procedures* (29 CFR 1607.14.B.(8).(b)) and *Standards for Educational and Psychological Testing* (American Educational Research Association, et al., 1999, pp. 80 and 82) require an investigation of the relationship between test scores and job performance for evidence of differential prediction by subgroup. The classical, regression-based model of test bias was used as the analytic framework by Young, Broach, and Farmer (1996) to evaluate the degree to which the written ATCS test battery differentially predicted performance in initial ATCS qualification training at the FAA Academy. After correcting correlations between test score, FAA Academy composite score, and gender for explicit and implicit restrictions in range due to prior selection of the sample on aptitude test score, the null hypothesis of a common regression line for the genders was rejected, suggesting the presence of some degree of test bias. Statistically significant differences by gender in regression slopes and intercepts were found in the step-down regression analysis, indicating the need for separate regression equations for men and women. The regression line for men slightly over-predicted the performance of women in the FAA Academy, as shown in Figure 9.4. The practical consequence of the apparent differential prediction was that women effectively needed a higher test score than men to have an equal likelihood of passing the initial qualification training at the FAA Academy.

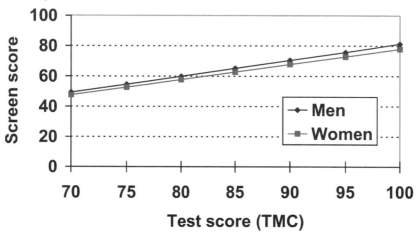

Figure 9.4   **Regression of test score (TMC) on FAA Academy score (SCREEN) by gender**

Race A formal adverse impact analysis on the basis of race was not technically feasible, as racial identifiers were not collected for ATCS job candidates. However, previous research (Rock, Dailey, Ozur, Boone, and Pickrel, 1984) found that the OPM test battery had adverse impact against African-Americans and Hispanics. In view of the previous research, Broach, Farmer, and Young (1997) investigated the relationship between test scores and performance in the FAA Academy for evidence of differential prediction on the basis of race. The classical, regression-based model of test bias was again used as the analytic framework to evaluate the degree to which the written ATCS test battery differentially predicted performance in initial ATCS qualification training at the FAA Academy for African-American and white controllers. After correcting correlations between test score, FAA Academy composite score, and race for explicit and implicit restrictions in range, the null hypothesis of a common regression line for African-American and white controllers was rejected, suggesting the presence of some degree of test bias. Statistically significant differences by race in regression slopes and intercepts were found in the step-down regression analysis, indicating the need for separate regression equations for African-American and white controllers. The regression line for whites over-predicted the performance of African-Americans in the FAA Academy. The practical consequence of the apparent differential prediction was that African-American controllers effectively needed a higher test score than white controllers to have an equal likelihood of passing the initial qualification training at the FAA Academy, as shown in Figure 9.5.

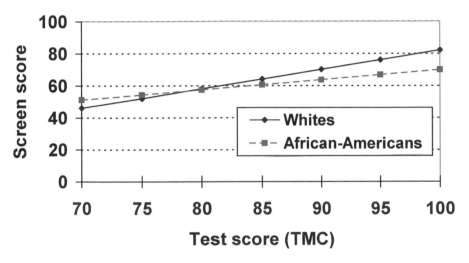

Figure 9.5   **Regression of test score (TMC) on FAA Academy score (SCREEN) by race**

Human Factors Issues in Future Controller Selection

On 21st February 1992, the FAA closed the process of hiring from the general population. No new applications to take the ATCS selection test were processed, except for certain veterans who had a statutory right to take the examination. Since that date, hiring has been substantially reduced due to a stable workforce with very low attrition and retirement rates (General Accounting Office (GAO), 1997) However, the FAA may increase its hiring in the not-too-distant future as the post-strike generation of controllers ages.

   Controllers hired in the 21st century will come to a very different FAA – and aviation industry – than their older, soon-to-retire post-strike predecessors. In April of 1996, the FAA was legally exempted from most of the civil service laws. The agency now has the authority to develop new and innovative personnel systems and, has been incorporating new procedures based on studies of the best practices in other government and private organizations. Based on those best practices and lessons learned in the recovery of the controller workforce between 1981 and 1992, the FAA is developing and implementing innovative processes to select and train the next generation of air traffic controllers. As the wave of retirement eligibility for post-strike controllers approaches, considerable resources will be required to plan for and initiate recruitment, selection, and training programs for the inevitable influx of new controllers. As a result, the FAA faces a number of personnel-related research and operational questions, including the aging of the controller workforce, training in a period of rapid technological change, and selection for future systems.

Aging of the ATCS Workforce

Future Retirements As the present controller workforce ages, and inexorably moves toward retirement eligibility, the issue of future selection to replenish the ATC system becomes critical. The vast majority (88 per cent) of the current ATCS workforce entered the FAA following the PATCO strike of 1981. The age distribution for the current terminal and en-route controller workforce is presented in Figure 9.6. The average age of the current workforce is 41 (SD = 7) with 17 years of experience as of the end of April, 2003. Just 10 per cent of the 15,531 controllers (as of the end of April, 2003) are age 50 or older. However, that proportion will grow as the controller workforce continues to age.

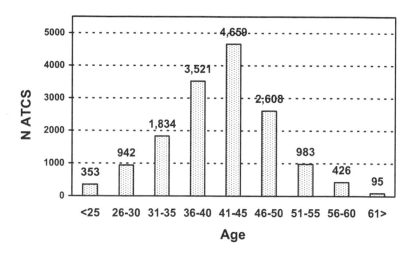

Figure 9.6   Age distribution for terminal and en-route controllers (4/2003)

Forecasting future retirements is critical to the strategic management of the human capital represented by the controller workforce. Given the time required to train a controller (historically, about 3 years in the en-route option), new controllers will have to be brought on board well in advance of expected retirements. Controllers may retire at an earlier age than most other federal employees. For example, they may take 'special optional retirement' if either (a) they have 25 years of service as a controller actively engaged in the separation and control of air traffic (see 5 USC 2901) without regard to age, or (b) at age 50 with at least 20 years of service as a controller (as defined by 5 USC 2901). US law also provides that a controller will be separated from service as a controller on the last day of the month in which he or she reaches age 56, unless exempted by the Secretary of Transportation as having 'exceptional skills and experience' until reaching age 61 (5 USC 8335). Expected and actual retirements under these rules are tracked closely by the Air Traffic Resource Management program at FAA Headquarters, as shown in Figure 9.7. Predicted retirements are expected to increase over the next several years as the post-strike generation of controllers' ages. New controllers will have to be hired to replace the retiring controllers.

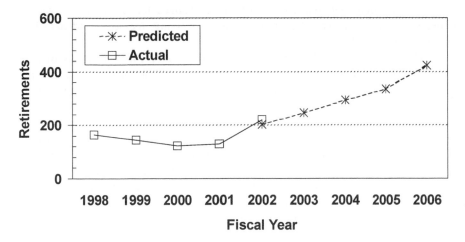

Figure 9.7   Actual (FY1998-2002) and predicted (FY2002-2006) retirements from terminal and en-route controller workforce

Age and Performance Another consideration associated with the overall aging of the terminal and En-route controller workforce during the first two decades of the 21st century concerns the potential effects of aging on cognitive functioning and the consequences of those changes for on-the-job performance. There are indications in the scientific literature that during the middle decades of life most individuals experience some decline in the speed of their cognitive functioning. In fact, Fozard and Nutall (1971), in their study of age-related changes in scores on the General Aptitude Test Battery (GATB), indicated that the average 60 year-old individual would be able to meet the occupational aptitude patterns for only 17 of 36 occupations. Changes are more frequently observed in the fluid or process aspects of cognition (speed of processing information, memory, and making decisions). Some scientists feel that overall performance on those tasks may start to decline at ages as early as in the 20s (Salthouse, 1990). However, what is often referred to as the crystallized measures of intelligence (vocabulary and general information) may improve with increasing age until the late 60s. When viewed in the context of most jobs, Schaie (1988) has argued that the reported age differences are generally too small to have meaningful consequences. However, little has been said about the potential implications of these changes for personnel in safety-related positions where speed of information-processing, memory, and decision-making are of critical importance.

Nearly all of the scientific information available refers to the effects of age-related declines in performance on somewhat novel laboratory tasks, compared with more familiar tasks or job performance. A series of studies by Trites, Cobb, and their associates (e.g., Cobb, Lay, and Bourdet, 1971; Trites, 1963; Trites and Cobb, 1964a; 1964b) consistently demonstrated an inverse relationship between age of entry and performance of ATCS trainees at the FAA Academy and in field training. Those studies led to Congressional action to restrict the age of entry into the FAA Academy training to age 30. A subsequent study (VanDeventer and Baxter, 1984) of 8,573 FAA Academy trainees where the age of entry was more restrictive (18-30) than that studied by Cobb and associates, demonstrated the typical inverse relationship between age and pass rate. Figure 9.8 presents the outcomes in the FAA Academy screening programs for 27,925 entrants from 1981 through 1992. Overall success in the FAA Academy declined from 66.6 per cent of those age 22 and younger to percentage who reached FPL status ranged from 76.9 per cent of controllers in the 22 years of age or younger group to 46.3 per cent of those in the 31 and older age group. Thus, in the more demanding terminal environments, there is 53.7 per cent of those aged 27-28, and 48.6 per cent of those in the 31 or older age category. Age-related changes were evident in both the failure rates as well as withdrawals.

Lower success rates were also evident in field training. Outcomes for the FAA Academy graduates initially assigned to an en-route ATC facility for training are presented in Figure 9.9. To be consistent with Manning's (1998) analysis of OJT outcomes, controllers still in training, or who left training for reasons other than performance, were excluded from this analysis. While 80.7 per cent of those age 22 or younger successfully reached the full performance level at their first en-route facility, only 52.8 per cent of those in the oldest age group (31+) did. From the younger through older age groups, there was a systematic increase in the percentages of trainees who switched to another option (Terminal or Flight Service Station) or were terminated from training.

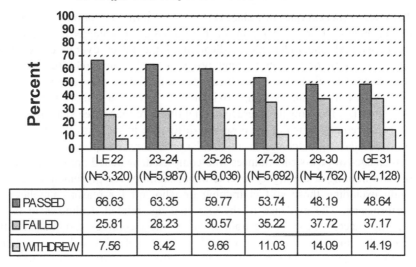

	LE 22 (N=3,320)	23-24 (N=5,987)	25-26 (N=6,036)	27-28 (N=5,692)	29-30 (N=4,762)	GE 31 (N=2,128)
■ PASSED	66.63	63.35	59.77	53.74	48.19	48.64
□ FAILED	25.81	28.23	30.57	35.22	37.72	37.17
□ WITHDREW	7.56	8.42	9.66	11.03	14.09	14.19

Figure 9.8   FAA Academy outcomes by age group for all entrants, 1981-1992

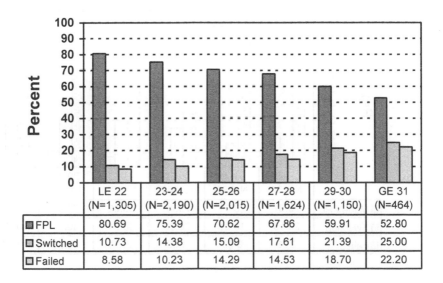

	LE 22 (N=1,305)	23-24 (N=2,190)	25-26 (N=2,015)	27-28 (N=1,624)	29-30 (N=1,150)	GE 31 (N=464)
■ FPL	80.69	75.39	70.62	67.86	59.91	52.80
□ Switched	10.73	14.38	15.09	17.61	21.39	25.00
□ Failed	8.58	10.23	14.29	14.53	18.70	22.20

Figure 9.9   OJT outcomes at first assigned facility by age group for FAA Academy graduates initially assigned to en-route centers (*N*=8,748)

The overall success rates of trainees entering the terminal option were considerably higher, ranging from 88 per cent for those aged 22 or younger to 78.6 per cent of those aged 31 and older. One indication of the lower overall difficulty of the terminal option on-the-job training program is the time required to reach full performance level (FPL) status. It ranges from slightly over one year at the low-level VFR-only terminals to approximately 2½ years at a more complex facility. The latter is still below the average of slightly more than three years required at the En-route centers. Much of the observed difference in success rates across ages can be attributed to outcomes from the larger, more complex terminal facilities. Percentages of post-strike trainees who completed, switched options, or washed out of field training at the higher level terminal facilities are presented in Figure 9.10, again excluding those still in training or who left for reasons unrelated to job performance. The evidence of the expected age-related decline is in on-the-job training success. Most of the trainees at the most complex terminal facilities are selected from the ranks of FPL controllers at other facilities. These outcomes are again consistent with historical data gathered from respective field facilities (Trites, 1963; Trites and Cobb, 1964a, 1964b). Thus, older adults experience greater difficulty completing the FAA Academy screening and on-the-job training programs for Air Traffic Control Specialists.

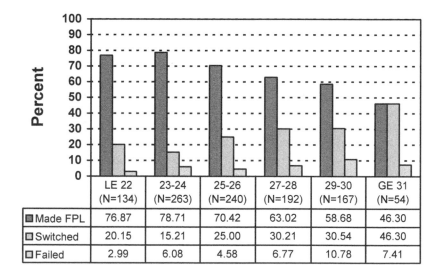

	LE 22 (N=134)	23-24 (N=263)	25-26 (N=240)	27-28 (N=192)	29-30 (N=167)	GE 31 (N=54)
■ Made FPL	76.87	78.71	70.42	63.02	58.68	46.30
▢ Switched	20.15	15.21	25.00	30.21	30.54	46.30
▢ Failed	2.99	6.08	4.58	6.77	10.78	7.41

Figure 9.10   OJT outcomes at first assigned facility by age category for FAA Academy graduates initially assigned to more complex, busy terminals (Level 4 and 5 in 1995) (*N*=1,050)

Some scientists have argued that age-related differences at work are not likely to be observed because many of the tasks are highly over-learned. It is difficult to assess the extent to which older employees may be less proficient at work due to the multitude of factors that influence attrition as employees age (workers performing poorly often drop out or move to other jobs, and the better performing workers are often selected for promotions or are given opportunities to switch to more challenging jobs). The effects of seniority on job assignments in many occupational settings are also often not well understood. Salthouse (1990) describes several mechanisms that may serve to preserve overall competency and job performance as adults experience age-related declines in the efficiency of basic processes: compensation, accommodation, elimination, and compilation. Rhodes (1983), in a review of age-related differences in work attitudes and behavior, concluded that the outcomes from studies of the relationship between age and job performance are mixed. Older employees did express higher levels of job satisfaction, satisfaction with the work itself, and job involvement.

Evidence for a possible moderating effect of experience on the typical effects of age on job-related tasks is mixed. General support was found by Salthouse for typing tasks (1990) and by Morrow, Leirer, Altieri, and Fitzsimmons (1994) on readback of visual or vocally-presented ATC communications. Evidence was more limited for a set of time-sharing tasks involving groups of pilots and non-pilots (Tsang and Voss, 1996). Salthouse, Babcock, Skovronek, Mitchell, and Palmon (1990) found little support among groups of practicing architects on spatial visualization tasks. Thus, while there is some laboratory-based evidence to support the critical role of expertise or experience in the performance of older workers on job-related tasks, there is insufficient job-related performance research available to clearly document the importance of various factors in determining the job performance of older workers, particularly in safety-critical occupations. However, Cobb and associates (Cobb, 1967; Cobb, Nelson, and Matthews, 1973) have demonstrated that combined ratings (supervisors, crew chiefs, and co-workers) of controller performance are lower for older than for younger controllers. This is consistent with data gathered earlier at the En-route centers. There are several factors (motivation, differences in aptitudes and abilities, and rating bias) that may have contributed to these differences, other than physiological aging. While we do not understand the full implications of the age-related changes on the performance of air traffic controllers, careful attention should be paid to what appears to be age-related performance changes as the overall age of the controller workforce increases during the next 15 years.

Training Implications

Based on the projected increase in retirement eligible personnel and the lead-time of around two to three years for individuals to complete training, the FAA is developing a strategic human capital management plan for the terminal and en-route ATCS workforce. Facilities are likely to face substantially greater training requirements over the next decade to equip new controllers to handle traffic in the future, as older controllers retire. However, this is not the only training burden faced by facility managers and the Air Traffic Service. Enhanced hardware/software ATC workstations, decision aids, and new policies and procedures will be introduced into the National Airspace System (NAS) with modernization to improve capacity and to increase safety and efficiency. During the first decade of the 21st century, new automation, including decision support systems (DSS), as well as updated controller workstations will be introduced. Emphasis on integrating 'free flight' concepts into the NAS may result in a number of significant changes in how air traffic is handled. Under some proposals there may be a shift in responsibility for separation of aircraft between flight crew members and controllers, as a part of 'free flight.' No matter which scenario is finally adopted, the introduction of these new systems and procedures into the NAS over the next decade will generate continuing requirements to train ATCSs to operate these new systems.

Since these changes will largely coincide with the influx of new controllers, it will present a significant logistical effort to retrain the existing cadre of full performance level controllers and simultaneously train new controllers. Concerns about the availability of sufficient staff within the facility to maintain normal operations and still provide on-the-job training are likely to be of special concern at facilities that have been identified as 'hard-to-staff,' where staffing levels have remained below projected levels. Thus, facility-specific plans will be needed to address the training requirements associated with the hiring of new controllers to replace retirees and the retraining of the existing workforce to function on the more automated systems of the future.

While the cognitive changes in performance capabilities may not pose a significant problem for the safe and expeditious flow of traffic under typical conditions, it may pose additional burdens as older ATCSs are required to transition to new technologies and new procedures. Evidence suggests that older employees experience greater difficulty acquiring new skills than do younger employees (Charness and Bosman, 1990). In the use of computer technology, there is research that suggests that older adults require longer training time and, in general, do not obtain the same level of skill as younger individuals (Czaja, 1996; Sharit and Czaja, 1994). The difficulty of this transition is, of course, dependent on the extent to which the job changes and the nature of those changes. At this point, not enough is known about the proposed future systems to determine the extent to which age-related factors should be considered in the development of the new procedures and training programs. Human Factors assessments are needed to determine the appropriateness of training approaches for transitioning to the new technologies.

Selecting Controllers for Future Systems

Research is currently ongoing with respect to developing and validating a 3rd generation computer-administered test battery to select controllers for the near term (2004 – 2008). But what are the cognitive and personal characteristics that may be needed by controllers operating future systems? The new technologies will involve greater automation and the availability of increased decision-aiding. The extent to which the new system and procedures impact on the nature of the job is highly dependent on the level of automation selected (Wickens, Mavor, and McGee, 1997). In his recent book on automation in aviation, Billings (1997) provides four scenarios regarding the nature of the future air traffic control system: management by delegation; management by consent; management by exception; and free flight. There are marked differences in the roles and responsibilities of ATCSs under these scenarios. Both the Automated En-route Air Traffic Control (AERA) and free flight concepts appear to raise a number of Human Factors concerns related to controller-machine interactions.

CAMI scientists have embarked on a program of research designed to provide an answer to those questions. The initial study was designed to assess controller perceptions of AERA and determine how perceptions of task activities under AERA would impact the requisite knowledge, skills, and abilities of ATCSs. Findings by Manning and Broach (1993) indicate that many of the same skills and abilities might be required. Since the FAA is no longer moving down an identical technological pathway, the question remains regarding the demands and job tasks associated with the future ATC systems. The infrastructure technologies that are being readied for introduction into the NAS during the remainder of this decade are not likely to dramatically alter the job duties and functions of ATCSs. However, the implementation of the initial and upgraded versions of decision-support systems, and the revised AERA system are likely to substantially modify the nature of the job. Given the time required to recruit, select, and train personnel, a new advanced selection system needs to be developed coincident with the development of the proposed future systems.

Traditional approaches to job analysis have focused on jobs that incumbents have already been performing. In planning for the future, an approach is needed that will allow us to identify the knowledge, skills and abilities required as the systems are under development. One strategy is to utilize the 'strategic job analysis' approach presented by Schneider and Konz (1989). Efforts are underway at CAMI to develop the methodology required to conduct a strategic job analysis for both air traffic control and airway facilities positions. Once developed, the approach will be used to assess the changes in knowledge, skills, and abilities as new systems are developed for the NAS. With the anticipated increased emphasis on coordination and shared decision-making, there may be additional requirements to identify personality dimensions that play a critical role in teamwork. Of course if the level of automation selected is one where the human operator is primarily a system monitor and will only take over in the advent of an automation-related failure, this presents a very different view of the nature of the job and the necessary knowledge, skills and abilities. Outcomes from the future job analysis assessments will be compared with the recent baseline job analysis data (Nickels, Bobko, Blair, Sands, and Tartak, 1995). The identification of the requisite knowledge, skills, and abilities that are sufficiently different from those required with current systems will be used to modify existing selection and training programs. Ideally, this will allow the FAA to develop an integrated recruitment, selection, training, and job performance data collection and analysis system. That system should be in place in time to select the personnel who will be operating those future systems. An outline for research to support selection of personnel for the future NAS architecture is described in Broach (1997).

Conclusions

The written ATCS aptitude test battery in use from 1981 through 1992 was developed to enable the agency to make reasonable predictions, for large numbers of applicants, about their future job performance as the basis for initial selection into a safety-related occupation. From a practical perspective, the primary function of the written ATCS test battery was to winnow a huge pool of applicants down to a smaller, manageable number suitable for further intensive, expensive evaluation. On one hand, the written ATCS aptitude test battery, as used from 1981 to 1992, achieved that organizational goal. The tests comprising the battery produced reliable scores. A composite of those reliable scores, used as the basis for selection decisions, was valid as a predictor of near-term (relative to date-of-hire) training outcomes. On the other hand, the composite score was not a valid predictor of far-term (2 to 3 years from hire) field training outcomes and may have had adverse impact against women and minorities.

Despite these limitations, the written test battery was a Human Factors success story. Growing out of research on man-machine interfaces and controller decision-making, the MCAT became the cornerstone of a valid selection battery suitable for mass testing. Continued applied research on practical problems resulted in the development of an objective assessment of prior experience. The FAA used these tools to reduce the applicant pool to a smaller, affordable number for more intensive evaluation of their aptitude for the controller occupation through screening at the FAA Academy. The relative costs of assessment methods must be carefully considered, particularly in view of the substantial numbers of job applicants. For example, the cost of administering the paper-and-pencil ATCS test battery was about $20 per examinee, compared with a cost of about $10,000 per person for initial training at the FAA Academy (Broach and Brecht-Clark, 1993). There was significant financial utility for the agency in using a valid test as the first step in a multiple-hurdle, sequential personnel selection system. Moreover, the test battery was fair, in that test scores over-predicted subsequent training performance of African-Americans and women. Overall, the written ATCS aptitude test battery, rooted in Human Factors research on human performance, was an inexpensive, practical, valid, fair, and invaluable tool in rebuilding the controller workforce between 1981 and 1992. In other words, the test battery was a singular success.

References

American Educational Research Association, American Psychological Association, & National Council on Measurement in Education (1999), *Standards for educational and psychological testing,* 6th ed., American Psychological Association, Washington, DC.

Aul, J.C. (1998), 'Employing air traffic controllers, 1981-1992', in D. Broach (ed.), *Recovery of the FAA Air Traffic Control Specialist Workforce, 1981-1995,* DOT/FAA/AM-98/23, pp. 3-6, Federal Aviation Administration, Washington, DC.

Billings, C.E. (1997), *Aviation automation: The search for a human-centered approach,* Lawrence Erlbaum Associates, Mahwah NJ.

Broach, D. (1991), *The impact of test preparation on ATCS aptitude test and screen scores.* Memorandum Report CAMI-MR-AAM523-91/1, Federal Aviation Administration, Civil Aerospace Medical Institute, Oklahoma City, OK.

Broach, D. (1997), *Designing selection tests for the future National Airspace System architecture,* DOT/FAA/AM-97/17, Federal Aviation Administration, Office of Aviation Medicine, Washington DC.

Broach, D., and Brecht-Clark, J. (1993), 'Validation of the Federal Aviation Administration air traffic control specialist Pre-Training Screen', *Air Traffic Control Quarterly,* Vol. 1, pp. 115-133.

Broach, D., Farmer, W.L., and Young, W.C. (1997), *Differential prediction of FAA Academy performance on the basis of race and written Air Traffic Control Specialist aptitude test scores,* DOT/FAA/AM-99/16, Federal Aviation Administration, Office of Aviation Medicine, Washington, DC.

Buckley, E. and Beebe, T. (1972), *The development of a motion picture measurement for aptitude for air traffic control,* FAA-RD-71-106, Federal Aviation Administration, National Aviation Facilities Experimental Center, Atlantic City, NJ.

Charness, N. and Bosman, E.A. (1990), 'Human factors and design for older adults', in J.E. Birren and K. W. Schaie (eds.), *Handbook of the psychology of aging,* 3rd ed., pp. 446-460, Academic Press, New York.

Cobb, B.B. (1967) 'Relationships among chronological age, length of experience, and job performance ratings of air route traffic control specialists' *Aerospace Medicine,* Vol. 39, pp. 119-124.

Cobb, B.B., Lay, C.D., and Bourdet, N.M. (1971), *Relationship between chronological age and aptitude test measures of advanced-level air traffic control trainees,* DOT/FAA/AM-71/36, Federal Aviation Administration, Office of Aviation Medicine, Washington DC.

Cobb, B.B., Nelson, P.L., and Matthews, J.J. (1973), *The relationships of age and ATC experience to job performance ratings of Terminal area controllers,* DOT/FAA/AM-73/7, Federal Aviation Administration., Office of Aviation Medicine, Washington DC.

Collins, W.E. (1970), *Study of air traffic controller selection approaches; PT-5 ltr of 5 Nov 1970,* Memorandum from Chief, Psychology Laboratory (AC-118) to the Associate Administrator for Manpower (PT-1) dated November 18, 1970, Federal Aviation Administration Civil Aerospace Medical Institute, Oklahoma City, OK.

Collins, W., Boone, J., and VanDeventer, A. (1984), 'The selection of air traffic control specialists: Contributions by the Civil Aeromedical Institute', in S. B. Sells, J. T. Dailey, and E.W. Pickrel, (eds.), *Selection of air traffic controllers,* pp. 79-112, DOT/FAA/AM-84/2, Federal Aviation Administration, Office of Aviation Medicine, Washington, DC.

Corson, J.J., Bernhard, P.W., Catterson, A.D., Fleming, R.W., Lewis, A.D., Mitchell, J.M., and Ruttenberg, S.H. (1970), *The career of the air traffic controller – a course of action,* Final report of the Air Traffic Controller Career Committee, Department of Transportation, Office of the Secretary of Transportation, Washington, DC.

Coulter, A.L. (1970a), *Evaluation of air traffic controller selection, PT-1 letter of 18 August 1970,* Memorandum from the Director, Aeronautical Center (AC-1) to Associate Administrator for Manpower (PT-1) dated September 3, 1970, Federal Aviation Administration Civil Aerospace Medical Institute, Oklahoma City, OK.

Coulter, A.L. (1970b), *Review of Phase II of study of controller selection techniques, PT-1 ltr dtd 9/25/70,* Memorandum from Director, Aeronautical Center (AC-1) to Associate Administrator for Manpower (PT-1) dated October 9, 1970, Federal Aviation Administration Civil Aerospace Medical Institute, Oklahoma City, OK.

Czaja, S.J. (1996), 'Aging and the acquisition of computer skills', in W.A. Rogers, A.D. Fisk, and N. Walker (eds.), *Aging and skilled performance: Advances in theory and applications,* pp. 201-220, Lawrence Erlbaum Associates, Mahwah NJ.

Dailey, J.T. (1972), *New ATC battery of selection tests,* Memorandum from Chief, Analytic Branch (AAM-310) to AMN-1 dated December 8, 1972, Federal Aviation Administration Civil Aerospace Medical Institute, Oklahoma City, OK.

Dailey, J.T. and Pickre, E.W. (1977), *Development of new selection tests for air traffic controllers,* DOT/FAA/AM-77/25, Federal Aviation Administration, Office of Aviation Medicine, Washington, DC.

Dailey, J.T. and Pickrel, E.W. (1984a), 'Development of the Multiplex Controller Aptitude Test', in S. B. Sells, J. T. Dailey, and E. W. Pickrel (eds.), *Selection of air traffic controllers,* pp. 281-298, DOT/FAA/AM-84/2, Federal Aviation Administration Office of Aviation Medicine, Washington, DC.

Dailey, J.T., and Pickrel, E.W. (1984b), 'Development of the air traffic controller Occupational Knowledge Test', in S. B. Sells, J. T. Dailey, and E. W. Pickrel (eds.), *Selection of air traffic controllers,* pp. 299-322, DOT/FAA/AM-84/2, Federal Aviation Administration, Office of Aviation Medicine, Washington, DC.

Della Rocco, P.S. (1998), 'Air Traffic Control Specialist screening programs and strike recovery', in D. Broach (ed.), *Recovery of the FAA Air Traffic Control Specialist Workforce, 1981-1995,* pp. 17-22, (DOT/FAA/AM-98/23, Federal Aviation Administration, Washington, DC.

Dille, J.R. (1970), *Comments on Dr. Colmen's report,* Memorandum from Chief, Civil Aeromedical Institute to Director of the Mike Monroney Aeronautical Center (AC-1),

Federal Aviation Administration Civil Aerospace Medical Institute, Oklahoma City, OK.

Education and Public Affairs, Inc. (1970), *Review and evaluation of present system for selection of air traffic controllers,* Report on Phase I, Task 1 for FAA Contract DOT-FA70WA-2371 dated July 31, 1970, Washington, DC.

Equal Employment Opportunity Commission (1978), *Uniform Guidelines on Employee Selection Procedures,* 29 CFR, Part 1607.

Fozard, J.L. and Nutall, R.L. (1971), 'General aptitude test battery scores for men differing in age and socioeconomic status', *Journal of Applied Psychology,* Vol. 55, pp. 372-379.

General Accounting Office (1997), *Aviation safety: Opportunities exist for FAA to refine the controller staffing process,* GAO/RCED-97-84, Washington, DC.

Ghiselli, E., Campbell, J., and Zedeck, S. (1981), *Measurement theory for the behavioral sciences,* Freeman, San Francisco.

Gottfredson, L.S. (1994), 'The science and politics of race-norming', *American Psychologist,* Vol. 49, pp. 955-963.

Harding, B.M. (1970), *Study of controller selection technique,* Memorandum from Associate Administrator for Manpower (PT-1) to Regional and Center Directors, Director, Air Traffic Service dated September 25, 1970, Federal Aviation Administration, Civil Aerospace Medical Institute, Oklahoma City, OK.

Henry, J.H., Ramrass, M.E., Orlansky, J., Rowan, T.C., String, J., and Reichenbach, R.E. (1975), *Training of U.S. air traffic controllers,* IDA Report R-206, Institute for Defense Analysis, Arlington, VA.

Kuder, G.F. and Richardson, M.W. (1937), 'The theory of the estimation of test reliability', *Psychometrika,* Vol. 2, pp. 151-160.

Lewis, M.A. (1978), 'Objective assessment of prior air traffic control related experience through the use of the Occupational Knowledge Test', *Aviation, Space and Environmental Medicine,* Vol. 49, pp. 1155-1159.

Lilienthal, M.G., and Pettyjohn, F.S. (1981), *Multiplex Controller Aptitude Test and Occupational Knowledge Test: Selection tools for air traffic controllers,* NAMRL Special Report 82-1, Naval Aerospace Medical Research Laboratory, Pensacola, FL.

Manning, C.A. (1998), 'ATCS Field training programs, 1981-1992', in D. Broach (ed.), *Recovery of the FAA Air Traffic Control Specialist Workforce, 1981-1995,* pp. 23-32, DOT/FAA/AM-98/23, Federal Aviation Administration, Washington, DC.

Manning, C.A., and Broach, D. (1992), *Identifying ability requirements for operators of future automated air traffic control systems,* DOT/FAA/AM-92/26, Federal Aviation Administration, Office of Aviation Medicine, Washington, DC.

Milne, A. M. and Colmen, J. G. (1977), *Selection of air traffic controllers for Federal Aviation Administration,* Final report under FAA contract DOT-FA70WA-2371, Education and Public Affairs, Inc., Washington, DC.

Morrow, D., Leirer, V., Altieri, P., and Fitzsimmons, C. (1994), 'When expertise reduces age differences in performance', *Psychology and Aging,* Vol. 9, pp. 134-148.

Nickels, B.J., Bobko, P., Blair, M.D., Sands, W.A., and Tartak, E.L. (1995), *Separation and Control Hiring Assessment (SACHA) final job analysis report,* Deliverable Item 007A under FAA contract DTFA01-91-C-00032, Federal Aviation Administration, Office of Personnel, Washington DC.

Rock, D.B., Dailey, J.T., Ozur, H., Boone, J.O., and Pickrel, E.W. (1982), *Selection of applicants for the air traffic controller occupation,* DOT/FAA/AM-82/11, Federal Aviation Administration Office of Aviation Medicine, Washington, DC.

Rock, D.B., Dailey, J.T., Ozur, H., Boone, J.O., and Pickrel, E.W. (1984a), 'Validity and utility of the ATC experimental tests battery. Study of Academy trainees, 1982', in S. B. Sells, J. T. Dailey, and E. W. Pickrel (eds.), *Selection of air traffic controllers,* pp. 459-502, DOT/FAA/AM-84/2, Federal Aviation Administration Office of Aviation Medicine, Washington, DC.

Rock, D.B., Dailey, J.T., Ozur, H., Boone, J.O., and Pickrel, E.W. (1984b), 'Conformity of the new experimental test battery to the Uniform Guidelines on Employee Selection Requirements' , in S. B. Sells, J. T. Dailey, and E. W. Pickrel (eds.), *Selection of air traffic controllers,* pp. 503-542, DOT/FAA/AM-84/2, Federal Aviation Administration Office of Aviation Medicine, Washington, DC.

Sackett, P.R. and Wilk, S.L. (1994), 'Within-group norming and other forms of score adjustment in pre employment testing', *American Psychologist,* Vol. 49, pp. 929-954.

Salthouse, T.A. (1990), 'Influence of experience on age differences in cognitive functioning', *Human Factors,* Vol. 32, pp. 551-569.

Salthouse, T.A., Babcock, R.L., Skovronek, E., Mitchell, D.R.D., and Palmon, R. (1990), 'Age and experience effects in spatial visualization', *Developmental Psychology,* Vol. 36, pp 128-136.

Schaie, K. W. (1988), 'Ageism in psychological research', *American Psychologist,* Vol. 43, pp. 179-183.

Schneider, B., and Konz, A.M. (1989), 'Strategic job analysis', *Human Resource Management,* Vol. 28, pp. 51-53.

Seymour, R.T. (1988), Why plaintiffs' counsel challenge tests, and how they can successfully challenge the theory of 'validity generalization', *Journal of Vocational Behavior,* Vol. 33, pp. 331-364.

Sharit, J. and Czaja, S.J. (1994), 'Aging, computer-based task performance, and stress: Issues and challenges', *Ergonomics,* Vol. 37, pp. 559-577.

Trites, D.K. (1963), 'Ground support personnel', *Aerospace Medicine,* Vol. 34, pp. 539-541.

Trites, D.K. and Cobb, B.B. (1964a), 'Problems in air traffic management: III. Implications of training-entry age for training and job performance of air traffic control specialists', *Aerospace Medicine,* Vol. 35, pp. 336-340.

Trites, D.K. and Cobb, B.B. (1964b), 'Problems in air traffic management: IV. Comparison of pre-employment, job-related experience with aptitude tests as predictors of training and job performance of air traffic control specialists', *Aerospace Medicine,* Vol. 35, pp. 428-436.

Tsang, P.S. and Voss, D.T. (1996), 'Boundaries of cognitive performance as a function of age and flight experience', *International Journal of Aviation Psychology,* Vol. 6, pp. 359-377.

VanDeventer, A.D. and Baxter, N.E. (1984), Age and performance in air traffic control specialist training, in A. D. VanDeventer, W. E. Collins, C. A. Manning, D. K. Taylor, and N. E. Baxter (eds.), *Studies of poststrike air traffic control specialist trainees: I. Age, biographic factors, and selection test performance related to Academy training success*, pp. 1-6, DOT/FAA/AM-84/6, Federal Aviation Administration, Office of Aviation Medicine, Washington DC.

Wickens, C.D., Mavor, A.S., and McGee, J.P. (1997), *Flight to the future: Human factors in air traffic control*, National Academy Press, Washington DC.

Wormser, E. (1970), *Letter from the Special Assistant to the Associate Administrator for Manpower to Dr. Joseph Colmen date December 14, 1970*, Federal Aviation Administration, Civil Aerospace Medical Institute, Oklahoma City, OK.

Young, W.C., Broach, D., and Farmer, W.L. (1996), *Differential prediction of FAA Academy performance on the basis of gender and written Air Traffic Control Specialist aptitude test scores*, DOT/FAA/AM-96/13, Federal Aviation Administration, Office of Aviation Medicine, Washington, DC.

Chapter 10

Implementation of Critical Incident Stress Management at the German Air Navigation Services

Jörg Leonhardt

Introduction

Disasters and other events causing devastating damage typically result in intensive media coverage and extended public discussions. Apart from reporting on the extent of the disaster and its victims and causes, the public has also increasingly focused its attention on helpers, members of rescue teams and other emergency service personnel. The consequences of such missions on the emotional state of members of rescue teams and the question of how to cope with such situations have increasingly gained importance. This topic has also been discussed within rescue organizations and, consequently, the self-understanding of these professional groups has changed. These discussions have concluded that a special 'method of treatment' for members of certain professional groups is required, efficient and different from measures commonly taken in psychotherapy. One of the methods fulfilling these criteria has been introduced in several organizations under the name of Critical Incident Stress Management (CISM).

Rationale for Critical Incident Stress Management

Air traffic controllers work in an environment that demands constant vigilance and involves situations requiring total attentiveness and concentration and are consequently exposed to a high level of professional stress. Controllers need to rely on the smooth functioning of the ATC system. Additional stress may arise if parts of this system do not interface smoothly. Tensions within the social and family life also have the effect of causing additional stress. These additional stress factors are referred to as cumulative stress, while occupational stress caused through the professional environment is referred to as general stress. Any malfunction of the air traffic system to operate smoothly and safely may cause critical incidents that lead to a specific form of stress, namely critical incident stress or critical incident stress reactions.

Stress

Stress is not always negative. Quite on the contrary, human beings need a certain healthy level of stress to operate effectively. The crucial factor is how people handle positive stress *(eustress)* and negative stress *(distress)*. Stress may be defined as a condition of physical and psychological arousal, and broken down into two categories, the *normal pathway* and the *pathological pathway*. The normal pathway includes *general stress* and *critical incident stress*. The pathological pathway includes *cumulative stress* and *post-traumatic stress disorder* (PTSD). The distinction made between these forms of stress hinges on whether the form of stress is normal or pathological. Note that critical incident stress is classed as normal stress meaning that its initial effects are not necessarily pathological.

Both forms of stress can have adverse affects in the long-term if not managed properly. Even the forms of stress in the normal pathway must be actively managed in order to avoid any adverse effects. Both normal pathways can develop into pathological forms of stress if not addressed adequately. General stress can, in the absence of adequate management and change strategies, cause cumulative stress and ultimately develop into burnout syndrome. A critical incident (CI) that is not adequately resolved and the associated critical incident stress can cause a PTSD.

The issue of stress and stress management is of crucial importance for the 'well-being' and professional development of air traffic controllers. Controllers should consequently be informed about stress and methods to cope with it. Stress management should be an integral part of air traffic controller training. Besides, refresher courses should be provided to controllers throughout their careers.

Critical Incidents and Critical Incident Stress

Throughout their private and professional lives individuals may continually be confronted with situations that are felt as critical, dangerous or threatening. Such situations might involve traffic accidents, natural disasters, illnesses or even the death of close members of the family. They might also involve critical situations occurring in a professional context.

Individuals have different ways of experiencing and coping with critical situations. However, we can assume that these situations may generally cause stress reactions that show through symptoms such as: increased pulse, excessive sweating, rapid breathing, adrenalin rushes, increased heart rate, anxiety, trembling, and a whole range of other physical reactions. The stress reactions will subside or cease over time, depending on their impact and the individual's capacity to manage stress.

If stress reactions persist, individual coping strategies are normally applied in order to deal with them. A person's capacity to cope with or manage stress reactions on his/her own depends on the critical incident itself and his/her available capacity to manage stress. If the individual concerned is psychologically and physically healthy and has a stable professional and private environment, he/she will have a greater capacity to deal with stress than in less favorable conditions. The individual capacity to deal with stress depends on the level of cumulative stress and the general ability to cope with stress. With professional self-care, active stress management can become part of daily life so that an individual's capacity to deal with critical incidents is significantly higher. Individuals have a major influence on their capacity to deal with stress.

There are, however, situations in which an individual's capacity to deal with stress is stretched or events occur with such a huge potential for trauma that the individual is unable to cope with them on his/her own. Since the inability to cope with stress can lead to a post-traumatic stress disorder and thus become pathological, it seems logical to offer professional support to individuals in comparable circumstances. To this end, Critical Incident Stress Management has been developed which is based on the work of a peer who helps the individual cope with the situation and deal with the critical incident stress reactions.

Occurrence of Critical Incidents and PTSD

A person might be confronted at any time throughout his or her life with a critical situation with the potential to lead to traumatic reactions. The International Critical Incident Stress Foundation (ICISF) has compiled statistics for the US concerning the probability of experiencing traumatic reactions to critical situations:

- More than 80 per cent of US citizens are exposed to trauma. About 9 per cent of those exposed develop PTSD.
- 40 per cent of US children are exposed to trauma.
- 14 per cent of violent crimes occur at work. Violence is directly associated with PTSD.
- 16 per cent of Vietnam veterans developed PTSD.
- 31 per cent of urban fire fighters report PTSD symptoms.
- 10-15 per cent of law enforcement personnel develop PTSD.
- For the police, the risk of suicide is 8.6 times greater than the risk of accidental death.
- Up to 35 per cent of disaster victims develop PTSD.
- An estimated 1.5 million people in New York City may need counseling post-September 11th (US Center for Mental Health Services).
- Over 50 per cent of disaster workers can be expected to develop significant post-traumatic distress (Wee and Myers, 2001).

Although these statistics pertain to the United States of America, they are relevant for Europe in many respects. Quick and efficient management of critical incidents and situations with potential for trauma is a form of secondary prevention, which stops symptoms from deteriorating or becoming chronic. CISM offers severely traumatized individuals the opportunity to cope with the stressful event through structured discussions and to cope with the associated critical incident stress reactions.

Critical Incidents in Air Traffic Management

The necessity to have a well-functioning Critical Incident Stress Management system in place has been recognized and acted upon only in the wake of a serious accident. [21] Although it is human nature that remedial actions are predominantly implemented in response to and not in prevention of disasters, it is quite important to have an operative CISM in place before disasters happen. In the case of an acute crisis, the only available resources are those that are already in place, well established and rehearsed. Programs and procedures must be designed, established, and tested well in advance – having to learn crisis management in the midst of a crisis will inevitably cause a deterioration of the situation. DFS (Deutsche Flugsicherung GmbH) saw the need to develop a CISM model without delay and decided in 1998 to develop and implement a CISM system, not to wait for a disaster to happen. Since 2000, a network of highly qualified CISM peers has been in place, offering professional assistance to colleagues at all DFS establishments.

Air traffic controllers belong to a professional group which works in a high safety-critical domain – a professional environment in which, due to the complexity of the working situation and the immediate and catastrophic consequences of errors or malfunctions, the likelihood of encountering critical incidents is substantially higher than for most other professional groups. ATC staff may be confronted with critical incidents in a variety of situations – loss of separation, air proximities, runway incursions, loss of overview and control, malfunctioning of technical equipment (radar, R/T communication), thunderstorms, aircraft in distress, emergency landings, crashes or even collisions between aircraft. Such critical incidents have the potential to cause the associated critical incident stress (CIS) reactions, e.g. excessive sweating, trembling hands, adrenalin rushes or the after-effects of nightmares, sleeplessness, loss of appetite or behavioral changes such as aggression, depression or other CIS reactions.

Even though some controllers are more prone to stress reactions than others, depending on their personality and their individual ability to cope with the critical incident stress most controllers are likely to encounter a critical incident at least once in their career and might develop critical stress reactions. In such a

---

[21] Canada initiated a CISM system after the DC 10 crash at Sioux City airport; the Netherlands started to develop CISM after a Boeing 747 crashed into an apartment building near Amsterdam and after a City Hopper aircraft crashed near Amsterdam airport; Finland started after an MD 80 landing in heavy fog failed to notice a maintenance car on the runway; after the mid-air collision over Lake Constance in July 2002 Skyguide built a CISM system in their organization.

professional environment, provision has to be made not only for continuous stress management to prevent or reduce cumulative stress, but also for a system to deal with critical incidents and associated critical incident stress reactions.

On account of the specific requirements of this profession, controllers have a personality profile that typically encompasses the following characteristics: target-, control- and action-orientation, commitment, high degree of stress tolerance, strong sense of responsibility, and ability to take decisions.

Such a personality profile contributes to the ability of air traffic controllers to execute their functions successfully. Typically, individuals with the described personality profile are accustomed to handling critical situations and if they detect reactions in themselves, which are contrary to the characteristics listed, they often encounter feelings of insecurity and irritation. CIS reactions such as apathy, aimlessness, or loss of control can cause individuals with the above personality profile to question their own professional image and self-esteem.

The importance of a functioning and supportive social environment, including family, colleagues, friends and acquaintances, in order to help concerned individuals to cope with any crisis must not be underestimated. However and without wanting to denigrate the assistance brought about by family and friends, they usually do not act on a professional basis. Additional assistance through a qualified professional is often necessary in order to effectively reduce and deal with critical incident stress reactions. Professionalism in this context means being well versed in stress and stress management and being aware of developments after a critical incident, thus being able to assess the CIS reactions triggered by such incidents. Professionalism also means that the discussion must follow a pre-defined fixed structure. The assistance through so-called peers, i.e. colleagues trained in CISM, has proven to be the most helpful approach.

The worst way of reacting to stress, less common now than in the past but still observable, is for the controller to speak to nobody about their situation and reactions and to 'deal with the crisis' unaided. This behavior is quite understandable, since the mentioned personality profile contributes to the feeling of insecurity and irritation when facing one's own stress reactions. Insecurity inhibits concerned individuals from contacting a peer and encourages them to withdraw and deal with the situation on their own. Their self-understanding is undermined, the mechanisms that previously served them well no longer work and they feel impotent in the face of their reactions.

In addition to the question of self-understanding, critical incidents and the associated stress reactions also cast doubt on their professional image, or at least the predominating professional image. Controllers are expected to be 'in control' and any reaction or event seemingly contradicting this facility can make them question their self-esteem and professional aptitude. At the same time they might register reactions and behavior patterns which either they do not recognize for what they are, or which they expect, in the light of earlier experiences, to pass after some time. A negative perception of such reactions and changes in behavior, either by the individual or his/her colleagues, can aggravate feelings of confusion or further deter them from seeking help, since they do not want to unfold their problems in front of their colleagues. Uncertainties in the course of operational duties and anxieties about losing his/her job become more acute and much energy is required to suppress such uncertainties and other undesirable reactions such as shaking hands or sweating. Frequently this has the potential of trapping the victims in a vicious circle and causing the CIS reactions to become chronic. In the worst case it may lead to post-traumatic stress disorder (PTSD) with the consequence of becoming unfit to execute the job.

Adequate professional support in assimilating critical incidents and coping with the stress reactions associated with the incident is the most effective way of effectively assimilating critical incidents and preventing PTSD. This presupposes that there is a recognized concept for support in the management of critical situations (such as CISM), that this management concept is supported and promoted by executive staff and that there are trained peers who are able to implement the CIS measures in a professional and committed way. Employees must be familiar with the CISM program and information about stress and critical incident stress. Such information and knowledge helps them to assess their own reactions and enables them, through self-assessment, to make the appropriate adjustments.

Critical Incident Stress Management

A psychological *crisis* may be considered as an acute reaction to a critical incident, trauma or disaster and is typically characterized by the following symptoms: (a) the individual's psychological balance is disrupted; (b) the usual coping mechanisms fail; and (c) there is evidence of significant distress, impairment, and dysfunction.

Critical incidents may be defined as 'situations faced by a person causing him or her to experience unusually strong emotional reactions' (EUROCONTROL, 1997). The definition emphasizes two important points: firstly, that *any situation* has the potential of causing unusual reactions; and secondly, that the critical incident is not defined in terms of the incident, but always in terms of the individual's *unusual reactions*.

Some events may, from an objective point of view, not appear particularly critical but still cause critical incident stress reactions in the individual concerned. Similarly, an event with obvious potential to cause significant distress may cause no unusual reactions. Whether CIS occurs generally depends on how the controller has experienced the situation. If the situation was critical, but he/she felt to have been 'in control of the situation' long-term CIS reactions may be completely absent since no feeling of helplessness or impotence prevailed. On the other hand, even relatively uncritical situations can cause feelings of helplessness or impotence and therefore lead to CIS reactions. Furthermore, the ability of any individual to deal with critical situations depends on the mental and psychological resources he/she possesses when the situation arises. Furthermore, past critical incidents that have not been adequately resolved adversely affect the situation and increase the potential for CIS reactions.

The second fundamental aspect in the above definition is the significance of unusual reactions. Everyone develops specific reactions in critical situations so that we all have a 'normal' reaction to stress with which we are fairly familiar. Reactions are unusual if they differ from the norm – someone who otherwise is rather active, seems withdrawn; a committed and motivated colleague becomes apathetic and de-motivated. It is important both for controllers and for their supervisors and colleagues to be on the lookout for unusual reactions or patterns of behavior in the wake of a critical situation.

Critical Incident Stress (CIS)

Critical incident stress may be defined as 'the psychological, cognitive, emotional and/or behavioral reaction to a critical incident. This reaction is a normal human reaction to an abnormal event' (EUROCONTROL, 1997).

It is difficult for individuals who experience unusual, persistent and intense reactions to interpret these reactions correctly. More often than not they will seek the causes in themselves and, even though perfectly understandable, this attitude is a significant obstacle to coping with the stress reactions. It is the responsibility of the peer to help concerned individuals to put their reactions into perspective by making it clear that these reactions are normal and a direct consequence of the critical incident. This will help them to reframe their self-perception: if reactions can be considered a normal and healthy response to an abnormal event they can be accepted more easily which is a prerequisite for starting to deal with the reactions.

The peer may use the following rationale to explain that stress reactions are perfectly normal: a critical incident triggers an instinctive and immediate physiological reaction, the *sympathetic* reaction or 'fight or flight' reflex with physical reactions beyond conscious control. Such sympathetic reactions include: higher pulse, increased blood pressure, expanded lung capacity, constricted blood vessels, slower digestion, and sweating. Thus, Critical Incident Stress reactions are biologically determined, perfectly normal and cannot be consciously controlled.

The converse of a sympathetic reaction is a *para-sympathetic* reaction, which triggers counter-reactions. Numerous overwhelming impressions in a critical incident flow through the sensory channels (hearing, sight, touch, taste) directly into our subconscious. Sometimes the para-sympathetic reactions may be insufficient to resolve the event due to the sheer intensity of the sensory and emotional impressions. In this case structured support will be necessary to resolve the critical incident as a traumatic event. Constant replaying of the situation or having nightmares about it are self-regulatory attempts to make sense of and resolve the sensory impressions related to the CI and should therefore also be considered as normal reactions.

Potentially Traumatizing Events and Stress Reactions

An event is has the potential of causing a trauma if one or more of the following criteria apply:

- feeling of helplessness/powerlessness;
- feelings of guilt;
- significant personal disruption;
- high degree of identification;
- high intensity of the event;
- threat to life and limb;
- deaths in connection with the exercise of a profession.

Stress reactions typically occur along different stages: *acute stress reactions* typically occur during and up to 24 hours after the event and cause massive stress reactions; typically, individual coping mechanisms are discernible. *Acute stress disorder* may be experienced between 24 hours and 4 weeks after the event during which massive stress reactions persist or recur with undiminished intensity and individual coping mechanisms do not seem to be working. If the stress symptoms persist for more than four weeks the stress disorder may become chronic, in which case the recovery process becomes significantly more demanding; in this case the term *chronic stress reaction* is used.

In the worst case, the stress reaction may develop into a *post-traumatic stress disorder* (PTSD) – 'the pathological end result of an unresolved critical incident stress reaction' (Mitchell and Everly, 1997). It is the most serious form of stress reaction and is typically characterized by chronic reactions such as flashbacks, agitation and recurring intense images and impressions. PTSD generally causes significant disruptions to one's personality and well-being and often results in the individual's inability to continue his/her profession.

The frequency of occurrence of PTSD has been estimated as 1-3 per cent (general population); 9 per cent (urban adolescents); 15-20 per cent (Vietnam veterans); 15-32 per cent (emergency services personnel).

Crisis Intervention and its Distinction from Psychotherapy

Crisis intervention must be clearly distinguished from psychotherapy. It offers rapid and intensive support in a crisis and in all phases of the respective CISM method are focused solely on the critical incident: 'as physical first aid is to surgery; crisis intervention is to psychotherapy' (Mitchell and Everly, 1993). CISM does not provide the series of follow-up discussions that is generally a feature of psychotherapy. If in the course of the discussion it appears that there is a need for further counseling in other areas of life, the peer will propose internal or external specialist services such as psychotherapy. CISM is provided by trained peers, is solution-oriented in terms of approach and has the ultimate goal of facilitating a positive attitude and overcoming an acute emotional distress. Crisis intervention is directed towards stabilizing, mitigating, mobilizing, normalizing, and restoring adaptive functions (Mitchell and Everly, 1997). To this end, crisis intervention is based on the principles of simplicity, brevity, innovation, pragmatism, proximity, immediacy, and expectancy.

CISM Methods

CISM measures should exclusively be provided to homogeneous groups since heterogeneous groups may traumatize individuals who were previously unaffected and may cause discussions to focus on the event and criticisms of persons or institutions, which makes the situation unmanageable for those responsible for implementing the measures. The basic principle underlying all measures is that they should help individuals to assess their reactions and deal with them. All CISM measures should be carried out by a peer and/or mental health professional (MHP) trained in the appropriate methods.

The Peer Model

Critical Incident Stress Management is based on the involvement of *peers* from the same professional environment. The peer model offers several key advantages:

- Colleagues are often accepted more readily than individuals from outside the profession such as psychologists, doctors, members of the clergy and pastoral workers.
- Peers from the same occupational group appreciate the individual's CIS reactions better, since they may understand them better – the peer may even have experienced similar reactions.
- Peers have a better understanding of the facts and circumstances of the event, since they are from the same work environment.
- Often, an action to address the situation is more credible and more easily accepted if proposed by a peer.
- The involvement of health professionals (medical doctors, psychologists) may cause the impression that the individual is 'ill', which is not the case with peers.
- Peers can often be called in more quickly and without complicated or bureaucratic procedures.

The use of peers can also involve certain disadvantages, e.g.:

- The identification with the colleague and the critical incident may trigger critical incident stress reactions amongst peers. In the worst case, the peer's professional aptitude may be impaired.
- The fact-finding phase might become overly comprehensive, with both parties becoming too involved in discussions of detail and procedure.
- Emotional phases in CISM, namely the *reaction phase*, may be too brief or even skipped entirely, since both parties may consider them as 'threatening'.
- People or procedures may be left unduly exposed to criticism.
- Too little consideration is given to self-motivation as a peer.
- The limits of the discussion are not respected – often it may be difficult to avoid straying into personal matters.

However, the advantages clearly outweigh the disadvantages and these can often be limited or even eliminated through good peer management.

SAFER Model

The SAFER model is a step-by-step model for supporting people in situations of acute crises. It is easy to apply, and very effective since it reduces the initial CIS reactions. The SAFER model is generally implemented immediately after the critical incident and at the location where the event took place. SAFER aims at stabilizing the individual concerned and identifying the need for further support through application of the following steps:

- *Stabilizing* involves distancing the individuals concerned from the direct impressions and stimuli of the critical incident, i.e. interrupting all visual, olfactory and auditory stimuli.
- *Acknowledging* involves making it clear to the individuals concerned that the experience they have just encountered was a critical incident that has triggered, or may trigger, CIS reactions.
- *Facilitating* involves helping the individuals concerned to understand and assimilate their experiences and reactions and promoting understanding in order to 'normalize' the CI-related reactions.
- *Encouraging* involves helping the individuals concerned to develop suitable coping strategies, to promote their own coping resources and drawing up a list of useful coping strategies and tactics.
- *Recovery or referral* involves determining either that the individuals concerned have regained their stability or that they need further assistance/support. In the latter case, further help can be provided or the necessary contacts can be made.

Defusing

Defusing is a shortened form of debriefing, conceived as a group exercise, and should take place on the same day as the event, no later than 8-12 hours after the end of the critical incident. Defusing is carried out by at least two people, e.g. two peers, with one peer leading the defusing. Defusing should alleviate reactions and tension and allow the individuals to resume normal functions. Defusing can help to identify individuals in need of further support, while at the same time advising them of the necessity for and availability of such support. Defusing typically comprises three phases: (a) introducing the team members; (b) exploring the experience; (c) providing information on how to manage stress. It has proven worthwhile at the end of the defusing session to address the possibility of and offer further support services and issue a handout with the telephone numbers of peers and information about stress and coping with CIS reactions.

Debriefing

The critical incident stress debriefing (CISD) forms the basis of CISM measures. It lasts approximately three hours and is designed as a group exercise led by a mental health professional (MHP). The debriefing may be described as 'a peer-driven and MHP led group discussion of a traumatic event' (Mitchell and Everly, 1997). The CISD team comprises one MHP and three or more peers. The debriefing should take place between four and six weeks after the event and, as with all measures, the group should be homogeneous and the participants should meet the criteria relating to the potential trauma. The debriefing has seven phases (Mitchell and Everly, 1997):

The *introduction phase* is used to

- introduce the leader and identify support staff;
- explain the purpose of CISD;
- describe the CISD process;
- motivate participants;
- lay out ground rules or guidelines;
- preview and answer first questions.

The *fact phase* addresses questions such as

- Who are you?
- What was your role in the incident?
- Please relate a brief description of your experience during the event!
- Who arrived first; what happened?
- Who came in next; what happened?

The *thought phase* addresses

- the first or most prominent thought that occurred;
- unusual or discomforting thoughts.

During the *reaction phase* the following questions are raised:

- What was the worst thing about the experience?
- Was there anything you wish that could be erased?
- Did anything occur that would have made the situation a bit easier to manage had it not happened?

The *symptom phase* addresses:

- How has this experience affected your life?
- Did you experience signs of distress at the time of the event?
- Did you experience signs of distress during the following days?
- Are you experiencing residual effects now?

The teaching phase intends to:

- discuss the signs of distress brought up by the group;
- teach stress management tactics according to the needs of the group;
- identify positive aspects or important lessons learned from the experience.

The *re-entry phase* serves to:

- integrate;
- explain;
- answer questions;
- summarize;
- thank, acknowledge, validate, encourage;
- provide a sense of closure to the acute phase.

The debriefing is a proven method since it provides the individuals concerned with a structured framework, which enables them to assess and assimilate their reactions and feelings. Communicating and sharing with the other participants brings relief, removes the feeling of isolation and helps the individual to resolve the event.

Demobilization

Demobilization is a post-incident discussion which marks the transition between an incident with the potential to traumatize and a return to work or leisure time.

Demobilization involves providing information to large groups and comprises two parts: in the first part, information about the critical incident stress is provided (approximately 10 minutes); the second part consists of a rest period with food and drinks (approximately 20 minutes). The demobilization should be conducted immediately after the incident, either on the next day or during the next shift. It is designed to help the people involved in the incident to better resolve the event and inform them about critical incident stress and its possible consequences. At the same time, information should be given about the availability of assistance, follow-ups and CISM peers who can be consulted. Mitchell warns that 'no one should return to disaster work on the same day as the demobilization. A minimum of six hours should pass before a unit is reactivated for service at the disaster site.'

Crisis Management Briefing

Crisis management briefing (CMB) is a relatively recent method developed for people with secondary and tertiary traumata, who were not directly involved in the incident but were nevertheless affected by the event. It is therefore different from demobilization, which is a post-incident discussion held directly for individuals involved in a critical incident. The crisis management briefing is a large group process that is used to provide practical and often anxiety-reducing information to large groups of people who have been exposed to traumatic events (Mitchell and Everly, 1997). The CMB team ideally comprises a CISM team member (peer), an MHP and a credible representative of the organization.

Family/Organizational Support

This involves all the measures the family and the organization need to be aware of in order to provide support. This may involve checklists, recommended behavior, information, care by peers, referral and general recommendations. It is quite important that peers cooperate with family members in order to complement measures taken within the organization with the support individuals receive through their family.

Critical Incident Stress Management at DFS

Specific Requirements and Conditions in Air Traffic Management

In contrast to rescue services, which are typically called in after an accident has occurred, air traffic controllers are usually directly involved in the critical incident, sometimes even as a factor contributing to the event. Therefore the critical incident is experienced not only in terms of its consequences but also through the controller's interaction through which the situation was defused or possibly even brought about. This is why the process of establishing joint responsibility and reducing individual feelings of guilt plays a key role for controllers in most CISM discussions. Even if such sessions do not address objective responsibility, it is hugely important to deal with the feelings of guilt individuals perceive. Feelings of guilt need to be considered and identified as critical incident stress reactions and thus as normal reactions, which should not be confused with guilt.

In contrast to tower and ground controllers, controllers working in arrival/departure or En-route control facilities perceive the event predominantly through the radar screen and R/T. Accordingly, mental images of the event may be distorted, although they are usually experienced by the individuals concerned as if they were real and consequently affect emotions and thought processes. Therefore it is absolutely essential to reconcile the mental images with the reality of what actually happened during CISM sessions.

A second important point is that a critical incident itself does not in most cases lead to an accident or incident. Fortunately losses of separation do not, in most cases, result in accidents or incidents but have the potential to be perceived as a critical incident. The general public, however, will not learn of the critical incident, since it will not be reported in the media – as far as the general public is concerned, there was no critical incident. Consequently family members and friends will not be aware of the incident and hardly understand the gravity of the situation for the controller concerned, and colleagues are often the only group who acknowledges the severity of the experience. Peers can understand the CIS reactions caused by a loss of separation, even if there is no accident, and can be extremely useful in normalizing stress reactions, given the credibility they have with the individual concerned.

Specific Requirements and Conditions at DFS

When DFS decided to implement a CSIM system, the challenge was to make peers available at all locations. Furthermore, the differences between tower and ground control on the one hand and arrival/departure and En-route control on the other hand had to be taken into consideration through independent staffing. At the 17 DFS units 65 peers have been trained so that the availability of peers is guaranteed at all times. In general, area control centers (ACC) are staffed with five peers and tower control centers with two peers. The peers are well trained and undergo regular refresher courses.

Peer Selection and Training

At DFS, all peers are selected by their colleagues to ensure they are fully accepted – no assessment-based selection is applied. In the implementation phase of CISM, emphasis is deliberately placed on the confidence controllers have in the peers. Confidence is the basis of acceptance and it was felt that only a high level of acceptance would ensure that the CISM facilities would be widely used. There is no financial remuneration for acting as a peer.

Peers initially attend basic training, which focuses on self-assessment, motivation and social skills. These key points are addressed, reflected on and discussed. Peers are introduced to the CISM task itself and provided with the necessary knowledge. A further aim of the basic training is to develop a CISM 'culture' among the peers, a collective understanding of the task and a form of system of ethics to govern their approach. The possibilities and limitations of their role as peers are examined and determined.

After the basic training the CISM basic course, based on the standards of the International Critical Incident Stress Foundation (ICISF), follows. The CISM Basic Course provides the essentials of CISM, i.e. the knowledge required regarding stress and critical incident stress. Course participants are familiarized with CISM methods, to enable them to work independently and in accordance with the SAFER model. They have the ability to carry out a CISM defusing and to take part in a CISM debriefing exercise as peers. A CISM Basic Course is a prerequisite for working as a peer. DFS opted to work and provide training on the basis of the ICISF standards since these offer a high level of standardization, are based on experience and a scientific background and provide convincing results.

Regular refresher courses are offered in order to maintain the level of training of peers and to increase their competence. In the refresher courses, integral parts of CISM are reviewed and enriched in the form of add-ons. Methods of leading discussions might, for example, be offered as an add-on.

An annual meeting of all peers takes places during a dedicated CISM forum; the purpose of this is to exchange experience and review CISM within the DFS and in other organizations that use CISM.

Three years after receiving the basic training, the peers at DFS were trained by Prof. J.T. Mitchell who led both the CISM Advanced Course and the Assisting Individuals in Crisis Course. The peers are highly trained and are awarded ICISF certificates in all courses.

Experiences at DFS

At DFS locations with a high number of operational staff, peers were more frequently called upon and therefore have more experience than peers at establishments with fewer air traffic controllers. Owing to the successful work of the peers to date, acceptance has increased both among colleagues and executive staff and has resulted in the fact that CISM is offered as the standard procedure throughout all locations. A blame culture, still present in the early stages, has disappeared entirely and both CISM and the peers are accepted and widely used.

The annual CISM forum is a three-day event aiming at an exchange of views and experiences between the peers. In addition to the peers the forum is also attended by external advisers with a wide range of experience in implementing CISM and CISM procedures in the various organizations or the application of CISM measures. Typically, a selected topic is discussed in detail during the CISM forum, such as 'leading discussions', 'special therapeutic procedures', or 'case studies from trauma centers'. This intends to provide peers with an insight into other procedures and alternative methods in order to broaden their professional horizons.

Peer Debriefing

Since by definition peers are active air traffic controllers, there is a certain risk that they may develop strong feelings of identification with the individuals they are aiding; in the long run an overly intense identification has the potential of compromising their professional abilities. Peers are offered debriefings on a regular basis in order to address this problem. Peer debriefings are conducted as group discussion and supervision exercises; the peers' own emotions are discussed and specific working methods are reflected. In addition, every peer may request a personal debriefing at any time. After major interventions, in particular after serious accidents, there are compulsory debriefings for the peers and all mental health professionals involved.

Role of the Watch Supervisor and Company Executives

Watch supervisors play a central role in the critical incident stress management system: they have to take the decision whether to replace the individual concerned and, instigate CISM measures and mobilize the peer. Furthermore, on account of their presence in situ, watch supervisor are usually the first person to speak to the concerned controller so that they have a key influence on the direction taken by the CISM measures. Watch supervisors lay the foundation groundwork for successful critical incident stress management.

The role of watch supervisors both as responsible decision-maker and as the first person to talk to a controller after a critical incident is not always an easy one. They have to take decisions that on the spur of the moment may not always be correct. They must reconcile their management responsibility (i.e. their duty of care towards the employee) with their responsibility for operational procedures. Furthermore, as the first person to talk to the individual concerned, they should effectively carry out the preparatory work for the ensuing CISM measures. It may be assumed that in direct succession to a critical incident a watch supervisor will be busy with his own crisis management and various internal and external demands. Therefore, it is vitally important to provide watch supervisors CISM training tailored to their requirements. They should also be provided with checklists and standardized deadlines and liaise closely with the peers. It is recommended that a compulsory CISM module be incorporated in the watch supervisor qualification.

Crisis Management Briefing

Within DFS, the use of Crisis Management Briefings (CMB) was very helpful after the mid-air collision over Überlingen on 1st July 2002. The CMBs were offered before the normal daily briefings at the control centers and proved an efficient way of helping individuals with secondary and tertiary involvement to resolve the traumatic event. It became apparent that the interaction of the CMB team is a crucial factor, in particular the close cooperation and the clear separation of roles between managers/executives and peers was experienced as very helpful.

The normalization of reactions was facilitated by the information on and understanding of the reactions and by removing the sense of isolation in the group process, which considerably helped the briefed air traffic controllers in resuming their work and retaining or regaining their self-confidence.

The Crisis Intervention Team

In the case of serious accidents, the intervention through peers may not be sufficient to alleviate the resulting stress upon controllers; these cases may require further CISM measures and professional crisis management. CISM is an established part of crisis management at DFS, and a crisis intervention team, consisting of three mental health professionals, offers a round-the-clock crisis hotline. The crisis intervention team may be mobilized at any time via this hotline in order to steer and coordinate all necessary CISM measures. The mental health professionals in the crisis intervention team have a background in psychology/psychotherapy and are trained and certified in CISM.

Benefits

When introducing CISM, DFS placed strong emphasis on the principle of confidentiality. In order to create a sound basis of trust, no data pertaining to peers, participants, and individual CISM interventions or their results were disclosed. An assessment of the benefits of CISM to date therefore cannot be backed up with statistics but has to be based on individual observations.

The high degree of trust in CISM and the peers at DFS is best reflected by the fact that, nowadays CISM is in most cases directly requested by the concerned air traffic controllers themselves. Word of mouth between operational staff and management regarding successful CISM measures has quickly removed the initial skepticism. Watch supervisors have been convinced of the benefits of CISM as a result of positive experience they have gathered and their own special training. CISM has become a standard procedure in the operational service and is widely accepted as such. Initial concerns that controllers requesting CISM measures might be frowned upon by their colleagues have fortunately proven entirely unjustified. Furthermore, CISM enjoys the full support of senior management at DFS who consider it an important and necessary measure. This is obvious through the fact that CISM has been offered to other ATC providers in the wake of serious accidents in a pre-emptive and non-bureaucratic way.

In many cases, CISM measures have led to discussions about specific safety procedures and the safety culture in general. The discussions were often continued during the CISM forum and resulting suggestions put forward in the concerned units and discussed with colleagues.

Since the basis of trust has been established, and this is not undermined through the collection of data, we are planning a first, anonymous survey of CISM in 2003.

Training

DFS has developed various CISM training curricula that are available for other institutions and organizations. All courses are delivered in accordance with ICISF standards and will result in the award of the ICISF certificate. The courses comprise:

- advice and support during concept development;
- information sessions for managing directors and managers;
- training for executive staff, special training for watch supervisors;
- training and qualification for peers;
- refresher training courses.

Conclusions

Air traffic controllers work in a professional environment in which a critical incident is a daily risk. These potential critical incidents trigger critical incident stress reactions, with the inherent danger of post-traumatic stress disorder. PTSD will inevitably lead to the inability to continue the job and PTSD symptoms will have a major negative impact on the quality of life of the individual concerned. Critical incident stress reactions must therefore be properly dealt with in order to eliminate the danger of PTSD.

On account of the specific requirements in the air traffic control domain, it is suggested to base CISM on the peer model. Only highly qualified peers can help controllers to deal with critical incident stress reactions quickly and effectively. Further CISM models for the ATM community should be based on established, proven and standardized procedures. In order to be able to draw on experience made by ATM providers, to further develop the model on ATM requirements and to provide mutual support where necessary, it would appear very desirable for CISM procedures to be comparable and uniform. The critical incident stress management of the International Critical Incident Stress Foundation is – in the experience of DFS – currently the most suitable model.

References

CISM (2000), *Advanced Group Crisis Intervention*, Training Manual, 2nd Edition, International Critical Incident Stress Foundation (ICISF), Baltimore, USA.

CISM (2003), *Basic Group Crisis Intervention*, Training Manual, 3rd Edition, International Critical Incident Stress Foundation (ICISF), Baltimore, USA.

European Organization for the Safety of Air Navigation (1997), *Human Factors Module Critical Incident Stress Management*, EUROCONTROL, Brussels, Belgium.

Mitchell, J.T. and Everly, G.S. (1993), *Critical Incident Stress Debriefing: An Operational Manual for the Prevention of Traumatic Stress*, Chevron Publishing Corporation, Ellicott City, MD.

Mitchell, J.T. and Everly, G.S. (1994), *Human Elements Training for Emergency Services, Public Safety and Disaster Personnel*, Chevron Publishing Corporation, Ellicott City, MD.

Mitchell, J.T. and Everly, G.S. (1997), *Critical Incident Stress Management: A New Era and Standard of Care in Crisis Intervention*, Chevron Publishing Corporation: Ellicott City, MD.

Shapiro, F. (1998), *EMDR: Eye Movement Desensitization and Reprocessing*, Junferman Verlag, Paderborn, Germany.

Van der Kolk B., Mc Farlane A. and Weisaeth L. (2000), *Traumatic Stress*, Junferman Verlag, Paderborn, Germany.

ICISF (2000), Workbooks of International Critical Incident Stress Foundation, Baltimore, USA.

Chapter 11

Team Resource Management in European Air Traffic Control: Results of a Seven-Year Development and Implementation Program

Michiel Woldring, Dominique Van Damme,
Ian Patterson and Patrícia Henriques

Introduction

Over the last 25 years airlines have been very successful in promoting the ideas of enhanced team working practices. Nearly every airline around the world applies the principles of *Crew Resource Management (CRM)* for pilots and other operational airline staff. CRM is a concept that involves the idea of optimizing not only the person-machine interface and the acquisition of timely, appropriate information but also interpersonal activities including leadership, effective team formation and maintenance, problem-solving, decision-making, and maintaining situation awareness (Wiener, Kanki and Helmreich, 1993). Thus, training in CRM involves communicating basic knowledge of Human Factors concepts that relate to aviation and providing the tools necessary to apply these concepts operationally. It represents a focus on crew-level (as opposed to individual-level) aspects of training and operations (Wiener, Kanki and Helmreich, 1993).

CRM on the flight deck gained much of its impetus from the fact that there had been a number of serious incidents and accidents in which poor communication, crew performance and inadequate behavior were seen as significant contributory, if not causal, factors. A growing body of anecdotal and incident analysis evidence has illustrated similar problems with team performance in ATC and has pointed to the need to complement the improvements made in the air, via CRM, with better team functioning within air traffic control. The point has been made elsewhere that 'it is somewhat surprising that "Controller Resource Management" did not develop in conjunction with Cockpit Resource Management' (Helmreich et al., 1993).

Although a great deal of effort and expertise is devoted to training individuals in the technical skills necessary for the ATC task, little, if anything, has been done to train these individuals to function as team members. Incidents and accidents, in which inadequate team work has been shown to be a factor, suggest that much more attention needs to be focused on this vital area; the adoption of the title 'Team Resource Management' is intended to reflect the importance of the team in the safe and efficient conduct of air traffic control.

Therefore, TRM concept emerges as training that is designed to improve the performance of air traffic control teams, by increasing the awareness and understanding of interpersonal behavior and human factor capabilities, with the goal of improving flight safety.

> Now is the time to apply the same emphasis and standards to teamwork skills that we currently apply primarily to technical skills (Biegalski, 1994).

When EUROCONTROL became manager of the European Air Traffic Management Programme (EATMP, former EATCHIP), a Human Resources Business Plan was presented to the first international EATMP Human Resources Team in 1994. The plan included a task: 'Develop an Air Traffic Services (ATS) Crew Resource Management Programme'.

Following on from this, a Study Group was created to investigate the possible benefits of, and requirements for, a Team Resource Management (TRM) Programme, official definition: *Strategies for the best use of all resources – information, equipment and people – to optimize the safety and efficiency of Air Traffic Services.*

A TRM Task Force was therefore established in summer 1995, including European Civil Aviation Conference (ECAC) State representatives from France, Germany, Switzerland, United Kingdom, EUROCONTROL Headquarters and the Institute for Air Navigation, resulting in a mixed team of active controllers, trainers and Human Factors experts, see Figure 11.1.

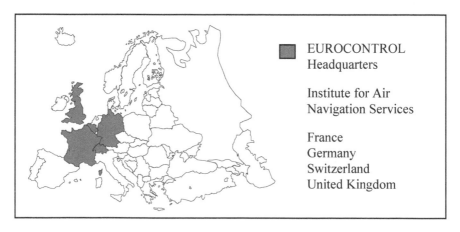

Figure 11.1   Members of the TRM Task Force in 1995

In 1996, the TRM Task Force submitted Guidelines for Developing and Implementing Team Resource Management, see Figure 11.2. Based on these a prototype TRM course was developed in 1997 – within the frame of the EUROCONTROL Human Factors Programme.

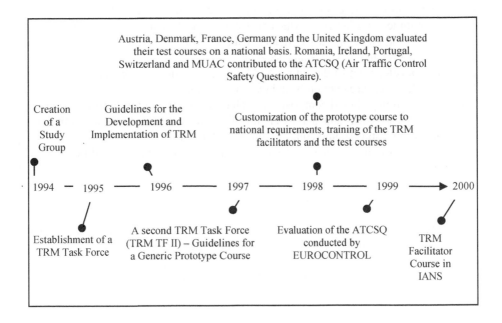

Figure 11.2 Timeline of the TRM Development and Implementation process

Objectives

TRM training seeks to ensure the effective functioning of operational staff, through the timely and proficient use of all available resources, and aimed at the safe and efficient flow of air traffic. Key objectives for TRM training are to develop team-member's attitudes and behavior towards enhanced teamwork skills and performance in Air Traffic Management.

The Guidelines for the development of TRM are shown in Figure 11.3.

1. The practical benefits of enhanced team performance for both management and operational staff should be communicated as early as possible. This will develop the necessary commitment to develop and reinforce TRM throughout the organization.
2. The main objective for TRM for operational staff should be the development of attitudes and behavior which will contribute to enhanced teamwork skills and performance in order to reduce team work failures as a contributory factor in ATM related incidents and accidents.
3. The development of the future ATM system should consider TRM principles in order to ensure continuity and teamwork stability.
4. TRM training should comprise three phases: an introductory/awareness phase, a practical phase and a refresher/reinforcement phase. Related training for operational staff should contain elements of TRM.
5. TRM should be mandatory elements in the selection, training and licensing of operational staff.
6. Situation awareness, decision making, communication, teamwork, leadership and stress management should form the mandatory subjects of a TRM training course.
7. The first phase of TRM training should be provided both to operational controllers and supervisors and should later be extended to other operational staff in ATM.
8. TRM instructors should be carefully selected and trained, and when possible should be current operational staff.
9. Scenarios for training purposes should be realistic, relevant to the course participants and regularly updated. The provision of a simulation environment should be considered such that participants can practice and reinforce TRM skills in both normal and emergency situations.
10. TRM training tools and methods should include lectures, examples, discussions, role-plays, videos on team related errors, hand outs, check-lists and simulator exercises.
11. The reinforcement of TRM in the operational environment should be ensured by management backup and support, team and individual (de)briefings, visual reminders and feedback from incident investigations.
12. The benefits of TRM should be maintained by continuously evaluating training courses and the changes in attitudes and behavior of operational staff in the work environment.
13. As TRM training evolves, an extension of the target population and refinement of TRM concept in the future ATM system should be considered.

Figure 11.3   Development guidelines of the TRM training

An Example: The TRM Prototype Course

The TRM prototype course has been developed with the support of operational controllers from several European countries and from extensive knowledge of human performance from the cognitive, psychological, social and physiological sciences. The content has been carefully developed and refined to be acceptable to the majority of the participating nations, but it is also realized that in the future not only will the content be modified to suit the changing air traffic control environment, but also to support national needs. It is hoped that the majority of the course will be delivered in all nations in a similar way, but it is also understood that modifications may be necessary; particularly to suit the learning needs of the participants and the addition of material with regard to incident examples that illustrate the course content.

The prototype course lasts three days (with an ideal number of course participants between eight and twelve) and has been prepared in eight separate modules: introduction, teamwork, team roles, communication, situation awareness, decision making, stress and a conclusion. There follows a description of what course participants should be able to do after completion of each module.

Teamwork

- determine typical characteristics of ATC related teamwork;
- identify behavior that has negative impact on teamwork and consequently develop and practice behavioral strategies that help to improve effective teamwork;
- identify the importance of recognizing different character types within teams and their influence on team work;
- understand the meaning and differences between team identity and corporate identity;
- identify safety-related issues concerned with teams and teamwork;
- recognize the importance of different individuals within teams and develop strategies which will improve teamwork skills and performance.

Team Roles

- understand the formal and informal hierarchical structures in the ATM system;
- understand how attitudes towards authority are formed and how you define your own authority;
- develop strategies to avoid errors due to misunderstandings arising from the roles of leadership and followership;
- develop strategies to deal with submissiveness, assertiveness and aggressiveness.

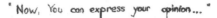

Communication

- identify the functions of communication;
- identify how communication is performed within teams and how it can affect safety;
- develop strategies on how to communicate effectively;
- develop strategies on how to intervene effectively in a typical ATM-related situation;
- develop ways to give and receive feedback and constructive criticism.

Situation Awareness

- better understand situation awareness;
- describe the effect of high and low workload on situation awareness;
- identify the symptoms of loss of individual and team situation awareness;
- develop appropriate strategies on how to prevent the loss of situation awareness;
- identify factors that may have a positive or negative influence on situation awareness.

Decision-Making

- establish the factors which contribute to effective decision-making;
- appreciate the importance of situation and risk assessment skills;
- appreciate the concepts of shared problem models and the use of resource management skills in team decision-making;
- identify an example of structured decision-making in special situations.

Stress Management

- define job-related stress situations;
- explain what stress is and how it affects your work;
- determine the strategies to help you cope with stress and its effects;
- explain how stress affects teamwork;
- develop skills to recognize and cope with stress situations in teams.

The TRM training includes evaluation material on both the course and the cultural differences between the states involved (in this context a mixture of national, organizational and safety culture). Incident reports have been placed in several modules. These are all real air traffic control-related incidents from various countries. Videos and video scenarios have been supplied within some of the modules. The videos have been carefully chosen to illustrate specific points within the modules whereas the video scenarios have been written to enable each country to create their own videos with local conditions and environmental framework. Several activities and exercises have been developed for each module.

There are two handbooks accompanying the TRM prototype course. These are the facilitator's handbook and the participant's handbook. The facilitator's handbook contains all the teaching materials, which include information regarding support for the facilitators, a summary of the slides found in each module, instructional notes, pages for the facilitator to take support notes and copies of all activities and exercises. The participant's handbook contains information which includes a selection of the most important slides, summaries with the main safety points regarding each module and pages for the participants to take notes throughout the course.

Customization

ECAC states were invited to participate in the customization and test phases of the TRM project. The prototype course should serve as a basis for developing customized TRM courses. Most of the participating states required external support for their customization. Some contracted companies with experience in Crew Resource Management (CRM), while others were supported by EUROCONTROL. In the TRM chapter there is a description of the method that EUROCONTROL followed and the lessons learnt while facilitating the TRM customization and facilitator training in Austria, Portugal, Ireland, Denmark and the EUROCONTROL Upper Airspace Centre in the Netherlands.

The subject of Human Factors is independent of culture (the experience up to now has illustrated the diversity of safety cultures within European ATC) and is often seen as a rather 'fuzzy' subject. It became apparent that a structured approach would certainly help to overcome any reluctance. There was also a benefit in applying a facilitation technique: discussions were needed to help the participants understand the relevance of the different topics. A standard method would also enable us to compare the customizations in the different countries. And, last but not least, time would be gained. Initial customization of all eight modules took only approximately four days, after which participants needed one more week to finalize the product (translation, creation of incident reports, exercise preparation, duplication, and rehearsal).

EUROCONTROL developed and applied a four-step method for the customization of the prototype course in the different test-sites for TRM (Woldring and Amat, 1998). The four steps are described in Figure 11.4.

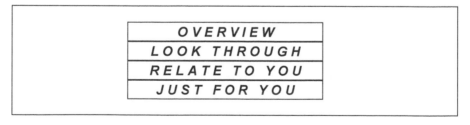

Figure 11.4   The four steps to customize the TRM prototype course

Step 1 – Overview ·
Step 1 was given the slogan 'overview'. Main objectives of step 1 are to introduce the issues and the scope of the expected discussions and to emphasize the structure of each module. This structure is obviously not compulsory and can be changed, but the predefined structures were successfully used in steps 1, 2 and even in 3. It was very important for the participants to keep enough distance from the material during steps 2 and 3 so that they could put things into context. We used one overhead projector slide per module to explain the most important messages. This slide corresponds to a table of content but is formulated in questions. For each question we briefly explained the underlying Human Factors issues and the different exercises and activities.

Step 2 – Look Through
Step 2 was given the slogan 'look through'. Objectives are to understand the aims of each module (content) and to understand the suggested techniques of facilitation or instruction (process). In this step we presented the whole module by leading the participants through the course material. We ran – where necessary – parts of the course, showed different games, looked at available videos, performed exercises and debated and debriefed case studies and incident reports.

Step 3 – Relate to You
Step 3 was given the slogan 'relate to you'. The main objective of step 3 is to assess the relevance of each part of the prototype course to the participants' culture – both in terms of content (is the message relevant for us?) and in terms of process (does this technique suit our culture?). In step 3 we discussed in detail the content and the relevance of the different topics and techniques, suggestions for alternative messages, questions, exercises, examples, pictures etc. In this phase we applied explicitly the different facilitation techniques that were taught at the beginning of the week.

Step 4 – Just for You
The slogan for step 4 was 'just for you'. The main objectives of this step are to screen the prototype material, and to modify the selected material that requires customization. Next to that a realistic schedule per module had to be decided. The modifications and the order in the prototype material were made immediately. Items were put on a 'to-do' list when the customization required much local information (local incidents, new developed exercises, relevant statistics).

Lessons Learned from the Customization

Facilitation At the beginning of each customization we trained the participants in facilitation techniques. The participants expressed current fears: how do I get a group to talk, and when they talk, how do I stop or steer them, what do I do with unexpected conclusions, how do I get rid of my tendency to teach, and how do I predict the outcome of a discussion?

First of all we restricted the number of techniques to self-presentation, mini-lessons, open and closed questions and introducing and summarizing discussions. Already during the facilitation training we used examples from the TRM prototype course. During the customization we could repetitively refer to these exercises in which the participants performed themselves. Facilitation techniques were explicitly applied by us, by doing so we functioned as a role model for the inexperienced facilitators. During step 4 the participants were encouraged to act as facilitators. Fears and concerns about facilitating gradually disappeared in the participants.

TRM Implementation Some important ingredients for the success of a TRM implementation come from the phase before the customization. Active management support and a carefully prepared information campaign exert critical influence over the attitudes towards TRM. Also, the involvement of incident/accident investigators was of high value in the choice of local examples for the different case studies and exercises. However interesting the outcome of an accident investigation on the other side of the planet is, the accident took place 'far away'. Analysis of a local incident brings the awareness that it can and does happen 'here' as well, so the learning experience is stronger.

Interactivity and Ownership The interactivity between facilitators and participants changed during the 4 steps of the customizations. At the start the communication was generally one way, us explaining and informing the participants. In the next step the participants became active, asking questions for clarification and seeking background information. In the exercises and case studies the participants were fully involved. Step 3 was characterized by discussions amongst the participants and exchange of ideas with the facilitators for alternatives for the prototype materials. In the end the facilitators of the customization were able to withdraw, the participants were in charge and they created the final product, with only little process oriented interferences from us, the facilitators. This swap in activity during the process of customization led to ownership of the end-product by the participants, and they considered the outcome as 'their' course.

Evaluation of the TRM Program

The objective of TRM is the use of all available resources – information, equipment and people – in order to achieve the safe and efficient movement of air traffic. This objective is obviously ambitious and as with the history of its counterpart on the flight deck – CRM – there will be difficulties in the measurement of its effectiveness. However, to ignore the challenge of this evaluation would be foolish. To this end the measurement and evaluation of the TRM program was undertaken.

The program evaluation should not only provide information on the effects of the training, but it should also provide direction for continued training. The most basic type of information comes from participant evaluations, usually collected by questionnaire at the end of the training course. Positive reactions to the training provide necessary, but not sufficient evidence of impact. That is, while a positive reaction to the training is not sufficient in itself to indicate a positive shift in behavior, by the same token a negative reaction to the training is an almost certain indication that positive behavior change is not going to occur.

A second source of information comes from the use of an evaluative questionnaire concerned with attitudes and behaviors, which can be administered before and after the training sessions. Often a more robust method of evaluating the changes of these attitudes and behaviors can be captured by administrating a third identical questionnaire some four to six months after the training course.

A third and more rigorous evaluation comes from the correlation of these attitudinal changes with observation or interview of the same personnel to gauge meaningful behavioral changes. From this methodology, measurable positive changes in interaction should be present following the training. Lastly, the ultimate validation can be found in the correlation of the training program and a reduction in the frequency of incidents within the system. The latter two methodologies are highly complex and take considerable time to achieve. It is for this reason that the test and evaluation phase of the TRM program used only the course evaluation and the monitoring of attitudinal and behavioral changes as an assessment of its effectiveness.

The Air Traffic Control Safety Questionnaire (ATCSQ)

The Air Traffic Control Safety Questionnaire (ATCSQ) was developed to enable the evaluation of the Team Resource Management (TRM) program. Several other questionnaires which have been developed for similar purposes were reviewed and the architecture of the ATCSQ reflected these developments within the flight deck and operating room environments. The questionnaire consists of four main sections. The first section concerns attitudes towards the quality of training, working conditions and documentation. The last section concerns demographic information. The second and third sections contain the main evaluative information, the second being concerned with attitudes and the third with those responses associated with behaviors.

Overall the results indicate a well-balanced and informative questionnaire which can be considered robust and reliable. Responses concerning the change in attitudes between the two courses are a little more difficult to determine. However, the results indicate that the questionnaire is sensitive to changes in attitude, and with a larger sample and strict adherence to data gathering a more meaningful database can be considered in the future. This will not only strengthen the use of such a questionnaire but will also help individual States to customize their individual needs in their TRM programs.

Seven Years Later

TRM facilitation training is now regularly available through the EUROCONTROL Institute of Air Navigation Services (IANS) in Luxembourg. Controllers from more than 20 European states have attended the facilitation courses. Also active interest is shown from Japan, Thailand, Australia, New Zealand, USA, Israel, Bahamas, Brazil and Canada.

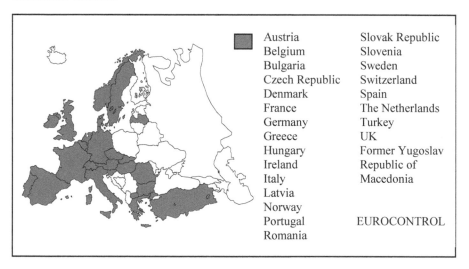

■	Austria	Slovak Republic
	Belgium	Slovenia
	Bulgaria	Sweden
	Czech Republic	Switzerland
	Denmark	Spain
	France	The Netherlands
	Germany	Turkey
	Greece	UK
	Hungary	Former Yugoslav
	Ireland	Republic of
	Italy	Macedonia
	Latvia	
	Norway	
	Portugal	EUROCONTROL
	Romania	

Figure 11.5   States represented at EUROCONTROL IANS TRM courses

In short, TRM has developed into a world-wide accepted safety concept. The main benefits of training are considered to be: the reduced teamwork-related incidents, enhanced task efficiency, improved use of staff resources, enhanced continuity and stability of teamwork in ATM, enhanced sense of working as a part of a larger and more efficient team and increased job satisfaction. Maintaining the momentum is the difficult part. A major difficulty of classroom training is to keep the new-learned attitudes and behaviors alive. Ideally TRM provides training in basic Human Factors issues and aims to shape controller attitude towards an open safety culture, but the industry is not yet there. Current new developments focus on simulation-observation training as a means of developing desirable on TRM-related behaviors in controller teams. Two specific initiatives are worth mentioning: TRM-Oriented ATC Simulator Training (TOAST) and Behavior Oriented Observation Method (BOOM).

TRM-Oriented ATC Simulator Training (TOAST)

In order to ensure that the knowledge acquired within the TRM initial training is brought to the operational environment, the next step was to allow controllers to apply and consolidate TRM in a risk-free environment. That was the main learning objective of TOAST. From the work conducted by a TOAST Working Group (2001) it was concluded that TRM-Oriented ATC Simulator Training is feasible. TOAST provided a strong foundation on which to develop TRM skills but further work had to be carried out at this stage. Taking the working group opinions on EUROCONTROL contribution and the EATMP HRS Programme (EATMP, 2000a) into account, the following recommendations were issued:

Table 11.1   Recommendations from the TOAST Working Group

1	TRM-related scenarios need to be introduced within the simulation exercises in order to trigger events that will allow participants to practice knowledge acquired during TRM training.
2	The TOAST facilitators should be provided with training on the briefing and debriefing philosophy and techniques that suit the TRM approach.
3	The simulation-exercise scenarios need to include specific scenarios/scripts for the pseudo pilots.
4	If enough resources are available one TOAST facilitator should be dedicated to the observation of TRM-related aspects within the simulation configuration.
5	The TOAST facilitator should be trained in behavioral-oriented observation method (BOOM).
6	A TOAST training program would allow the TOAST training objectives to be structured in order to ensure all TRM-related aspects are dealt with at the end of the training.
7	It is important within the exercises to have a well-balanced combination of technical and non-technical aspects.

Behavior-Oriented Observation Method (BOOM)

BOOM provides a method by which trained observers, usually TRM facilitators, can objectively observe the behaviors of controllers and provide feedback on the non-technical (TRM) aspects of their performance. The need to develop such a behaviorally oriented observation and debriefing method was essential if there was to be some consistency/homogeneity of approaches to simulation training in European ATM. Logically it also provides a framework by which feedback is given, the ultimate objective of which is to provide controllers (and other professional groups) with information by which they can improve their Non-Technical Skill (NTS) or teamwork skills.

Underlying BOOM are a number of principles that have driven its development:

- Only observable behavior is used as a basis for the BOOM feedback.
- The evaluation must exclude reference to a controller character trait or emotional capacities and should be based only on observed behaviors and discussions arising from these.
- For a Non-Technical Skill (NTS) to be considered it should have an operational significance or safety should be actually (or potentially) compromised.
- Open discussion of observations.
- Every interpretation, judgment or opinion on the observed behavior must be discussed within the context of its operational relevance.
- The trainer/observer shall be knowledgeable about TRM and should ideally have experience as an ATCO.
- The observer should be motivated to use TRM concepts and knowledge in the operational environment.
- The observed person will be fully informed.
- The observed person should be provided with clear explanations concerning the process, the NTS concept and the learning objectives of BOOM.
- The method and training should be simple and as jargon-free as possible.
- Culturally sensitive; the method and training should be flexible enough to fit in with the local culture.
- Keep it simple.

BOOM helps the observer build a good observation frame and assists the analysis of observed behaviors within the context of the TRM domains. The feedback will help the observed controller become more aware of how their behaviors influence team functioning within operational environment.

The expected outcome (or benefit) for the 'BOOMEE' (observed person) is that they become aware of their own practices and safety-related behaviors and develop their NTS; making the bridge between the 'theory' and the practice of the TRM concepts.

Why this approach in BOOM? The BOOM approach relies heavily on the judgment of the trained facilitators/observers. Initially, it had been intended that a simple checklist of desirable non-technical skills and their associated behaviors might be produced. However, early attempts at developing the BOOM method using TRM facilitators and controllers to generate a list of desirable non-technical skills (embedded in the TRM framework) and the behavioral descriptions associated with theses were unsuccessful. Such prototype lists were shown to be not comprehensive to be useful in all contexts and practical experience showed that the experts could not agree on the items that should be included, or the interpretation of these.

The resulting method is therefore reliant on the BOOMER (observer) observing behavior and the context in which occurred, then discussing this with the BOOMEE (observed person) with a view to refining the non-technical skills of the BOOMEE.

BOOM has therefore developed as a more qualitative feedback process than was initially intended.

BOOM Process
BOOM has three simple steps:

- 1st step: Observation of the BOOMEE by a trained observer (BOOMER). Observation in the real situation, or in simulation, of the BOOMEE behaviors and the contexts in which these occur;
- 2nd step: Debriefing Preparation, i.e. listing of questions to discuss with the BOOMEE, the interpretation of these behaviors in the context of the underlying NTS they represent;
- 3rd step: Debriefing with the BOOMEE.

It is anticipated that BOOM brings TRM a step closer to daily operations.

EUROCONTROL planned developments in TRM

New TRM modules on the management of error and violation and the impacts of automation are being considered. The Study Group of 1994 based its recommendations on a literature survey, an 'ATCO attitudes to teamwork' survey, a training survey and last but not least a teamwork-related ATC incident survey.

In terms of the results concerning the attitudes towards the professional training and working environment, the ATCSQ questionnaire clearly indicated acceptable satisfaction within most of the areas of training, but some aspects with respect to operation and safety manuals, handling of emergency traffic and feedback in daily operations were not as positive.

The time and TRM maturity are now there to directly link the debriefing of incident investigations to TRM, which would help optimize safety and efficiency of Air Traffic Services.

References

Barbarino, M. (1997), 'Team Resource Management in European Air Traffic Services', *Proceedings of the 9th International Symposium on Aviation Psychology*, Ohio State University, Columbus, Ohio, USA.

Biegalski, C. S. (1995), The critical factor in CRM training effectiveness (the management factor), in N. McDonald, N. Johnston, & R. Fuller (eds.), *Aviation psychology: Training and selection*, pp. 23-31, Avebury Aviation, Aldershot.

EATCHIP (1996), *Guidelines for Developing and Implementing Team Resource Management, HUM.ET1.ST10.1000-GUI-01*, EUROCONTROL, Brussels, Belgium.

EATCHIP (1998), *Proceedings of the second EUROCONTROL Human Factors Workshop: Teamwork in Air Traffic Services,* HUM.ET1.ST13.000-REP-02, EUROCONTROL, Brussels, Belgium.

EATMP (2000), *Proceedings of the fourth EUROCONTROL Human Factors Workshop: Team Resource Management in European ATM,* EUROCONTROL, Brussels, Belgium.

Isaac, A.R. and Barbarino, M. (1998), 'Development of Team Resource Management in European Air Traffic Control', *Proceedings of the 10th International Symposium on Aviation Psychology,* Auckland, Australia.

Wiener, E.L., Kanki, B.G. and Helmreich, R.L. (1993), *Cockpit Resource Management,* Academic Press, San Diego.

Woldring M. and Amat A.L. (1998), 'Team Resource Management In European Air Traffic Control: Customisation of a Prototype Course', *Proceedings of the 23rd Conference of the European Association for Aviation Psychology,* Vienna, Austria.

Woldring, M. and Isaac, A.R. (1999), *Team Resource Management, Test and Evaluation,* HUM.ET1.ST10.2000-REP-01, EUROCONTROL, Brussels, Belgium.

Chapter 12

Shiftwork and Air Traffic Control: Transitioning Research Results to the Workforce

Pamela S. Della Rocco and Thomas E. Nesthus

Introduction

Everyone who works shifts understands. Among other things, shiftwork makes you tired. Tired people are prone to making errors. Personnel in safety-related occupations, like Air Traffic Control Specialists (ATCSs), must staff facilities 24 hours per day, 7 days per week.

Shiftwork presents an interesting Human Factors challenge. Researchers have, for many years, reported the disruptive effects of shiftwork on sleep, performance, circadian rhythms, social and family relations, and longer-term health status (US Congress, Office of Technology Assessment, 1991; Schroeder and Goulden, 1983). With advances in our understanding of the circadian clock and the importance of sleep, researchers and practitioners have begun to focus on the challenge to mitigate the undesirable effects and to minimize conditions that are conducive to error. Over the past decade, a number of empirically developed coping strategies and fatigue countermeasures have been deployed in various operational environments (e.g., Graeber, Rosekind, Connell, and Dinges, 1990; Neri, Oyung, Colletti, Mallis, Tam, and Dinges, 2002; Rosekind, Graeber, Dinges, Connell, Roundtree, Spinweber, and Gillen, 1994).

It was in this context that, in 1990, the Federal Aviation Administration's (FAA) Civil Aerospace Medical Institute (CAMI) revived a program of research on shiftwork in the FAA's ATC facilities. The program built upon several CAMI studies from the 1970s that had focused on shiftwork and stress. Research in the 1990s replicated and extended the early findings to understand how the shiftwork issues manifested in the ATC environment and to target the fatigue countermeasures for transition from the laboratory to the workforce.

An impetus for facilitating the transition came in Fiscal Year (FY) 1999 when the FAA's congressional appropriations mandated a study of ATC shiftwork issues. Coincidentally, Article 55 of the National Air Traffic Controllers Association

(NATCA) 1998 Collective Bargaining Agreement (CBA) with the FAA also called for a CAMI study of fatigue and shiftwork issues. The Air Traffic Service (ATS) established an 'Article 55 Human Factors Workgroup' to address the CBA requirements. Controllers' work schedules and conditions were a negotiated item in Articles 32-34 of the 1998 CBA. CAMI researchers, therefore, had a forum within which to work, with the goal of moving the research findings from the laboratory into practical Human Factors solutions.

This chapter describes the research from the CAMI Shiftwork and Fatigue Research Program and the activities the authors undertook to transfer the findings to the ATC workforce.

Background

Shiftwork affects many aspects of an individual's life. An extensive literature exists on the disruptive effects of working variable shifts (Folkard, 1989; US Congress OTA, 1991; Waterhouse, Folkard, and Minors, 1992). Notable effects include disruptions of circadian rhythms, sleep, and social aspects of an individual's life (Comperatore and Krueger, 1990). Disruptions due to shiftwork have been termed 'shiftlag' because the effects are analogous to jetlag. Shiftlag may also result in physical symptoms, such that the individual does not feel well and, therefore, may frequently use sick leave in an attempt to cope. These effects have implications for safety-related occupations as well as organizational and individual costs. The disruptions may result in a situation whereby an individual is not at peak performance during duty hours. The Human Factors goal is to manage the shiftwork-related disruptions and to develop countermeasures to mitigate the potential adverse outcomes. An in-depth discussion of these issues is beyond the scope of this chapter; however, a high level outline of the topics follows.

Circadian Rhythms

Circadian rhythms are physiologically-based rhythms found in many measures of human biology (Costa, 1999; Dinges, Graeber, Rosekind, Samel, and Wegmann, 1996; Minors and Waterhouse, 1985). Circadian means 'about a day,' so these rhythms tend to cycle about every 24 hours. Body temperature is a commonly studied example. For day-oriented individuals, core body temperature rises over the course of the day until evening, when it falls until about 0300 hours (or 3 am) in the early morning. If you have stayed up all night or you wake to find the bed covers fell off around the early morning hours, you may have felt very cold. Body temperature reaches its lowest point at about that time of night for many people. Cognitive functions, performance, sleep, and alertness also demonstrate circadian variations (Monk, Folkard, and Wedderburn, 1996). These rhythms tend to track the day/night environmental cycle for day-oriented individuals. Thus, when it is daylight, an individual feels awake and alert. When it gets dark, the individual feels sleepy and less alert. Day-oriented individuals working a night shift must not only fight sleep but must also function when their performance rhythms are at the circadian low point or nadir. This makes them prone to errors and accidents just due to normal circadian variations (Monk, 1990). Costa (1999) notes that the circadian decline in alertness and cognitive performance is often aggravated by disruption of circadian rhythms, sleep deficit, and fatigue.

Circadian rhythms are maintained by both endogenous and exogenous factors (Aschoff, 1965; Costa, 1999). One effective exogenous factor or 'Zeitgeber' is daylight exposure. Shiftwork and travel across time zones result in disrupted rhythms (i.e., shiftlag or jetlag) when exposure to daylight is different than the times to which the body clock is adjusted (Comperatore and Krueger, 1990; Minors and Waterhouse, 1985). Disruptions in the rhythms may result in physiological symptoms, such as gastrointestinal complaints and/or cardiovascular effects (Comperatore and Krueger). Moreover, cognitive resources become unpredictable. So, until the rhythms adjust, a person may feel groggy or sleepy during times when they were expecting to be alert. The circadian rhythms will reset to a new time zone over time with consistent daylight exposure at the new times. Some rhythms reset quickly. However, it can take up to 2 weeks for other measurable rhythms to reset. Thus, the circadian system is relatively slow to change. So, when an individual changes shifts, the new exogenous cues may result in shiftlag or desynchronized rhythms. With stable exogenous cues, the rhythms will take several days to adjust. Chronic circadian disruption can have other long-term health implications, such as cardiovascular and other gastrointestinal problems (Costa, 1999).

Sleep

Sleep, like eating, is a physiological need (Dinges et al., 1996). Sleep is required to maintain alertness, concentration, and performance, as well as health (Anch, Browman, Mitler, and Walsh, 1988; Naitoh, 1992; Naitoh, Kelly, and Englund, 1990). People differ in the amount of sleep they need as individuals. However, the common 'rule of thumb' is that 8 hours is average; although, there is a wide variability (Dinges et al., 1996). Shiftwork disrupts sleep in a number of ways (Tepas, 1982; Webb, 1982). It may shorten sleep duration, disturb the sleep architecture, and lead to fragmentation (Tilley, Wilkinson, Warren, Watson, and Drud, 1982). If a person needs 8 hours and only gets 6, they experience sleep loss. Consecutive days of restricted sleep periods will result in cumulative sleep debt. Sleep-deprived individuals find it difficult to fight sleep and may 'doze off' unintentionally. Dinges et al., suggest that losing as little as 2 hours of sleep will degrade performance and alertness. Only sleep effectively fights sleep loss.

Shiftworkers, particularly night-shift workers, demonstrate increased sleepiness during work, with most sleepiness occurring during the last half of the shift (Akerstedt, Torsvall, and Gillberg, 1982). Researchers have reported that the increased sleepiness leads to a decrease in performance (Mitler, Carskadon,

Czeisler, Dement, Dinges and Graeber, 1988), an increase in accidents (Ribak, Ashkenazi, Klepfish, Avgar, Tall, Kallner and Noyman, 1983), and an increase in spontaneous sleep periods on the job (Torsvall and Akerstedt, 1987).

Other Individual Factors

Other factors affect how shiftworkers adapt to working shift schedules. As employees age, research suggests that changes in sleep patterns and biological rhythms may also increase difficulty adapting to different shift schedules that result in partial sleep loss or circadian disruption (Akerstedt and Torsvall, 1980; Monk and Folkard, 1985). Shiftwork also disrupts an employee's family and social life. These social and family disruptions can also affect the individual's job performance (Penn and Bootzin, 1990).

Schedules

There is no single 'optimum shift system.' However, shift schedule designs can either minimize or exacerbate disruptions experienced by a shiftworker (Knauth, 1993). Schedules directly determine how much time is available for sleep. For example, if an employee gets off from an evening shift at 2300 hours and is scheduled to return for an 0700 shift the next day, the employee only has 8 hours between shifts. Factor in the time to commute home and back to work, unwind from the evening shift, and prepare for bed, available time to sleep can be substantially reduced between shifts. Schedules also interact with the factors that stabilize circadian rhythms. Because daylight is a primary 'Zeitgeber,' then a shiftworker's variable daylight exposures may tend to desynchronize the circadian rhythms. Shift schedules can be designed to facilitate stability and mitigate desynchronizing effects. Therefore, schedule design is a very important Human Factors consideration.

Particular features of schedule design include *permanent versus rotating schedules, slow versus rapid rotations, advancing (counterclockwise) versus delaying (clockwise) rotations*, as well as *the number of night shifts* and *early morning shifts* included in the schedule (Knauth, 1993). Rotating schedules involve a change of shift start times. This change can be either slow (e.g., 2 weeks of the same shift before changing) or rapid (e.g., 2 days of the same shift before changing to another shift start time). In rotating schedules, the direction of rotation of the start times is important to consider. If start times of the rotating shifts are progressively earlier, the schedule is termed 'advancing' or counterclockwise. If the start times are progressively later, the schedule is termed 'delaying' or clockwise. Finally, the length of the shift (e.g., 8 hours versus 10 hours) is a consideration.

Less direct information is available about the merits of the different speeds of rotation and the direction of rotation. Researchers have debated the speed and direction of rotating shift schedules in the literature (Folkard, 1992; Wilkinson, 1992); however, very few actual data from controlled studies were available when we revived the study of ATC schedules (Turek, 1986). Researchers have generally argued that delaying systems, with clockwise rotations, are preferable to advancing systems with counterclockwise rotation (Barton and Folkard, 1993).

Of all the features of schedule design, night shifts are considered to be one of the most disruptive for most workers. Knauth and Rutenfranz (1982) concluded from a review of the literature that re-entrainment to night shift work remains incomplete even on permanent night shift schedules. As European researchers have long recognized, early morning shifts also have drawbacks (Knauth, Landau, Droge, Schwitteck, Widynski, and Rutenfranz, 1980). Simons and Valk (1999) demonstrated that start times prior to 0900 resulted in shortened total sleep times, reduced sleep quality, decreased daytime alertness, and significant vigilance performance impairment in short-haul pilots. Taub and Berger (1976) reported a series of studies, in which they concluded that maintaining stable sleep schedules was of equal or greater importance than sleep duration to maintain performance. Thus, stability in sleep and daylight exposure is important to promoting optimal performance and adaptation to new schedules.

ATC Shiftwork

ATC shift schedules are interesting and unique in the context of shiftwork research. In general, many ATC facilities use counterclockwise, rapidly rotating schedules. That means that ATCSs work different shifts (e.g., afternoon, morning, and midnight) within one workweek (rapidly rotating), and start times for the different shifts are progressively earlier (counterclockwise). These features act to compress the workweek, providing more time-off between weeks.

A specific counterclockwise, rotating shift schedule, the '2-2-1,' worked in ATC facilities was a primary focus of CAMI research in both the early and the more recent years of the program (Della Rocco, Cruz, and Schroeder, 1996). The '2-2-1' involves a rapid, counterclockwise rotation through two afternoon shifts, two morning shifts (usually early morning), and one midnight shift. Table 12.1 presents an example of a 'typical' 5-day schedule. This scheduling practice was reported as early as the early 1970s (Melton and Bartanowicz, 1986).

Table 12.1   A 'typical' 2-2-1 shift schedule

Shift	2-2-1	Hours Between
1	1500-2300	15
2	1400-2200	9
3	0700-1500	15
4	0600-1400	8
5	2200-0600	

Note: Shift 5 actually begins at 2200 on Day 4.

This schedule compresses a 40-hour workweek into a total of about 88 hours, as opposed to the typical 104 hours required to work 40 hours on a typical straight-day schedule. The transitions from the afternoons to mornings and from the mornings to the nights involve 'quick-turn-arounds' with as few as 8 hours off between shifts. The 2-2-1-type schedule results in about 80 hours off between workweeks compared to approximately 64 hours off on a straight-day schedule. Thus, ATCSs report that they prefer the 2-2-1-type schedules because they have more time off between workweeks.

Melton and Bartanowicz (1986) argued a case in favor of the 2-2-1. They suggested that one advantage of the schedule was that four of the five shifts were worked during normal waking hours. Therefore, employees could maintain relatively stable sleep/wake cycles for the 4 days of the schedule in order to remain day-oriented in their circadian rhythms. They suggested that the 2-2-1 may be less disruptive than other schedules that require weekly shift rotations in which an employee works one shift for a 5-day workweek, and then rotates to a new schedule, such as the 'straight-5.' In the FAA version of the straight-5, employees work 5 days on the same shift, have 2 days off between workweeks, and then rotate to a different shift for 5 straight workdays. Melton and Bartanowicz argued that the straight-5 schedules could result in relatively continuous circadian disruption because, during each 5-day workweek, the body's circadian rhythms may not entrain to any given shift until about the time the employee rotates to a different shift. Thus, the employee's circadian rhythms would be constantly trying to adapt. Another advantage of the 2-2-1 may be that it includes only one night shift per week. This minimizes employee exposure to problems associated with the midnight shift in addition to minimizing circadian disruption. The placement of the single midnight shift at the end of the week precludes the circadian disruption, due to the midnight shift, from adversely affecting work on the following day.

Certain characteristics of the counterclockwise 2-2-1 schedule, however, are counter to the prevailing recommendations in the literature. The 2-2-1, on its face, is a multiple phase-advancing schedule. Although it could be argued that it is both an advancing and delaying mix. The change between the afternoon and early morning shifts is an advance; and, the change between early morning to the midnight within the same day, in combination with late asleep and arise times on off days before the afternoons, has delaying properties (Boquet, Cruz, Nesthus, Detwiler, Knecht, and Holcomb, 2004). Depending upon the employee's sleep/wake cycle and schedule of daylight exposure, their circadian clock might not obtain the regular timing cues required to maintain stable day-orientation, and therefore, their rhythms may desynchronize. The quick-turn-arounds from one shift to the next offer as little as 8 hours off between shifts. This arrangement has the potential to result in cumulative partial sleep loss during the week, as well as circadian rhythm disruption. During the transition from the morning shift to the midnight shift, employees' sleep is essentially a nap period during the afternoon sleep period. This is likely to result in poor quality sleep for a variety of reasons, including the circadian characteristics of afternoon sleep, social and family activities, and possibly, the sun's bright light in the sleeping environment. CAMI researchers hypothesized that the placement of the midnight shift at the end of the week, following the phase advances and possible resulting shiftlag earlier in the week, may interact with sleep loss to exacerbate the problems associated with the midnight shift (i.e., the circadian trough) effects on performance. We set out on a program of research to investigate these issues.

Over the course of the shiftwork and fatigue research program, we discovered a variety of schedules in use at ATC facilities. Based upon the previous CAMI research and the existing literature, we adopted five shift designations to examine schedules: early morning (start times before 0800), day (start times between 0800 and 0959), midday (start times between 1000 and 1259), afternoon (start times between 1300-1959), and midnight (start times between 2000 and 0100). The term *midnight* was used to avoid confusion with an ATCS convention of calling shifts starting between 1300 and 1900 'night shifts.'

Table 12.2 presents samples of schedules we found in other studies with shift types: A=Afternoon, D=Day, E=Early Morning, N=Mid-night and M=Mid-day.

In 1999, we surveyed the watch schedules assigned to ATCSs for FAA Terminal facilities to determine the prevailing scheduling practices (Della Rocco, Dobbins, and Nguyen, 1999). We examined 2,893 schedules from 36 Terminal facilities with 24-hour operations. The schedule study demonstrated that ATC facility schedulers used a large variety of shift start times to cover staffing needs. Shift start times for employees were staggered at most facilities. Unlike hospitals or industries where all employees arrive for a shift at the same time, a typical ATC schedule would involve a couple of employees scheduled to arrive in their area every hour or half-hour between 0500 and 0700. The study found that schedule patterns with a 2-2-1-like rotation (counterclockwise, rapid rotation with one or more midnight shifts) represented approximately 25 per cent of the schedule types in terminal facilities. Other counterclockwise rotations without midnight shifts represented approximately 36 per cent of the schedules in use. In addition, some of the facilities worked schedules with rotating days off. This schedule type involved working, for example, 4 straight afternoon shifts followed by 2 days off, 4 early morning shifts followed by 2 days off, and finally, 4 midnight shifts followed by 2 days off to complete the rotation. These schedules represented approximately 9 per cent of this sample. From this and the field studies, it was apparent that 24-hour facility staffing did not require all employees to actually work midnight shifts. So, in some facilities, a majority of schedules did not include midnight shifts. Scheduling practices varied widely among the facilities to reflect staffing requirements for covering typical air traffic patterns. The schedules in ARTCCs may have included a greater percentage of 2-2-1 type schedules.

Table 12.2   ATC facility work schedules from CAMI studies

	Day 1	Day 2	Day 3	Day 4	Day 5
2-2-1 Schedule					
Shift Type	A	A	E	E	N
Sample Schedule	1430-2230	1330-2130	0700-1500	0600-1400	2230-0630
Range of Start Times	1330-1600	1000-1600	0600-0800	0600-0620	2200-2400
Range of End Times	2130-2400	1800-2400	1400-1600	1400-1420	0600-0800
Hours Off Between Shifts	16	8	16	8	
2-1-2 Schedule Field	**(Miami)**				
Shift Type	A	A	M	E	E
Sample Schedule	1430-2230	1330-2130	1030-1830	0700-1500	0700-1500
Range of Start Times	1330-1500	1250-1400	0955-1100	0630-0700	0600-0745
Range of End Times	2130-2400	2250-2300	1755-2000	1430-1600	1400-1545
Hours Off Between Shifts	16	12	12	16	
10 Hour Schedule (Minneapolis)					
Shift Type	A	A	E	E	
Sample Schedule	1400-2400	1200-2200	0800-2000	0600-1600	
Range of Start Times	1200-1400	0600-1400	0600-1000	0600-0700	
Range of End Times	2200-2400	1600-2400	1600-2000	1600-1700	
Hours Off Between Shifts	12	10	10		
Straight Early Morning Schedule					
Shift Type	E	E	E	E	E
Sample Schedule	0730-1530	0700-1500	0700-1500	0630-1430	0630-1430
Range of Start Times	0630-1000	0630-0900	0630-1000	0630-0700	0630-0645
Range of End Times	1430-1900	1430-1800	1430-1900	1430-1600	1430-1445
Hours Off Between Shifts	16	16	16	16	

The 1998 FAA/NATCA CBA addressed only a few items related to the work schedules. It defined a basic watch schedule in terms of days of the week, hours of the day, rotation of shifts, and changes in regular days off. It stated that permanent/rotating shifts and/or permanent rotating days off were options for scheduling (Article 32 Section 1). A basic workday was defined as 8 consecutive hours. A basic workweek was defined as 5 consecutive days (Article 34, Section 1, FAA/NATCA CBA, 1998). Finally, the CBA specified, 'employees shall not be required to spend more than 2 consecutive hours performing operational duties without a break' (Article 33 Section 1, FAA/NATCA CBA, 1998). The CBA addressed overtime to a limited extent in Article 38. However, total time on duty was not limited. Finally, controllers could earn non-overtime 'credit hours' at the election of the employee and with the employer's approval (Article 34 Section 8).

ATC Studies: 1970-1985

Several CAMI studies investigated ATCS shiftwork in the decade spanning the late 1960s through the late 1970s. C. E. Melton, in collaboration with a number of CAMI researchers, conducted a series of studies at the request of the Air Traffic Service to identify and quantify sources of physiological stress in ATCSs. Among these studies were several on shiftwork. Melton and Bartanowicz (1986) reported on two basic work shift rotation patterns had evolved during that period of time: 1) the straight-5 shift schedule and 2) the 2-2-1. The straight-5 schedule involved working 1or 2 weeks of shifts, beginning at the same time of day, for 5 days straight, with 2 days off between workweeks. Generally, shifts progressed from two weeks of day shifts (0800-1600) to two weeks of afternoons (1600-0000) to two weeks of night shifts (0000-0800), respectively. The 2-2-1 schedule required different start times each of the 5 days of the schedule beginning with 2 evening shifts (1600-0000 and 1400-2200), followed by two day shifts (0800-1600 and 0600-1400), and ending on the midnight shift (0000-0800) on the night of the fourth day.

Their studies compared a number of physiological markers of stress, sleep, and subjective measures in employees working these two schedule rotations. Researchers collected data on the sleep patterns of participants as well as a mood scale data. In initial studies comparing the 2-2-1 to the straight-5 schedule, Melton and his colleagues (Melton et al., 1973; Melton, Smith, McKenzie, Saldivar, Hoffman, and Fowler, 1975) reported that controllers found their jobs fatiguing, but measures of stress levels showed only slight stress differences between the 2 schedule patterns. In addition, they documented the decreasing sleep durations over the course of the 2-2-1, as well as an afternoon sleep period prior to the midnight shift.

Saldivar, Hoffman, and Melton (1977) administered a survey to 185 ATCSs to compare differences in reported sleep for ARTCC controllers working the two different schedules. Saldivar found no significant differences in the average amount of sleep reported over a 7-day period. The greatest amount of sleep was associated with the evening shift and the least amount of sleep was associated with the night shift in both schedules. Fatigue, weakness and somnolence were the most frequently reported complaints on the midnight shift. Controllers on the 2-2-1 reported better quality sleep than controllers working the straight-5 on their days off. They suggested that readjustment to night sleep consumed most of the straight-5 controllers' days off and, consequently controllers reported returning to work tired.

Finally, Melton (1985) compared data from the Miami International Flight Service Station (FSS) from non-rotating 'steady' shifts to the 2-2-1 schedule.

Fatigue checklist results showed that the 2-2-1 workers reported significantly more pre-work subjective fatigue than steady shift workers. In general, the 2-2-1 group also reported significantly greater levels of fatigue after 8 hours of work. Evaluation of the sleep logs found that sleep declined almost linearly over the 2-2-1 rotation from 8.3 to 5.4 hours. He concluded that while there were minimal biochemical indicators of elevated stress levels due to the 2-2-1, controllers reported greater fatigue levels.

Summary The early CAMI program investigated research questions that are still relevant today about the amount of fatigue and stress experienced by ATCSs on different shift schedules. Melton and his colleagues conducted several studies with ATCSS suggesting that the 2-2-1 might be less stressful than the straight-5 schedule. The authors argued that the 2-2-1 caused less circadian disruption because the majority of shifts were during normal working hours. The phase shift studies of other CAMI researchers, Higgins and his colleagues (Higgins et al., 1975; Higgins, Chiles, Mckenzie, Funkhouser, Burr, Jennings and Vaughan, 1976) supported the time differences in rephasal times of various rhythms, which would occur on the straight-5 schedule. Melton's group, however, did not provide the empirical data to demonstrate their speculation on the circadian effects. They also reported that ATCSs generally preferred the 2-2-1 rotation because of the 80 hours off between workweeks. In his 1985 study, Melton noted that straight, non-rotating schedules might be an 'obvious' remedy to the problems. However, he documented several objections to steady shifts in ATC. The primary reason ATC objected to non-rotating, steady shift schedules was because traffic was characteristically light on the midnight shift. Managers were concerned that controllers working only non-rotating midnight shifts would lose their proficiency to handle high traffic loads. Thus, the 2-2-1 schedule and similar scheduling patterns have survived to the present day.

ATC Studies: 1990-2003

For a variety of reasons, CAMI researchers revived the shiftwork research program in the early 1990s and conducted additional laboratory, survey, and field studies. We focused primarily on performance, sleep, mood, and circadian rhythms and how to develop countermeasures to shiftwork-induced disruptions.

10-Hour Field Study In the early 1990s, the FAA approved the use of compressed and flexible work schedules for employees. This implementation of alternate work schedules (AWS) in part rekindled interest in shiftwork research. Following an agreement with controllers to implement 10-hour workdays on a trial basis, the ATS requested that CAMI conduct a study to examine the possible effects of the longer workday on performance (Schroeder, Rosa, and Witt, 1998). The purpose of the study was to compare measures of performance and alertness of ATCSs working 10-hour shift schedules versus the more traditional 8-hour 2-2-1-schedule types. The facility schedulers designed the 10-hour schedules with a backward rotation from two afternoon shifts followed by two morning shifts and sometimes a midnight (in place of the second early morning).

Schroeder et al. (1998) collected data from 52 volunteers at the Minneapolis ARTCC. Of these, 26 ATCSs worked the 2-2-1 schedule and 26 worked four 10-hour shifts. Researchers collected performance data using the choice reaction time, mental arithmetic, and grammatical reasoning tests from the National Institute of Occupational Safety and Health (NIOSH) Fatigue Test Battery (Rosa and Colligan, 1988). Participants also completed questions about sleep, somatic complaints, and mood. Researchers administered three test sessions during each workday: at the beginning of the workday, 2 hours prior to the end of the workday (at the end of 6 hours for the 8-hour shift workers and at 8 hours for the 10-hour shift workers), and at the end of the workday.

The researchers did not find any statistically significant performance differences between the 8-hour and 10-hour groups. However, the analyses did reveal performance effects associated with day of the workweek and changes across the course of the workday. In general, choice reaction times decreased across the workday and workweek, but errors were found to be significantly higher at the end of the workday and workweek. The midnight shift performance tests resulted in decrements on each of the measures across the shift for the 8-hour group.

For both groups, the sleep data revealed a decline in sleep duration from an average of 8.35 hours on the night prior to the first workday (usually an afternoon shift) to an average of 5.75 hours on the night prior to the fourth day of the workweek. ATCS ratings of 'feeling rested' tracked the decline in sleep durations with a mean high rating (on a scale of 1-5) of 3.46 for the 10-hour group and 3.28 for the 8-hour group before the first workday of the week to 2.89 and 2.70, respectively, prior to the fourth day of the workweek. The researchers used the Naval Psychiatric Research Unit's (NPRU) mood scale (Johnson and Naitoh, 1974) to assess participant's mood. Positive and negative mood ratings remained relatively stable for the first 3 and 4 workdays, respectively. Again, the greatest changes were seen across the midnight shift. This group of participants reported very few somatic complaints.

Schroeder et al. (1998) concluded that the results of this study did not show evidence of differential fatigue between the 8-hour and 10-hour shift schedules. The results indicated performance decrements on the NIOSH battery tests across the workweek and across the workdays in both groups. The researchers pointed out that the declines in test performance did not necessarily indicate changes in actual job performance because, in the operational tasks, ATCSs often have more time for analyses than was provided for the battery tasks. However, they did note that the results reflected some general effects on the participants' readiness to perform.

Miami Field Study In 1993, we had the opportunity for the second field study when the Miami ARTCC 'Quality Through Partnership' (QTP) team requested a study of sleepiness on the midnight shift. Because the QTP program was a labor-management effort, the study was well supported. The facility management provided one hour of administrative time for potential volunteers to attend a briefing and complete the questionnaire. Management and labor representatives jointly briefed employees that the QTP team had requested the study. To accommodate the largest number of employees possible, the briefing was offered daily for 1 week. As a result of the labor/management support, 225 of the approximately 400 ATCSs at the ARTCC completed a shiftwork questionnaire and 95 ATCSs completed logbooks.

Logbook data presented an opportunity to compare the sleep patterns of ATCSs working different shift schedules, as well as sleepiness ratings (Cruz and Della Rocco, 1995a). We adapted the logbooks from those used by the National Aeronautics and Space Administration in their aircrew fatigue studies (Gander, Myhre, Graeber, Andersen, and Lauber, 1989). We were able to examine three specific shift schedules from the sample: 1) a 2-2-1, 2) a '2-1-2' (2 afternoons, 1 midday, and 2 early mornings), and 3) straight early mornings. For total sleep time (or TST), the data revealed: 1) a consistently low average TST for the straight early morning schedule across the workweek (M ranged from 6.3 hours, $sd=1.3$ to 5.6 hours, $sd=1.1$); 2) longer TSTs for the two afternoons and midday shift on the 2-2-1 and 2-1-2, respectively (M ranged from 8.3, $sd=1.0$ to 7.3 hours, $sd=1.5$); 3) two significant decreases on the quick-turn-rounds into the early mornings and midnight shift, on the 2-1-2 and 2-2-1, respectively. Over the course of the workweek, the 2-1-2 resulted in the greatest mean TST ($M=7.0$ hours) compared to both the 2-2-1 and the straight early morning shifts means of 5.7 hours and 6.0 hours, respectively. The asleep times remained relatively stable for each of the schedules, except for the sleep period during the quick-turn-around afternoon prior to the 2-2-1's midnight shift. Thus, even though the shifts changed, the participants generally maintained a stable asleep time. The ATCSs working early mornings reported asleep times around 2300, approximately 1 hour before the midnight times reported by the 2-2-1 and 2-1-2 participants. ATCSs reported awake times between 2-3 hours earlier on average before the early morning shifts than either the afternoon or midday shifts. The Stanford Sleepiness Scale (SSS) (Hoddes, Zarcone, Smythe, Phillips, and Dement, 1973) data did not reveal any differences between the schedule types as predicted. However, as in the 10-hour study, ATCS's rated their sleepiness higher at the end of each workday than at the beginning.

In a second paper, we investigated the turn-around time as a factor in the amount and quality of sleep obtained on the counterclockwise rapidly rotating schedules (Cruz and Della Rocco, 1995b). We compared short turn-around (STA) times of less than 10 hours between shifts to medium turn-around (MTA) times of between 10 and 15 hours. Both of these comparison groups had employees reporting to an early morning shift. The results suggested that the STA restricted sleep more than the MTA.

To determine how widespread shiftwork-related adaptation issues might be among the Miami ATCSs, we chose a general shiftwork survey (Cruz, Della Rocco, and Hackworth, 2000). The arguments posed in the literature against the counterclockwise rotation of shifts included the following: 1) greater disruption of circadian rhythms, 2) shortened sleep periods as a result of reduced time off between shifts, 3) increased fatigue, and 4) physical complaints from shiftworkers (Czeisler, Moore-Ede, and Coleman, 1982; Folkard, 1989. The NIOSH General Health and Adjustment Questionnaire (Tasto, Colligan, Skjei, and Polly, 1978), a measure of general adjustment to the shift schedules, served to assess employee health, mood, sleep patterns, eating patterns, and general lifestyle for the purpose of examining adaptation to shiftwork. The 225 participants represented the general demographics of the center, (with 78 per cent males, 78 per cent Caucasian, 13 per cent Hispanic, and an average age of 33 years, *sd*=6 years).

The results indicated that the population was indeed generally healthy because they must pass an annual medical examination. The five most commonly reported health symptoms in this sample were colds and sore throats, back pain, periods of severe fatigue or exhaustion, gas or gas pains, and bloated or full feeling. However, just over half of the ATCSs (55 per cent) reported periods of severe fatigue or exhaustion at least occasionally. Of those, 7 per cent reported fatigue frequently or constantly. About 50 per cent of the sample reported gastrointestinal symptoms occasionally or more frequently.

ATCSs responses suggested that they were experiencing elevated sleepiness during work hours. Over half of the sample reported experiencing the following: 1) feeling tired or sleepy at work at least two to three times per week (56 per cent), 2) catching themselves about to doze off at work in the last year (68 per cent), and 3) taking naps while at work (52 per cent). In addition, about one-third of the sample reported falling asleep while driving home from work. In addition, a higher percentage of ATCSs working 9-hour shifts (83 per cent) than 8-hour shifts (60 per cent) reported catching themselves about to doze off at work. By comparison to NIOSH Questionnaire data from two other shiftworking populations, food processors (Smith, Colligan, and Tasto, 1982) and ferry operators (Sparks, 1992), the ATCSs generally reported being healthier. ATCSs responses were similar to the comparison populations on measures of sleep. However, a higher percentage of ATCSs reported being tired or sleepy at work at least two times per week (56 per cent) compared with the food processors working straight night shifts (43 per cent).

Laboratory Study – Comparison of 2-2-1 to Straight Day Shifts In 1992, we initiated a laboratory-based study to gain more experimental control than was available during a field study. The study compared the 2-2-1 schedule to straight-day shifts to examine the effects of working the 2-2-1 on sleep, performance, neuroendocrine measures, and circadian rhythms in two different age groups (Della Rocco, 1994). Twenty male participants comprised the 'younger' group (n=10, age=30-35 years) and the 'older' group (n=10, age=50-55). Because they were not ATCSs, participants were matched on characteristics of the ATCS population including medical status and cognitive abilities. During the A-B-A research protocol, participants worked three consecutive weeks in the laboratory as follows: 1) 1 week of day shifts (0800-1600), 2) 1 week of the 2-2-1 schedule (1600-2400;

1400-2200; 0800-1600; 0600-1400; 2400-0800), and 3) 1 week of day shifts (0800-1600). Participants recorded data on sleep times, duration, and quality in daily logbooks (Della Rocco and Cruz, 1995). The core body temperature and activity data were verified using an ambulatory physiological monitor (Vitalog model HMS-5000, Redwood City, CA). Finally, urine samples were collected five times over the course of the day for neuroendocrine measures.

Participants provided performance measures by working on the computerized CAMI version of the Multiple Task Performance Battery (MTPB) (Chiles, Alluisi, and Adams, 1968; Della Rocco and Cruz, 1995). The MTPB tasks included monitoring and information-processing tasks involving mental arithmetic, complex visual discrimination, and problem solving. The tasks measured basic psychological or cognitive functions relevant to control of complex systems, in general, and ATC tasks, in particular (Chiles, Jennings, and West, 1972). The MTPB provided a motivating synthetic work environment in which participants completed three 2-hour MTPB sessions per 8-hour shift.

The results from analyses of the MTPB performance data revealed that the only significant decrement in performance occurred on the midnight shift (Della Rocco and Cruz, 1996). Participants evidenced significant decrements in both Active and Passive Composite scores during the third session of the midnight shift. Planned multiple comparisons revealed that performance on the passive tasks (both monitoring tasks) declined significantly over the course of the night shift. Planned multiple comparisons on the active tasks revealed that performance on the night shift was also significantly reduced compared to the previous day shift. Performance decrements ranged between 0-12 per cent.

The TST for the day shifts during this study were relatively short (M=6.8 hours, sd=.69). During the 2-2-1 schedule, TST demonstrated a decline across the 2-2-1 workweek, from the longest before the afternoon shifts (M=7.6 hours, sd=1.5), a shorter TST before the day (M=5.8 hours, sd=.9) and early morning shifts (M=6.6 hours, sd=1.0), and the shortest TST before the midnight shift (M=3.7 hours, sd=1.6). During the 2-2-1 work week in the laboratory study, asleep times were delayed by 2-hours prior to the second afternoon and advanced prior to the day shifts. In addition, the asleep time for the daytime sleep prior to the midnight shift was also significantly earlier than the previous sleep period because it occurred during the day. Within both the 2-1-2 schedule and straight early morning schedules in the Miami study, no significant differences in asleep time were found from one sleep period to the next. ATCSs on the early morning shifts appeared to have consistently earlier asleep times than employees on the other two schedules. Unlike asleep times during the 2-2-1 workweek in the Miami study, awake time was significantly advanced between the second afternoon shift and the first day shift. Likewise, awake time was advanced on the daytime sleep before the midnight shift. These differences between the studies demonstrate the importance of verifying laboratory findings by conducting field trials in the operational environment of the ATCS.

Circadian rhythm data (body temperature amplitude and acrophase) were compared for seven participants for the last 2 days of the day shift week and the 2 days following the 2-2-1 workweek (Della Rocco, Hackworth, and Cruz, 2000). Participants' core temperature rhythms revealed a significant delay of approximately 103 minutes between the acrophase during the day shift week (acrophase=1603 hours) and the end of the 2-2-1 workweek (acrophase=1746 hours). Likewise, the time of the peak cortisol and EPI levels appeared to be delayed. No significant differences were found in the mesor or amplitude of these participants.

The Congressionally Mandated Studies During the late 1990s, the National Transportation Safety Board, NASA, and researchers from other organizations were actively raising the awareness of the importance of addressing fatigue in safety-related occupations (Dinges et al., 1996). In the wake of this increased public awareness, the 1998 Department of Transportation Appropriations bill cited the CAMI program of research with ATCSs' shiftwork and fatigue. In 1999, Congress provided funding for CAMI to conduct an agency-wide comprehensive survey of ATC personnel to determine the extent of fatigue among the workforce and effects of current shift patterns and rotation practices on health, well-being, and performance. We adopted a multiphase research plan to be responsive to this mandate and to Article 55, Section 2 of the FAA/NATCA CBA, which also called for CAMI studies of fatigue, shiftwork, and stress-related issues. The first phase involved a comprehensive survey of the ATCS workforce. The second phase involved a more in-depth follow-up field study using objective measures to validate survey findings. The third phase involved a controlled laboratory study to directly and empirically compare the counterclockwise, rapidly rotating, 2-2-1 schedule to a clockwise rapid rotation recommended by the scientific literature.

To ensure the best scientific approach, CAMI assembled an internationally recognized group of scientists with expertise in shiftwork and fatigue research to advise on the study. In addition, recognizing a variety of coordination issues associated with the survey, we requested a group comprising FAA headquarters and management, union (NATCA and National Association of Air Traffic Specialists (NAATS)), and CAMI representatives. The ATS was forming workgroups to address several issues in the 1998 FAA/NATCA CBA at the time of our request. The ATS named the 'Article 55 Human Factors Workgroup' to serve as the administrative advisory group on this project. The group comprised eight NATCA and FAA representatives. The scientific advisory group discussed each item on the survey, made recommendations for additions, deletions, and modifications, and provided guidance for the study protocol. Both groups provided extensive input into the survey questions and methodology and continued to provide scientific and administrative oversight for the study.

Survey of the ATCS Workforce This phase provided our first opportunity to examine the extent of the shiftwork issues in the entire workforce. To ensure that the data collected in the congressionally mandated study could be compared with other shiftworking populations, we selected the Standard Shiftwork Index (SSI) (Barton, Spelten, Totterdell, Smith, Folkard, and Costa, 1995) as the survey instrument. The SSI was a battery of standard questionnaires assessing the respondents' shift schedules, job satisfaction, sleep patterns, fatigue, physical and mental health, and social and domestic life. The battery was easy to administer, and the psychometric properties of the subscales have been reported (Barton et al., 1995). The SSI has been used to examine a number of aspects of shift rotation systems including: advancing (counterclockwise) and delaying (clockwise) shift rotation systems (Barton and Folkard, 1993), permanent night and rotating shift systems (Barton, Smith, Totterdell, Spelten, and Folkard, 1993; Spelten, Totterdell, Barton, and Folkard, 1995), 8- and 12-hour shifts (Iskra-Golec, Folkard, Marek, and Noworol, 1996), 6 a.m. and 7 a.m. start times for 8- and 12-hour shifts (Tucker, Smith, Macdonald, and Folkard, 1998), regular and irregular shift rotations (Bohle and Tilley, 1998), and the number of consecutive night shifts for permanent night and rotating shift systems (Barton et al., 1995).

For purposes of this study, the SSI was modified based upon recommendations of the scientific advisory group (Della Rocco, Ramos, McCloy, and Burnfield, 2000). Experience with the SSI and other similar instruments, knowledge of ATC schedule characteristics, and the requirement to have a survey instrument that was suitable for a large-scale administration were all factors in decisions associated with the content of the survey. In some cases, the Article 55 Workgroup provided input to phrasing questions to improve the assessment in the ATC environment.

CAMI researchers mailed surveys to representatives at ATC facilities for distribution to all ATCSs with an FAA occupational code of 2152. A cover letter, signed by the Office of Air Traffic and the respective union presidents, informed each participant about the purpose, benefits, and risks of the study and provided a CAMI point of contact for questions. A total of 22,958 surveys were distributed during late November 1999. The useful return rate was 28.7 per cent.

In general, this combined sample was over age 35 (84 per cent) and reported an average of 19 years of shiftwork experience. This group was slightly older and had between 4-8 years more of shiftwork history than the nurses (age 33, shiftwork experience 12 years) and the Industrial/service (age 39, shiftwork experience 16) comparison groups in the Barton et al., (1995) normative sample. Compared to the normative sample, however, the ATCSs were similar to or reported higher levels of psychological well-being. Consistent with previous CAMI research (Smith, 1980; Collins, Schroeder, and Nye, 1991) which indicated that controllers report low levels of state and trait anxiety, controllers in this study reported lower levels of cognitive anxiety than the normative group. Levels of cardiovascular problems, digestive disorders, and chronic fatigue were low and comparable to the normative group. Compared to the normative group, the ATCSs reported higher general job satisfaction. The ATCSs did, however report greater disruption in social and domestic life than the normative group.

In 1995, Costa et al. reported SSI data from a group of 572 Italian air traffic controllers. All were male with a mean age of 43 and a mean shiftworking experience of nearly 20 years. This sample was very similar to the present FAA sample, with the exception of a 17 per cent gender difference. The Italian controllers worked shift schedules with rapid, counterclockwise shift rotations. Mean scores for the male FAA and Italian ATCSs were comparable on the measures of digestive complaints, psychological well-being, and cardiovascular complaints. However, the FAA ATCSs scored lower in cognitive anxiety and neuroticism.

To examine the effect of shift scheduling patterns on the outcome variables, we identified four prevalent scheduling patterns for analysis. We identified a comparison group of controllers working straight shifts (SS) (n=731), which included straight early morning, day, midday, or evening shifts. The second group of individuals worked a counterclockwise, rapidly rotating schedule with *no* midnight shifts (CR) (n=1,994). The CR grouping included schedules that generally started a workweek with afternoon shifts and rotated, with progressively earlier start times, to end the week with early morning shifts. The third schedule group was a counterclockwise, rapidly rotating shift schedule with midnights (CRM) (n=1,486). This group included the traditional 2-2-1 shift. The final schedule group was the straight-5 (S5) (n=313), the schedule examined by Melton and Bartanowicz, 1986, involving one week of the same shift and advancing to an earlier shift in the following week.

The results from analyses of the modified SSI data revealed that, as anticipated, the two schedule groups with midnight shifts, the CRM and straight-5, had the poorer outcome measures. Specifically, chronic fatigue, sleepiness, and domestic outcome measures were significantly different from the straight shift group. Likewise, both the CRM and straight-5 groups reported significantly higher levels of digestive disorders than did the SS group. Sleep quality ratings were significantly worse for the CRM group than the SS group. The CR group, however, also tended to show some adverse effects of the rapid rotations. The CR group reported significantly greater chronic fatigue and sleepiness scores than the SS group. All of the rotating shift groups showed poorer general psychological scores on the General Health Questionnaire (GHQ) than the SS group. Of note in these data is the fact that the SS group did have consistently better outcomes than the groups with shift rotations despite the fact that the majority of the SS group reported typically working straight early morning shifts. This finding would suggest that the rotations result in worse outcomes, even without midnight shifts, than the sleep loss associated with the straight early morning shifts. Thus, the stability provided by the straight shifts is likely important to the better outcomes. We should note that, while the differences were statistically significant, they were small.

Examination of some performance indicators revealed that decrements after the midnight shift were the worst for mental sharpness and automobile driving. More than one third of the respondents in this sample reported that they had experienced falling asleep or experiencing lapses driving home from the midnight shift during the previous 12 months. Nesthus, Cruz, Boquet, and Hackworth (2004) reported that the number of commuting miles (>20 miles); type of roadway (highway, city traffic, country roads); and subjective mental sharpness (high mental sharpness vs. low mental sharpness) contributed to greater odds ratios for the outcome variables, including lapses of attention, falling asleep, near misses, and accidents during *early-mornings*, particularly for nightshift workers returning home. These results also support the findings from a previous study (Cruz, Della Rocco, and Hackworth, 2000). Similarly, the cognitive performance measures from laboratory studies (Della Rocco and Cruz, 1996; Schroeder et al., 1998) had previously pointed to the midnight shift as a vulnerable time during the CRM schedules. In this sample, 5-6 per cent reported having operational deviations or operational errors in the past year. Of all of those cases, 48 per cent reported that fatigue was a factor. The CR and CRM group reported fatigue as a factor in 49 per cent and 46 per cent, respectively. However, the straight-5 group reported fatigue as a factor in 58 per cent of the cases.

From these analyses, the straight-5 resulted in worse outcomes than the CRM schedules. An assessment of the CRM schedules in comparison with the European standards for schedule design reveals that, from an ergonomic standpoint, there were some positive features. The CRM schedules minimize exposure to the midnight shift. Days off are provided after the midnight shift. The down sides are the counterclockwise rotations, the early morning start times, and quick-turn-arounds.

Air Traffic Shiftwork and Fatigue Evaluation – AT-SAFE Field Study Researchers designed this second phase of the congressionally mandated examination of the effects shift scheduling practices on ATCSs to provide empirical data to supplement the self report data collected by the national survey. The Administrative Steering Group (Article 55 Human Factors Workgroup), a retired controller, and data acquisition team manager provided invaluable assistance in gaining agency approvals, facilitating logistical support, and coordinating with the facility management, union and volunteers. Participants included 70 full-time Full Performance Level (FPL) ATCSs working at either a Terminal Radar Approach Control Tower (TRACON, n=19) or an ARTCC (n=51). Participants completed a 3-week protocol that included measures of both cognitive performance during the first 10 days as well as actigraphy (sleep/wake) and logbook reports of sleep duration and sleep quality, and various cognitive, subjective mood, and sleepiness scales during the entire 21 days.

Researchers selected CogScreen© Aeromedical Edition, a computerized test battery that was developed by the Advance Resource Development Corporation and Georgetown University (Kay, 1995) for performance measures because it has been shown to detect subtle clinical changes in the cognitive function. Analyses of the CogScreen© data revealed a significant effect of age on performance with differences on tests involving information-processing speed and divided attention. Researchers calculated composite scores in order to compare to data from the available normative samples (i.e., general aviation, the military, and commercial aviation pilots). We found the greatest effect for the Speed/Working Memory factor, such that younger participants (<40 yrs) were significantly faster and produced greater throughput than the older participants (>40 yrs). So, age was significantly associated with speed and working memory, as was the alertness rating. Older controllers performed less efficiently than younger controllers; and, those with greater subjective alertness also performed better. But there was no significant interaction between chronological age and subjective alertness. As subjective alertness increased, performance improved, and as age increased, performance fell. Also, participants' performance demonstrated a dip in subjective alertness in the morning at approximately 1000 hours, which coincided with approximately 3-4 hours of time on duty (for the early morning shifts). A secondary analysis revealed a significant relationship between time on shift and performance. In summary, the analysis of the performance data with respect to age, suggested that ATCSs may need to become more aware of their internal state of alertness as they become older. Although all of the performance scores were well within normal limits, it appeared that those individuals over the age of 45 should attend carefully to their subjective alertness. If subjective alertness is low, these individuals would likely benefit more from an intervention to maintain alertness and to counter the effects of fatigue (Becker, Nesthus, Caldararro, and Luther, in press).

The researchers also analyzed sleep durations, sleep quality, mood, and subjective sleepiness (SSS) to address the effects of shift start times, shift schedules, and the quick-turn-around shift rotations. Previous research (Folkard, 1989) has indicated that some shift start times promote greater sleep reduction/disruption than others. Reduced sleep and poor quality sleep generally have a negative affect on mood and sleepiness and are, therefore, of interest with this sample of data. Controllers working the early morning shifts (i.e., beginning before 0800) showed 5.8 hours of TST. Day, mid-day, and afternoon shifts (starting between 0800-1000, 1000-1259, and 1300-1959, respectively) promoted longer TSTs of 6.5, 7.7, and 7.6 hours. Controllers napped an average of 2.3 hours before the midnight shift and slept following the midnight shift for an average of 4.5 hours. Their regular days off (generally 2 consecutive days) produced sleep durations of just less than 8 hours. The sleep quality ratings associated with these shift start times followed a consistent pattern with more sleep producing better quality ratings. The data revealed higher positive mood ratings for the day, mid-day, and afternoon shift start times also associated with greater sleep duration. The higher sleepiness ratings associated with post-shift times occurring in the evening (M=3.1-3.4) and the early morning (M=4.6) shifts revealed a time of rating (i.e., time of day) effect apparent in these data. Sleepiness ratings during the drive home after the midnight shift were the highest at M=4.9.

To address the question concerning the effects of current schedules on these measures, we evaluated four different shift types from our limited field study sample, including: 2-2-1, 2-1-2 (2 afternoons, 1 mid-day, 2 early mornings), 2-3 (2 afternoons, 3 early mornings), and straight early mornings (Nesthus, Holcomb,

Cruz, Dobbins, and Becker, 2002). A day by shift-type interaction for sleep duration showed that the amount of sleep controllers experienced was clearly a function of shift start time as previously discussed. A day main effect was found for the sleep quality data revealing a general decline in ratings across the workweek for all but the 2-1-2 shift type. On the 2-1-2 schedule, sleep quality ratings improved slightly across the week. We found positive mood to favor the 2-1-2 schedule with as much as a 24 per cent improvement in ratings over the week compared with the other schedules. For the pre-shift sleepiness scale ratings, controllers generally reported the least sleepiness on the 2-1-2 schedule and the 2-3 and 2-2-1 schedules reported increased sleepiness across the workweek. The post-shift ratings of sleepiness suggested that working an 8-hour shift affected each daily report. The most notable change occurred following the midnight shift for the 2-2-1 schedule. Here, a significantly increased rating of sleepiness was reported (M=4.75). The time this measure was taken was in the early morning, just before their drive home, which makes this elevated sleepiness report noteworthy. In response to the national ATCS shiftwork and fatigue survey, more than a third of the respondents indicated having fallen asleep or experienced a lapse of attention during their drive home following the midnight shift during the prior 12 months.

We conducted a third evaluation of the sleep measures to determine the effects occurring over the quick-turn-around rotations from afternoon to early morning shifts (A/EM) and from early morning to midnight shifts (EM/N). Comparisons were also made to determine the potential effects of the length of time off between rotations (i.e., 8-, 9-, or 10-hour quick-turn-around rotations) found in this field study sample (Nesthus, Cruz, Boquet, and Holcomb, 2003). Generally speaking, afternoon shifts allowed for later waking times and resulted in sleep durations of 7-8 hours or more, whereas early morning shifts typically restricted sleep to 5-6 hours for normally day-entrained individuals. They are required to get up shortly after their biological nadir, while the need for sleep is still somewhat unfulfilled. Attempts to obtain more sleep by going to bed earlier are often unsuccessful because our circadian rhythms are entrained by daylight exposure and tend to be set to maintain activity and alertness into the evening. This pattern does not promote sleep until a later hour (e.g., 2200 or later). Specific, measures would need to be taken to shift the circadian rhythms to an earlier orientation. In the present study, sleep duration fell from M=7.2 hours before the afternoon shifts to M=5.4 hours before the early morning shifts. Other studies have shown similar influences and results (Cruz, Boquet, Detwiler, and Nesthus, 2002; Cruz et al., 2000; Cruz and Della Rocco, 1995b; Della Rocco et al., 1999).

Likewise, the reduced sleep duration experienced before the midnight shift is due to the combined difficulty in sleeping and decreased quality of sleep during the day with the social and domestic demands controllers typically experience before the midnight shift (Della Rocco, Comperatore, Caldwell, and Cruz, 2000). In the early morning to midnight shift quick-turn-around in this study, TST averaged 3.3 hours. Most controllers in the study reported attempting this nap before the midnight shift; however, not all controllers were able to get a nap during the quick-turn-around to the midnight shift. PANAS (mood) ratings for the afternoon to early morning shift quick-turn-around rotations indicated that positive affect (PA) declined, whereas negative affect (NA) increased. These changes may reflect trends associated with the amount of sleep controllers got before their shifts began. The reported SSS values for the beginning of the shift showed a 23 per cent increase in sleepiness for the early morning shift (having had less sleep), compared with the afternoon shift (with more sleep). Åkerstedt (1996) reported similar (i.e., higher) levels of sleepiness for the early morning shift (4.2 at 0700) compared with the afternoon shift (2.6 at 1500) for nuclear power station shiftworkers on an 8-point Karolinska Institute sleepiness scale. Even more pronounced for ATCSs in this study was the reported 30-36 per cent increase in sleepiness at the end of the shift, during the drive home, and after arriving at home following the midnight shift. These changes reflected the cumulative effects of working at night when normally asleep and the resultant sleep loss, itself.

The quick-turn-around time-off periods between shift rotations required controllers to compress the time available for commuting, personal hygiene, sleep, eating, and returning to work. The quick-turn-around time-off periods for the afternoon to early morning shift rotations were 8 hours, 9 hours, or 10 hours. The quick-turn-around time-off periods for the early morning to midnight shift rotations were mostly 8 hours with a few at 9 hours. In general, 9-hour time-off periods between both quick-turn-around rotations resulted in better outcomes than 8-hour time-off periods. Specifically, the 9-hour time-off period resulted in significantly better mood, reduced sleepiness, and better sleep quality in the afternoon to early morning shift rotation along with higher PA and better sleep quality in the early morning to midnight shift rotation. While 9-hour time-off periods were most common in the afternoon to early morning shift rotations, 8-hour time-off periods were more typical in the early morning to midnight-shift rotations. These data suggest that even one additional hour for all of the activities required between shifts resulted in better outcomes.

Laboratory Study – Comparison of 2-2-1 Counterclockwise to Clockwise Rapid Rotation As we have discussed, much commentary in the literature suggested that a clockwise (CW) rotation of shift schedules would be better than a counterclockwise (CCW) rotation (Czeisler et al., 1982; Folkard, 1989). Researchers argued that the clockwise rotation took advantage of the inherent slight delay of the circadian system. However, not much empirical data existed. The Congressional funding provided CAMI with the opportunity to directly compare schedules with the two different rotations (Boquet, Cruz, Nesthus, Detwiler, Knecht, and Holcomb, 2004; Cruz, Detwiler, Nesthus, and Boquet, 2003a; Cruz, Boquet, Detwiler, and Nesthus, 2003b). As in the first laboratory study, researchers used a 3-week protocol. Twenty-eight participants completed a week of day shifts (0800-1600), followed by 2 weeks of one of the rotation conditions. For each of the rotations, the two afternoon shifts were scheduled between 1400-2200. The two early morning shifts were between 0600 and 1400. Finally, the midnight shift was scheduled between

2200-0600. Fourteen participants (7 males and 7 females) with an average age of 40.6 years (*sd*=9.4) completed the CW rotation. Fourteen participants (5 males and 9 females) with an average age of 41.9 years (*sd*=9.0) completed the CCW rotations.

As in the first laboratory study in 1992, the researchers used the MTPB as a motivating, synthetic work environment for one of the performance measures. However, they also used the Bakan Vigilance Task (Dollins, Lynch, Wurtman, Deng, Kischka, Gleason, and Lieberman, 1993) at the beginning and end of each shift. Participants completed a logbook to document sleep parameters, sleep quality and sleepiness ratings, mood, as well as other daily activities. Researchers used an ambulatory physiological monitor (Series 2000 Minilogger, MiniMitter Co., Inc., Sunriver, OR) to measure core body temperature, wrist activity, and ambient light. Researchers collected saliva samples at the end of the baseline week for melatonin and cortisol assays, as well as at the same times of day during the 2 shiftwork weeks.

There was no effect of rotation condition for any of the sleep measures. A main effect for sleep period indicated that sleep duration before the two early morning shifts (*M*=5.0h and 5.6h; *sd*=1.0 h for both) was significantly shorter than before the two afternoon shifts (*M*=7.9 h and 7.5 h; *sd*=1.3 h and 1.1 h). The difference in the sleep onset and awakening times for the shifts was noteworthy. Sleep onset time before the second early morning shift (*M*=22:12, *sd*=01:06) was significantly earlier than before the afternoon shift (*M*=23:33 and 23:53, *sd*=01:23 and 00:47). Also, the awake time was significantly earlier before the two early morning shifts (*M*=04:39 and 04:40; *sd*=00:24 and 01:05). This is of particular interest for the CCW condition because, when combined with the weekend nights, the sleep onset and awake times for the first 2 afternoon shifts reveal that there are four consecutive opportunities for delayed sleep. This might be analogous to traveling west and aligning one's activity and sleep to a later clock time, which could give delaying signals to the circadian clock.

On the performance measures, results indicated that effects of rotation condition were modulated by shift type, such that on particular shifts, performance in the CCW rotation was actually better than in the CW rotation. (Cruz et al., 2003b). For the Bakan task, performance during the first week of shiftwork was significantly better (*M*=101.7, *sd*=22.6) than during the second week of shiftwork (*M*=97.5, *sd*=25.6). In addition, performance was significantly better at the beginning of shifts (*M*=102.8, *sd*=23.2) than at the end (*M*=96.4, *sd*=25.0). While it appeared that the CCW group performed consistently better than the CW group across all shifts, results of the simple effects analyses indicated a significant difference only on the first afternoon shift, where the CCW group (M=113.5, sd=14.9) performed significantly better than the CW group (M=90.2, sd=28.9). For the MTPB, the active task composite score showed that on Session 1, there was no significant effect of rotation condition. On Session 2, there was a significant effect for rotation condition on the first early morning shift, with the CCW rotation (*M*=525.9, *sd*=32.5) performing better than the CW rotation (*M*=483.0, *sd*=70.7). On Session

3, there was a significant effect for rotation condition on the first afternoon shift, with the CCW rotation (*M*=534.1, *sd*=22.5) performing better than the CW rotation (*M*=464.8, *sd*=111.8). The authors note that the first afternoon shift was the first day of the CCW schedule and the third day of the workweek for the CW schedule. Analysis for the Overall and Passive task composites revealed that performance was better at the beginning of the midnight shift than at the end, and that performance at the end of the afternoon shift was better than the start of the midnight shift.

Analyses of the cortisol and melatonin data found results consistent with describing a normal circadian rhythm for these two hormones, including a suppression of melatonin during the midnight shifts for both groups, relative to their baseline values. The analyses revealed relationships between levels of hypothalamic-pituitary axis (HPA) activity and cognitive performance: high cortisol was associated with poorer performance on passive monitoring tasks, active problem solving tasks, and the Bakan vigilance task (Detwiler, Boquet, Cruz, and Nesthus, 2002). No rotation group differences were found for cortisol; however, the clockwise group had a significantly greater increase in melatonin during the early morning shift than the counterclockwise group (Boquet et al., 2004).

Finally, analyses of the core body temperature data for the last 72 hours of each week revealed that the amplitude for the counterclockwise rotation (*M*=0.3°C, *sd*=0.06°) was significantly lower than the amplitude for the clockwise rotation (*M*=0.5°C, *sd*=0.13°). In addition, an 84-minute delay of the acrophase was found for the counterclockwise rotation (*M*=1808, *sd*=1.5h) relative to the clockwise rotation (*M*=1644, *sd*=1.9h). This delay was evidenced in relation to the baseline value for the counterclockwise rotation as well. Indeed baseline acrophase values for the clockwise (*M*=1610, *sd*=1.5h) and counterclockwise (*M*=1613, *sd*=2.1h) rotations were not significantly different from each other (Boquet et al., 2004). Researchers suggested that the delaying effects of the later sleep onset and awake times for the weekend sleep periods and the afternoon shift sleep periods early in the week combined with the extended activities for the last day of the schedule (i.e., an early morning shift followed by the quick-turn-around to the midnight shift) may have contributed to this effect (Nesthus, Cruz, Boquet, Detweiler, Holcomb, and Della Rocco, 2001). Also, an evaluation of these data with MTPB performance, indicated that a group of subjects with lower bathyphase values (i.e., the lowest point on the temperature rhythm) performed significantly better on the midnight shift than the higher bathyphase value group on both the active and passive MTPB composite scores (Boquet, Cruz, Nesthus, 2003).

In summary, results of this rotation comparison study indicated that for the most part, performance was similarly affected on each shift for both rotation conditions, and that if anything, performance was actually better during the counterclockwise rotation. Results of the study were similar to past research on the 2-2-1 schedule performed by Della Rocco and Cruz (1996) indicating that performance is maintained on early morning and afternoon shifts and drops more dramatically across the midnight shift. Again, Monk (1990) and others (Akerstedt, 1988; Klein, Bruner, and Holtman, 1970) have consistently found that performance decrements on the midnight shift are to be expected in any shift schedule configuration. A significant benefit of the 2-2-1 schedule worked by many ATCSs, then is that it generally includes only one midnight shift, which is placed at the end of the workweek and is followed by at least two days off. This is in keeping with European recommendations that no more than two to four midnight shifts should be worked in succession (Wedderburn, 1991).

Summary of Recent Research

The body of evidence on ATCS schedules revealed the following disruptive influences: 1) sleep loss occurred toward the end of the week on counterclockwise, rapidly rotating schedules as employees rotated into early morning and midnight shifts, 2) decrements in cognitive performance were primarily evidenced on the midnight shift, although some changes were observed on the early morning shift, 3) performance was not found to differ between the 10-hour and the first 4 days of the 8-hour 2-2-1 schedules, 4) straight early morning shift schedules resulted in as much sleep loss as the counterclockwise, rapidly rotating shift schedule with a midnight shift, 5) on surveys over half of the participants reported at least some fatigue and shiftwork maladaptation, and 6) circadian rhythms were disrupted. Scheduling pattern data suggested that rotating schedules resulted in worse outcomes than stable patterns, and that quick rotations were better than straight-5 schedules. These data argued for fatigue countermeasures, particularly on the midnight shift, as well as scheduling manipulations.

Development of the Countermeasures

The scientific community had begun exploring methods for managing the effects of sleep loss and shiftwork. Strategies included exercise, caffeine, circadian rhythm management, and napping (Dinges et al., 1996). The goal of these strategies was to manage alertness and cognitive resources and/or to develop strategies to protect against those times when alertness could be predicted to be degraded.

The napping literature was developing data to suggest that strategic and prophylactic naps were very effective at maintaining alertness during circadian rhythm troughs, as well as under conditions of sleep loss (Della Rocco, Comperatore, Caldwell, and Cruz, 2000). NASA researchers were exploring naps for pilots in the cockpit (Rosekind, et al., 1994). Using electroencephalography, the NASA researchers found microsleeps in pilots during flights. They noted that sleepiness is very difficult for a sleep-deprived individual to fight and that only napping effectively counters sleep deprivation. Because the findings from ATCSs were primarily that controllers were experiencing sleep loss associated with the shift schedules, and that performance and alertness degradation was primarily on the midnight shifts, napping strategies uniquely addressed a number of the shiftwork-related fatigue observed in previous ATC studies. Napping would address the sleep loss, tiredness, and could work to maintain performance across the shift. It might help with the drive home from the midnight shift. Finally, a nap on the midnight shift might serve as an 'anchor' sleep and may work to help stabilize circadian rhythms.

CAMI, in collaboration the US Army Aeromedical Research Laboratory (USAARL), conducted a study to examine the effectiveness of 1 or 2 hour naps during a midnight shift at maintaining performance and alertness (Della Rocco, Comperatore et al., 2000). Sixty ATCSs participated at the USAARL sleep laboratory. Researchers assigned the participants to one of three midnight shift napping conditions: a long nap of 2 hours, a short nap of 45 minutes, and a no nap condition. ATCSs completed a 4-day protocol during which they worked three early morning shifts (0700-1500) followed by a rapid rotation to the midnight shift (2300-0700). Subjects completed three 1.5 hour test sessions (one session before the nap and two sessions after the nap) during the midnight shift involving two computer-based tasks: 1) the Air Traffic Scenarios Test (ATST), a task developed for selection of ATCSs (Broach and Brecht-Clark, 1994), and 2) the Bakan, a test of vigilance (Dollins et al., 1993). Both cognitive performance and subjective measures of sleepiness supported the use of naps during the midnight shift. In fact, both the long nap of 2 hours and the short nap of 45 minutes resulted in better performance than no nap on the Bakan test at the end of the midnight shift. A doze-response relationship existed such that the long nap also resulted in better performance than the short nap. The ATST, on the other hand, was much less sensitive to differences in napping condition and even to the expected circadian trough, which would have been expected to affect all groups. Sleepiness ratings on the SSS suggested that sleepiness increased across the midnight shift for all groups, but ratings were lower for the long nap condition and were lower for males in the short nap condition, when compared with the no nap condition.

These results and those of a recent unpublished CAMI report suggested that naps taken during the midnight shift could be useful as a countermeasure to performance decrement and sleepiness on the midnight shift. However, a number of important issues remained to be explored before we would recommend that napping be adopted as a countermeasure in the field. One very significant issue concerns sleep inertia, the period of grogginess experienced upon awakening, that may pose a problem in implementing a napping policy. If nappers reach deeper stages of sleep, there is a period of time after awakening (10-20 minutes) during which performance and alertness may be compromised (Della Rocco, Comperatore, et al., 2000). This is termed sleep inertia. In a sleep-deprived population like shiftworkers, it may be more likely that nappers would reach the deeper stages of sleep more quickly than a rested population. CAMI proposed a study to examine these issues. Napping presents a number of problems for field facilities, also. First, it is a logistical problem in terms of a facility's physical accommodations. Second, policies need to be developed to manage personnel issues and ensure fairness.

Policies would need, for example, to ensure that the individual was not experiencing sleep inertia prior to returning to controlling traffic. Finally, sleeping on duty is currently not permitted by the Office of Personnel Management. Thus, much more work needs to be accomplished, both scientifically and through coordination with agency operational personnel.

Napping was not the only available countermeasure, however. As we have discussed, scheduling can work to mitigate or exacerbate disruptions due to shiftwork. The laboratory study associated with the congressional project (Boquet, et al., 2004; Cruz et al., 2003a and b) compared the counterclockwise, rapidly rotating shift rotation to the clockwise rotation. Contrary to the recommendations in the literature, the clockwise rotation did *not* provide measurable improvements, at least in a short-term study. Longer-term studies might find otherwise; however, this particular countermeasure did not appear to provide a benefit. During the course of the research program, a number of findings pointed to areas where improvements could be made. For example, longer times between shifts on quick-turn-arounds can improve outcomes (e.g., sleep duration, mood) (Cruz and Della Rocco, 1995b; Nesthus, Cruz, Boquet, and Holcomb, 2003). Finally, stability in scheduling patterns may be more important to better outcomes than sleep deprivation due to early morning start times (Della Rocco, Ramos et al., 2000).

Transitioning the Results to the Workforce

After the napping study, we sought to transition state-of-the-art knowledge from the literature and our research to the workforce through development of an educational brochure. We included information on disruptions due to shiftwork, management of sleep circadian rhythms and alertness, particularly with respect to counterclockwise, rapidly rotating shift schedules. The Air Traffic Service, for a variety of reasons, failed to approve distribution of the brochure to the workforce.

Although this was unfortunate, it was instructive. We quickly became aware, that while we may have known the problems of shiftwork, we did not adequately coordinate the research findings with Air Traffic. The Congressional funding provided an opportunity to improve the coordination. We established the two advisory groups to maximize this opportunity from the beginning. The internationally renowned group of scientists on the project helped to ensure that the research was 'state-of-the-art.' We sought the administrative group, the Article 55 Workgroup, to optimize the logistics of the Congressionally mandated research and provide a mechanism – a labor-management forum which could facilitate transitioning the findings. This group consisted of representatives from FAA Headquarters Air Traffic Service, field facility managers, NATCA, and CAMI scientists. Because this group was to address Article 55 (Human Factors) in the 1998 FAA/NATCA CBA, it had an official status which provided a better forum for the study results. Both of the groups were instrumental in key phases of the project.

In June 2000, we convened a meeting of both the Scientific and Administrative Groups to collectively review the shiftwork survey data. The scientists made several recommendations, including education of the workforce on problematic areas associated with shiftwork and its effects on performance, as well as fatigue countermeasures and coping strategies. Because the problems identified were primarily sleep loss and fatigue, they also recommended that we institute periods of controlled rest or napping as a primary fatigue countermeasure. Finally, they recommended that the Agency look for opportunities to improve the current scheduling practices based upon known ergonomic principles, quality of life concerns, and operational constraints. Dr. Giovanni Costa provided a list of general guiding principles, including minimizing exposure to midnight shifts, providing rest the day after midnight shifts, rotating shift start times in a clockwise rotation, provide long intervals between cycles (at least 2 days off), avoiding quick-turn-arounds (8 or less hours between the end of a shift and start of the next), avoiding early morning start times, providing weekend days off, scheduling 8-hour shifts instead of longer 12 hours, and providing predictability in scheduling for the employee.

Based upon the data and recommendations of the scientists, the Article 55 Human Factors Workgroup identified the following four recommendations as the highest priority for the Agency to consider (Article 55 Human Factors Workgroup, 2001). The workgroups' co-leads presented these recommendations during two meetings with the Associate Administrator for Air Traffic and the NATCA National Executive Board.

1. Feedback the survey results to the workforce.
2. Educate the workforce on shiftwork, its effects on performance, and management of fatigue.
3. Allow napping during break periods as a tool to mitigate the effects of fatigue and sleep loss resulting from shiftwork in order to increase alertness and maintain performance.
4. Encourage facilities at the local level, in collaboration with NATCA, to evaluate their current schedule for opportunities to apply the ergonomic principles identified in this study in order to improve the adaptation to shiftwork.

The efforts resulted in development and delivery of the survey results summary brochures and an educational multi-media CD ROM titled 'Shiftwork Coping Strategies' to all 23,000 air traffic control specialists (with the work code 2152). This CD was developed for CAMI under contract with Key Multi-Media Solutions (Contract No. DTFA-02-98P80593). Since this time, the CD has been modified for the FAA Airways Facilities workforce, duplicated and delivered to some of the trucking industry workforce through collaborative efforts with department of transportation (DOT) VOLPE. Also, in collaboration with the US DOT Research and Special Programs Administration, and in particular, the Human Factors

Coordinating Committee Operator Fatigue Management Team efforts, some materials from the CD have been incorporated in the Battelle-developed 'Commercial Transportation Operator Fatigue Management Reference' released July 31, 2003 (Transaction Agreement No. DTRS56-01-T-003).

Discussion

This program was an example of successful development of research into products and information for both the individual controller and managers. The program benefited from longevity and a little luck. The longevity allowed the concepts to mature both in the scientific community and within the FAA's understanding of the ATCS's specific environment.

We were able to investigate shiftwork-related fatigue as it was manifested in the ATC environment. The patterns are different in different operational communities. By examination of the variety of scheduling practices, we were able to confirm patterns due to specific schedules. Then we could target specific countermeasures. It is very important for researchers to first examine the basic scheduling patterns within an operational community, as well as the existing coping strategies of the personnel. While the principles of sleep and circadian rhythms have been studied for a long time, countermeasures must be targeted for specific circumstances. Misapplication of these principles has the potential to result in worse disruption than the coping strategies employees might adopt left to their own strategies.

Within the ATCS population, napping strategies appeared to uniquely address the issues of sleepiness and performance decrements, which were so clearly present in the data. The data pointed to the midnight shift for, at least initially, targeting the napping countermeasures. Napping can reverse the observed performance decrements. Sleep could reverse some of the sleep loss that we found accumulating over the course of the workweek. A nap placed a few hours before the end of the midnight shift can improve performance at the end of the shift when traffic levels begin to escalate in many facilities. Finally, an extended nap on the midnight shift can serve as an 'anchor' sleep which might help provide some stability for the circadian clock (Minors and Waterhouse, 1981). However, prior to the implementation of any countermeasure, both laboratory and field trials should be conducted to reveal unforeseen complications. The napping countermeasure, for example, involves a 'side-effect' if you will, called sleep inertia. Sleep inertia is a transient period of 'grogginess' experienced after a nap. The duration of this period varies over the course of the day or night and depends on the stage of sleep reached upon awakening. The time-course of sleep inertia is an important addition to the total time of the napping break, since clear thinking is required in the performance of most jobs.

Research on schedule design holds a lot of potential. Researchers have begun developing computer-based programs for evaluating schedules (Schoenfelder and Knauth, 1993; Hursh, 2003). These could be of great help to ATC schedulers if we could tailor them to the ATC operational environment. Also, schedule design must be carefully considered locally with each facility because of a variety of reasons, including local traffic patterns, geographical considerations which influence circadian rhythms. If the literature shows that starting a shift at 0800 results in 30 minutes of additional sleep for an employee as compared to a 0700 start time but the commute to arrive at work at 0800 requires the employee to leave at the same time, the countermeasure is not going to work. The literature does suggest, however, that small changes can result in improvements worth considering.

While the recognition of the importance of sleep loss to health and performance has been in the literature for some time, American society is just recently becoming aware. A recent study provided a focus that many people can understand. Dawson and Reid (1997) conducted a study comparing driving performance of participants impaired with a blood alcohol concentration of 0.10 per cent (legally drunk in many states) to 24-hour sleep loss. They found the same levels of impairment in judgment and reaction time in both groups. In addition, the degree to which sleepiness is an impairment is difficult for people to accurately identify. Researchers have demonstrated that people were unable to accurately recognize when they had nodded-off for a short period of time (Gastaut and Broughton, 1963).

Adopting innovative strategies to counter the adverse effects of shiftwork benefits both the organization, as well as the individual. By adopting strategies that maintain performance during times when performance would otherwise be degraded (i.e., midnight shifts), we increase the safety margin. It is possible that with better adaptation strategies (e.g. locally improved schedule design) other benefits could be seen, such as reduced use of sick leave. Finally, the long-term health and mood of the ATCS could be improved over the course of his/her career.

We as researchers must understand and acknowledge the issues that arise in transitioning our results to the field. Even though we may think we have solved a problem, with supporting empirical data, we may be creating other problems for managers or employees. The extent to which we as researchers work to understand all of the issues in the operational environment will greatly influence our success transitioning our findings from the laboratory to the field.

References

Åkerstedt, T. (1988), 'Sleepiness as a consequence of shiftwork,' *Sleep*, Vol. 11, pp. 17–34.

Åkerstedt, T. (1996), *Wide awake at odd hours, Shift work, time zones, and burning the midnight oil*, Swedish Council for Work Life Research, Stockholm, Sweden.

Åkerstedt, T. and Torsvall, L. (1980), 'Age, Sleep, and Adjustment to Shift Work,' in W.P. Koella (ed.), *Sleep*, pp. 190–194, Karger, Basel.

Åkerstedt, T., Torsvall, L. and Gillberg, M. (1982), 'Sleepiness and Shift Work: Field Studies,' *Sleep*, Vol. 5, pp. 95-106.

Anch, A.M., Browman, C.P. Mitler, M.M., and Walsh, J.K. (1988), *Sleep: A Scientific Perspective*, Prentice Hall, Englewood Cliffs, NJ.

Aschoff, J. (1965), 'Circadian Rhythms in Man,' *Science*, Vol. 148, pp. 1427-1432.

Barton, J. and Folkard, S. (1993), 'Advancing versus Delaying Shift Systems', *Ergonomics*, Vol. 36, pp. 59-64.

Barton, J., Smith, L., Totterdell, P., Spelten, E., and Folkard, S. (1993), 'Does individual choice determine shift system acceptability?', *Ergonomics*, Vol. 36(1-3), pp. 93-99.

Barton, J., Spelten, E., Totterdell, P., Smith, L., Folkard, S. and Costa, G. (1995), 'The Standard Shiftwork Index: A Battery of Questionnaires for Assessing Shiftwork-related Problems', *Work and Stress*, Vol. 9(1), pp. 4-30.

Becker, J.T., Nesthus, T.E., Caldararro, R. and Luther, J. (in press), 'Shiftwork and Age-Associated Performance Variation Among Air Traffic Control Specialists', DOT/FAA/AM, Federal Aviation Administration, Office of Aviation Medicine, Washington, DC.

Bohle, P. and Tilley, A. (1998), 'Early experience of shiftwork: Influences on attitudes', *Journal of Occupational and Organizational Psychology*, Vol. 71, pp. 61-79.

Boquet, A., Cruz, C., and Nesthus, T. (2003), 'The Bathyphase Value of the Temperature is related to Improved Performance on the Midnight Shift.' Abstract, *Aviation, Space, and Environmental Medicine*, Vol. 74(4), p. 414.

Boquet, A., Cruz, C.E., Nesthus, T.E., Detwiler, C.A., Knecht, W.R. and Holcomb, K.A., (2004), 'A Laboratory Comparison of Clockwise and Counter–clockwise Rapidly Rotating Shift Schedules. Effects on Temperature and Neuroendocrine Measures', *Aviation, Space, and Environmental Medicine, Vol. 75(10)*.

Broach, D. and Brecht-Clark, J. (1994), *Validation of the Federal Aviation Administration Air Traffic Control Specialist Pre–training Screen*, DOT/FAA/AM–94/4, Federal Aviation Administration, Office of Aviation Medicine, Washington, DC.

Chiles, W.D., Alluisi, E.Z., and Adams, O.S. (1968), 'Work Schedules and Performance During Confinement', *Human Factors*, Vol. 10(2), pp. 143-196.

Chiles, W.D., Jennings, A.E., and West, G. (1972), *Multiple Task Performance as a Predictor of the Potential of Air Traffic Controller Trainees*, DOT/FAA/AM-72-5, Federal Aviation Administration, Office of Aviation Medicine, Washington, DC.

Collins, W.E., Schroeder, D.J., and Nye, L.G. (1991), 'Relationships of anxiety scores to academy and field training performance of Air Traffic Control Specialists', *Aviation, Space, and Environmental Medicine*, Vol. 62, pp. 236-240.

Comperatore, C.A. and Krueger, G.P. (1990), 'Circadian Rhythm Desynchronosis, Jetlag, Shiftlag, and Coping Strategies', *Occupational Medicine: State of the Art Reviews*, Vol. 5(2), pp. 323-341.

Costa, G. (1999), 'Fatigue and Biological Rhythms', in D.J. Garland, J.A. Wise, and V.D. Hopkin, (eds.), *Handbook of Aviation Human Factors*, pp. 235-255, Lawrence Erlbaum Associates, London.

Costa, G., Shallenberg, G., Ferracin, A., and Gaffuri, E. (1995), 'Psychophysical conditions of Air Traffic Controllers Evaluated by the Standard Shiftwork Index', *Work and Stress*, Vol. 9, pp. 281-288.

Cruz, C., Boquet, A., Detwiler, C., and Nesthus, T. (2003b), 'Clockwise and Counterclockwise Rotating Shifts: Effects on Vigilance and Performance', *Aviation, Space, and Environmental Medicine,* Vol. 74(6), pp. 606-614.

Cruz, C., Della Rocco, P., and Hackworth, C. (2000), 'Effects of Quick Rotating Shift Schedules on the Health and Adjustment of Air Traffic Controllers', *Aviation, Space, and Environmental Medicine,* Vol. 71(4), pp. 400-407.

Cruz, C., Detwiler, C., Nesthus, T., and Boquet, A. (2003a), 'Clockwise and Counterclockwise Rotating Shifts: Effects on Sleep Duration, Timing, and Quality', *Aviation, Space, and Environmental Medicine,* Vol. 74(6), pp. 597-605.

Cruz, C.E. and Della Rocco, P.S. (1995a), *Sleep Patterns in Air Traffic Controllers Working Rapidly Rotating Shifts: A Field Study,* DOT/FAA/AM-95/12, Federal Aviation Administration, Office of Aviation Medicine, Washington, DC.

Cruz, C.E. and Della Rocco, P.S. (1995b), 'Investigation of Sleep Patterns Among Air Traffic Control Specialists as a Function of Time Off Between Shifts in Rapidly Rotating Work Schedules', in R. Jensen and L. Rakovan (eds.), *Proceedings of the Eighth International Symposium on Aviation Psychology,* Vol. 2, pp. 974-979.

Cruz, C.E., Detwiler, C., Nesthus, T., and Boquet, A. (2002), *A Laboratory Comparison of Clockwise and Counter-clockwise Rapidly Rotating Shift Schedules, Part 1: Sleep,* DOT/FAA/AM-02/8, Federal Aviation Administration, Office of Aviation Medicine, Washington, D.C.

Czeisler, C., Moore-Ede, M., Coleman, R. (1982), 'Rotating Shift Work Schedules that Disrupt Sleep are Improved by Applying Circadian Principles', *Science,* Vol. 217, pp. 460-463.

Dawson D., and Reid K. (1997), 'Fatigue, Alcohol, and Performance Impairment', *Nature,* Vol. 388(6639), pp. 235-236.

Della Rocco, P., Cruz, C., and Schroeder, D. (1996), 'Fatigue and performance in the air traffic control environment,' *Proceedings of the Aerospace Medical Panel Symposium of the Advisory Group for Aerospace Research and Development (AGARD), Neurological Limitations of Aircraft Operations: Human Performance Implications,* p. 579.

Della Rocco, P.S. (1994), *Shiftwork, Age, and Performance: Investigation of a Counterclockwise, Rapidly Rotating Shift Schedule Used in Air Traffic Control Facilities,* Unpublished Dissertation, The University of Oklahoma Health Sciences Center, Graduate College.

Della Rocco, P.S. and Cruz, C.E. (1995), *Shift Work, Age, and Performance: Investigation of the 2–2–1 Shift Schedule Used in Air Traffic Control Facilities I. The Sleep/Wake Cycle,* DOT/FAA/AM-95/19, Federal Aviation Administration, Office of Aviation Medicine, Washington, DC.

Della Rocco, P.S. and Cruz, C.E. (1996), *Shiftwork, Age, and Performance: Investigation of the 2–2–1 Shift Schedule Used in Air Traffic Control Facilities II. Laboratory Performance Measures,* DOT/FAA/AM-96/23, Federal Aviation Administration, Office of Aviation Medicine, Washington, DC.

Della Rocco, P.S., Comperatore, C., Caldwell, L., and Cruz, C. (2000), *The Effects of Napping on Night Shift Performance*, DOT/FAA/AM-00/10, Federal Aviation Administration, Office of Aviation Medicine, Washington, DC.

Della Rocco, P.S., Dobbins, L.P. and Nguyen, K.T., (1999), 'Shift Schedule Sampling from FAA Air Traffic Control Towers', presented at the *70th Annual Scientific Meeting of the Aerospace Medical Association*, Detroit, MI.

Della Rocco, P.S., Hackworth, C.A., Cruz, C.E., (2000), 'Circadian Rhythm Disruption on a Counterclockwise, Rapidly Rotating, Shift Schedule', presented at the *71st Annual Scientific Meeting of the Aerospace Medical Association*, Houston, TX.

Della Rocco, P.S., Ramos, R., McCloy, R.A. and Burnfield, J.L. (2000), *Shiftwork and Fatigue Factors in Air Traffic Control Work: Results of a Survey*, Report submitted to the Chief Scientist for Human Factors, Federal Aviation Administration, Washington DC.

Detwiler, C., Boquet, A., Cruz, C., and Nesthus, T. (2002), 'The Relationship between Glucocorticoid Activity and Cognitive Performance on the Bakan Vigilance Task.' Abstract, *Aviation, Space, and Environmental Medicine*, Vol. 73(3), p. 282.

Dinges, D.F., Graeber, R.C., Rosekind, M.R., Samel, A., and Wegmann, H.M. (1996), *Principles and Guidelines for Duty and Rest Scheduling in Commercial Aviation*, National Aeronautics and Space Administration, Ames Research Center, Moffett Field, CA.

Dollins, A., Lynch, H., Wurtman, R., Deng, M., Kischka, K., Gleason, R., and Lieberman, R. (1993), 'Effects of Pharmacological daytime doses of Melatonin on Human Mood and Performance,' *Psychopharmacology*, Vol. 122, pp. 490-496.

Folkard, S. (1989), 'Shiftwork – A Growing Occupational Hazard', *Occupational Health*, Vol. 41(7), pp. 182-186.

Folkard, S. (1992), 'Is there a "Best Compromise" Shift System?', *Ergonomics*, Vol. 35(12), pp. 1453-1463.

Gander, P., Myhre, G., Graeber, R., Andersen, H., and Lauber, J. (1989), 'Adjustment of Sleep and the Circadian Temperature Rhythm after Flights Across Nine Time Zone', *Aviation Space, and Environmental Medicine*, Vol. 60, pp. 733-743.

Gastaut, H. and Broughton, R. (1963), 'Paroxysmal psychological events and certain phases of sleep', *Perceptual and Motor Skills*, Vol. 17(2), pp. 362.

Graeber, R., Rosekind, M.R., Connell, L.J. and Dinges, D. (1990), 'Cockpit Napping', *ICAO Journal*, Vol. 45, pp. 6-10.

Higgins, E., Chiles, W., McKenzie, J., Funkhouser, G., Burr, M., Jennings, A., and Vaughan, J. (1976), *Physiological, biochemical, and multiple–task–performance responses to different alterations of the wake–sleep cycle*, DOT/FAA/AM-76-11, Federal Aviation Administration, Office of Aviation Medicine, Washington, DC.

Higgins, E., Chiles, W., McKenzie, J., Iampietro, P., Winget, C., Funkhouser, G., Burr, M., Vaughan, J., and Jennings, A. (1975), *The effects of a 12–hour shift in the wake–sleep cycle on physiological and biochemical responses and on multiple task performance* DOT/FAA/AM-75-10, Federal Aviation Administration, Office of Aviation Medicine, Washington, DC.

Hoddes, E., Zarcone, V., Smythe, E., Phillips, R. and Dement, W.C. (1973), 'Quantification of Sleepiness: A New Approach', *Psychophysiology*, Vol. 10(4), pp. 431-436.

Hursh, S. (2003), 'Fatigue Risk Management Using Performance Modeling', Abstract: *Aviation Space, and Environmental Medicine*, Vol. 74, pp. 364.

Iskra-Golec, I., Folkard, S., Marek, T., and Noworol, C. (1996), 'Health, well-being and burnout of ICU nurses on 12- and 8-h shifts', *Work and Stress*, Vol. 10(3), pp. 251-256.

Johnson, L.C. and Naitoh, P. (1974), *The Operational Consequences of Sleep Deprivation and Sleep Deficit*, AGARDograph AF-193, NATO Advisory Group for Aerospace Research and Development. London, England.

Kay, G.G. (1995), *CogScreen Aeromedical Edition Professional Manual*, Psychological Assessment Resources, Inc., Odessa, FL.

Klein D., Bruner, H., and Holtman, H. (1970), 'Circadian rhythm of pilot's efficiency and effects of multiple time zone travel', *Journal of Aerospace Medicine*, Vol. 41, pp. 125-132.

Knauth, P. (1993), 'The design of shift systems', *Ergonomics*, Vol. 36 (1-3), pp. 15-28.

Knauth, P., and Rutenfranz, J. (1982), 'Development of criteria for the design of shiftwork systems', *Journal of Human Ergology*, Vol. 11 (supplement), pp. 155-164.

Knauth, P., Landau, K., Droge, C., Schwitteck, M., Widynski, M., and Rutenfranz, J., (1980), 'Duration of Sleep Depending on the Type of Shift Work', *International Archives of Occupational and Environmental Health*, Vol. 46, pp. 167-177.

Melton, C.E. and Bartanowicz, R.S. (1986), *Biological Rhythms and Rotating Shiftwork: Some Considerations for Air Traffic Controllers and Managers*, DOT/FAA/AM-86-2, Federal Aviation Administration, Office of Aviation Medicine, Washington, DC.

Melton, C.E., McKenzie, J.M., Smith, R.C., Polis, B.D., Higgins, E.A., Hoffman, S.M., Funkhouser, G.E., and Saldivar, J.T. (1973), *Physiological Biochemical, and Psychological Responses to Air Traffic Control Personnel: Comparisons of the 5–day and 2–2–1 Shift Rotation Patterns*, DOT/FAA/AM-73-22, Federal Aviation Administration, Washington, DC.

Melton, C.E., Smith, R.C., McKenzie, J.M., Saldivar, J.T., Hoffman S.M., and Fowler, P.R. (1975), *Stress in Air Traffic Controllers: Comparison of Two Air Route Traffic Control Centers on Different Shift Rotation Patterns*, DOT/FAA/AM-75-7, Federal Aviation Administration, Washington, DC.

Minors, D.S., and Waterhouse, J.M. (1981), 'Anchor Sleep as a Synchronizer of Rhythms on Abnormal Routines,' *International Journal of Chronobiology*, 7(3), pp. 165-188.

Minors, D.S., and Waterhouse, J.M. (1985), 'Introduction to Circadian Rhythms', in S. Folkard and T.H. Monk (eds.), *Hours of Work: Temporal Factors in Work Scheduling*, pp. 1-14, John Wiley and Sons Ltd., New York.

Mitler, M., Carskadon, M., Czeisler, C., Dement, W., Dinges, D., and Greaber, R. (1988), 'Catastrophes, Sleep, and Public Policy: Concensus Report', *Sleep*, Vol. 11(1), pp. 100-109.

Monk and Folkard S. (1985), 'Individual Differences in Shiftwork Adjustment', in S. Folkard and T.H. Monk (eds.), *Hours of Work: Temporal Factors in Work Scheduling*, pp. 227-237, John Wiley and Sons Ltd., New York.

Monk, T.H. (1990), 'Shiftworker Performance,' *Occupational Medicine: State of the Art Reviews* Vol. 5(2), pp. 183-198.

Monk, T.H., Folkard, S., and Wedderburn, A.I. (1996), 'Maintaining Safety and High Performance on Shiftwork', *Applied Ergonomics*, Vol. 27(1), pp. 17-23.

Naitoh, P. (1992), 'Minimal Sleep to Maintain Performance: The Search for Sleep Quantum in Sustained Operations', in C. Stampi (ed.), *Why We Nap: Evolution, Chronobiology, and Functions of Polyphasic and Ultrashort Sleep*, pp 199-216, Burkhauser, Boston.

Naitoh, P., Kelly, T.L. and Englund, C. (1990), 'Health Effects of Sleep Deprivation', *Occupational Medicine: State of the Art Reviews* Vol. 5(2), pp. 209-237.

Neri, D.F., Oyung, R.L., Colletti, L.M., Mallis, M.M., Tam, P.Y., and Dinges, D.F. (2002), 'Controlled Breaks as a Fatigue Countermeasure on the Flight Deck', *Aviation Space and Environmental Medicine*, Vol. 73(7), pp. 654-664.

Nesthus, T., Cruz, C., Boquet, A., and Holcomb, K. (2003), 'Comparisons of Sleep Duration and Quality, Mood, and Fatigue Ratings During Quick-Turn Shift Rotations for Air Traffic Control Specialists', Abstract, *Aviation, Space, and Environmental Medicine,* Vol. 74(4), p. 381.

Nesthus, T.E., Cruz, C. Boquet, A., Detwiler, C., Holcomb, K., and Della Rocco, P. (2001), 'Circadian Temperature Rhythms in Clockwise and Counter-Clockwise Rapidly Rotating Shift Schedules' *Journal of Human Ergology* Vol. 30(1-2), pp. 245-249.

Nesthus, T.E., Cruz, C., Boquet, A., and Hackworth, C. (2004), 'Risk Factors for Air Traffic Control Specialists Commuting to and From Early Morning and Midnight Shifts' Presentation, *75th Annual Scientific Meeting of the Aerospace Medical Association,* Anchorage, AK.

Nesthus, T.E., Holcomb, K., Cruz, C., Dobbins, L., and Becker, J. (2002), 'Comparisons of Sleep Duration, Subjective Fatigue, and Mood among Four Air Traffic Control Shift Schedule-Types', Abstract: *Aviation, Space, and Environmental Medicine,* Vol. 73(3), p. 272.

Penn, P. and Bootzin, R. (1990), 'Behavioral Techniques for Enhancing Alertness and Performance in Shift Work', *Work and Stress,* Vol. 4(3), pp. 213-226.

Ribak, J., Ashkenazi, I., Klepfish, A., Avgar, D., Tall, J., Kallner, B., and Noyman, Y. (1983), 'Diurnal Rhythmicity and Air Force Flight Accidents Due to Pilot Error', *Aviation, Space, and Environmental Medicine*, Vol. 54(12), pp. 1096-1099.

Rosa, R.R. and Colligan, M.J. (1988), 'Long Workdays versus Restdays: Assessing Fatigue and Alertness with a Portable Performance Battery', *Human Factors*, Vol. 30, pp. 305-317.

Rosekind, M.R., Graeber, R.C., Dinges, D.F., Connell, L.J. Roundtree, M.S., Spinweber, C.L., and Gillen, K.A. (1994), *Crew Factors in Flight Operations IX: Effects of Planned Cockpit Rest on Crew Performance and Alertness in Long-Haul Operations*, Technical Memorandum No. 108839, NASA Ames Research Center, Moffett Field, CA.

Saldivar, J., Hoffman, S., and Melton, C. (1977), *Sleep in Air Traffic Controllers*, DOT/FAA/AM-77-5, Federal Aviation Administration, Office of Aviation Medicine, Washington, DC.

Schoenfelder, E. and Knauth, P. (1993), 'A Procedure to Assess Shift Systems Based on Ergonomic Criteria', *Ergonomics*, Vol. 36(1-3), pp. 65-76.

Schroeder D.J., Rosa, R.R., and Witt, L.A. (1998), 'Some Effects of 8- vs. 10-hour Work Schedules on the Test Performance/Alertness of Air Traffic Control Specialists', *International Journal of Industrial Ergonomics*, Vol. 21, pp. 307-321.

Schroeder, D.J. and Goulden D.R. (1983), *A Bibliography of Shift Work Research: 1950– 1982*, FAA/AM/83-17, Federal Aviation Administration, Office of Aviation Medicine, Washington, DC.

Simons, M. and Valk, P.J. (1999), *Sleep and Alertness Management during Military Operations: Review and Plan of Action*, DTIC 425137, ADA372731.

Smith, M., Colligan, M., and Tasto, D. (1982), 'Health and Safety Consequences of Shift work in the Food Processing Industry', *Ergonomics*, Vol. 25(2), pp. 133-144.

Smith, R. (1980), *Stress Anxiety, and the Air Traffic Control Specialist: Some Conclusions for a Decade of Research*, DOT/FAA/AM-80-14, Federal Aviation Administration, Office of Aviation Medicine, Washington DC.

Sparks, P. (1992), 'Questionnaire Survey of Masters, Mates, and Pilots of a State Ferries System on Health, Social, and Performance Indices Relevant to Shift work', *American. Journal of Industrial Medicine*, Vol. 21, pp. 507-516.

Spelten, E., Totterdell, P., Barton, J., and Folkard, S. (1995), 'Effects of age and domestic commitment on the sleep and alertness of female shiftworkers', *Work and Stress*, Vol. 9(2/3), pp. 165-175.

Tasto, D., Colligan, M., Skjei, E., and Polly, S. (1978), *Health Consequences of Shiftwork*, DHEW Publication No. 78-154, U.S. Government Printing Office, Washington, DC.

Taub J.M. and Berger, R.J. (1976), 'The Effects of Changing the Phase and Duration of Sleep', *Journal of Experimental Psychology: Human Perception and Performance* Vol. 2(1), pp. 30-41.

Tepas, D.I. (1982), 'Work/Sleep Time Schedules and Performance', in W. Webb (ed.), *Biological Rhythms, Sleep, and Performance*, pp. 175-204, John Wiley and Sons Ltd., Chichester, England.

Tilley, A., Wilkinson, R., Warren, P., Watson, W., and Drud, M. (1982), 'The Sleep and Performance of Shift Workers', *Human Factors*, Vol. 24(6), pp. 629-641.

Torsvall, L., and Åkerstedt, T. (1987), 'Sleepiness on the job: Continuously measured EEG Changes in Train Drivers', *Electroencephalography and Clinical Neurophysiology*, Vol. 66, pp. 502-511.

Tucker, P., Smith, L., MacDonald, I., and Folkard, S. (1998), 'The impact of early and late shift changeovers on sleep, health, and well-being in 8- and 12-hour shift systems', *Journal of Occupational Health Psychology*, Vol. 3, pp. 265-275.

Turek, F. (1986), 'Circadian principles and design of rotating shift work schedules', *American Journal of Physiology*, Vol. 251, pp. 636-638.

U.S. Congress, Office of Technology Assessment (1991), *Biological rhythms: Implications for the Worker*, OTA-BA-463, U.S. Government Printing Office, Washington, DC.

Waterhouse, J.M., Folkard, S. and Minors, D.S. (1992), *Shiftwork, health and safety: An overview of the scientific literature 1978-1990*, HMSO, London.

Webb, W.B. (1982), 'Sleep and Biological Rhythms', in W.B. Webb (ed.), *Biological Rhythms, Sleep, and Performance*, pp. 87-110, John Wiley and Sons Ltd. Chichester, England.

Wedderburn, A. (1991), 'Guidelines for Shiftworkers', *Bulletin of European Shiftwork Topics,* Vol. 3.

Wilkinson, R.T. (1992), 'How fast should the night shift rotate?', *Ergonomics*, Vol. 35(12), pp. 1425-1446.

PART IV
HUMAN FACTORS
METHODOLOGIES

Human Factors aims to allow for informed decision-making based on findings that are scientifically sound, and therefore depends on the development and application of reliable and valid methods. Since Human Factors issues by their very nature are often open to discussion and expression of opinions, it helps if concrete measurements of human performance lend objectivity, clarity, and precision to decisions involving Human Factors. This section therefore considers the methodological and measurement considerations necessary to ensure the sound application of Human Factors to ATM. Carefully chosen and applied methods and measures can often bring insight to complex issues, and sometimes show that prior assumptions were in fact unfounded. This section addresses three important areas of Human Factors methods and measurement in ATM: measuring performance when observing behavior, predicting performance using computer modeling techniques, and understanding human error and recovery in current and future systems. The weight of this section is indeed on understanding human performance for the future, especially given the significant changes to ATM systems that are envisaged in the next decade and a half.

The first chapter by Manning and Stein focuses on the measurement of actual human performance, whether in a laboratory, in a simulator, or in the field. The first half of the chapter lays out a range of techniques, including both objective and subjective measures, the latter of which are particularly important in ATM. This first part also includes a discussion on the usefulness and drawbacks of using Operational Errors (OE) as a measure of human performance. The second half of the chapter considers examples of application of such measures, and shows how these measurement techniques have been used together to yield insights and answers to real ATM issues. This part of the chapter reveals the depth of forethought, planning, and discipline that is required to deliver scientific results that give an objective answer, not simply the one stakeholders (including sometimes controllers and even Human Factors professionals) expect. It considers commonly-asked questions such as 'is there a best set of measures?' and 'what are the relative benefits of using 'generic' versus 'realistic' airspace in simulations?' Lastly, it raises a critical issue in this area of Human Factors, namely the challenge of effectively communicating the scientific results of human performance measurement studies to operational controllers and management.

The second chapter in this section by Corker focuses on modeling human performance, an alternative to measurement of actual observed behavior. This may be particularly useful in the design and development stages, allowing one to understand if a new procedure or interface works without the cost of development and testing to generate findings that may indicate a change is required, but may be prohibitively expensive to make once initial development has proceeded. First, Corker develops an overall model of the ATM system, showing where human performance modeling fits. The chapter then considers the different types of modeling approaches, from early to more recent ones, and then focuses in more depth on those modeling approaches most relevant to ATM. These include computational models, risk models, and models developed to support the design process. A case study of one of these (Air-MIDAS) is given at the end of the chapter to show how such models can be used to support operational decisions. The chapter indicates towards the end that often models are built for a specific purpose, and then cannot be used more generally, even though some of these models require substantial and protracted development cycles. There is therefore clearly a need for more generic models that will not only answer specific issues today, but will remain useful for years to come. Corker also notes the need for modeling approaches that are less 'piecemeal', and that can be used to address issues at the right level of 'granularity', thereby linking issues of human performance to system performance.

The third and final chapter in this section by Shorrock et al. deals with the issue of human error prediction, linking Human Factors clearly to risk assessment and safety management. The chapter considers approaches borrowed from other industrial domains with a more mature and explicit risk management tradition, such as nuclear power generation. Two techniques in particular (TRACER-Lite and HAZOP) are shown to have been useful and reliable, with a number of short case studies highlighting their application. The chapter addresses the importance of the area of Human Error Identification, then outlines the two techniques themselves, using six case studies to show both the 'mechanics' of the techniques and the impacts achieved. Then, because these approaches are relatively new to ATM, a comparative validation study is briefly discussed, showing how the two techniques performed when used independently on the same ATM projects. This chapter therefore shows straightforward approaches to finding future vulnerabilities in systems, whether for risk or safety assessments, or for the purpose of making design more robust and reliable.

In summary, human performance measurement is key to understanding the efficiency of user interactions with a task or environment and can therefore contribute to identifying required changes to the system and assessing if an intervention has made a difference or not. Gaining unbiased and appropriate measurements related to such objectives is never an easy task in all but the simplest situations. Certainly ATM is no exception to his point and may be one of the most challenging areas to apply human performance assessment. Human performance measurement allows a degree of objectivity and analysis to enter into any decision-making situation about Human Factors, and so helps ensure the right decision is made, even if, as occasionally happens, the original Human Factors experts themselves are shown to be wrong by their own methods and results. Work in this difficult but essential area of human performance measurement, and in the area of 'prospective' performance modeling via computer simulations and human error modeling, therefore needs to continue, and to remain a fundamental backbone of Human Factors. The measurement techniques themselves, effectively the Human Factors practitioner's and researcher's 'toolkit', remain a reservoir for achieving effective and sustainable impacts in ATM.

Chapter 13

Measuring Air Traffic Controller Performance in the 21st Century

Carol Manning and Earl Stein

Introduction

Why Measure Human Performance?

There are many reasons to measure human performance in occupational settings. For example, organizations need to assess their employees' performance to determine whether they are doing their jobs correctly. Measuring performance can document the contribution of their employees' work products to the organization's goals. Further, US regulations require organizations to be able to demonstrate that procedures used to select employees (selection tests, performance standards, and promotion criteria) are directly related to job duties. Organizations also need to determine whether job changes (for example, the introduction of new technologies or procedures) might affect their employees' performance. It is also necessary to measure system performance (which encompasses the interaction of human and equipment) to assess the effect of job changes on system output. There are other reasons to measure job performance, including identifying employees who should receive higher compensation and determining employees who should receive training (Landy, 1989, p. 92).

To assess performance effectively, researchers and practitioners must develop valid, reliable measures related to the job being performed. This chapter describes the development of human performance measures for the job of Air Traffic Control Specialist (ATCS) in the US Federal Aviation Administration (FAA). Air traffic control (ATC) is a dynamic job performed in a constantly changing environment that produces intangible products (maintaining the safe, orderly, and expeditious flow of air traffic) through mostly unobservable (cognitive) activity. ATCSs perform more than one task concurrently and may either work alone or as part of a team. While efficiency is important, the most important factor in this job is safety because the output contributes to the safety of those who use the service. At the same time, the products are intangible, and there may be many similar ways to accomplish the same result. If a job with intangible products is performed in a dynamic environment where activity occurs constantly, measurement of job performance is further complicated because it must be accomplished in such a way as to not interfere with the job being performed.

The nature of the ATCS job determines how difficult it is to develop appropriate performance measures. In some jobs, the product is tangible and production activity can be observed easily, thus, the employee's performance can usually be measured fairly well by counting the number of products, assessing their quality, or evaluating the production process. As products become more intangible, as in the job of ATCS, and the production activity is less observable (for example, because most of the activity involves thinking), then performance measurement becomes more difficult to accomplish because the number and quality of products may be more difficult to assess, and the activities used to produce the product are not clear. The ATCS job produces intangible results and is performed in a dynamic environment; thus, measurement of job performance is further complicated because it must occur in a way as to not interfere with the job being performed. It is the measurement of complex job performance in the ATCS occupation that is described here.

The Job of the Air Traffic Controller

Before discussing the development of performance measures for ATCSs, it is helpful to first describe the ATCS job. ATCSs assure the safe, orderly, and expeditious flow of controlled aircraft between a departure point and a destination. Three types of ATCSs control traffic in the US: Tower controllers are responsible for aircraft departing from or arriving at an airport, Terminal Radar Approach Control (TRACON) controllers are responsible for aircraft moving in a terminal area outside the immediate location of an airport, and En-route controllers are responsible for controlled aircraft outside of terminal areas.

ATCSs plan to assist a pilot in following the flight plan route in the most efficient way possible, given that other aircraft flight plans may, in the future, be in conflict. ATCSs (depending on their specific job assignment) must organize multiple aircraft so they fly conflict free in a limited amount of airspace, and sequence a series of arriving aircraft with different performance characteristics so they can land on a limited number of runways at an airport. They must also comply with letters of agreement with other facilities that require aircraft arriving over a certain fix to fly at a certain speed and altitude, and must separate high speed, high altitude crossing or transitioning traffic.

A radar display assists controllers in visualizing the relative positions of aircraft in two dimensions and presents altitude as a number in a data tag linked to an aircraft's position on the display. A controller using radar must take aircraft altitudes into account, either by forming a mental 'picture' of all the aircraft moving through the airspace or by systematically comparing aircraft pairs to determine if they will conflict. In locations where radar is not available, controllers separate aircraft using paper flight progress strips, which estimate the time when an aircraft will cross a fix. This non-radar separation has a more lengthy separation requirement than is in effect when radar is available.

One factor that must be considered when measuring ATCS performance is that most of what they do is cognitive and, thus, is unobservable. Observable behaviors associated with ATC include talking to pilots and other controllers, entering information into the computer, and (sometimes) marking flight progress strips. However, most of their activity involves problem identification and resolution, which are not easily observed. Moreover, controllers' observable activities often occur concurrently, for example, simultaneously talking to a pilot or other controller while entering data in the computer. This combination of activities is commonly called 'multitasking.' Though all acknowledge that multitasking is important, it is a complex construct to measure.

The lack of consistency in the output also complicates the measurement of ATCS performance. There is no 'best way' to control a group of aircraft – there are usually multiple ways to accomplish the same goal, and controllers seldom agree that one solution is better than another. One reason for the lack of consistency in performance standards is the influence of situation-specific factors. ATC situational variables include airspace, terrain, weather, adjacent facilities, and facility letters of agreement regarding how aircraft are to be controlled. Variations in operational concepts and emphases that occur between and within large facilities can also influence both system and individual performance.

Some skills are measured against an absolute standard of performance, which is clearly defined and is easily recognized by anyone in the trade or occupation. In contrast, a relative standard assumes that there are many ways of evaluating performance. Relative standards make performance measurement very complicated.

Measurement of Performance in Air Traffic Control

The *Uniform Guidelines on Employee Selection Procedures* (1978) attempted to prevent discrimination in employment on the basis of race, color, religion, sex, or national origin yet still allow employers to use tests to select their employees. If a selection test resulted in disparate impact against a protected group, it was unlawful unless the employer could show that it fairly predicted performance on the job. Thus, organizations had to be able to demonstrate that their selection tests could be linked to criterion measures of job performance, which necessitated developing measures that adequately measured job performance. This section discusses several types of performance measures that have been used for ATCSs.

Subjective Ratings

Traditional assessment techniques for dynamic, safety-related occupations, such as pilots and controllers, have placed a heavy emphasis on the observational and rating skills of subject matter experts (SMEs). For example, 'check airmen' evaluate military and civilian pilot performance using checklists containing items that describe performance of specific job-related tasks.

Subjective performance ratings have been frequently used in the ATC environment. For example, Trites, Miller, and Cobb (1965) used subjective rating scales to assess the validity of some experimental Air Traffic Control Specialist (ATCS) selection procedures. They used 14 items that described desirable ATCS characteristics (e.g., 'ability to make decisions required by his position,' 'aptitude for ATCS activities,' and 'potential to perform journeyman duties'). Supervisors responded on a 1-5 point scale ranging from 'unsatisfactory' to 'excellent.' They rated the controllers' performance about a year after the ATCSs completed their FAA Academy training. Trites et al. (1965) found that a set of ratings made during their Academy training predicted these job performance ratings better than did their class or lab scores.

In the advent of the *Uniform Guidelines on Employee Selection Procedures* (1978), organizations have had to demonstrate that their selection procedures predict job performance in a fair manner. While easy to obtain, traditional rating scales were problematic because they were often vague. For example, like the rating scale developed by Trites et al. (1965), many scales contained endpoints ranging from 'Poor' to 'Excellent' or 'Unsatisfactory' to 'Superior.' The use of such vague rating scales and lack of rater training can contribute to rater judgment biases. These biases include differential interpretation of the terms used to anchor the scales, increased tendencies to provide harsh, lenient, or moderate ratings for all employees, and increased tendencies to provide all positive or negative ratings for employees they either like or dislike, respectively. Until the adoption of the *Uniform Guidelines*, the consequences of using generic subjective ratings were not particularly important because the results of faulty decision-making based on such measures in industrial jobs could be fixed without serious consequences.

Although subjective ratings and checklists are still sometimes used to evaluate ATCS performance (e.g., evaluation of trainees), efforts have been made to replace vague subjective ratings scales with better, more specific behaviorally-anchored rating scales, situational judgment tests, objective measures, and work sample tests. These newer performance measurement methods are based on more specific information about employees' performance and, to the extent possible, control for extraneous variability. It is important to control the setting in which performance measurement occurs, because in an uncontrolled environment, factors outside an employee's control could affect the resulting product or outcome in an unrealistically positive or negative way. The next section describes several newer types of performance measures and how researchers and trainers use them to evaluate air traffic controller performance.

Behaviorally-Anchored Rating Scales

According to Anastasi (1988), designers can reduce the magnitude of commonly reported rating errors by carefully constructing rating scales to include specifics about what is being rated. She suggested providing a behavioral anchor on scale increments wherever possible and emphasized the importance of rater training. She also recommended training to enhance raters' observational skills.

Behaviorally-anchored rating scales (BARS) contain intermediate and end points that are tied to observable behaviors. To develop a BARS, an SME panel identifies critical incidents (Flanagan, 1954) that might be faced by an employee. A second SME panel identifies different ways employees might handle the incidents and maps these choices to specific points on a rating scale. The panel assigns values to each identified point on the scale. Designated raters receive specialized training for using the BARS to rate employee performance. The result is a set of scales that can be used to more specifically describe employee performance and a supervisor who is better trained to use the methodology.

BARS have been used by researchers to look at both ATC selection and system development. Borman, Hedge, Hanson, Bruskiewicz, Mogilka, Manning, Bunch, and Horgen (2001) developed a set of BARS as a criterion measure against which the Air Traffic Control Selection and Training (AT-SAT) battery was validated. Based on a job analysis developed for the Selection and Control Hiring Assessment (SACHA) project (Nickels, Bobko, Blair, Sands, and Tartak, 1995) and extensive input from several SME controller teams, Borman et al. developed ten Behavior Summary Scales with anchors describing different controller effectiveness levels. Inter-rater reliabilities for these scales, using combined peer and supervisor ratings, ranged between .50 and .69 (n=1227; Borman et. al., 2001, p. 89).

Bruskiewicz, Hedge, Manning, and Mogilka (2000) subsequently developed a similar set of BARS (called Over-the-Shoulder or OTS Rating Scales) for a high-fidelity simulation study that evaluated the criterion measures used for the AT-SAT validation. Using ten extensively trained raters, median interrater reliabilities for these scales ranged from .83 to .95 (Borman et. al., 2001, p. 93.)

Sollenberger, Stein, and Gromelski (1997) developed a set of BARS for research examining the effects of new technologies on TRACON controllers. SMEs used twenty-four rating scales to assess observable controller actions across different areas of the controller's performance domain. SMEs initially received no training and rated videotaped sessions of an ATC simulation study, with minimal inter-rater reliability. After raters were trained, inter-rater reliability increased to between r=.70 to r=.90 for most of the rating scales. Sollenberger et al. (1997) found that continuing to reinforce the basics of objective observation and facilitating communication between observers until they came to use a common mental model of what to look for and how to rate provided the greatest gains in reliability. The combination of specific rating scales and training contributed to the improvement in results. Similar sets of scales were designed for use in the En-route (Vardaman and Stein, 1998) and tower cab environments (Sollenberger and Della Rocco, 2002).

Advantages Associated with BARS The advantages of using BARS to assess controller performance are that, first, they capture important aspects of the job because they are based on formal job analyses and systematic input from SMEs. Detailed descriptions of anchors located at different points on the scales provide good examples for raters to use to decide which performance rating to assign. BARS can also be used in a variety of settings, from a general annual performance assessment to a more specific work sample simulation test. Depending on the circumstance, BARS can be used to assess either controllers' typical performance (Landy, 1989; p. 116), based on supervisors' memories of how a controller performed over a period of time, or their maximum performance, such as their best attempts to perform a task as demonstrated during simulation testing. Measuring typical performance allows a supervisor to assess how an employee behaves on a day-to-day basis, while measuring maximum performance allows an observer to assess how the employee behaved while aware that he/she was being observed.

Disadvantages Associated with BARS Disadvantages of using BARS are that the scales are very difficult to develop and require extensive input from several independent groups of SMEs. Extensive rater training is also required to use the scales correctly and minimize rater biases.

Situational Judgment Tests

Situational judgment tests (described in Hanson, Borman, Mogilka, Manning, and Hedge, 1999) sample an employee's judgment about specific work situations. Realistic job-related situations are presented to employees who use their judgment to choose the best option from a set. SMEs develop the situations to represent important situations that occur on the job. They are then presented in a part-task format to job incumbents who respond by answering multiple-choice items. Response alternatives for each item have been assigned different values based on the collective judgment of an SME panel. An incumbent's score is the sum of values assigned to the selected response alternatives.

The FAA has used situational judgment tests in the ATC environment for a number of years. Buckley and Beebe (1972) developed the Controller Decision Evaluation Technique (CODE) test that presented an evolving ATC situation. The authors originally presented each situation as a motion picture film, then later as slides. Each subsequent slide showed a slight movement in the progression of traffic. The timed test included items that evaluated applicants' abilities to project a displayed air traffic situation into the future. The CODE was later converted to a paper-and-pencil format that was administered to candidate controllers as an ATCS selection test. Called the Multiplex Controller Aptitude Test (MCAT; Rock, Dailey, Ozur, Boone, and Pickrel, 1981), the Office of Personnel Management used this test for more than ten years as a major part of the initial selection procedure for large groups of controller applicants. Scores on the MCAT were found to be correlated with field training performance (r=.12 before adjustment for restriction in range, but r=.35 when adjusted for restriction in range (N=402); Manning, 1991).

Controller Skills Tests (CSTs; Pickrel and Greener, 1984; Tucker, 1984) administered at the end of ATCS initial qualification training were another example of situational judgment testing. CSTs required controller trainees to complete timed multiple-choice items describing air traffic situations. The questions included responses that tested a candidate's ability to assess the situation, predict what would happen, and formulate a reasonable response. Pickrel and Greener (1984) reported that CST scores distinguished between developmental and full performance level controllers and CST scores were highly related to laboratory assessment scores (p. 211). CST scores were found to be correlated with field training performance for En-route developmentals (r=.16 before adjustment for restriction in range but r=.26 after adjustment for restriction in range (N=402); Manning, 1991).

Borman et al. (2001) developed a situational judgment test as a criterion measure for the validation of the AT-SAT selection battery. The Computer Based Performance Measure (CBPM; Hanson et al., 1999) included items developed by ATC SMEs in numbers that proportionally represented important En-route job tasks from the SACHA job analysis (Nickels et al., 1995). SME teams evaluated the effectiveness of response alternatives. During the AT-SAT concurrent validation study, operational controllers observed re-creations of short traffic samples occurring in a generic sector that included controller-pilot voice communications and a set of relevant flight progress strips. Each traffic sample was followed by a group of items. Controllers were required to respond to each item during a restricted time period by choosing what they thought was the most effective response. The CBPM score was the sum of the SME-assigned effectiveness ratings associated with the responses chosen for each item. The CBPM had a corrected correlation of .70 with the AT-SAT predictor composite (Waugh, 2001, p. 121). The CBPM also had a correlation of r=.32 with the En-route non-radar CST administered to 212 controllers 10-15 years earlier when they attended the ATC Initial Qualification Training Program (Manning and Heil, 2001, pp. 139-140).

Advantages Associated with Situational Judgment Tests The advantage of using situational judgment tests to assess ATCS performance is that the testing environment is dynamic but can also be controlled systematically. All incumbents view the same dynamic ATC traffic samples and can choose from among the same responses. When developed according to recommended practices, several groups of SMEs determine the response alternatives and their effectiveness ratings according to a systematic procedure. Another advantage is that situational judgment tests are timed, placing the controller under some pressure to complete the task.

Disadvantages Associated with Situational Judgment Tests Disadvantages of using situational judgment tests to assess controller performance are that controllers choose what they consider to be the best response from a limited set of options. Controllers do not formulate or execute the responses they have chosen. Thus, if a controller earns a high score on this test, it demonstrates that he/she can identify the most appropriate response from a list of available responses but does not guarantee that he/she would have identified the best response in the absence of any options, nor does it guarantee that he/she would have executed the chosen response properly.

Objective Measures

Monitoring employee output or analyzing routinely recorded data are methods used to derive objective performance measures. The primary characteristic of these measures is that the incumbent is unaware of their collection. Supervisors can use objective or unobtrusive measures to assess 'typical' rather than 'maximum' performance (Landy, 1989; p. 116). Several types of objective measures have been developed for the ATC environment. These measures are usually based upon recordings of ATCS actions or relative positions or movements of aircraft. However, tabulations of controller activities made by SME observers are also considered here. It may be argued that counts of events made by observers are subjective because observers may under- or over-record events or display bias in their classifications. However, if the events are well defined and sufficient training is provided on making the observations, there should be no question about whether a recorded event occurred or was classified properly. Such tabulations may be considered 'nearly' objective, and so are discussed here.

This section will describe several types of objective ATC performance measures. The section will start with a discussion of specific measures often included in a catalog of objective measures, such as ATC Operational Errors, system effectiveness measures, and POWER measures (see below). Later, methods for collecting, replaying, and analyzing objective ATC performance measures are discussed. These include simulation replays, SATORI re-creations, training simulations, and baseline studies.

Operational Errors In air traffic control, the minimum separation allowed between aircraft under positive radar control is one of the most fundamental standards against which all controllers are judged. Operational errors (OEs) occur when controllers allow an aircraft to get too close to another aircraft or an obstacle by an amount that varies depending on the altitude and type of airspace being worked. When an OE is identified, a controller is removed from position, the available materials describing the situation are reviewed, and, if a violation is found to have occurred, then the controller is decertified until remedial training is provided and the controller can recertify. As such, everyone in ATC meets the standards or risks being removed. Thus, occurrences of OEs are clearly important, but using them as performance indicators is problematic for several reasons.

The most notable reason that OEs are inadequate performance measures is that they occur infrequently, both in the operational system and to individual controllers. The overall OE rate is about 0.16 per 1,000,000 operations. The rare occurrence of OEs in the operational environment makes them a non-viable measure to assess controller performance for the purposes of system development or evaluation, though they may be useful in a training context.

When using simulation tests, it is possible to develop scenarios that are sufficiently busy or complex that one or more OEs may occur. One must consider, though, that if researchers designed scenarios to include extremely high workload conditions, then any resulting OEs would be based upon events that are extremely unlikely to occur in the operational environment. Depending on the reason for measuring controller performance, it may or may not be appropriate to include such unusual events in a scenario.

Another problem with the use of OEs as performance measures is that only a small part of the distribution of aircraft separation statistics is available for analysis. An En-route OE only occurred if an algorithm in the National Airspace System (NAS) software determined that separation standards were violated. But most controllers perform at above minimum level, so, most of the time, they maintain more than the required lateral En-route separation between aircraft of 5 nautical miles when they are at the same altitude. However, NAS software does not identify events that almost occurred, such as when a controller who normally keeps aircraft separated by 7 nautical miles allowed a pair to come within 5.25 or 6 nautical miles. While events that result in actual violations are heavily studied, no information is available for situations that came close to but did not actually become violations. The truncated distribution of OEs necessarily results in a skewed analysis of separation violations and factors contributing to their incidence.

For evaluating new systems or personnel who have already achieved journeyman status, Operational Errors are a crude measure. However, they continue to be used because they have face validity for the ATC community. If one chooses to use OEs as an objective performance measure, they may be analyzed in a variety of ways. Typical measures include frequency of OEs and how long they lasted. However, additional factors may be taken into account, such as the minimum distance between aircraft at the time of the event (Rodgers, Mogford, and Mogford, 1998).

Paul (1989, 1990) created a unique tool for use in ATC simulation research called the Aircraft Proximity Index (API). Instead of simply counting aircraft conflicts, the API provides a graded severity scale ranging from 0 to 100. As long as it is 0, there is no conflict and, as the numbers rise, so does the severity of the conflict. An API of 100 means a collision is imminent. Instead of assuming that all conflicts are alike, this tool takes into consideration horizontal and vertical separation and the slant-range distance between aircraft. 'API is not linear. A linear decrease between aircraft, either vertically or laterally, increases the API exponentially' (Paul, 1990, p. 8). Research personnel now routinely use the API in ATC simulations at the William J. Hughes Technical Center.

Two severity classifications to describe OEs are being used operationally by the FAA. The FAA Office of Investigations uses a classification algorithm that allocates 100 points across several variables, including vertical and horizontal separation distances, flight paths and cumulative closure rates, and the level of air traffic control involvement in radar environments (FAA, 2002). The distribution of severity scores is then divided into High, Moderate (uncontrolled), Moderate (controlled), and Low categories. The FAA Runway Safety Program uses a different method for classifying severity of runway incursions. Incursions are classified into groups (A, B, C, or D) by a panel of SMEs. Five interdependent parameters are used: available reaction time, evasive or corrective actions required, environmental conditions, speed of aircraft and/or vehicle, and proximity of aircraft and/or vehicle (FAA, 2001).

System Effectiveness Measures If it is possible to display either simulated or real ATC data graphically, it should also be possible to count, measure, and otherwise analyze these data. Buckley, DeBaryshe, Hitchner, and Kohn (1983) originally developed the idea to derive objective ATC performance measures from simulation output. Buckley et al. (1983) developed ten performance measures by identifying basic dimensions for measuring ATC functions in real time. They conducted two experiments to study the interaction of sector geometry and density and the use of simulation for performance evaluation. The authors found significant effects of sector geometry and traffic density for most of the ten system performance measures. The second experiment collected an extensive amount of data over time using repeated measures. Thirty-nine controllers participated in 12 one-hour runs using the same sector with the same traffic level. The authors used the resulting database to compute a factor analysis that produced four meaningful factors: Confliction, Occupancy, Communication, and Delay. 'Confliction' included measures of 3-, 4-, and 5-mile conflicts. 'Occupancy' included measures of the amount of time an aircraft was controlled, distance flown, fuel consumption, and time within sector boundary. 'Communications' included path changes, number and duration of ground-to-air communications. 'Delay' included total number of delays and total delay times. Two other measures, 'Number of Aircraft Handled' and 'Fuel Consumption,' were also relevant. These experiments conducted by Buckley et al. (1983) served as building blocks for most of the objective controller performance measurement research that followed.

Objective measures obtained from data collected during simulations are often used to answer relevant research questions. Buckley and Stein (1992) identified a set of measures called System Effectiveness Measures (SEM) that were derived from the output of simulations and based on the Buckley et al. (1983) study. These measures fall into a number of categories including: Conflict, Complexity, Error, Communications, Taskload, and Workload. See Hadley, Guttman, and Stringer (1999) for a concise summary and a listing of all the variables. The William J. Hughes Technical Center Research Development and Human Factors Laboratory

(RDHFL) included this toolkit of measures in their simulation system to address research questions in many of their studies.

POWER Measuring controller taskload and performance using routinely recorded ATC data is an offshoot of the Buckley et al. (1983) idea of deriving objective measures of controller performance from simulation data.

The Performance and Objective Workload Evaluation Research (POWER) software (Mills, Pfleiderer, and Manning, 2002), which followed the development of SATORI (Rodgers and Duke, 1993; discussed below), analyzes information extracted from files produced by the NAS Data Analysis and Reduction Tool (DART; FAA, 1993) program. Software called the NAS Data Management System (NDMS) consolidates and organizes the output of these files into Microsoft Access tables to facilitate extraction and computation of the POWER variables.

After the DART program extracts the data, POWER computes a set of taskload measures (Mills et al., 2002), including number of aircraft controlled, maximum aircraft controlled simultaneously, control duration, handoffs (both accepted and initiated), handoff latencies (time to accept both initiated and accepted handoffs), number of data entries and data entry errors (by position), specific data entry measures (including distance reference indicator (DRI) requests, route displays, pointouts, datablock offsets), pairs of aircraft in conflict, assigned altitude changes, number, duration and amount of aircraft altitude and heading changes, visual clustering (maximum number of aircraft simultaneously within 10nm laterally), and average distance (distance between aircraft in lateral and vertical dimensions and in Euclidean distance). The set of POWER measures increases as new measures are identified and decreases as some measures are found to overlap with others.

Simulation Replays The simplest way of using objective measures to analyze ATCS performance is to replay recorded ATC simulations. Instructors originally used replays to provide feedback to developmental controllers after they finished running simulated scenarios when training. The FAA Academy replays recorded ATC training simulations in their Radar Training Facility (RTF) laboratory. En-route centers replay recorded ATC simulations in their Dynamic Simulation (DYSIM) training facilities. Reviewing recorded air traffic scenarios has also been used to provide feedback or compute performance measures during experiments.

SATORI Re-creations Rodgers and Duke (1993) developed a sophisticated method of re-creating operational ATC traffic samples. Systematic Air Traffic Operations Research Initiative (SATORI) software graphically displays routinely recorded operational ATC data that is synchronized with recorded voice files on an independent computer workstation not linked to a host computer or simulator. Investigators use SATORI to re-create operational errors and other important ATC events for review. SATORI has also been used to display traffic samples to SMEs to obtain their reactions and ratings (e.g., Endsley and Rodgers, 1997; Manning,

Mills, Fox, Pfleiderer, and Mogilka, 2001; Manning, Mills, Fox, Pfleiderer, and Mogilka, 2002).

Replays and re-creations of air traffic scenarios supported performance measurement by allowing an observer to review events as they had occurred and then provide feedback to a controller about his or her performance. Alternatively, an experimenter could replay an experimental scenario and count events as they occurred in near-real time (Galushka, Frederick, Mogford, and Krois, 1995) or discuss a controller's reasons for taking certain actions. Thus, the playback of re-created ATC traffic samples provided observers with an opportunity to see what had occurred but did not, by itself, support improved measurement of controller performance.

Training Simulations When the FAA's RTF laboratory was developed, the plans called for computerized measurement of student performance (Boone, Van Buskirk, and Steen, 1980). The measures under consideration were: number of aircraft in the sample, ideal aircraft time-in-system (based on flight plan), ratio of ideal aircraft time-in-system to number of aircraft, number of completable flights, arrivals, departures, arrival/departure ratio, scheduled arrival and departure rate, En-route and terminal conflicts, number of delays (holding, arrivals, departures), delay time (holding, arrivals, departures), aircraft time in system, aircraft handled, completed flights, arrivals and departures achieved, achieved arrival and departure rate, air-ground contacts, air-ground communication time, heading, speed, altitude changes, path changes, and handoffs. Boone et al. (1980) found that using computerized measures increased the reliability of student grades as compared with using grades based solely on over-the-shoulder ratings. Although the computerized measures appeared promising, Boone (1984) later indicated that system limitations and lack of research had prevented the Academy from using the measures operationally.

Manning, Mills, Mogilka and Pfleiderer (1998) compared computerized performance measures with those counted by SMEs. They used data collected during the AT-SAT high fidelity simulation study and compared the number of operational errors (OEs) identified by software with the number of OEs identified by trained observers. Discrepancies were resolved by an SME who reviewed the recorded simulation data. The SME found that 63 per cent of raters' OE classifications were correct, but only 37 per cent of the computer's classifications were correct. Raters identified some OEs that the computer could not identify because they occurred in non-radar airspace and thus, were not recorded. When only radar OEs were considered, the computer was correct 53 per cent of the time and the raters were correct 47 per cent of the time for those errors on which they disagreed. About 55 per cent of raters' errors involved failing to identify OEs that actually occurred (false negatives) while more than 2/3 of the computer's errors involved incorrectly identifying OEs that did not occur (false positives). The authors concluded that the computer could not identify all radar errors that the raters identified correctly, although the computer could become more accurate if additional resources were invested in software development.

Part of the AT-SAT validation study involved comparing the BARS and situational judgment criterion measures with performance in a high-fidelity simulation exercise conducted at the FAA Academy's RTF laboratory (described in Borman et al., 2001). Besides the OTS Rating Scales (discussed above) and a behavioral checklist, Manning, Mills, Pfleiderer, Mogilka, Hedge, and Bruskiewicz (2000) recorded data from the simulations and retrieved them for analysis. They extracted a set of measures that included times of first and last aircraft within a group to pass a fix, number of aircraft in the group, delays and holds issued, heading and altitude changes, and number and duration of operational errors. They used a subset of the objective measures and items on the behavioral checklist to predict the overall performance rating for a difficult scenario. They found that the error checklist was the best predictor of overall performance (R=.66), but the objective measures were found to be somewhat predictive (R=.32).

Baseline Studies A wide variety of objective and subjective measures have been used in baseline studies designed to describe ATCS activity and performance for three En-route systems. Galushka et al. (1995) developed a set of baseline measures for the then-current En-route Plan View Display (PVD) console and M1 system. They developed objective and subjective measures according to six constructs: Safety, Capacity, Performance, Workload, Usability, and System Fidelity. They ran simulations and collected measures that quantified traffic volume, flight duration, and traffic characteristics and assessed how the controller used the system (via input/output messages). The authors did not discuss the decision-making involved in assigning individual measures to the global constructs. They concluded that the number of simulation scenarios used (and their durations) should be increased and the simulation parameters should be changed. They also identified additional variables to represent each construct.

Allendoerfer, Galushka, and Mogford (2000) conducted a baseline simulation of En-route air traffic control operations using an experimental version of the En-route Display System Replacement (DSR) system that eventually replaced the M1 system. The simulation collected objective data based on the same operational constructs used by Galushka et al. (1995). The DSR Baseline used the same airspace, traffic scenarios, and controller participants as the Galushka et al. (1995) PVD Baseline study. The simulation used four Washington Air Route Traffic Control Center (ARTCC) sectors and two traffic scenarios that represented a 90th percentile day for traffic volume. Although there were some methodological differences between the baselines (e.g., simulation platforms, communication equipment, and pseudo-pilots), the DSR Baseline used the same data collection and analysis techniques as the PVD Baseline. Human Factors researchers collected objective data from the output of the simulation platform and the communication system. They collected subjective data using controller and expert observer questionnaires and measured subjective controller workload using the Air Traffic

Workload Input Technique (ATWIT). The review team used the data to generate some recommendations for the DSR program, and for improving the baselining process.

Marsden and Krois (1997) conducted a third baseline study that compared Eurocontrol's Operational Display and Input Development (ODID) IV system with the FAA's En-route PVD system. The simulation was intended to compare FAA's then current PVD technology with the ODID graphical interface for airspace and scenarios derived from routinely recorded ATC data from the Washington ARTCC. Measures used were computed from ODID system recordings. The researchers collected aircraft profile information and sector occupancy, ghost pilot inputs to the system in response to controller instructions, timing of all communications, and controller inputs to the system and related system responses. These comparisons allowed identification of changes in system performance that might occur with new ATC technologies.

Other Objective Measures Checklists may be considered 'nearly' objective measures when completed by SMEs formally trained to use them to observe controller activities. The difference between observation checklists and rating scales is that the SME observers who complete the checklists do not provide a subjective rating, but simply record events they observed. Each time a controller takes an action or makes an error, either during a simulation or while working operationally, an SME observer tallies it. For example, SMEs used an observational checklist to record flight strip markings (e.g., issued, coordinated, and planned clearances, incoming/outgoing radar/communications, non-clearance coordinations, information updates) and actions (e.g., point, move, offset; Durso, Batsakes, Crutchfield, Braden, and Manning, 2004). SMEs who participated in this study worked with researchers to develop a checklist and a standardization guide that provided guidance on how they should code all markings and actions made using flight strips.

Bruskiewicz et al. (2000) developed the Behavior and Event Checklist (BEC) used by trained raters to record errors observed during the high-fidelity simulation study conducted in support of the AT-SAT validation (Borman et al., 2001). Items included in the BEC were failure to a) accept handoffs, b) issue weather information, or c) coordinate pilot requests. Additionally, instances of d) Letters of Agreement (LOA)/directive violations, e) readback/hearback errors, f) unnecessary delays, g) incorrect information entered into the computer, and h) making late frequency changes were recorded. Raters also recorded OEs, Operational Deviations, and Special Use Airspace (SUA) violations. As in the Durso et al. (2004) study, the SME observers developed a standardization guide to document decisions about how errors would be identified and classified.

Non-SMEs also use observational checklists to record specific events. Several studies used checklists to assess 'compensatory behaviors,' that is, activities occurring in today's ATC environment that may be expected to increase or decrease as a result of changes to equipment or procedures. For example, Albright,

Truitt, Barile, Vortac, and Manning (1995) counted compensatory behaviors to assess controllers' performance when they controlled traffic with and without flight strips. They counted Flight Plan Readouts (FPRs), Route Displays, J-rings (also known as Distance Reference Indicators or DRIs), and Conflict Alerts. They also measured the amount of time the controller spent looking at the situation display. They found that controllers compensated for the lack of strips by requesting more FPRs and spent more time looking at the situation display. Truitt, Durso, Crutchfield, Moertl, and Manning (2000) looked at similar compensatory behaviors when controllers could use an optional strip posting/marking procedure that allowed strips to be removed early.

Advantages Associated with Objective Measures Since they are based on observable events, objective measures are derived directly from a controller's activities rather than opinions about the controller's effectiveness. Rater bias does not influence objective measures. With the exception of the checklists completed by SMEs, whether or not an event occurred is not subject to interpretation when using objective measures, and events will not be 'missed' because software failed to identify them. If a standardization guide exists, classification of events is subject to little interpretation. The use of objective measures, if obtained unobtrusively, also allows for the measurement of typical performance.

Disadvantages Associated with Objective Measures Landy (1989, pp. 113-114) cites three problems with the use of what he calls 'basic production data' as job performance measures. The first problem is that the observation period needs to be long enough to provide a reasonable assessment of an individual's performance. The second problem is that the output of the system may not be observably related to the actions of the employee. Third, there may not be any good objective measures available. Determining the effectiveness of the extracted measures requires knowing what they mean and how they relate to job performance.

There are some problems associated with the use of objective measures. First, there may be ethical problems associated with the analysis of performance measures obtained from recorded data if the employee is unaware the analysis is taking place. This may be dealt with by ensuring informed consent of all involved parties.

A larger problem involves understanding what the measures mean. It is easy to compute numbers that describe what is happening in an ATC traffic sample but more difficult to determine which numbers are meaningful. Several researchers have tried to determine the meaning of some of these measures. For example, Galushka et al. (1995) and Allendoerfer et al. (2000) linked their objective measures to descriptive categories (Capacity, Performance, Workload, and Usability) but provided no evidence about the validity of the linkages. Marsden and Krois (1997) indicated that, in spite of the objective data collected, it was difficult for them to assess the effects on safety of using ODID.

Hennessy (1990) suggested that, in an operational environment, expert observational techniques are more relevant and useful than measures recorded using some sort of automation. He proposed that ratings are based on many assumptions about human operators, including how they learn and how that learning is implemented in performance. He felt that routinely recorded measures cannot take these factors into account.

A similar argument suggests that taskload measures, such as SEMs and POWER, capture observable variations in ATC activity, but do not take into account the controller's subjective reaction to events in the environment (Stein, 1998). Stein contended that controllers' individual differences influenced their perception of the effects of a particular taskload. Thus, subjective workload ratings are affected by a component that cannot be derived simply by analyzing recorded data. However, other research found significant correlations between objective taskload and subjective workload measures (Stein, 1985; Manning et al., 2001), suggesting that taskload measures alone may allow researchers to evaluate the effects of new systems.

Standardized Work Sample (Simulation) Tests

Standardized work sample tests (Landy, 1989, p. 118) are designed to measure important job tasks in a controlled environment. While the situations may be similar to those examined by the situational judgment tests previously described, work sample tests are unique because they require incumbents to both decide on and take an action rather than simply choosing among a set of available response options. Because they know they are being observed and their performance is being measured, work sample tests assess an incumbent's maximum performance.

Both trainers and researchers use work sample simulation tests to evaluate the performance of air traffic controllers. In the air traffic environment, this type of testing is typically called simulation testing, and the work samples are called scenarios. Simulation testing can be used to assess the controller by measuring performance in initial or recurrent training. Simulation testing can also be used to assess new operational concepts, equipment, or procedures by examining the effects of system changes on the performance or workload of the controller. This section will describe uses of simulation testing, examples of simulation tests, issues concerning the inclusion of specific performance measures in simulation tests, and advantages and disadvantages of using simulation tests to measure the performance of air traffic controllers.

Use of Simulation Tests to Assess Training Performance Some ATC simulation tests consist of laboratory scenarios that developmental controllers run during different stages of their training. For example, the 'Non-radar lab' scenarios were administered to En-route and terminal controller trainees during the last stage of the former pass/fail Initial Qualification Training course. Laboratory scenarios were used because they allowed trainees to demonstrate that they could execute actions required of a controller in a non-radar environment, in addition to remembering important job knowledge. Trainees used non-radar procedures (with only flight strip displays and no radar equipment) to assess an air traffic situation and issued clearances to 'ghost' pilots and 'remote' controllers that they believed would ensure separation between aircraft in the scenario. After observing the scenario, an instructor would assign each trainee a Technical Assessment (TA; a numerical score based on the number of errors made) and an Instructor Assessment (IA; based on a rating scale used to assess the student's future potential). The TA and IA were averaged together to produce a Laboratory Score for the scenario. The average score for the best five of six graded laboratory scenarios comprised 65 per cent of the final grade in the pass/fail Initial Qualification Training course. The average Non-radar Technical Assessment had a correlation of .21 with a variable indicating field training status (r=.30 when adjusted for restriction in range; Manning, 1991) and a correlation of .35 with the CBPM situational judgment test, which was administered some 10-15 years later to 212 incumbent En-route controllers (Manning and Heil, 2001, pp. 139-140). Similarly, the average Non-radar Instructor Assessment had a correlation of .22 with field training status (r=.37 when adjusted for restriction in range) and a correlation of .29 with the CBPM. Instructors administered a corresponding set of laboratory scenarios to ATC trainees (and used similar scoring techniques) during the Radar Training Program (Boone et al., 1980).

En-route controller trainees also ran graded ATC scenarios during field training that occurred after the Initial Qualification Training course. Depending on the phase of field training, the scenarios addressed the functions of either the Radar Associate controller or the Radar controller. Unlike the non-radar scenarios administered at the Academy, field training scenarios were designed specifically for the airspace where the trainees actually worked and were more difficult than those used in Initial Qualification Training.

Use of Simulation Tests in Concept Exploration Research While trainers often used simulation tests to assess performance in En-route controller training, researchers also used them in a variety of ATC research settings. In particular, researchers have often used simulation testing to assess the feasibility of new technology concepts and the usability of specific systems. Several examples are presented below.

Ladue, Sollenberger, Belanger, and Heinze (1997) conducted a field study using recorded communications of both analog and digital radios with varied background noises from different types of aircraft. The results indicated that vocoded (i.e., digitized, transmitted, and re-synthesized) speech was both understandable and acceptable to controllers. Sollenberger, Ladue, Carver, and Heinze (1997) conducted a simulation test with digital and analog communications alternated in random order over a series of moderate and high traffic level scenarios. The authors found no differences in performance and workload; however, controller ratings of acceptability and intelligibility were higher for analog radio than for two versions of the vocorder.

Hadley, Sollenberger, D'Arcy, and Basset (2000) conducted simulation testing to examine the impact of inter-facility dynamic re-sectorization on controllers' performance, workload, communication, situation awareness, and control strategies. The study focused on the impact of lateral boundary changes between adjacent En-route facilities. Results indicated that dynamic re-sectorization, as defined in the study, was feasible. Moreover, dynamic re-sectorization did not reduce controller situation awareness and did not increase operational errors.

Yuditsky, Sollenberger, Della Rocco, Friedman-Berg, and Manning (2002) investigated the effectiveness of visual coding enhancements for special use airspace (SUA) activation, transitioning aircraft, and destination airports in a simulation study. They examined the effects of visual enhancements on controller performance, workload, and efficiency. Researchers hypothesized that the automated visual cues would reduce memory demands and cognitive workload, and would increase airspace efficiency. Results indicated that there were significant effects on controller performance with each enhancement tested. However, when all of the enhancements were presented simultaneously during a scenario, they did not find the beneficial effects that occurred when the enhancements were tested individually. It appeared as though the enhancements were subtractive, rather than additive, effectively canceling each other out.

Since 1991, researchers have conducted a program to investigate the use of flight progress strips by En-route controllers. Simulation tests were conducted at the FAA Academy and in field En-route DYSIM laboratories to investigate a number of research questions related to flight progress strips. In general, the studies compared two competing hypotheses: The interaction hypothesis predicted that automation of flight progress strips should result in poorer performance and impaired cognitive processing, because controllers would no longer be able to interact physically with the strips. The workload hypothesis predicted that using electronic flight data would improve controller performance and cognitive processing, because the computer would take over much of the housekeeping activities involved with updating and maintaining information on the strips. To test the interaction hypothesis, Vortac, Edwards, Fuller, and Manning (1993) glued strip holders together and prevented controllers from writing on the strips while they ran scenarios. They found that controllers who had restricted access to strips granted significantly more prospective requests and did so sooner than did controllers who had normal access to strips.

Vortac, Barile, Albright, Truitt, Manning, and Bain (1996) tested a control condition of normal strip usage against two versions of electronic flight data, a full automation condition and a partial automation condition. Results suggested that the full automation condition improved both performance and prospective memory while the partial automation condition was not always superior to the normal condition. The authors speculated that board management functions required by the normal and partial automation conditions resulted in a negative effect.

Several other simulation studies involving manipulations of flight strip conditions have been conducted (Albright et al., 1995; Durso, Truitt, Hackworth,

Albright, Bleckley, and Manning, 1998; Truitt et al., 2000; Vortac, Edwards, Fuller, and Manning, 1993). Before these studies were conducted, it was thought that automating strips might interfere considerably with the control of air traffic (Hopkin, 1990). Research on flight strips suggested that eliminating the requirement to write on strips might compensate, at least in part, for any negative effects on controller cognitive processing and performance.

What Events or Behaviors should be Measured during a Scenario? While some may view the scenario itself as a form of performance measurement, another level of performance measurement can occur within the scenario, that is, the specific measures obtained from the controllers during and after the scenario.

In general, the choice of measures obtained from a scenario should depend on the reason for conducting the simulation test. In other words, the measures chosen should address the circumstances being investigated by the simulation test. For example, if an employer is conducting a simulation test to determine if a candidate is an appropriate choice for a job, then performance measures should be obtained from the scenario that assess the candidate's aptitude. If, on the other hand, a trainer or evaluator is conducting a simulation test to determine whether a candidate has learned a set of ATC skills, performance measures obtained from the scenario should provide information about whether the candidate demonstrated that he/she can perform the job task just trained under a variety of circumstances.

If a Human Factors specialist is conducting a simulation test to assess the utility of new equipment or procedures, then the performance measures obtained from the scenario should provide information about the potential effects of these changes on controller performance, workload, situation awareness, information-processing, or other factors related to the aspects of the job likely to be affected by the change. Usability ratings (Lund, 2001) are often used to infer potential system and individual operator performance. Again, the effects should be measured under a variety of circumstances that potentially reflect the anticipated use of new equipment or procedures and those circumstances should be programmed into the scenario.

What Kinds of Measures are Useful for Assessing Controller Performance during Scenarios? Depending on the purpose of the study, any of the second-generation performance measures described above are appropriate. For example, researchers might use BARS when an SME's perception of a controller's performance (or workload) is considered relevant to answering the question addressed by the simulation test. Although situational judgment tests are infrequently used during simulations, researchers could use them during part-task simulations conducted when a full fidelity environment is not available (because a prototype version of the proposed system has not yet been built.) Researchers often include objective measures in simulations, although, as indicated above, they may not fully understand their meaning. Observation checklists are often useful if designed specifically for the purpose of the simulation. Finally, researchers or trainers can evaluate whether the elements of the overall ATC job were performed correctly during a scenario by determining if the steps were executed correctly, determining the amount of time required to complete them, and/or evaluating the quality of the execution.

Frequencies and timing of events are the most widely used dependent variables in simulation testing. These may be discrete or cumulative and are based on a specific time period, in this case, the time encompassed by the scenario. One might hypothesize, for example, a change in conflict frequency and duration based on the amount of time a controller has been on position. In this case it would be important to have the capability to compute statistics across predetermined time blocks. Researchers have used frequency measures in numerous studies over the years to evaluate concepts and systems. For most simulation research, investigators have found that using multiple measures (both objective and subjective) provides the necessary sensitivity and discriminability needed to learn something of value from a simulation test. However, it has been argued, with some justification, that researchers cannot always clearly define the difference between *systems* and *individual* performance measures because the two often mesh.

Examples Researchers have used a wide variety of task-specific performance measures in simulation tests conducted to explore ATC concepts. A few examples are described below.

In the multisector planner Airspace Coordinator study, Willems, Heiney, and Sollenberger (2002) measured visual scanning, communications, performance, workload, and situation awareness (SA). An eye tracking system collected visual scanning data for the Airspace Coordinator. They used push-to-talk (PTT) software to examine landline and ground-to-air communications. They looked at the number of conflicts and length of time in sector. Researchers collected workload ratings from a Workload Assessment Keypad (WAK), the NASA Task Load Index (TLX), and self-reports. They evaluated SA using the SA Verification and ANalysis Tool (SAVANT), self-report measures, and over-the-shoulder (OTS) ratings made by ATC SMEs. Post-scenario questionnaires (PSQ) provided self-report data from the controllers, and OTS ratings provided subjective performance data.

In the inter-facility dynamic re-sectorization study, Hadley et al. (2000) used a suite of objective measures produced by the DESIREE laboratory simulation software, observational measures, and over-the-shoulder ATC BARS-type rating forms discussed previously. Workload estimates were obtained from the NASA TLX and ATWIT. Participants completed situation awareness questionnaires and self-ratings on a variety of dimensions. Willems and Truitt's (1999) study on increases in automation measured visual scanning, workload, situation awareness, system performance, SME over-the shoulder-ratings, and participant responses to PSQs.

Vortac et al. (1993), Vortac et al. (1996), Durso et al. (1998), Albright et al. (1995), and Truitt et al. (2000) designed a series of dependent measures for various flight strip studies. These studies compared the effects of alternate versions of flight progress data on controller workload, performance, and cognition. The authors developed multiple measures of controller cognition, performance, compensatory behaviors, and opinions for the studies. Initially, they used a battery of cognitive tests that included measures of attention, recall, prospective memory (remembering to perform future actions), and traffic planning. They retained the measures of attention, which were based on embedded secondary tasks, and later supplemented them with subjective workload assessments to assess changes in workload that might result from different flight data configurations.

The performance measures used in the flight strip studies included simulated position relief briefings and SME ratings made using the OJT evaluation form (FAA, 1998). Durso et al. (1998) modified the standard OJT evaluation form to assess strip-marking activities specifically whereas Vortac et al. (1993) developed a measure of efficiency called Remaining Actions that counted the number of control actions that needed to be taken to move the remaining aircraft from the sector at the end of the scenario, as determined by an SME. Researchers also used a checklist to count compensatory behaviors that they expected would be affected by unavailability of flight strips. For example, some of the compensatory behaviors examined were Flight Plan Readout (FPR), other Computer Readout Display (CRD) entries, pilot requests for information, time spent looking at the Plan View Display (PVD), time spent talking, and number of times strips were pointed to or examined. Some behaviors specific to the location and circumstances were also counted (e.g., number of Coast Tracks seen when running non-radar scenarios).

For a comprehensive compilation of all known measures available at that time for ATC performance and workload assessment, see Hadley et al. (1999). Their ATCS performance measurement database, which describes most of these measures in more detail, is available online at http://acb220.tc.faa.gov/atcpmdb/default.htm.

Advantages Associated with Work Sample Simulation Tests Advantages of using work sample tests are that they allow conducting realistic, high-fidelity simulations of tasks controllers perform regularly. Scenarios can be created that begin and end at exactly the same time for each participant, and include scripts where aircraft enter or leave the sector at exactly the same time. Using a simulation test instead of the operational environment allows a researcher to construct ATC situations that may not occur naturally and/or might be undesirable in the 'real world.' For example, a researcher can introduce events into a scenario such as different numbers and types of aircraft, situations that have been determined to be 'complex' (i.e., 19 factors including climbing and descending aircraft, frequency congestion; see Mogford, Murphy, Roske-Hofstrand, Yastrop, and Guttman, 1994). Other events that can be scripted in include aircraft emergencies, pilot requests for flight plan changes, aircraft that will conflict if the controller does not issue a clearance, and pilot failures to respond to ATC instructions. Activities can also be built into scenarios that could never be introduced into the operational environment, such as stopping the scenario to assess situation awareness or asking a controller to provide information unrelated to the traffic situation while controlling traffic.

Disadvantages Associated with Work Sample Simulation Tests There are also some disadvantages associated with using work sample tests in the ATC environment. First, once a controller takes action during a scenario, he/she will never again see exactly the same situation as other controllers who run the same scenario even though the scenario will begin the same way for each participant and will include all the same aircraft. This lack of consistency occurs because no two controllers control aircraft in exactly the same way. Once a controller issues a clearance to a pilot during a scenario, the position of that aircraft in relation to other aircraft will be different than the position of those aircraft when other controllers run the same scenario. Thus, work sample testing allows the same initial stimuli to be provided to multiple controllers and the same outcome measures to be computed, but the outcome measures will depend on different traffic situations experienced by the participants. This result is more realistic than other methods used to measure controller performance, but it provides the researcher with less experimental control. Unfortunately, if we tried to artificially increase experimental control, then realism or fidelity would inevitably decline.

Second, work sample simulations may be somewhat different than the operational environment in ways that may seem inconsequential to researchers but may be notable to the controllers running them. For example, controllers running simulations are aware that they are not working in a real environment. If they commit an operational error during the scenario, it is not real, and thus, they know it will not be counted against them. This knowledge may allow participants to take chances they would never take in the operational environment. They may also take the scenario less seriously than they would if it were real. However, more often than not, they will get caught up in the scenario and behave as they would during normal operations.

Third, there are additional concerns associated with airspace used in a simulation, the amount of training provided on the airspace, as well as the amount of training provided for experimental equipment or procedures being tested. Airspace is important because controllers work traffic in a limited number of geographical areas. For example, En-route controllers are certified on between five and ten 'sectors' of airspace in a single 'area of specialization.' Six to eight areas of specialization are found in any En-route center, and each controller works in only one area. Sectors use detailed maps, procedures, and letters of agreement that must all be memorized. Controllers are very familiar with the sectors on which they are certified and are less familiar with other controllers' airspace. Thus, if they were to control traffic at a sector other than one of their own, controllers may not fully understand the traffic flows, choke points, and other potential problems associated with the sector.

It might seem desirable to develop scenarios based upon a known sector to assure a participant's familiarity. However, if one sector is selected for a simulation, then only those controllers who work that airspace could participate, severely reducing the available subject pool. Furthermore, we are not sure to what extent the results obtained from a simulation test conducted on only one sector can be generalized to another. One alternative is to use a generic airspace instead of a real one. Generic airspace is not real, but is constructed to provide a fairly realistic airspace while minimizing some of the complications of a real airspace. Use of generic airspace allows any controller to participate in a study instead of limiting the subject pool to controllers familiar with a specific sector. Controller participants will be equally unfamiliar with the generic airspace so there should be no differential effect of airspace familiarity. The use of generic airspace can simplify and reduce the cost of training and selection of participants if controllers are able to perform about as well using a generic sector as they can in when using a known sector. However, controller performance in a generic airspace will depend on the amount of training they receive relative to the complexity of the airspace and procedures used.

Guttman and Stein (1997) evaluated the feasibility of using generic airspace that participating controllers had not seen before. They collected data on ATCS performance, workload, system effectiveness, and self-assessment during simulated scenarios conducted at both a generic and a familiar sector. Three of four performance categories showed high and consistent correlations between the generic and home sectors. These categories were ATWIT workload ratings, system effectiveness measures, and controller self-ratings of performance. These correlations suggest that controller workload, communication, and task management were basically the same regardless of the sector used. Workload was also highly correlated between the home sector and the fourth block of scenarios run on the generic sector. This result suggested that, once the controllers learned the sector, the workload was basically the same, regardless of the sector configuration. The results also indicated that system performance, as measured by system effectiveness measures, was very similar in both sector configurations.

While generic sectors are designed to be easier to understand and remember than real sectors, a certain amount of training is still required for a controller to develop some amount of expertise on the generic sector. The amount of sector training provided can also create problems with running scenarios. If insufficient training is provided, then it will be difficult to determine whether results of a study occurred because of the experimental treatment or because of lack of familiarity with the sector, perhaps interacting with an unfamiliar procedure or new piece of equipment. The amount and content of training for using an experimental procedure or equipment is also important. Without sufficient training, a realistic assessment of effects of procedures or equipment on controller workload or performance cannot be performed.

Additional issues surrounding the use of simulation tests will be discussed in the Issues/Questions section below.

Issues/Questions

Several questions remain about the use of measures to assess the performance of air traffic controllers. These include:

- Are certain controller performance measures better than others?
- How can we ensure that controller performance measures are valid?
- How do ATC performance measures relate to operational goals?
- How do we explain the results obtained using controller performance measures to organizational decision-makers?

Are Certain Controller Performance Measures Better Than Others?

The discussions in this chapter may suggest that the authors believe that objective measures are best, records of observed behavior are good, and subjective ratings are less valuable. However, we wish to emphasize that the best measures to use for any particular circumstance are those that address the question being asked. Through personal experience, we have learned that some measures worked better than others under specified conditions, but these experiences may be incomplete. On the other hand, some measures seem to produce insufficient variability (e.g., ceiling effects on certain evaluations or infrequent events such as OEs) while others produce too much unrelated variability (e.g., recall of aircraft speeds).

Harvey, Buondonno, Kopardikar, Magyarits, and Racine (2003) produced a document that recommended 'best practices' for conducting simulation tests. One section of the document described measurement. The authors' answer to the question of which measure is best to use is, generally, 'it depends,' an answer usually unpopular with sponsors of ATC research. They said, 'The choice of measures (dependent variables) and experimental factors (independent variables) will depend on the objectives, and in turn the objectives should depend on the existence of suitable variables' (page 25). Before specific measures are identified, it is necessary for sponsors to acknowledge the importance of systematically collecting performance measures, especially during concept exploration and validation.

Subjective vs. Objective Measures If we accept the proposition that the choice of measures used should depend on the question asked, are there still some types of measures that are better than others? More specifically, how effective are subjective ratings? There is little doubt that subjective ratings have a place in the overall scheme of ATC performance assessment. However, a number of problems arise repeatedly with their continued use. Some user groups see such measures as ends in and of themselves, even if other, more objective, data are available. 'If controllers like it, the system must be good, and if they do not, then the system is invariably flawed,' in the eyes of the user. Ratings are, by their very nature, only as good as those who designed rating forms/items, the descriptions of the anchors, and the training of the raters. Unless rater training can provide behavioral anchors and reduce or eliminate rater bias, the ratings are not worth the paper they are printed on.

Harvey et al. (2003) questioned the need to distinguish between objective and subjective measures. 'In many cases these terms are misused to express the difference between qualitative and quantitative data. In the particular case of Human-in-the-Loop studies, it can be useful to distinguish between the objectively measured system variables and those derived from the subjective opinions of the participants. An alternative terminology could be to define perception measures (subjective) and observed measures (objective). This definition helps to clarify that the subjectivity is that of the human participant and not that of the observer' (p. 25).

System vs. Individual Measures In Human Factors we often find ourselves working almost at cross purposes. We want to measure human performance within the system, and yet many of the measures used are essentially system measures. We infer human performance based on system changes when the human operator is the primary variable in the system, and we have controlled as much of the remainder of the variability as we can.

Changes in human performance can cascade throughout the system. 'The primary objective of performance measurement is to provide a better understanding of NAS critical elements and to help to diagnose and solve system performance issues. From a Human Factors research standpoint, one important question is how to establish the link between ATCS performance and system effectiveness' (Hadley et al., 1999, p. 9.)

Yet, not all measurement attempts to make these inferences. There are many examples of models and research that simply look at the system and do not focus on the human components. We are barraged by measures that are used to evaluate facilities such as terminals (e.g., Cocanower, 2002). One of the most frequently cited measures in the media is delays, theoretically induced by a less-than-efficient air traffic control system and, by implication, the controller workforce themselves. Reducing delays has been practically a mantra in the FAA. However, from a practical standpoint, political pressure to reduce delays may have a downside in that one way to accomplish the task is to encourage controllers to reduce the buffer space beyond the minimum separation standard with which some feel more comfortable. Less buffer space means less time to correct potential problems such as an aircraft speeding up while in trail behind another. Thus, fewer delays may lead to more OEs.

Types of Measures Used Another issue to be considered is which measures should be used in any given investigation. We recommend measuring multiple constructs and using multiple measures of each construct. The linkage between a measure, such as a scale that rates perceived activity level, and a construct, such as subjective workload, usually appears sensible but is difficult to establish conclusively. The convergence of several variables proposed to measure the same construct is more credible than an effect based on only one measure of that construct. Any research study or system evaluation that is based solely on one type of measurement such as subjective ratings will likely be flawed. If we are to get a good handle on human performance in the sense of true variance, then multiple methods and multiple measures are necessary.

How can we Ensure that Controller Performance Measures are Valid?

First, we must keep in mind that reliability is an upper bound to validity. Thus, reliability of measures (especially subjective ratings) must be ensured before we can be confident of their validity. Earlier discussions pointed out the lack of reliability in subjective ratings that can occur unless extensive training is provided to raters concerning potential rater biases, as well as the use of the rating scales. The use of BARS also helps raters associate their evaluations with calibrated performance descriptions.

By controlling extraneous variability as much as possible, even in the applied environment, and by attempting to use scientific procedures such as randomization, then, to the extent possible, we can draw conclusions about the relationship between variables and thus may be able to infer internal validity (Campbell and Stanley, 1966). Threats to validity may result from lack of randomization (which may occur when using participants selected by the union or other volunteers), low sample sizes (inherent in research where bureaucratic resistance makes it difficult to get access to participants), and the Hawthorne effect (Mayo, 1933) in which changes in behavior occur simply because of the observation and not the experimental treatment. Using within-subjects designs may reduce some variability related to individual differences, though examination of individual differences is also an important question to consider.

We believe that important evidence about the validity of measures can be obtained from the use of multiple measures associated with multiple constructs.

How do ATC Performance Measures Relate to Operational Goals?

External validity is also an important consideration. External validity is the extent to which the results of a study can be generalized to other, similar studies or similar environments. If a simulation is conducted and a new ATC tool is found to reduce controller workload, we want to be sure that the same result will occur for other controllers, other traffic configurations, and other airspace.

In the world of applied psychology, researchers do not develop methods and measures for the general advancement of science, but rather to address questions and issues important to the research sponsor. Although important to researchers, the internal validity of applied measures and methodology is of little apparent value to research sponsors. The development of valid performance measures is tied to specific systems development goals. Measurement and the tools that support it can have a very positive long-term impact on systems that are developed and fielded.

Over the years, applied researchers have attempted to produce value-added measures. Some measures were actually developed as an end goal of funded research projects. However, more were side products in that sponsors are more interested in funding system-specific HF research, and we attempt to leverage off that work to enhance our collective measurement tool kits.

A related issue concerns the relationship between performance measures and operational requirements. The concern is especially relevant when considering objective events such as controller behaviors. What does it mean to operational decision-makers if we say that controllers made significantly fewer heading, speed, or altitude changes during Condition 1 than during Condition 2? Are those results good or bad?

Harvey et al. (2003) provided the following perspective: 'The fact that statistical significance is found does not necessary mean that results translate to meaningful operational differences... The statistical significance will tell you if there is a difference but it will not indicate how meaningful the difference is from the operational perspective. Therefore, statistical significance must be treated with caution. Often, perception (subjective) responses gathered from questionnaires, interviews, and debriefings help to sort out the operational relevance of observed (objective) results' (p. 30). On the other hand, lack of statistical significance may occur because of reduced power due to a small number of participants or a limited treatment effect relative to the experimental design. In this case, researchers should not automatically refrain from considering how their results may relate to operational meaningfulness.

Again, the use of multiple measures may help interpret the meaning of results. Another recommendation from the Harvey et al. (2003) report is that a simulation test should be designed to address high-level objectives related to operational considerations such as feasibility, safety, and benefits. If the relationship between the components of the simulation (scenario elements, dependent measures, etc.) is established before the study is conducted, then the relevance of the results should be easier to interpret than if no such a priori linkage was established.

How do we Explain Results Obtained using Controller Performance Measures to Organizational Decision-Makers?

If researchers collaborate with operational decision makers to clearly establish the purpose of the study in advance, then the design and performance measures should be easily linked to the study's purpose. Results should be understandable in the context established for the study and so the explanation of results should be reasonably easy to convey.

While there are few certainties when it comes to human performance measurement, the following statement by Bailey (1982) is accurate, '. . . people do not perform consistently and available measurement devices are imperfect' (p. 554). Despite these admonitions and the difficulty of obtaining effective performance ratings, such evaluations are very popular and continue in business, industry, and government. Subjective ratings have face validity for many decision makers, even when they fail to meet basic criteria for reliability and criterion-related validity.

Long (2000) commented that universities turning out HF professionals may be spending most of their instructional time on technique, rather than dealing with underlying concepts in measurement. He suggests that we may have technicians, not scientists, who do not integrate measurement with the philosophy underlying it. The philosophy, if we continued to embrace it, would force us, he notes, to look beyond the surface. This is similar to what Hopkin (1980) said many years ago. Long and Hopkin were correct in that customers do not care about, nor do they want to support, research that is not directly related to their projects. Philosophy and the development of measurement based on theory takes second place or no place at all in much of applied research.

The good news is that we do not have to sell philosophy or higher levels of science to sponsors who are developing or acquiring systems. What we do have to sell over and over again is ourselves, our profession, and the tools we have that will help them put better and safer systems into operation. We need to continually market a systems model of development that emphasizes measurement under controlled conditions until we are ready to go to the field. Organizations talk a good game about systems thinking, but in practice they often shortcut the model in an effort to save time and money. We have to show them that with effective systems planning, measurement improves the product and lowers potential long-term costs. Change will only occur when we can demonstrate, in their own terms, the payoff for using quality performance measurement as part of their overall development process.

Conclusions

Developing and using ATC performance measures that are reliable and valid is a non-trivial activity. It requires people, material, money, and time. Program managers often find it easier and cheaper to use either a systematic but subjective evaluation or else fall back on a non-systematic technique that relies on the personal preferences and biases of whomever is looking over the system they are procuring. These are often user groups who generally believe they know what they like and what will work for them.

Systems developers are often under time pressure. Time and money are generally used as the rationale for shortcutting an effective evaluation that uses substantive performance measures. We tend to reward program managers for getting systems out the door and into the field. This approach does not factor in the long-term costs of sub-optimal systems that contain design errors that could have been caught with effective measurement during development or prototype testing. We have all seen or experienced equipment and software that did not live up to its potential, yet we paid top dollar for it.

There is an old mechanic's dictum: 'You can pay me now or you can pay me later.' In the mechanic's world this refers to doing preventive maintenance now or facing a breakdown later. In the world of systems acquisition, you do not get many chances to get it right. Once a system hits the field, reengineering it is often prohibitively expensive and, instead, the operators are tasked to somehow make it work. Such a system takes more time or effort, increases the potential for errors, and reduces efficiency. It may well take detailed cost accounting to demonstrate what most Human Factors professionals view as common sense. Good measures used early and often during the evolution of a system that has human operators improve the probability of a safe, efficient system. In most applied settings, the developer does not actually even have to create performance measures. He/she can draw on the expertise of others and on previous studies that created various tools. Some valid measurement, especially during development and evaluation of prototypes, can determine if the new system is superior or inferior to the one it replaces. One hopes the evaluation and baseline studies were accomplished under controlled conditions. Allendoerfer et al., (2000) clearly documented that there is no substitute for good research design and the best measures will not help you if you do not have control over confounding variables.

Testing and measurement are neither fast nor cheap. However, if planned as part of the overall development program, they will be part of the developmental flow and will serve to highlight small problems before they become major design errors.

When we look at the success stories associated with the measurement of controller performance, we can consider that FAA researchers have attempted to systematically apply scientific principles to the examination of applied problems related to air traffic control. To the extent that we have been able to identify potential problems with new systems, and system designers have begun to acknowledge that we contribute something useful to system development (and even occasionally ask for our opinions), it appears that we are beginning to achieve some success. As designers realize that the systematic application of performance measures to all aspects of systems design (especially concept development and validation) can significantly improve the resulting products, they will increasingly collect reliable performance measures to assess their systems.

References

Albright, C.A., Truitt, T.R., Barile, A.L., Vortac, O.U., and Manning, C.A. (1995), 'Controlling Traffic Without Flight Progress Strips: Compensation, Workload, Performance, and Opinion', *Air Traffic Control Quarterly*, Vol. 2, pp. 229-248.

Allendoerfer, K.R., Galushka, J.J. and Mogford, R. (2000), *Display System Replacement Baseline Research Report*, DOT/FAA/TN00/31, William J. Hughes Technical Center, Atlantic City International Airport, NJ.

Anastasi, A. (1988), *Psychological Testing*, 6th Edition, Macmillan, New York.

Bailey, R.W. (1982), *Human Performance Engineering: A Guide for System Designers,* Prentice Hall, Englewood Cliffs, NJ.

Boone, J.O., Van Buskirk, L., and Steen, J. (1980), *The Federal Aviation Administration's Radar Training Facility and employee selection and training,* DOT/FAA/AM-80/15, FAA Office of Aviation Medicine, Washington, DC.

Boone, J.O. (1984), 'The FAA Air Traffic Controller Training Program, with Emphasis on Assessment of Student Performance', in S. B. Sells, J. T. Dailey, and E. W. Pickrel (eds.), *Selection of Air Traffic Controllers,* DOT/FAA/AM-84/2, pp. 211-214, FAA Office of Aviation Medicine, Washington, DC.

Borman, W.C., Hedge, J.W., Hanson, M.A., Bruskiewicz, K.T., Mogilka, H., Manning, C., Bunch, L.B., and Horgen, K.E. (2001), 'Development of Criterion Measures of Air Traffic Controller Performance', in R.A. Ramos, M.C. Heil, C.A. Manning, (eds.), *Documentation of Validity for the AT-SAT Computerized Test Battery, Volume II,* DOT/FAA/AM-01/16, FAA Office of Aviation Medicine, Washington, DC.

Bruskiewicz, K.T., Hedge, J.W., Manning, C.A., and Mogilka, H.J. (2000), 'Measuring the Performance of Air Traffic Controllers Using a High-Fidelity Work Sample Approach', in C.A. Manning (ed.) *Measuring Air Traffic Controller Performance in a High-Fidelity Simulation,* DOT/FAA/AM-00/2, FAA Office of Aviation Medicine, Washington, DC.

Buckley, E.P. and Beebe, T. (1972), *The Development of a Motion Picture Measurement Instrument for Aptitude for Air Traffic Control,* DOT/FAA/RD-71/106, National Aviation Facilities Experimental Center, Atlantic City, NJ.

Buckley, E.P., DeBaryshe, B.D., Hitchner, N., and Kohn, P. (1983, April), *Methods and Measurements in Real-time Air Traffic Control System Simulation,* DOT/FAA/CT-83/26, Federal Aviation Administration Technical Center, Atlantic City International Airport, NJ.

Buckley, E.P., and Stein, E.S. (1992), *Simulation Research Variable Specifications,* unpublished Manuscript, William J. Hughes Technical Center, Atlantic City International Airport, NJ.

Campbell, D.T., and Stanley, J.C. (1966), *Experimental and Quasi-Experimental Designs for Research,* Rand McNally, Chicago.

Cocanower, A. B. (2002), 'Terminal Airspace Metrics for Forty Airports', *The Journal of Air Traffic Control,* Vol. 44(2), pp. 16-19.

Durso, F.T., Batsakes, P.J., Crutchfield, J.M., Braden, J.B., and Manning, C.A. (2004), 'The Use of Flight Progress Strips While Working Live Traffic: Frequencies, Performance, and Perceived Benefits', *Human Factors,* Vol. 46(1), pp. 32-49.

Durso, F.T., Truitt, T.R., Hackworth, C.A., Albright, C.A., Bleckley, M.K., and Manning. C.A. (1998), *Reduced Flight Progress Strips in En Route ATC Mixed Environments,* DOT/FAA/AM-98/26, FAA Office of Aviation Medicine, Washington, DC.

Endsley, M.R. and Rodgers, M.D. (1997), *Distribution of Attention, Situation Awareness, and Workload in a Passive Air Traffic Control Task: Implications for Operational Errors and Automation,* DOT/FAA/AM-97/13, FAA Office of Aviation Medicine, Washington, DC.

FAA (1993), *Multiple Virtual Storage (MVS) User's Manual; Data Analysis and Reduction Tool (DART),* NASP-9247-PO2, Washington, DC.

FAA (1998), 'Air Traffic Control Specialist On-the-Job Training and Position Certification' in *Air Traffic Technical Training*, FAA Order 3120.4J, Federal Aviation Administration, Washington, DC.

FAA (2001), *FAA Runway Safety Report: Runway Incursion Severity Trends at Towered Airports in the United States, 1997-2000*, FAA Office of Runway Safety, Washington, DC.

FAA (2002), *Air Traffic Quality Assurance*, Order 7210.56C, Washington, DC.

Flanagan, J.C. (1954), 'The Critical Incident Technique', *Psychological Bulletin,* Vol. 51(4), pp. 327-358.

Galushka, J., Frederick, J., Mogford, R.H., and Krois, P. (1995), *Plan View Display Baseline Report,* DOT/FAA/CT-TN95/45, William J. Hughes Technical Center, Atlantic City International Airport, NJ.

Guttman, J.A., and Stein, E.S. (1997), *En Route Generic Airspace Evaluation*, DOT/FAA/CT-TN97/7, William J. Hughes Technical Center, Atlantic City International Airport, NJ.

Hadley, G.A., Guttman, J.A., and Stringer, P.G. (1999), *Air Traffic Control Specialist Performance Measurement Database,* DOT/FAA/CT-TN99/17, William J. Hughes Technical Center, Atlantic City International Airport, NJ.

Hadley, J., Sollenberger, R.L., D'Arcy, J-F., and Bassett, P. (2000), *Interfacility Boundary Adjustment,* DOT/FAA/CT-TN00/06, William J. Hughes Technical Center, Atlantic City International Airport, NJ.

Hanson, M.A., Borman, W.C., Mogilka, H.J., Manning, C.A., and Hedge, J.W. (1999), 'Computerized Assessment of Skill for a Highly Technical Job', in F. Drasgow, and J.B. Olson-Buchanan, (eds.), *Innovations in Computerized Assessment*, Lawrence Erlbaum Associates, Mahwah, NJ.

Harvey, A., Buondonno, K., Kopardikar, P.K., Magyarits, S., and Racine, N. (2003) 'Appendix 1: Best Practices for Human-in-the-Loop Exercises', in *FAA/Eurocontrol Cooperative R&D Action Plan 5: Operational Concept Validation Strategy Document,* EATMP Info Centre, Reference 030303-1, Eurocontrol, Brussels, Belgium.

Hennessy, R.T. (1990), 'Practical Human Performance Testing and Evaluation', in H. Booher (ed.), *MANPRINT: An Approach to Systems Integration,* Van Nostrand Reinhold, New York.

Hopkin, V.D. (1980), 'The Measurement of the Air Traffic Controller', *Human Factors*, Vol. 22(5), pp. 547-560.

Hopkin, V.D. (1990), 'Automated Flight Strip Usage: Lessons from the Functions of Paper Strips', in *Book of Abstracts from the Symposium Challenges in Aviation Human Factors: The National Plan*, pp. 64-64, American Institute of Aeronautics and Astronautics, Inc., Vienna, VA.

Ladue, J., Sollenberger, R.L., Belanger, W., and Heinze, A. (1997), *Human Factors Evaluation of Vocoders for ATC Environments: Phase 1 Field Evaluation,* DOT/FAA/CT-TN97/11, William J. Hughes Technical Center, Atlantic City International Airport, NJ.

Landy, F.J. (1989), *Psychology of Work Behavior*, Brooks-Cole Publishing Company, Belmont, CA.

Long, G.M. (2000), 'Dynamic Visual Acuity: A Promising New Vision Test with an Old History', *Individual Differences in Performance NEWS*, Vol. 13(1), Villanova University.

Lund, A. (2001), 'Measuring usability with the USE questionnaire', *Usability Special Interest Group Newsletter*, Vol. 8(2), pp. 3-7.

Manning, C. (1991), 'Procedures for Selection of Air Traffic Control Specialists', in H. Wing and C. Manning (eds.), *Selection of Air Traffic Controllers: Complexity, Requirements, and Public Interest*, DOT/FAA/AM-91/9, FAA Office of Aviation Medicine, Washington DC.

Manning, C.A., and Heil, M.C. (2001), 'The Relationship of FAA Archival Data to AT-SAT Predictor and Criterion Measures,' in R.A. Ramos, M.C. Heil, C.A. Manning, (eds.), *Documentation of Validity for the AT-SAT Computerized Test Battery, Volume II,* DOT/FAA/AM-01/16, FAA Office of Aviation Medicine, Washington DC.

Manning, C.A., Mills, S.H., Fox, C., Pfleiderer, E., and Mogilka, H.J (2001), *Investigating the Validity of Performance and Objective Workload Evaluation Research (POWER),* DOT/FAA/AM-01/10, FAA Office of Aviation Medicine, Washington, DC.

Manning, C.A, Mills, S.H., Fox, C.M., Pfleiderer, E.M., and Mogilka, H.J. (2002), *Using Air Traffic Control Taskload Measures and Communication Events to Predict Subjective Workload,* DOT/FAA/AM-02/4, FAA Office of Aviation Medicine, Washington, DC.

Manning, C.A., Mills, S.H., Mogilka, H., and Pfleiderer, E.M. (1998), 'A Comparison of Rater Counts and Computer Calculations of Operational Errors Made in En route Air Traffic Control Simulations' *Proceedings of the 69th Annual Scientific Meeting of the Aerospace Medical Association*, Seattle, WA.

Manning, C.A., Mills, S.H., Pfleiderer, E.M., Mogilka, H.J., Hedge, J.W., and Bruskiewicz, K.T. (2000), 'Prediction of Subjective Ratings of Air Traffic Controller Performance by Computer-Derived Measures and Behavioral Observations', in C.A. Manning (ed.), *Measuring Air Traffic Controller Performance in a High-Fidelity Simulation,* DOT/FAA/AM-00/2, FAA Office of Aviation Medicine, Washington, DC.

Marsden, A., and Krois, P. (1997), *En route baseline comparison simulation final report,* EEC Report No. 311, Eurocontrol Experimental Centre, Brétigny sur Orge, France.

Mayo, E. (1933), *The Human Problems of an Industrial Civilization,* chapter 3, Macmillan, New York.

Mills, S.H., Pfleiderer, E.M., and Manning, C.A. (2002), *POWER: Objective Activity and Taskload Assessment in En route Air Traffic Control,* DOT/FAA/AM-02/2, FAA Office of Aviation Medicine, Washington, DC.

Mogford, R.H., Murphy, E.D., Roske-Hofstrand, R.J., Yastrop, G., and Guttman, J.A. (1994), *Research Techniques for Documenting Cognitive Processes in Air Traffic Control: Sector Complexity and Decision Making*, Report No. DOT/FAA/CT-TN94/3, Federal Aviation Administration Technical Center, Atlantic City International Airport, NJ.

Nickels, B.J., Bobko, P., Blair, M.D., Sands, W.A., and Tartak, E.L. (1995), *Separation and Control Hiring Assessment (SACHA) Final Job Analysis Report*, Deliverable Item 007A under FAA contract DTFA01-91-C-00032, Federal Aviation Administration, Office of Personnel, Washington, DC.

Paul, L.E. (1989), *The Evaluation of Conflicts in Air Traffic Control Simulation.* unpublished manuscript, William J. Hughes Technical Center, Atlantic City International Airport, NJ.

Paul, L. (1990), *Using Simulation to Evaluate the Safety of Proposed ATC Operations and Procedures,* DOT/FAA/CT-TN90/22, William J. Hughes Technical Center, Atlantic City International Airport, NJ.

Pickrel, E.W. and Greener, J.M. (1984), 'Controller Skills Tests', in S.B. Sells, J.T. Dailey, and E.W. Pickrel (eds.), *Selection of Air Traffic Controllers,* DOT/FAA/AM-84/2, pp. 211-214, FAA Office of Aviation Medicine, Washington, DC.

Rock, D.B., Dailey, J.T., Ozur, H. Boone, J.O., and Pickrel, E.W. (1981), *Selection of Applicants for the Air Traffic Controller Occupation,* DOT/FAA/AM-82/11, FAA Office of Aviation Medicine, Washington, DC.

Rodgers, M.D., and Duke, D.A. (1993), 'SATORI: Situation Assessment through Re-creation of Incidents', *The Journal of Air Traffic Control,* Vol. 35(4), pp. 10-14.

Rodgers, M.D., Mogford, R.H., and Mogford, L.S. (1998), *The Relationship of Sector Characteristics to Operational Errors*, Report No. DOT/FAA/AM-98/14, FAA Office of Aviation Medicine, Washington, DC.

Sollenberger, R.F., and Della Rocco P.S. (2002), *Tower Cab Metrics,* DOT/FAA/CT-TN02/03, William J. Hughes Technical Center, Atlantic City International Airport, NJ.

Sollenberger, R.L., Ladue, J., Carver, B. and Heinze, A. (1997), *Human Factors Evaluation of Vocoders for Air Traffic Control (ATC) Environments Phase II: ATC Simulation,* DOT/FAA/CT-TN 97/25, William J. Hughes Technical Center, Atlantic City International Airport, NJ.

Sollenberger, R., Stein, E.S., and Gromelski, S. (1997), *The Development and Evaluation of a Behaviorally Based Rating Form for the Assessment of Air Traffic Controller Performance,* DOT/FAA/CT-TN96/16, William J. Hughes Technical Center, Atlantic City International Airport, NJ.

Stein, E.S. (1985), *Air Traffic Controller Workload: An Examination of Workload Probe,* DOT/FAA/CT-TN-84/24, William J. Hughes Technical Center, Atlantic City International Airport, NJ.

Stein, E.S. (1998), 'Human Operator Workload in Air Traffic Control', in M.W. Smolensky and E.S. Stein (eds.), *Human Factors in Air Traffic Control,* pp. 155-184, Academic Press, San Diego, CA.

Trites, D.K., Miller, M.C., and Cobb, B.B. (1965), *Problems in Air Traffic Management: VII. Job and Training Performance of Air Traffic Control Specialists – Measurement, Structure, and Prediction,* DOT/FAA/AM 65-22, Federal Aviation Agency, Office of Aviation Medicine Washington, DC.

Truitt, T.R., Durso, F.T., Crutchfield, J.M., Moertl, P.M., and Manning, C.A. (2000), 'Test of an Optional Strip Posting and Marking Procedure', *Air Traffic Control Quarterly,* Vol. 8(2), pp. 131-154.

Tucker, J.A. (1984), 'Development of Dynamic Paper-and-Pencil Simulations for Measurement of Air Traffic Controller Proficiency', in S.B. Sells, J.T. Dailey and E.W.

Pickrel (eds.), *Selection of Air Traffic Controllers,* Report No. FAA-AM-84-2, pp. 215-241, FAA Office of Aviation Medicine, Washington, DC.

Uniform Guidelines on Employee Selection Procedures. (1978), *Federal Register,* Vol. 43, No. 166, pp. 38290-38309.

Vardaman, J.J. and Stein, E. (1998), *The Development and Evaluation of a Behaviorally Based Rating Form for the Assessment of En route Air Traffic Controller Performance,* DOT/FAA/CT-TN-98/5, William J. Hughes Technical Center, Atlantic City International Airport, NJ.

Vortac, O.U., Barile, A.B., Albright, C.A., Truitt, T.R., Manning, C.A., and Bain, D. (1996), 'Automation of Flight Data in Air Traffic Control', in D. Hermann, C. McEvoy, C. Hertzog, P. Hertel, and M.K. Johnson, (eds.), *Basic and Applied Memory Research,* Volume 2, Lawrence Erlbaum Associates, Mahwah, NJ.

Vortac, O.U., Edwards, M.B., Fuller, D.K., and Manning, C.A. (1993), 'Automation and Cognition in Air Traffic Control: An Empirical Investigation', *Applied Cognitive Psychology,* Vol. 7, pp. 631-651.

Waugh, G. (2001, March), 'Predictor-Criterion Analyses', in R.A. Ramos, M.C. Heil, C.A. Manning, (eds.), *Documentation of Validity for the AT-SAT Computerized Test Battery, Volume II,* DOT/FAA/AM-01/16, FAA Office of Aviation Medicine, Washington DC.

Willems, B., Heiney, M. and Sollenberger, R. (2002), *Study of an Air Traffic Control (ATC) Baseline for the Evaluation of Team Configurations: Information Requirements,* William J. Hughes Technical Center, Atlantic City International Airport, NJ.

Willems, B., and Truitt, T.R. (1999), *Implications of Reduced Involvement in En route Air Traffic Control,* DOT/FAA/CT-TN99/22, William J. Hughes Technical Center, Atlantic City International Airport, NJ.

Yuditsky, T., Sollenberger, R.L., Della Rocco, P.S., Friedman-Berg, F., and Manning, C.A. (2002), *Application of Color to Reduce Complexity in Air Traffic Control,* DOT/FAA/CT-TN02/13, William J. Hughes Technical Center, Atlantic City International Airport, NJ.

Chapter 14

Computational Human Performance Models and Air Traffic Management

Kevin Corker

Introduction

Complex socio-technical systems, for example power provision, production, communication and transportation systems, are generally designed through a process of iterative simulation and optimization. Safety critical and consequential systems are, by necessity, designed as fused systems of humans and technology (Hollnagel, 2001). In such systems, which are referred to here as joint cognitive systems, the human(s) and the technological components share control responsibility. They do so both by fixed assignment and through flexible shared modes dependent on system and operator state (Sheridan, 1992). The design of such systems requires efficient and effective methods for analysis and prediction of future system performance. Such methods often rely heavily on mathematical models that describe the non-human parts of the system and the environment in which they operate. There is therefore a need to describe the human portion of the system in a manner consistent with the majority of the system, that is to say, through some computational representation of the joint cognitive system (Baron and Corker 1989). Otherwise, the human 'component' may be left out of key decision-informing material. This chapter addresses the development and implementation of such computational models of human and joint cognitive performance as they apply to the processes of air traffic management.

Modeling Human Performance

Why Model the Human Components of ATM Systems?

The assessment and prediction of human performance in air traffic management (ATM) systems has been a central concern for system designers throughout the history of their development (Hopkins, 1995; Mavor, Parasuraman and McGee, 1998). The increased rate of development of ATM system enhancements, the tremendous economic pressure to implement and reap immediate benefits from

ATM technologies, and the significant complexity and cost of large-scale distributed air-ground tests have provided impetus for identification of efficiencies in system development and deployment.

This chapter reviews efforts to model the human operators and the systems with which they interact. These models are usually computational (that is they are implemented as programs or simulations), though some are descriptive or explanatory. The process of modeling the human-system function occurs at several levels of representation of those systems, and is undertaken for several purposes. This range in scale/scope and purpose has led to a diversity of models. However, the basic issues that they are intended to address are fairly straightforward.

Levels of Simulation or Modeling

First, it is necessary to establish what designers of future systems are trying to do, at several functional levels (individual operator; multi-operator; and system management). Designers and analysts of ATM systems are initially concerned with providing information to the (individual) human operators of systems in ways that are effective, and that reduce or restrict information loss or misinterpretation. Information should also result in improved coordination among the many operators in ATM systems (hence 'multi-operator' perspective). Second, they are interested in providing methods of responding to that information that are correspondingly efficient, correct and effective. Representations of the operator and the system at this level of performance are focused in perceptual, cognitive and motor-response processes. Figure 14.1 illustrates the basic components of this type of modeling.

The basic model in Figure 14.1 is implemented in a simulation that can produce output from each of the components represented. So predictions could be made as to whether the equipment provided information to the human operators that was both perceivable, and useful to the tasks at hand. Time required to perform the tasks and estimates of the load on the operator to perform the tasks are available. The control represented in Figure 14.1 includes the human operators and their perception, attention, memory and decision-making processes. It also includes some detail of their equipment and the procedures that these operators are required to carry out in response to their task.

Broader issues in human-system integration that can be modeled are those associated with how the action taken by individuals in an ATM control system affect other operators in that system. In trying to model these types of issues, a more extensive representation is needed. Figure 14.2 provides a representation of this level of model or simulation.

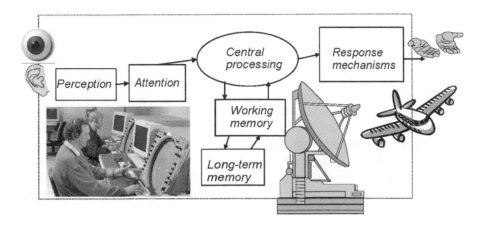

Figure 14.1   Perceptual, cognitive, and motor models

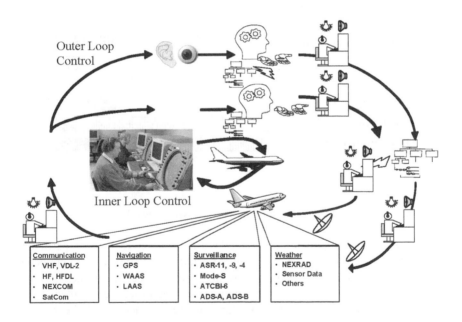

Figure 14.2   Multi-operator control systems

Figure 14.2 illustrates an expansion of the control model to include multiple operators and to include other control processes and information processes. The system represented in this way can be used to answer questions about interconnections among centers, the effect of local and remote knowledge in control, control timing and information requirements analysis.

Models of this second type, because of their scale, can also be used to get a broader sense of system performance with respect to issues of airspace capacity, interconnection, and safety. Questions about the impact of system-wide influences like weather, airspace reconfiguration, route structure and airspace structures changes can also be studied in such a wider level of system aggregation. The kinds of issues studied at the 'higher' resolution (such as perceptual motor demands) are typically not addressed at this broader scale of simulation.

Finally, issues of policy and large-scale structural changes in the system can be investigated by representing the elements of control and the procedures for that management at higher levels of abstraction. Figure 14.3 represents the ATM system modeled at a level of abstraction that allows policy and general management practice to be examined along with the assumptions or assertions about the technologies needed to support these systems.

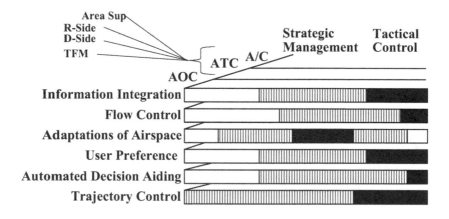

Figure 14.3   The ATM system modeled at an abstract level

In all cases, the outputs of the models are intended to answer particular questions about the ATM system. The level of detail and the type of data collected vary as a function of the questions that are being asked about the system's performance. The use of models of human-system performance is intended to provide the designers of equipment, procedures and requirements, with feedback as to what is feasible and what changes might be expected as a function of those implementations. The models are intended to be tools in a design loop that is represented in Figure 14.4.

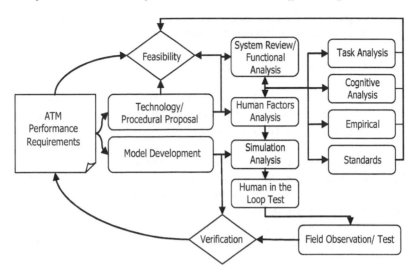

Figure 14.4   Modeling and simulation design loop

The loops in Figure 14.4 illustrate the intended use of models as a dual path of analysis and feedback as to the feasibility and validity of technological and procedural changes to meet changing performance requirements.

What kinds of data and information feed back from models to aid design depends on the purpose of the analysis and, to some extent, the knowledge base associated with the system under analysis. For instance, task sequences and timing of tasks is a common output from a human system model in ATM. How useful or reliable those task sequences are as metrics of the system's performance varies considerably. For a system that is well known and whose performance is well understood (for example the use of current switch technology for inter-facility voice communications), predictions of the effect of changing the system interface and infrastructure should be accurate and immediately useful in evaluation. However, if it is assumed, in another analysis, that traffic will be automatically separated by computer algorithms and that digital data link communication among aircraft and to ground stations will exist, then the task sequences are less useful as an evaluative tool and certainly less reliable in prediction. Output that predicts error in the phone system evaluation is likely to be fairly accurate. Output that predicts error in the process of automated airspace is considerably less likely to have an absolute level of accuracy.

The above therefore gives an outline of three different levels of modeling or simulation. As usual, the answers given depend on the questions asked, and the level of detail used to answer the question. This introductory preamble sets out how it works in theory, and the following sections detail what exists in practice. In the next sections, therefore, models are reviewed that are at varied levels of detail, with different assumptions of the representation for the ATM system and with different goals and purposes for their output. First, however, the need for models is briefly re-visited, now that what is meant by models has been clarified.

The Need for Models – ATM Paradigm Shifts

The motivation for these new paradigms of human performance analyses in air traffic lies in the goals espoused by the international aviation community for the next twenty-five years. These goals suggest a radical revamping of the practice of aircraft and air traffic management. This new vision finds expression in both the European and United States commissions and offices (Busquin, 2001; NASA Aerospace Technology Mission Vision, 2001; FAA NAS Modernization Plan, 2000). These visions assert advanced technological capabilities, and without exception these technologies change the fundamental process of the work of air traffic and air space management. The 'advancements' require a redistribution of information and control among the humans and the automation systems in airspace operations. These changes alter decision modes, execution modes and optimization processes among all participants in the aerospace transportation process.

Advancement in air traffic control processes motivates concern for human-system integration. Researchers in Europe and the United States (Bresolle et al., 2000; Billings 1996; and Wickens et al. 1998) are defining issues for tightly coupled automation and human systems. Such systems require attention to issues of system stability, human task load, information currency, control authority/responsibility and coordination. It is to address these issues in a dynamic and adaptive socio-technical system that model-based paradigms for consideration of the human operator are being developed. These methods seek to represent the human/system so as to account for the following phenomena:

- The information available has changed, and so have the roles and responsibilities in human and work-system operation.
- Local adaptation by air traffic controllers (and flight crews) occurs, beyond what is anticipated by the designers of these systems.
- These socio-technical systems are tightly linked – changes brought on by simple modifications of communications practice have been observed to have propagated goal effects across the entire human-system (Corker 2002). So large-scale simulations are needed to predict the systems response.

Human-system modeling can provide structures and methods to represent and to understand the interactions of the whole range of individual and system response associated with evolving air traffic management.

In this chapter, therefore, a review of the structure and process of such models is provided, and a review of select applications (as illustrative of general modeling type) and their effect is then provided. Finally, a summary of the current state-of-issues in this type of computational performance modeling in air traffic management is attempted.

Review of Models

The review of models is split into two major categories, structural/organizational, and computational, as defined below.

Structural/Organizational Models Modeling ATM is the process of representing the relationships among elements of the air traffic management process. One family of models relies on a description of the information-processing capabilities of the human operators that comprise the ATM system and the tasks that the operator(s) must perform. These models also focus on the interconnectedness of system elements and the exchange of information and meaning across the air traffic management system. Such models of the ATM system are developed to address its operability, to explain its potential failure modes and to guide its overarching architecture in design. These models of structures and processes represent not only the physical components but also the informational and social/organizational aspects of ATM (Issac and Pounds, 2001; Corker, 2002; Orasanu and Fisher, 1997). These modeling approaches include network and system state representations. Portrayals of specific human performance capabilities such as signal detection or multi-criteria decision-making, taxonomies of performance error and schema for procedure, influence and value are represented. The use of these models in analysis of system error and for diagnosis of system vulnerabilities will be reviewed.

Computational Models A second family of models focuses on capturing the time varying characteristic of the process of air traffic management. These models have as their focus the computational representation and prediction or prognostication of the performance of multi-agent systems. This paradigm is termed computational human performance modeling (Pew and Mavor, 2001; Corker, 2000; Laughery and Corker, 1998). In these approaches, the human(s) and any other system elements of interest are represented as computational entities or functions or aggregation of functions (e.g. teams). These computational agents interact as the system elements would in actual field operations. The benefit accrued from such models (assuming well-developed and validated models) is that the system and human characteristics can be quickly varied (e.g. based on an assumed technology change). Then the impact of that change (and any secondary effects) identified in the full-system context can be studied. Such models help focus expensive, complex simulations and field tests. In addition, performance at the edge of system safety can be safely explored in the computational human performance-modeling paradigm. Finally, these models can provide support to the structural insight of human-system interaction provided by the above-mentioned organizational/ structural models. The application of such models in ATM follows from successful use of human performance models in control systems (control theoretic modeling) in which the theory and techniques of control system design, analysis and evaluation are applied to human performance in control.

Both of these modeling approaches overlap in purpose. Each is intended to support inquiry to determine whether a system design is likely to achieve its purposes in efficiency, safety and economy; to assure appropriate consideration of Human Factors principles in functional allocation, performance loading, information communication and control; to support safety assessment, and to predict unanticipated effects of technology on procedure, training and selection.

Structural and Organizational Models

Task Decomposition Models

In this family of models, the focus of effort is in representing the task/procedure requirements for ATM according to either information-processing structures, hierarchic structures, process structures, or with respect to organizational structures (Kirwan and Ainsworth, 1992; Kallus, Van Damme and Dittmann, 1999; Endsley and Rodgers, 1997). The researchers use methods of task decomposition to view ATM as an interactive process of goal setting and information seeking on the part of the human operators, while sustaining sufficient resources to carry out ATM tasks in an accurate and timely fashion. The models of these interactive processes have taken several forms depending on the purpose of the analysis, e.g., allocation of function, determination of error process, maintenance of situation awareness. Most of these models are supported by a process of cognitive task analysis. A cognitive task analysis is a method of task/procedure decomposition that describes the physical tasks and cognitive plans required to accomplish a particular goal. The cognitive task analysis traces the information exchange, transformation and action patterns of an individual, a team and an organization as well as the technologies that support those individual in the performance of their work (Hollnagel, 1999). The process of cognitive task analysis is usually a walkthrough of the work process with explicit enumeration of the assumptions about an operator's informational state and the change in that state over time. The support for the work to be performed is examined and the flow of control through the system is analyzed. This type of analysis has been exceptionally useful in understanding the apparent emerging complexity in human-automation interaction in systems that are intended to 'support the human' operator (Vincente, 1999).

Function Allocation and Automation Integration Models

The issue addressed by these models is the appropriate allocation of tasks among human and automated systems in air traffic management. An increasingly prominent school of thought holds that simple substitution of human effort for machine effort does not work in practice; this is also known as 'The Substitution Myth', see Woods (2002) or Woods and Tinapple (1999). In order to address this issue an approach known as 'functional congruence' (Hollnagel, 1999) has been proposed. This view does not see human and machines as essentially interchangable information-processing units and, furthermore, considers the Joint Human-Machine Cognitive System (JCS) to be a non-decomposable unit of analysis. Cacciabue (1998) has developed such a functional allocation model with respect to flight crew and controller behavior. The model 'COSIMO' represents pilot checklist behavior, controller behavior and varied assumptions about automation.

The emphasis in analyses, therefore, becomes on how the JCS as a whole manages the process. Contextual dynamics become important in this view, as does the ability of the joint cognitive system (i.e., humans and machines viewed together as a cognitive unit) to redistribute or reorganize work in response to changing (and evolving) contextual requirements. An example of an effective representation process for human automation is provided in a state space representation described by Degani, Shafto and Kirlick (1996). Their approach maps human and automation states along with control and display methods to provide an effective diagnostic process evaluating human automation interaction. It also provides some insight into the system-human informational state and exchange process.

The increasing reliance of systems on modes of control (interleaving system behavior contingent on system/environment state) has proven problematic to the human operators of such systems. By examining crew interactions with flight deck automation and anticipating air traffic controller automation developments, Degani and his colleagues settled on a language that captures the nature of both the human and machine systems with respect to information and control state residence and state transition. The representation of the state of the system (which included the flight crew's possible actions in control) was a finite state machine and operator function model (Jones, Chu, and Mitchell, 1995). In implementation the team followed a method of state charts in which a hierarchy of goal-task relationships is specified to include the logic of goal satisfaction and the timing of task performance (sequential, concurrent, etc.). The team used these methods to embody: an environment which drives task requirements; human operator functions and tasks; controls (methods for undertaking the human-intended control); the physical plant that responded to control; and the feedback in the form of displays to the human operator executing the tasks. With this structure the team examined and explained several 'mode confusion' accidents in flight deck operation. And, perhaps more significantly, provided design and certification guidelines for mode-based automation so as to decrease the likelihood of mode confusion in multi-mode system operation (Degani, in press). The state-based representation method has also been used to predict human performance in the face of automated system support. Specifically, Ludtke and Modus (2004) provide a model of pilot behavior that simulates a learned procedural non-compliance as a function of lack of consequence for behavior (procedurally incorrect behavior). This model of 'learned carelessness' uses a state space method to predict what Mosier and Skitka (1995) would call 'automation over-reliance or bias'.

There are some shortfalls in a state space approach to modeling human automation integration. The dynamics of temporal load are not represented and information-processing constraints are implicit in the state transition process. There is currently no internal world representation for the human operator model. This means that the model cannot reference its past state to guide behavior, and it means that the analyst, when examining the model's behavior, cannot make a comparison between what the model 'knew' of the world and what a human operator in the same situation might know of the world. These structural characteristics aside, it is interesting to see that several air traffic models find similar structural support in the state-network representation: Traffic Organization and Perturbation Analyzer (TOPAZ) and a portion of the HERA-JANUS process both employ a backbone of world and human information states to either trace error evolution or to assign error causal factors.

Error Models

Several models focus on the explanation and prediction of human-system error. The premise of these models is that given appropriate taxonomies of human error (Reason, 1990) and an understanding of the work performance context (Rodgers, Mogford and Mogford, 1998), the organizational context in which the human-system performance occurs (Shappel and Weigmann, 1997), and an appropriate human information-processing model structure (Wickens, 1992), the analyst can explain and predict the sources of human system error in ATM incidents and accidents. This approach has been formalized in several systems that have successfully demonstrated their ability to provide useful and unique insight into system error. For example, TRACEr (Shorrock and Kirwan, 1988), POWER (Mills, Pfeiderer and Manning, 2002) and HERA-JANUS, which represents a harmonization of European and United States supported error modeling, (Isaac, Shorrock, Kennedy, Kirwan, Andersen and Bove, 2003; Issac and Pounds, 2002).

The questions asked of the human operator model in these paradigms are of the form provided by Reason (1990 p. 125): 'What kind of information-handling device could operate correctly for most of the time, but also produce the occasional wrong response characteristic of human behavior?' The models of human performance in systems then look to three basic interactions to answer that question. First, there are models of the basic information-processing elements and functions that the human operator brings to the task (Card, Moran and Newell, 1983). In these models the issue of limitations (perceptual, memorial, cognitive or motor) are provided to anticipate how humans might be overloaded by task requirements. Humans do not have infinite 'bandwidth' and, in fact, have some unique attributes that determine what types of tasks we can and cannot do simultaneously. There are established theories that address these issues such as Wicken's (1992) Multiple Resource Theory. Second, there are models of the requisite modification of those functions in response to environmental stressors (either internal or external to the operator). These models have played a large role in human reliability analyses such as 'performance-shaping functions' (Swain and Guttmann, 1983). Finally, context (Hollnagel, 1992) and organizational impacts (Shappel and Weigmann, 1997) are accounted for in these frameworks. Ultimately, these factors are shown to be interactive in the HERA framework for error with internal, external and psychological error modes modeled as influenced by context, stressors and other performance-shaping factors as well as the flow of the activities in an air traffic mission.

Decision-Making and Time-Sharing Models

The human operator's function in the distributed air/ground ATM system includes visual monitoring, perception, spatial reasoning, planning, decision-making, communication, procedure selection and execution. The level of detail to which each of these functions needs to be modeled depends upon the purpose of the prediction of the performance model. For example, in a detailed modeling of a specific aspect of controller performance, conflict detection and resolution, Masalonis and Parasuraman (2003) investigated the impact of fuzzy signal detection logic on decision support systems for controller identification of conflict. Their reasoning was that to better model the controller's mental process a fuzzy (degree of belonging) assessment of separation was more appropriate than a 'crisp' (separated or not) logic. In a simulation, 11 experienced controllers were asked to rate their confidence in the automated conflict detection system. The simulation consisted of 45 short traffic scenarios, in which controllers had to indicate whether or not actions needed to be taken to avoid a conflict. The fuzzy signal detection was better able to model how controllers in busy situations would attend to conflicts than did a standard signal detection theory. The fuzzy system also had a higher sensitivity than the traditional alerting logic and resulted in fewer false alarms relative to hits than did the standard. In this example a detailed model of the signal detection process of the controller was undertaken to address a specific design issue, the alerting logic in the User Request Evaluation Tool (URET).

Human operators sharing control and information with both automated aiding systems and other operators in the control of complex dynamic systems will, by the nature of those systems, need to perform several tasks within the same time frame or within closely spaced time frames. Multiple operators performing multiple tasks are a challenge to the state of the art in human performance representation.

Newman, Tattersall and Warren (1994) explore a framework for characterizing the cognitive performance of En-route air traffic controllers. They developed a set of activity types that included: radio transmission and communications, conflict detection and resolution, coordination among ATC team members, attention to the radar and flight data management. In their modeling study, the researchers tried to track cognitive strategies by observing the number and type of activities in the controller team. They found significant differences between the two controller roles.

Helbing and Eyferth (1994) also attempted to capture the structure of the information that the air traffic controllers used to develop their 'picture'. These early structural models were then expanded to include a computational representation (Niessen, Eyferth and Bierwagen, 1999) that will be reviewed later in this article. Powell (1998) also developed a model of controller performance in the context-based management of tasks in En-route performance. He too isolated differences in the controllers' information structures as a function of controller role.

It is essential to model multi-tasking and task management because there are limits to human bandwidth, information storage/retrieval capacity, and our ability to focus attentional resources. These limitations clearly affect human performance within the ATC system. For example, Isaac (1994) investigated the component elements of the short-term memory requirements for current ATC operations and projected the impact of technology on those memorial structures, suggesting that reduced affordance[22] will have a negative impact on encoding and retrieval. The resulting increase in memory load can lead directly to safety critical performance by the operators. Endsley, Allendorfer, Snyder and Stein (1997) performed similar analysis on the possible impact of free routing and separation aircraft operations on controller situation awareness. Hansman and Davidson (2000) also studied the relative differences in information access and information state as a function of varied technologies in air-ground integration. The conclusion of these models of informational state is that memory, attention and situation awareness factors are critically dependent on the technology and assumed role of the controller and flight crew. This is a theme that will be reinforced as we look at computational human performance models.

Contextual and Ecological Models of Human Reliability

The issues associated with error, erroneous behavior and its prediction are difficult to address in structural models. The reason for this is the requirement to 'know in advance' all possible paths for performance at each task junction, and to have effective mechanisms in place to choose the next task and predict the next erroneous performance. In response to human, environment, and organizational factors already mentioned, some newly developed structural models of human performance attempt to represent the human's adaptability and responsive restructuring of their task environment (to essentially make it different from that anticipated by the task analyst). In order to capture that type of flexibility 'contextual or ecological' models of human-system performance are developing.

There is an interesting set of models being developed that take a radically different view of the human-system integration process and consequently a different approach to modeling that interaction. This set is perhaps best represented by

---

[22] i.e. the linkage between cognitive processes and the external event triggering these.

Hollnagel's (1993) Contextual Control (COCOM) model and subsequent derivatives. The human and the task environment are considered not to be linear components in an information-processing system, but to be interactive and inter-responsive. The human operator modifies the process of task performance as an active adaptation to his/her perceived capabilities and the task environment demands. These adaptations include an understanding of context and a requirement to include context and its potential impact explicitly in the system performance model. The problems with defining and delimiting the boundaries of context in aeronautical applications are formidable, but an explicit inclusion of context responsive behavior in human performance models was recently demonstrated by Corker and his colleagues (Verma and Corker, 2001; Corker, Gore, Fleming and Lane, 2000) and is reviewed below.

Computational Models

There is a growing interest and expertise in the development of computational simulation of air traffic management systems, and including in these simulations explicit models of human performance including cognitive performance as the air space operations are dynamically simulated. For instance, the reorganized air traffic control mathematical simulator (RAMS) developed by EUROCONTROL was invested with a rule-set to emulate controller performance in conflict resolution (Mondolani, 1998) and had a process of accumulating 'task load' to identify possible controller workload issues given airspace and traffic loads. The following sections review the developments of such computational controller and flight crew models.

Control Theoretic Models applied to ATM

There is a long history of the use of human performance models based on a combination of engineering and psychological principles in dealing with complex aeronautical systems. Craik (1947) performed seminal work in human control of large inertial systems (in this case anti-aircraft gunnery systems) and characterized that control through models. In performing such experiments, data in tracking control studies led Craik to conclude the human operator behaves basically as an intermittent correction servo. This formulation was further refined by McRuer and Krendall (1957) and summarized by McRuer and Jex (1967). The resultant description of the human operator is that of a good servo with bandwidth constraints and a cross-over frequency response characteristic. Traditional engineering models of human performance in aeronautical systems have considered the human operator as a transfer function and remnant in a continuous control. They have concentrated on the interaction of one pilot and an aircraft with concern for system stability, accuracy of tracking performance, information-processing of displays, and ability to handle disturbances. They are intended to provide design guidance as to in design that determines whether the information provided, and the control system through which the operator performs their functions, allows successful performance with an acceptable level of effort (Baron and Corker, 1989). These models assume a closed loop control in which the human operator observes the current state of the system, constructs a set of expectations based on his/her knowledge of the system. This internal model is modified by the most recent observation, and based on expectations the operator assigns a set of control gains or weighting functions that maximize the accuracy of a command decision. In this loop the operator is also characterized as introducing observation and motor noise (effector inaccuracies) and time delay smoothed by an operator bandwidth constraint. Such a model is represented in Figure 14.5: the human operator is assumed to observe a display of system state and to compare that display to an internal model of the system, represented as a Kalman estimator and predictor. The operator then chooses an action that will offset any observed error between current and desired system state and acts through his neuro-motor processes, which include a noise and bandwidth limit, to effect the control.

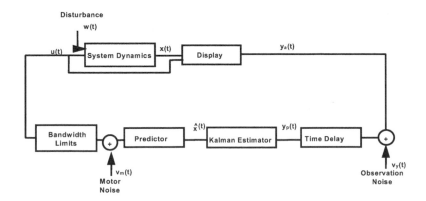

Figure 14.5   Optimal Control Model

The legacy that these control theoretic models provided will be shown to be important in the development of more fully articulated, multi-agent and cognitively focused models of air traffic management that will be reviewed below. From this early work, three principles for model development can be derived:

1. description of human and systems and their collaboration in the same mathematical, structural and dynamic terms;
2. analytic capability to define what information should be displayed to the human operator in the human-machine system as a consequence of his/her sensory/perceptual and cognitive characteristics in control; and

3. a fundamental paradigm shift in which man-machine systems could be conceived as a joint entity coupled to perform a specific task or set of tasks.

A new level of abstraction was introduced and systematized by Craik and subsequent developers of operator control models to guide the machine design. The control theoretic description served well for predictive modeling of flight crew, and flight crew procedures, However, despite some early attempts to extend the information-sampling characteristics of the OCM to air traffic control (Govindaraj, Ward, Poturalski and Vikmanis, 1985), there were few representations of the reactive and problem-solving aspects of air traffic management by optimal control theoretic models. What was needed was another level of complexity addressing the human cognitive process *with* monitoring and supervisory behavior guiding intervention strategies.

Supervisory Models Applied to ATM

As the human operators came to be served by automation that operated at remote sites in semi-autonomous modes, a new framework was developed, led by Sheridan's work in Supervisory Control (Sheridan and Ferrell, 1969). In these models, the operator stands back from the direct manual control of the systems and performs managerial functions, setting goals, training, observing performance, and intervening. The requirement for supervisory control by remote human operators and for local autonomy of a function was originally associated with distance/time relationships, bandwidth limits, or efficiencies gained by removing the human from the direct critical path of control. In current views of air traffic management the motivations to assign the human a supervisory role are concern for the cognitive complexity of full manual control and remoteness brought on by layers of automation (Parasuraman, Sheridan and Wickens, 2000). This more recent model of human supervisory control details levels of automation from low (fully manual) to high (fully automated) control across the range of information acquisition, information analysis, anddecision-making and action implementation.

In terms of system safety analyses, the research developments in modeling the human as a supervisor concentrated on two areas. First, what type of error would an operator make when operating in a supervisory function? Second, what system sampling functions might be considered optimal and what was the potential for error if the system sampling functions were not followed. The research considered how could that model be supported when correct or corrected when inaccurate. This line of thinking led to a large body of work intended to understand the human operator as an information-processing component in a system. Significant developments in what was coming to be called cognitive psychology were also being made in examining human information-processing capabilities and limitations. The convergence of increasingly automated systems and empirical models of humans as information processors led to considering the human as an information-processing element in a complex dynamic system. However, the consideration of the human in supervisory control has also led to models of the human operators in air traffic management systems put in 'double-binds' between monitoring system performance under some level of automated control and intervening at cost incurred by the operator in cognitive effort and by the system inefficiencies (Dekker and Woods, 1999; Mooij and Corker, 2002).

Task-Based Models

Callentine (2002) has developed a multi-agent architecture to represent the performance of air traffic controllers based on the Crew Activity Tracking System (CATS, see also Callentine, Mitchell, and Palmer, 1999). His agent-based representation applies a hierarchically decomposable set of task structures, beliefs about the state of the world, rules, roles and skills, and activities to capture controller performance. His system is built on a combination of operator function model development and task decomposition. Figure 14.6 provides a schematic representation of his system.

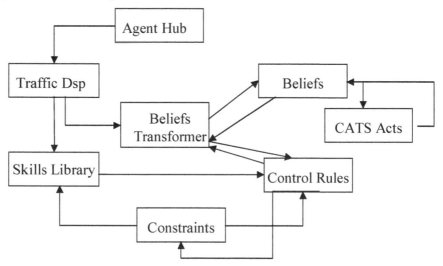

Figure 14.6   The CATS-based ATC model (Callentine, 2002)

The CATS-based ATC model assumes several interactive structures and produces behavior as the result of traffic patterns in the target airspace. The model revolves around a process of belief transformation that coordinates performance on the bases of skills, constraints, action libraries and control rules. The belief transformation and belief structure maintain a set of beliefs as follows: current beliefs (referencing current action and control rules), retrospective beliefs (referencing actions and rules applied previously), and prospective beliefs (beliefs about intended action into the future, plans). It is important to note that in the case of both future plans and with respect to plans already undertaken the prospective and retrospective belief systems provide a necessary hysteresis to maintain stability in a dynamic control environment. That is to say, the controller model needs to have a memory of action taken to account for the fact that action may have been taken – but the inertial characteristics of the system may not yet show the effects of that action. And the controller model needs to know what action is planned in order to select appropriate control rules. This requirement of some internal state reference, or self-awareness, is common to dynamic models of ATC and may provide some insight as to what information is needed to be provided to controllers in order to effect air traffic control (and to judge when that information is modified or removed by new procedures and or new technology).

These beliefs trigger a set of skills in the skills library that are organized into three categories, similar to the decompositions of controller tasks provided by framework models provided by Newman, Tattersall and Warren (1994). These skills are focused on classes of controller actions as follows:

- maintenance of situation awareness;
- determination as to which aircraft to attend to and service next;
- management of the airspace in terms of handoffs, descents, separation and non-compliance.

The selection of which action to take next is performed on a priority-based scheme and all action is initiated as soon as the conditions and constraints that determine its efficacy are met. The aircraft in the system have roles associated with them to guide both their behavior and the control rules provided by the controller. Plans are developed to handle specific control situations based on the management types above. Separation by speed and by vectoring is provided, as is an ability to meet time-over-fix constraints.

Callentine has reported a performance assessment of his model in the following manner. Two high altitude sectors and one low altitude approach feeder sector were modeled in the Dallas Fort Worth ARTCC. One low altitude sector supporting approach to Dallas airport was also modeled. The simulation was run in two modes one with only the descent rule-set activated and one with the full set of rules implemented. The results indicate improved performance in terms of fewer separation violations as a function of the full-rule set being employed. Callentine proposes that flexibility and dynamic updating of the agent's role bindings, plans and perceptual thresholds would provide a better representation of the ATC response to dynamic operations in their environment.

This performance analysis points to another of the issues associated with human performance modeling in ATC – that is validation. In this case the model in its full form is compared to its operation in reduced form and the increment of performance is taken to indicate that the fuller model is a better representation of human performance. In discussion of other models of controller performance this issue will reemerge.

Another example of a controller performance model that is basically structured by task is a model of En-route controller behavior developed by Leiden (2003).

This model is being used to determine the impact of controller-pilot data link communications (CPDLC) on controller communication loads. The simulation modeled En-route ATC using MicroSaint task network of the air traffic controllers.

The study represented the radar controller and the data controller tasks for En-route and approach traffic. The radar controller tasks included situation monitoring which subsumed tasks of detecting conflicts, detecting metering and spacing violations, initiating transfer and acceptance of handoffs, monitoring radio, and other tasks including conflict resolution, spacing conformance, metering conformance, radio communication with pilot. The Data controller managed flight strips, assisted with pilot readback and coordinated with other sectors. The study was focused on the potential gain in airspace capacity as a function of reduced handoff communication load assuming a CPDLC function to automate sector entry and exit communication requirements.

The simulation is based on a MicroSaint architecture that represents human information-processing functions called by the task network that is used to describe the required performance. The task network architecture organizes the simulation of a system based on the goals and tasks required to carry out target missions. The human performance of each task is further decomposed to a level chosen by the analyst (and to a level for which performance data are available). Figure 14.7 represents a standard monitoring task (Laughery, Archer and Corker, 1999). Each level of a task can be further decomposed to expand what is needed to 'compare' or to 'open'.

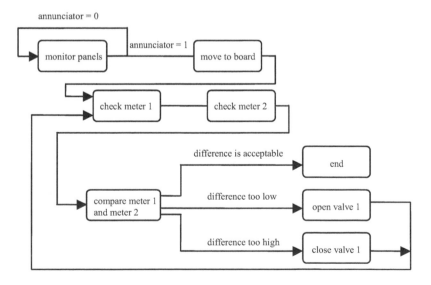

Figure 14.7   A task network for a monitoring and control task

The study concludes that the introduction of CPDLC will reduce the controller workload in En-route operation significantly (perhaps by as much as fifty-eight per cent) and that spare human capacity can be used to increase the capacity of En-route airspace (Leiden, 2003). This type of model has been extensively reviewed in Laughery, Archer and Corker (2001) and will not be detailed here.

However, several points are brought to issue in this study. First, the model is very sensitive to the assumed procedure that CPDLC will bring to the controller (as are all of these computational models). Small differences in the assumptions of required monitoring and modeling will lead to large changes in a result of the level of workload. The model, as implemented does not, in fact, calculate workload. Rather, it calculates communication time and the investigators make a post hoc coordination of communication time to workload to extended capability in controller performance. Such extrapolation requires that the scheduling mechanisms of the model be considered in order to understand the model's assumptions in task performance. In short, as we move into more sophisticated and interactive models, the results and conclusions that can be drawn from the model's performance need to be clearly based on the model's computational characteristics and assumptions about human performance.

Risk Assessment Models

We will review three types of human performance causal modeling for risk assessment in air traffic management. These are an event-tree based analysis (Spouge, 2002), a 'bow-tie' Bayesian network human-system reliability analysis (Roelen et al., 2002) and a dynamically colored Petri-Net simulation (Blom, 2002). These studies are good representatives of the range of approach that is taken in risk assessment analyses with explicit inclusion of human performance.

Event-tree Based Analysis Probability event-trees have been used to access probabilities of propagated failure in systems in which the likelihood of failure at some node is determinable based on mean-time-between failure data on component elements. In this study the component elements are individual behavioral elements associated with human performance in the system. Historically, other techniques such as Human Reliability Assessments (HRA) have been used to attempt to identify human performance transition probabilities. A problem with HRA and the causal side approach is that the units of behavior ('flight crew fails to respond to ATC warning') are arbitrarily assigned and no empirical data exists to support its possible failure rate – other than expert opinion. So the basic unit of causal propagation is not well founded. A positive feature of the approach is the development of a consequences model as a result of the risk-side model coming to an accident or incident. In representing the consequences of risk, the same event-tree probability risk assessment (PRA) is used to estimate the impact, both proximal and distal, of the accident event. The analysis then narrows the risk to the event set that produces an accident, then expands the consequence set as a function of an assumed accident to understand the impact in system and social context (the narrowing then expanding event set provides the name 'bow-tie' to this analysis paradigm).

'Bow-tie' Bayesian Network Analysis The Bayesian network approach takes into account explicit and directional conditional probabilities of one event influencing another. The Bayesian network suffers from the somewhat arbitrary definition of behavioral chunk and truncation of the propagation path. However, other Bayesian network approaches (e.g. Hudlicka, Zacharias and Psotka 2000) have attempted to overcome this shortcoming (truncation of conditional influence) by including the network as part of a dynamic simulation in which the conditional probabilities are calculated as a function of simulation process not fully pre-calculated. This solution is in fact the one implemented in the proposed approach.

Both approaches are acyclic which implies that the influence and probability accumulation process is unidirectional so that changes in behavior as a function of the simulated event propagation are not supported. Both approaches are essentially 'hardwired' once the original network structure is established (the branching is probabilistic, but the branches are fixed). Blom et al. (2000) have produced a simulation-based Bayesian-structured dynamically colored Petri-Net that directly addresses the issue of acyclic influence through Petri-Net rather than unidirectional network structures.

Dynamically Colored Petri-Net Simulation The Traffic Organization and Perturbation AnalyZer (TOPAZ, Klompstra, Bakker and Blom, 2000) is a simulation of accident evolution. The method begins with the identification of operations and possible hazards associated with those operations. After a set of operations has been defined a dynamically colored Petri-Net (DCPN) representation of the scenario to be simulated is developed. In this development both the sources of the cause of the hazards identified in step one and the consequences of those hazards (e.g. collision risk) are developed. Values are assigned to the parameters that define transition probabilities among the system states defined by the net and values are associated with the risk events that are triggers for hazard. Once the simulation is developed, tested and refined, the simulation is run to determine what the accident risk is going to be. This is undertaken by first making estimates of event sequence probabilities, based on the simulation runs. Conditional probability density functions are defined to assess the continuous safety state of the ATM system – based on hybrid state Markov process theory. Accident risk is then determined under conditional risk assessment. Finally, a consequence modeling effort is undertaken to determine the impact of an accident on human safety and economic and social consequences.

The TOPAZ process explicitly includes a model of human performance based on Hollnagel's (1993) Contextual Control (COCOM) model. That model moves the DCPN to different transition states as a function of the operating mode of the human operators in the simulation. The human operator is also modeled using the

DCPN process and the effect of training, fatigues and conditional impacts of current ATM state determine the human operators' performance (Daams, Nijhuis and Blom, 1999). The TOPAZ model has been successfully used to model operations in parallel and convergent approaches (Blom, Klompstra and Bakker, 2001), surface operations (Blom, Corker, Stroeve and van der Park, 2003), and free flight operations (Blom, Bakker, Blanker, Daams, Everidij and Klompstra, 1998). More recently the TOPAZ process and human performance modeling (Air MIDAS) have attempted a more extensive link between the risk assessment DCPN process and a more elaborated human performance model (Air MIDAS), which serves to provide transition probabilities and reaction times for several of the flight crew activities modeled in the DCPN process. This integration is reviewed in Blom, Corker, Stroeve and van der Park (2003). The hazard assessment function of human operator representation uses some model of human performance to determine the likelihood of specific action sequences in ATM. There are several modeling efforts that try to produce such action sequences for the operation of component and functional elements of human performance in general, and then tune those performance specifications to air traffic operations.

General Models of Human Performance in ATM

One very interesting and complete human-system representation is provided by Lindsey and Connelly (2001) who developed a model following from EUROCONTROL's work on controller behavior analysis and modeling. Kallus et al. (1999) and Planaque et al. (1999) applied Petri-Nets to modeling the effect of data link technology on controller performance, and Johnson (1997) applied these analyses to accident modeling. Lindsay and Connelly (2001) provide an En-route air traffic controller model developed using Hoares' Communicating Sequential Processes techniques. The model includes scanning, problem identification, memory, projection of action, experience-based monitoring, prioritization and decision to act (rules-based and knowledge-based) and validation of decision through further monitoring of action with feedback modification of performance. The model has been simulated with air traffic models to produce several types of human error. The methods of the model provide emergent behavior that represents failures in scanning, persistent misclassification, persistent mis-prioritization, planning error with action deferred beyond critical performance windows and indecision resulting in an inability to take action. The authors have not described any verification processes, but the research provides a unified representation and a sophisticated model of cognitive performance in air traffic management.

Niessen, Eyferth and Bierwagen (1999) provide a model using ACT-R (Anderson and Lebierre (1998) which takes advantage of their earlier work in the development of the air traffic 'picture' and implements the picture development process in computational simulation. The simulation efficacy was not reported relative to actual control of traffic, but was cited as useful for situation awareness training of controllers.

Another model that provides theory-based emergent task performance with the possibility of error is APEX (Freed and Remington, 1997). In this model Cognitive Perceptual Motor – Goals, Operators, Methods and Selection Rules (CPM-GOMS) based architecture provides selection rules guiding resource-constrained execution. Controller adaptive behavior is simulated by delay, interrupt, and choices of alternative methods using both transient and long-term knowledge. The controller decision strategy in action selection is modeled based on estimates of resources required for performance, expected other-task workload demands, expected urgency or performance, and likelihood that current default action selection values will be accurate. The team has provided a demonstration of 'habit capture' in ATC performance in TRACON operations. Assuming a rare event, e.g. runway closure, the model eventually migrates to 'most likely' runaway assignment if there are no perceptual or memorial cues to reinforce the rare event state. This model benefits from the extended body of research that supports the GOMS body of models (Johns and Kieras, 1998; Johns and Vera, 1992). It is being expanded to include a wider set of ATM tasks and the methods associated with GOMS task analysis, which are fairly work-intensive, are being supported by intelligent automated support.

Air MIDAS: An Integrated Human-System Model

Finally there is a set of air traffic management studies associated with the Air MIDAS (Man-machine Integrated Design and Analysis System). This model of individual and multiple operator behaviors has had an extensive application to flight deck operations. This set of research and supporting validation efforts are reviewed in Corker (2002) and will not be presented here. We will, however provide a review of the model's recent application in air traffic management, discuss issues in validation and introduce approaches to hybridization. The basic functional elements of the MIDAS Architecture for Human Representation in Complex Systems are presented in Figure 14.8. Each of the modules represented in this figure is a functional model of human performance. They are linked together into a closed-loop simulation of operator performance. This basic structure is replicated to account for multiple crew member operations.

The components of the model represent a working memory which is the store that is susceptible to interference and loss in the ongoing task context. We have modeled human memory structures as divided into long-term (knowledge) and working memory (short-term store). We have implemented working memory, described by Baddeley and Hitch (1974), as composed of a central control processor (of some limited capacity), an 'articulatory loop' (temporary storage of speech-based information) and a 'visio-spatial scratch pad' (temporary storage of spatial information). Long-term memory structure is provided via a semantic net. The interaction of procedure with memory is provided by a goal decomposition method implemented as a form of cognitive schema. The internal updateable world representation (UWR) provides a structure whereby simulated operators access their own tailored or personalized information about the operational world. Attention capture functions are represented through a 'pre-attentive' filter mechanism that responds to physical characteristics of environmental stimuli (e.g. color, blinking, auditory characteristics, etc.). The model generates action through the decomposition of goals, establishment of priority and management of performance and interruption queues. This management process is contingent on assessment of the operator's state and its interaction with the world state, or context.

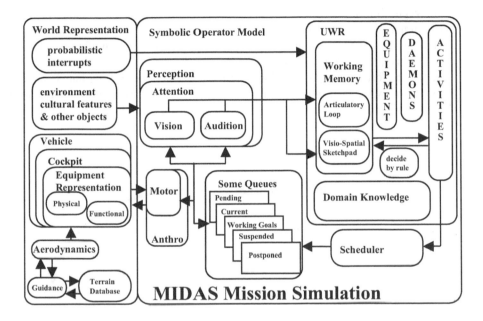

Figure 14.8   MIDAS – The Man-Machine Design and Analysis System

Modeling the effects of context in Air MIDAS involved a computational implementation of Hollnagel's (1993) contextual control model (Verma and Corker, 2001). The earlier versions of MIDAS did not have a context-control switching mechanism. The human agent's scheduler behavior simply performed tasks based on their priority, if resources were available. If resources were not available then activities were deferred or dropped depending on time passage. The availability of resources is assessed by using the task loads assigned to each activity. The resources are not limited to availability of time; the model instead employs a set of demand values for different tasks ranging from zero to seven. These values were generated by expert pilots and compiled by Aldrich, Craddock and McCracken (1984).

The model has also been expanded to represent the interaction of many operators (up to forty-six flight crews and six to ten air traffic controllers in the studies of ATM to be described below). The assessment of human performance in the system and of the system's safety risk is ultimately based on behavior (system and human) in time and environmental context. The behaviors represented have either safety neutral, safety enhancing or safety reduction consequences and it is a measurement of those consequences that our predictive models should provide. The dimensions of the output of the models can be scaled from individual and team performance to overall system response to a range of demands, perturbation and environmental influences. Behavior of multiple operators requires the development of computational structures to deal with the kinds of performance complexity that emerges from the interaction of individuals in joint co-operation of a system. Figure 14.9 illustrates the types of structures that should be in place to examine multiple human performance representations and their interactions. Setting human agents in interaction with each other requires the computational capability to have conflicts and variations of worldview and performance expectation emerge and track the consequences of these.

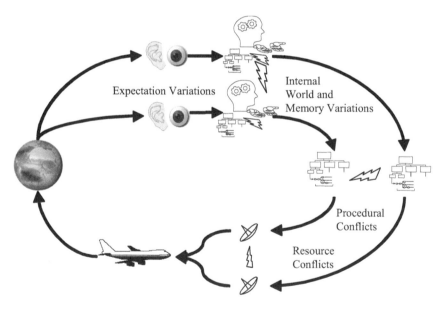

Figure 14.9   Structures for the examination of multiple human performance representations and their interactions

Air MIDAS has been used, recently, to try to predict the impact of procedural changes and advanced technologies on air traffic control performance. Corker, Gore, Fleming and Lane (2001) addressed the issue of control workload as a function of the mix of aircraft self-separating in the controllers' airspace. The study combined part-task simulation and follow-up human performance modeling to understand the mechanisms of impact of a procedural change without any corresponding change to the controllers' technical support. The model predicted and the part-task study found an increase in certain aspects of controller workload in such mixed equipage settings.

In keeping with this focus of considering the impact on controllers from flight deck technologies, two recent studies have been performed to investigate the impact of a clear air turbulence sensor on board the flight deck with respect to the air traffic controller. The study simulated sector airspace and several anticipated 'look ahead' times for the sensor system (Abkin et al., 2002). The simulation of the controller found specific parameters of look ahead that adversely affected controller performance based on the operational concepts anticipated procedures (Corker, 2002). The second of these studies examined the impact on the controller of a highly automated airspace concept in which the controller simply monitored the operation of an automated air traffic system and then managed control of aircraft, as they may need to exit from automatic control for various reasons. This concept is described in Erzberger (2001). In this study, the controller was predicted to have a very high punctuate workload at the point of transition of an aircraft from automatic to standard manual control. On the basis of these findings, alternative procedures are being explored for the fail-safe process in highly automated air traffic management.

Finally, most recently in collaboration with the ATAC Corp the Air MIDAS system has had a unique opportunity to explore the impact on controllers of different spacing paradigms on approach, and to have that simulation validated with actual air traffic performance data (Corker, Verma and Jadhav, 2003). The simulation links a simulation engine (Reconfigurable Flight Simulator) developed by Georgia Tech (Pritchett, 2002) to the Air MIDAS human performance model via run time interface through the DoD-developed high-level architecture (HLA).

The study was performed to explore the differences in workload resulting from time-based-metering (TBM) (in which the controller manages traffic to specific times to a fix which consists of a three-dimensional point in space and a time) and miles-in-trail (MIT) functions (in which the controllers manage traffic along a route with specific aircraft intervals to be maintained) for traffic on approach to Los Angeles airport through the Southern California TRACON and the Los Angeles ARTCC. Procedures for both types of approach control were developed for two center sectors and one TRACON sector. Air traffic was recorded for four days under each traffic control condition and this served as the validation traffic base. The simulation of controller performance was run using one hour of selected traffic with fifty-four aircraft in both approach and over-flight regimes.

An analysis was performed on the full data sets to support a comparison between the actual traffic, a 'nominal' set of traffic (sample over four days) and the traffic produced by the Air MIDAS simulation. The comparison was intended to identify if there were any statistical differences between the actual controllers performance and that of the model on specific measured variables (time the aircraft was in the sector, and miles flown in sector).[23] An examination shows a close correspondence between model and actual performance. This study provided one of the unique opportunities to validate air traffic management performance models against operational data.

The analysis of controller workload did illustrate some difference in pattern of activity. More maneuvers were undertaken in the center sectors under TBM control than in the TRACON. The overall workload, however, was evenly distributed between the two control performances. The source and timing of the load appeared to be redistributed as a function of different actions performed, but the overall workload for each metering performance remained the same. Total throughput in the hour observed and modeled favored the TBM approach. As this study was just completed (October 2003) further analysis and research is to be undertaken and will be published.

Conclusions

The process of modeling human performance in air traffic systems is maturing to the point of providing a valuable and viable tool for both large-scale and small-scale analysis of ATM. The process of performance modeling requires careful and detailed analysis of the air traffic management processes for each individual performance component in that system, and an equally detailed analysis of the interactions among those entities. In itself this level of careful insight provides information important to the understanding of air traffic management as it is actually performed (as opposed to how it is presumed to occur or how it is proscribed to occur).

In trying to provide some assessment of the level of effort needed to provide a successful analytic and predictive model of the processes of air traffic management, we faced several challenges. The range of coverage of the air traffic phenomena being studied is not uniform. Few, if any, model systems have had the luxury of uninterrupted funding and consistent direction from funding sources in the problem areas studied. Most model analyses have started from some level of either developed models and/or standing data and expertise; so the 'model development' process rarely starts from square one. Finally, there are no standards and little community consensus as to what actually constitutes a 'successful' analytic model effort. Despite these issues, however, some estimate is needed to provide a sense of scale and level of effort for these methodologies. We have conducted a set of

---

[23] The F test provided an $F = 0.113_{(2,42)}$ for the difference among the sources of the data tracks in time in sector and distance flown as dependent variables. Further analysis showed an $F = 1.648_{(1,42)}$ for the differences between MIT and TBM on those same variable. An R^2 of 0.996 was found in the analysis of the differences of variances among the actual and simulated variance. In short, there were no significant differences between the model generated flight tracks or actual flight tracks in terms of distance and time.

interviews with current and past practitioners in this field to try to provide that estimate. The result is illustrated in Figure 14.10[24] which presents the timeline of model development and analysis and attempts to give a sense of the amount of effort (and the steps for many model development exercises) for an analysis cycle. Two features of this timeline should be noted. First, most of the steps in the analysis overlap each other in time and this suggests that model analyses are both highly iterative and performed by teams of individuals with a mix of skills, and knowledge. Second, early steps are often revisited and modified as a function of constraints in later model analysis stages. The feedback and forward design information process is critical to the development of successful analyses. Not all model types require all steps, but the process depicted covers a majority of the major modeling efforts described in this chapter. As noted the time line of person/months of effort is a composite and not intended to predict or prescribe any one analysis.

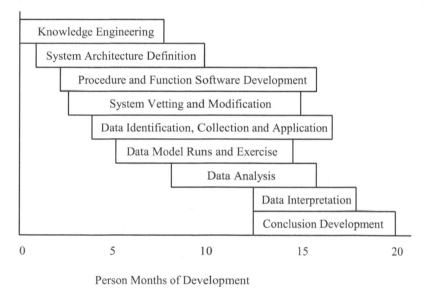

Person Months of Development

Figure 14.10 Estimated timeline for development and analysis of different models

There are open issues in air traffic management performance. First, a description is needed for what constitutes an appropriate level of detail in human performance, and what behaviors or behavioral categories are of interest to the

---

[24] The reader is advised that this represents the author's attempt at consolidation and consensus of opinion and is not intended to be canonical.

analyst. The range is from micro models of attentional or cognitive performance to models of aggregate behavior in the operation of a section of the national airspace. This provides a challenge for the appropriate selection of models to match analytic needs.

Second, there are a large set of issues associated with validation and verification of these models. The purposes of these models are to a large extent the determinant of their verification and validation process. Whether to provide a psychologically and technologically plausible explanation of why the predicted behavior was observed or to provide a validated platform from which to predict the impact of changes in the human technical system; each demands a much different strategy in response to verification requirements.

Finally, there is a clear interaction between the structure of the airspace and the conformance of both the human controller and the model to those constraints. The system is limited in the degrees of freedom afforded the operator in responding to the system constraints. Both the computational model and the human controller respond to those constraints in a similar way. That does not detract from the validation of the model – but suggests that a combination of factors including the constraint-based response needs to be considered in predicting performance. In this view then the appropriate unit of analysis is the airspace, the operator and the goals of the system.

References

Aldrich, T.B., Craddock, W., and McCracken, J.H. (1984), 'A Computer Analysis to Predict Crew Workload during LHX-scout-attack Missions, Vol. 1', *MDA903-81-C-0501/ASI479-054-1-84(B)*, U.S. Army Research Institute Field Unit, Fort Rucker, AL.

Aldrich, T.B., Szabo, S.M. and Bierbaum, C.R. (1989), 'The Development and Application of Models to Predict Operator Workload During System Design', in G. Macmillan, D. Beevis, E. Salas, M. Strub, R. Sutton and L. Van Breda (eds.) *Applications of Human Performance Models to System Design,* pp. 65-80, Plenum Press, New York.

Anderson, J. and Lebierre, C. (1998), *Atomic Components of Thought,* Erlbaum, Hillsdale, NJ.

Baron, S. and Corker, K. (1989), 'Engineering-Based Approaches to Human Performance Modeling', in G. McMillan, D. Beevis, E. Salas, M. Strub, R. Sutton, and L. Van Breda (eds.), *Applications of Human Performance Models to Systems Design,* Plenum Publishing, New York.

Baxter, G.D. and Ritter, F. (1999), 'Towards a Classification of State Misrepresentation', in D. Harris (ed.), *Proceedings of the 2nd International Conference on Engineering Psychology and Cognitive Ergonomics,* Oxford, England, pp. 35-42.

Bier, V.M., Caldwell, B.S., Gibson, N.M. and Kapp, E.A. (1996), *The effects of Workload, Information Flow, and Shift Length on Coordinated Task Performance and Human Error,* Final Report 133-Z572, Center for Human Performance in Complex Systems, University of Wisconsin.

Blom, H, Stoeve, S. Everidj, M., van der Park, M. (2003), 'Human Cognitive Performance Model to Evaluate Safe Spacing in Air Traffic', *Human Factors and Aerospace Safety,* Vol. 3 (1), pp. 59-82.

Blom, H.A.P., Bakker, G.J., Blanker, P.J.G., Daams, J., Everdidj, M.H.C. and Klompstra, M.B. (1998), 'Accident Risk Assessment for Advanced ATM', *Proceedings of the 2nd USA/Europe ATM R&D Seminar,* FAA/EUROCONTROL.

Blom, H.A.P., Corker, K., Stroeve, S.H. and van der Park, M.N.J. (2003), *Study on the Integration of Air MIDAS and TOPAZ,* NLR, Amsterdam, The Netherlands.

Bresolle, M.C., Benhacene, R., Boudes, N. and Parise, R. (2000), 'Advanced Decision Aids for Air Traffic Controllers: Understanding Different Working methods from a Cognitive Point of View', *Proceedings of the 3rd USA/Europe ATM R&D Seminar,* FAA/EUROCONTROL.

Broach, D. (1999), 'An Examination of the Relationship between Air Traffic Controller Age and En Route Operational Errors', *Proceedings of the 10th International Symposium on Aviation Psychology,* Columbus, OH.

Busquin, P. (2001), *European Aeronautics: A vision for 2020,* EUROCONTROL, Brussels, Belgium.

Cacciabue, P.C. (1998), *Modeling and Simulation of Human Behavior in System Control,* Springer, Amsterdam.

Callentine, T. J. (2002), *CATS-based ATC Agents,* NASA CR 2002-211856, NASA Ames Research Center, Moffett Field, CA.

Callentine, T.J., Mitchell, C. and Palmer, E. (1999), *GT-CATS: Tracking Operator Activities in Complex Systems,* NASA Technical Memorandum 208788, NASA Ames Research Center, Moffett Field, CA.

Corker, K (2003), 'Requirement for a Cognitive Framework for Operation in Advanced Aerospace Technologies', in E. Hollnagel (ed.), *The Handbook of Cognitive Task Design,* pp. 417-436, Erlbaum, Hillsdale, NJ.

Corker, K. (2000), 'Cognitive Models and Control: Human and System Dynamics in Advanced Airspace Operations', in N. Sarter and R. Amalberti (eds.) *Cognitive Engineering in the Aviation Domain,* pp.12-14 Erlbaum, Hillsdale, NJ.

Corker, K. (2001), 'Air-ground Integration Dynamics in Exchange of Information for Control', in L. Bianco, P. Dell'Ormo and A. Odani (eds.), *Transportation Analysis,* Springer Verlag, Heidleberg.

Corker, K. (2002), 'Hazard, Risk and Performance Prediction in Large-Scale Airspace Simulations: Human Performance Compared to Human Model Prediction', *Proceedings of the International Symposium on Air Traffic Management.*

Corker, K., Gore, B., Fleming, K. and Lane, J. (2000), 'Free Flight and the Context of Control: Experiments and Modeling to Determine the Impact of Distributed Air-Ground Air Traffic Management on Safety and Procedures', *Proceedings of the 3rd USA/Europe ATM R&D Seminar,* FAA/EUROCONTROL.

Craik, K.J.W. (1947) 'Theory of the Human Operator in Control Systems: I. The Operator as an Engineering System', *British Journal of Psychology,* Vol. 38, pp. 56-61.

Daams, J. Nijhuis, H.B. and Blom, H.A.P. (1999), *Accident risk assessment with a human cognition model using TOPAZ: Modeling detail*, NLR-Memorandum LL-99-030, NLR, Amsterdam, The Netherlands.

Degani, A., Shafto, M. and Kirlick, A. (1996), 'Modes in Automated Cockpits: Problem Data Analysis and a Modeling Framework', *Proceedings of the 36th Israel Annual Conference on Aerospace Sciences*, Haifa, Israel.

Dekker, S.W.A., Woods D.D. (1999), 'To Intervene or not to Intervene: The Dilemma of Management by Exception', *Cognition, Technology and Work*, Vol. 1, pp. 86-96.

Endsley, M. and Rodgers, M. (1997), *Distribution of attention, situation awareness and workload in a passive air traffic control task: implications for operational errors and automation*, DOT/FAA/AM-9713, Federal Aviation Administration, Washington, DC.

Endsley, M.R. and Rodgers, M. (1994), *Situation awareness information requirements for En route air traffic control*, DOT/FAA/AM-94/27, Federal Aviation Administration, Washington, DC.

Endsley, M.R., Mogford, R.H., Allendoefer, K.R., Snyder, M.D., and Stein, E. (1997), *Effect of Free Flight Conditions on Controller Performance, Workload, and Situational Awareness*, FAA technical report, Federal Aviation Administration, Washington, DC.

Ericsson, K.A., and J.A. Smith (eds.) (1991), *Toward a General Theory of Expertise: Prospects and Limits*, Cambridge University Press, Cambridge.

FAA (1997), *A Concept of Operations for the National Airspace System in 2005*, Revision 1.3., Federal Aviation Administration, Washington, DC.

Freed, M. and Remington, R. (1997), 'Managing Decision Resources in Plan Execution', *Proceedings of the 15th Joint Conference on Artificial Intelligence*, Nagoya, Japan.

Govindaraj, T, Ward, S., Poturalski, R. and Vikmanis, M. (1985), 'An Experiment and a Model for the Human Operator in Time-constrained Competing-task Environment', *IEEE Transactions on Systems Man and Cybernetics*, Vol. SMC-15, No. 4, pp. 496-503.

Hansmann, R.J., Davison, H. (2000), 'The Effect of Shared Information on Pilot/Controller and Controller/Controller Interactions', *Proceedings of the 3rd USA/Europe ATM R&D Seminar*, FAA/EUROCONTROL.

Harper, K., Guarimo, S., White, A., Hanson, M., Billamoria, K. and Mulfinger, D. (2002), *An Agent-Based Approach to Aircraft Conflict Resolution with Constraints*, AIAA, American Institute of Aeronautics and Astronautics, Reston, VA.

Helbing, H. and Eyferth, K. (1994), 'Structuring Information in ATC Mental Models', in N. Johnston, R. Fuller and N. McDonald (eds.), *Proceedings of the 21st Conference of the European Society for Aviation Psychology*, pp.113-117, Ashgate, Surrey, UK.

Hoare, C.A.R. (1985), *Communicating sequential processes*, Prentice-Hall, Oxford, UK.

Hollnagel, E. (1993), *Human Reliability Analysis, Context and Control*, London Academic Press, London.

Hollnagel, E. (1998), *Cognitive Reliability and Error Analysis (CREAM)*, Elsevier Science Offices, Amsterdam, The Netherlands.

Hollnagel, E. (1999), 'From Function Allocation to Function Congruence', in S. Dekker and E. Hollnagel (eds.), *Coping with computers in the cockpit*, Ashgate, Surrey, UK.

Hooey, B.L. and Foyle, D.C. (2001), 'A Post-Hoc Analysis of Navigation Errors during Surface Operations: Identification of Contributing Factors and Mitigating Strategies', *Proceedings of the 11th Symposium on Aviation Psychology*, Ohio State University.

Hopkins, V.D. (1995), *Human Factors in Air Traffic Control,* Taylor and Francis, London.

Hudlicka, E., Zacharias, G., and Psotka, J. (2000), 'Increasing Realism of Human Agents by Modeling Individual Differences', *Proceedings of the AAAI Symposium Simulating Human Agents*, TR FS-00-03, AAAI Press, Menlo Park, CA.

Isaac, A. (1994), 'Short-term Memory and Advanced Technology: the Use of Imagery in Air Traffic Control', in N. Johnston, R. Fuller and N. McDonald (eds.), *Aviation Psychology: Training and Selection,* pp. 107-111, Ashgate, Surrey, UK.

Isaac, A. and Pounds, J. (2001), 'Development of an FAA-EUROCONTROL Approach to the Analysis of Human Error in ATM', *Proceedings of the 4th USA/Europe R&D Seminar,* FAA/EUROCONTROL.

Isaac, A., Shorrock, S. T., Kennedy, R., Kirwan, B., Andersen, H. and Bove, T. (2003), *The Human Error in Air Traffic Management Technique (HERA-JANUS),* EUROCONTROL Info Ref. # 021217-01 HRS/HSP 002-REP-03, EUROCONTROL, Brussels, Belgium.

John, B.E. and Kieras, D. (1998), 'The GOMS Family of user Interface Analysis Techniques: Comparison and Contrast', *ACM Transactions of Computer-Human Interaction.*

Johns, B.E. and Vera, A. (1992), 'A GOMS Analysis of a Graphic, Machine-paced Highly Interactive Task', *Proceedings of the International Conference on Computer-Human Interaction (CHI),* section ACM, pp. 251-258.

Johnson, C. (1997), 'Reasoning about Human Error and System Failure for Accident Analysis' in S. Howard, J. Hammond and G. Lindgaard (eds.) *Human-computer interaction INTERACT'97*, pp. 331-338, Chapman-Hall, London.

Jones, P., Chu, R.W. and Mitchell, C. (1995), 'A Methodology for Human-Machine Systems Research: Knowledge Engineering, Modeling and Simulation', *IEEE Transactions on Systems Man and Cybernetics,* Vol. SMC-25(7), pp. 1025-1038.

Kallus, K, Van Damme, D. and Dittmann, A. (1999), *Integrated Task and Job Analyses of Air Traffic Controllers – Phase 2: Task Analysis of En Route Controllers,* EUROCONTROL Note HUM.ET1.STO1.1000.REP-004, EUROCONTROL, Brussels, Belgium.

Kirwan B, and Ainsworth L.K. (1992), *A Guide to Task Analysis,* Taylor and Francis, London.

Kirwan, B. and Hollnagel, E. (1998), 'The Requirements of Cognitive Simulations for Human Reliability and Probabilistic Safety Assessment' in E. Hollnagel and H. Yoshikawa (eds.) *Cognitive Systems Engineering in Process Control.*

Klompstra, M.B., Bakker, G.J. and Blom, H.A.P. (2000), 'TOPAZ Dynamically Coloured Petri Net (DCPN) specification for double missed approaches on parallel runways', *NLR Report NLR-TR-2000-471,* NLR, Amsterdam, The Netherlands.

Laughery, K.R., Archer, S. and Corker, K., (2001), 'Modeling human performance in complex systems', in G. Salvendy, (Ed.), *Handbook of Industrial Engineering,* 3rd. edition, pp. 2409-2444, Wiley, New York.

Leiden, K. (2003), 'Workload-based Airspace Capacity Predictions for NASA Distributed Air/Ground Traffic Management', *NASA/FAA Human Performance Modeling Workshop*, NASA Ames Research Center.

Lindsay, P. and Connelly, S. (2001), 'Conference in Research and Practice in Information Technology' Vol. 7, in John Grundy and Paul Calder (eds.), *Publications of the Australian Computer Society.*

Ludtke, A. and Mobus, C. (2004), 'A cognitive pilot model to predict learned carelessness', *(in review for) HCI-Aero 2004,* Toulouse, France.

Masalonis, A.J. and Parasuraman, R. (2003), 'Fuzzy Signal Detection Theory: Analysis of Human and Machine Performance in Air Traffic Control, and Analytic Considerations', *Ergonomics*, Vol. 4, pp. 5-20.

Mayberry, P.W., Kropp, K.V., Kirk, K.M., Breitler, A.L. and Wei, M.S. (1995), *Analysis of operational errors for air route traffic control centers*, CNA Corporation, Alexandria, VA.

Mills, S., Pfleiderer, E. and Manning, C. (2002), *POWER: Objective Activity and Taskload Assessment in En Route Air Traffic Control*, Federal Aviation Administration, Washington, D.C.

Mondolani, S. (1998), *Development of an En route conflict resolution rulebase for the reorganized air traffic control mathematical simulator*, NARIM report A09008-01, Washington, D.C.

Mooij, M. and Corker, K. (2002), 'Supervisory Control Paradigm: Limitations in Applicability to Advanced Air Traffic Management Systems', *Digital Avionics System Conference*, Long Beach, CA.

Mosier, K. and Skitka, L. (1996), 'Human decision makers and automated aids: Made for each other?' in R. Parasuraman, and Mouloua, G. (eds.), *Automation and Human Performance: Theory and Application*, Earlbaum, New Jersey.

Newman, M., Tattersall, A., Warren, C. (1994), 'Modeling Cognitive Processes in Air Traffic Control Operations', in N. Johnston, R. Fuller and N. McDonald (eds.), *Aviation Psychology: Training and Selection, 21st Conference of the European Society for Aviation Psychology,* pp. 118-123, Ashgate, Surrey, UK.

Niessen, C., Eyferth, K. and Bierwagen, T. (1999), 'Modeling Cognitive Processes of Experienced Air Traffic Controllers', *Ergonomics*, Vol. 42 (11), pp. 1507-1520.

Orasanu, J. and Fisher, U. (1997), 'Finding Decisions in Natural Environments: The View from the Cockpit', in C. Zambok and G. Klein (eds.), *Naturalistic Decision Making,* Lawrence Earlbaum.

Palanque, P., Bastide, R. and Paterno, F. (1997), 'Formal specification as a tool for objective assessment of safety critical interactive systems', in S. Howard, J. Hammond and G. Lindgaard (eds.), *Human-computer interaction INTERACT'97*, pp. 323-330, Chapman-Hall, London.

Parasuraman, R., Sheridan, T. and Wickens, C. (2000), 'A Model for Types and Levels of Human Interaction with Automation', *IEEE Transactions of Systems Man and Cybernetics,* Vol. 30, pp 286-297.

Pew, R.W. and Mavor, A.S. (1998), *Modeling Human and Organizational Behavior: Applications to Military Simulations*, National Academy Press, Washington D.C.

Powell, J. (1999), 'Ecological Modeling of Air Traffic Management Tasks', *Proceedings of the 9th European Conference on Cognitive Ergonomics*, pp. 109-116, Limerick, Ireland.

Rasmussen, J. (1983), 'Skills, Rules, and Knowledge; Signals, Signs and Symbols, and Other Distinctions in Human Performance Models' *IEEE Transactions on Systems, Man, and Cybernetics*, Vol. SMC-13, No. 3, pp. 257-266.

Rasmussen, J. and Vicente, K. (1989), 'Coping with human error through system design: implications for ecological interface design', *International Journal of Man-Machine Studies*, Vol. 31, pp. 517-534.

Rodgers, M. (1993), *An Examination of the Operational Error Database for Air Route Traffic Control Centers*, DOT/FAA/AM-93/72, Federal Aviation Administration, Washington, DC.

Rodgers, M., Mogford, R. and Mogford, L. (1998), *The Relationship of Sector Characteristics to Operational Error*, Final Report, DOT/FAA/AM-98/14, Federal Aviation Administration, Washington, DC.

Roelen, A.L.C., Wever, R., Hale, A.R., Goossens, L.H.J., Cooke, R.M., Lopuhaä, R., Simons, M., Valk, P.L.J. (2002), *Demonstration of an Integrated Risk Evaluation Causal Tool, Phase 2 Report*, Memorandum of Cooperation MoC-AIA/CA-52, Annex 8, between the United States Federal Aviation Administration FAA and Divisie Luchtvaart.

Sanderson, P., Naikar, N., Lintern, G. and Gross, S. (1999), 'Use of cognitive work analysis across the system lifecycle: requirements to decommissioning', *Proceedings of the Human Factors and Ergonomics Society's 43rd Annual Meeting*, Human Factors Society, Santa Monica, CA.

Shappell, S.A., and Wiegmann, D.A. (2000), *The Human Factors Analysis and Classification System-HFACS*, Tech. Rep.DOT/FAA/AM-00/7, Office of Aviation Medicine, Washington, DC.

Sheridan, T.B. (1992), *Telerobotics, Automation and Human Supervisory Control*, MIT Press, Boston, MA.

Shorrock, S.T. and Kirwan, B. (1988), 'The development of TRACEr Technique for Retrospective Analysis of Cognitive Errors in ATM', in D. Harris (ed.), *Proceedings of the 2nd International Conference on Engineering and Cognitive Ergonomics*, Oxford, England, Ashgate.

Spouge, J.R. (2002), *Causal Modeling of Air Safety*, Final Report CM-DNV-016, Det Norske Veritas.

Stein, E. and Garland, D. (1993), *Air Traffic Controller Working Memory: Considerations in Air Traffic Control Tactical Operations*, Tech. Rep. DOT/FAA/CT-TN93/37, Federal Aviation Administration, Washington, DC.

Tenney, Y.J. and Spector, S.L. (2001), 'Comparisons of AMBR models with human-in-the-loop performance in a simplified air traffic control simulation with and without HLA protocols: Task simulation, human data and results', *Proceedings of the 10th Conference on Computer Generated Forces and Behavioral Representation*, Vol. 10, pp. 15-26.

Van Lehn, K. (1990), *Mind Bugs: The origins of procedural misconceptions*, MIT Press, Cambridge, MA.

Verma, S. and Corker, K. (2002), 'Introduction of context in a human performance modeling to predict performance for new air traffic management initiatives', *Proceedings of the Advanced Simulation Technologies Conference*, San Diego, CA.

Vicente, K. (1999), *Cognitive Work Analysis: Toward safe, productive, and healthy computer-based work*, Erlbaum, Hillsdale, NJ.

Weigman, D.A. and Shappell, S.A. (1997), 'Human Factors analysis of past accident data: Applying theoretical taxonomies of human error', *International Journal of Aviation Psychology*, Vol. 7(1), pp. 67-81.

Wickens, C., Mavor, A. and McGee, J. (1997), *Flight to the Future: Human Factors in Air Traffic Control*, National Academy Press, Washington, D.C.

Wickens, C., Mavor, A., Parasuraman, R. and McGee, J. (1998), *The Future of Air Traffic Control: Human Operators and Automation*, National Academy Press, Washington, DC.

Wickens, C.A. (1992), *Engineering Psychology and Human Performance*, 2nd edition, Harper Collins, N.Y.

Woods, D. D. and Tinapple, D. (1999), 'W^3: Watching human factors watch people at work', *Presidential address, 43rd annual meeting of the Human Factors and Ergonomics Society*.

Woods, D.D. (2002), 'Steering reverberations of technology change on fields of practice: Laws that govern cognitive work', *Plenary address, 24th annual meeting of the Cognitive Science Society*, George Mason University, Fairfax, VA.

Chapter 15

Performance Prediction in Air Traffic Management: Applying Human Error Analysis Approaches to New Concepts

Steven Shorrock, Barry Kirwan and Ed Smith

Introduction

Air Traffic Management (ATM) is in a period of major change, with a variety of new concepts under consideration to help air traffic controllers cope with projected increases in traffic levels. New technology will enable considerable changes to older methods of operation, and automation to help the controller is now widely acknowledged to be one of the only ways to meet future capacity demands. Automation, or cognitive support tools, in aviation has been subject to increasing investigation and analysis over recent years, particularly since the introduction of glass cockpit systems. In ATM, full automation is highly unlikely in the foreseeable future, but automation of some functions has been introduced since the 1960s, for example flight data gathering and processing (e.g. radar data processing and flight data processing) and conflict alerting. This trend is continuing with tools such as Electronic Flight Strips and Final Approach Spacing Tools. In the future, communication tools such as datalink and decision support tools including 'risk' display tools and conflict resolution tools may be implemented, and perhaps changes to the way that ATM goals are achieved, such as free flight, time-based separation and increased delegation of controller tasks to the pilots.

One of the key issues that has emerged from these studies is the effect that technology might have on 'human error'. 'Human error reduction' was once seen as a natural consequence of automation; system designers thought that if human operators could be moved to the fringes, the risk of human error would decrease. But this premise was questioned (Wiener and Curry, 1980), and the real situation is still being revealed. On the basis of experimental studies and operational experience, many commentators asserted that computers, and automation in particular, produce new error forms or additional sources of errors (Wiener, 1988; Sarter and Woods, 1995). Another view is that 'Computers do not produce new sorts of errors. They merely provide new and easier opportunities for making the old errors' (Kletz, 1988b). It is also a widely held view that the consequences of errors that do occur are likely to be more serious (Wiener, 1985; Billings, 1988; Leroux, 2000). Whatever the case, it is clear that human error is an issue that still needs to be managed, particularly in the context of new automation. This chapter offers two alternative methods for tackling this difficult but important area.

The chapter is laid out in three main sections. The first is the more scientific or academic part, showing that Human Error Analysis (HEA) has credibility in other sections, and furthermore establishing the 'pedigree' of the two main approaches selected for ATM (called HAZOP and TRACEr-lite), and how they work. The second part then describes a series of short case studies that demonstrate the usefulness of these two approaches in ATM situations. The third part contrasts the use of the two techniques, to see whether they can be used inter-changeably or whether they have relative advantages and disadvantages for ATM applications.

Scientific Background to Human Error Analysis Techniques

Various means exist to assist in the process of identifying potential human errors. One family of approaches called 'Human Error Analysis' (HEA) has seen increasing use in other industries (e.g. nuclear power, chemical process and petrochemical) since the 1980s (see Kirwan 1998a and 1998b for a review of the area). These methods include two types of approaches. First, *group-based* approaches utilize a multi-disciplinary team of individuals to help brainstorm and analyze potential failures. These approaches stem from the Hazard and Operability Study (HAZOP: Kletz, 1974, 1988a) and 'What-if' query-based or checklist-based methods. Second, *analyst-led* approaches generally utilize a task analysis and a classification system to probe potential errors and their psychological and contextual origins. These approaches include SHERPA (Embrey, 1986), CREAM (Hollnagel, 1998) and TRACEr (Shorrock and Kirwan, 1999, 2002). Such methods, when supplemented with other techniques, can help to provide an integrated risk-informed approach to 'designing for safety' via explicit consideration of human involvements. Whilst HEA may be relatively new to ATM and to aviation, it is not new to other areas. An earlier review of HEA techniques in other domains revealed a large number of approaches (Kirwan, 1998a). Five broad classifications have been used to show the techniques' general orientation or form:

1. Taxonomies: Many techniques (e.g. Technique for Human Error Rate Prediction - THERP, Swain and Guttmann, 1983) tend to be taxonomic to some degree (i.e. they contain error taxonomies), but some techniques are solely taxonomic in nature. These taxonomic techniques offer checklists of error modes, and the reliance is placed on the analyst to interpret them in the context of interest.
2. Psychologically-based tools: These are tools that rely on an understanding of the factors affecting performance. This group is particularly characterized by tools that consider error causes (Performance Shaping/Influencing Factors,

PSFs/PIFs) and/or error mechanisms (Psychological Error Mechanisms, PEMs). The classic technique here is the SRK (skills, rules and knowledge) framework, parent to a number of techniques which have borrowed from its conceptual framework to some degree.

3. Cognitive modeling tools: these are tools that try to model cognitive aspects of performance, either in terms of relationships between knowledge items relating to symptoms of events (for diagnostic reliability assessment) (Influence Modeling and Assessment System - IMAS), or in terms of how various factors will affect cognitive performance aspects of the task (CREAM). This domain is perhaps the least mature of the human error analysis approaches, but also perhaps the most interesting, as it is an attempt to combine cognitive psychology, the currently dominant paradigm in engineering psychology, with a human reliability attitude.

4. Cognitive Simulations: These are generally computer simulations of operator performance. This is the most sophisticated human error identification area, often relying on expert system-type frameworks to predict performance and error. Most of these tools are aimed at modeling and predicting cognitive performance rather than psycho-motor performance, as it is felt that the former is both the more important and more dominant contributor to risk in complex systems.

5. Reliability-oriented tools: These stem from the reliability approaches that have proven their worth with non-human reliability problems: principally Hazard & Operability Study (HAZOP), Failure Mode and Effect Analysis (FMEA), and Event Tree Analysis (see Green, 1983). The HEA tools in this domain are therefore either super-imposed on their reliability parent framework or they are adaptations of the original concept. They may also be clearly focused by quantitative risk assessment concerns onto specific human error issues required for risk assessment integrity and insight. For example, in the nuclear power domain, the main concern now is the less predictable type of error, in which the operator does something unrequired (called an error of commission), due to unusual circumstances. This area of human error interest is likely to become a focus in ATM also, as will be discussed later.

This structure is inevitably subjective, but is hopefully useful to the reader wishing to further research and to group what is otherwise a large and somewhat confusing set of disparate techniques and approaches.

A large number and variety of approaches for HEA has been developed since the 1980s, initially in response to the Three Mile Island nuclear power plant accident in 1979. New approaches or ideas have appeared, which add to the breadth of HEA as a whole, e.g. affordance approaches; tools aimed at error of commission analysis; and violation assessment tools. Many HEA techniques have been influenced heavily by Rasmussen's (1981) Skill-, Rule-, and Knowledge-based (SRK) behavior framework and Reason's (1990) classification of slips, lapses, mistakes, and violations. Certain approaches have remained popular, such as the SHERPA-style of approach, and more generally the taxonomic approach. In practice only a small number have seen widespread use, and many are confined to history as funding is discontinued or where techniques are not published and made generally available.

However, regardless of their orientation, many HEA techniques work in the same way, requiring a task model, consideration of the context the human operators find themselves in, a taxonomy of error types, and an analyst or facilitator and one or more task domain experts. Typically, the taxonomy of error types operates on the task model, with direct or indirect consideration of contextual factors, often known as Performance Shaping (or Influencing) Factors (PSF or PIF).

A review of available approaches in Shorrock and Kirwan (2002) compared six approaches of potential value to ATM. These were SHERPA (Embrey, 1986), CREAM (Hollnagel, 1998), Generic Error Modeling System - GEMS (Reason, 1990), TAFEI (Baber and Stanton, 1991, 1994), HEIST - Human Error Identification in Systems Technique (Kirwan, 1994), and PHEA - Predictive Human Error Analysis (Embrey et al., 1994). The approaches were compared on the 12 criteria below. Table 15.1 summarizes these comparative evaluation ratings using the following criteria:

- comprehensiveness: the degree to which a technique can discriminate and classify a comprehensive range of errors and influencing factors at an acceptable level of mutual exclusivity;
- consistency: the degree to which a technique leads to consistent analyses between different users and with the same user over time;
- lifecycle applicability: the degree to which a technique can be used throughout the formative and summative phases of system design lifecycle;
- predictive validity: the degree to which a technique accurately predicts errors that could or actually do occur;
- theoretical validity: the degree to which a technique is based on a human performance model or framework, with a plausible internal structure;
- contextual validity: the degree to which a technique incorporates contextual and domain-specific information;
- flexibility: the degree to which a technique enables different levels of analysis according to the project needs, known information or expertise of the user;
- usefulness: the ability of a technique to suggest or generate practical error reduction or mitigation measures;
- resource efficiency (training): the time taken to become proficient in the use of a technique;
- resource efficiency (usage): the amount of time required to collect supporting information and apply a technique, including task-domain experts;
- usability: the practicality and ease of use of a technique in the applied setting; and
- auditability: the degree to which a technique lends itself to auditable documentation.

The evaluation in Table 15.1 is updated from the original evaluation, incorporating HAZOP and TRACEr, and taking into account additional applied experience and validation evidence. This evaluation took account of applied experience, validation evidence (Kirwan, 1992b; 1998a and b), and other papers that have reported on the use of the various techniques. Table 15.1 summarizes the evaluation of the techniques, and these results led to the conclusion that TRACEr and HAZOP would be the best starting point for ATM applications.

Table 15.1 Comparative evaluation of HEA techniques

	SHERPA	CREAM	GEMS	TAFEI	HEIST	PHEA	HAZOP	TRACEr
Comprehensiveness	M	H	H	L	H	L	M	H
Structure and consistency	H	L-M	L	H	L-M	M-H	M-H	M-H
Lifecycle stage applicability	M	M-H	L	L	M	M	H	M
Predictive validity	M	M	L	M-H	M	M	M	H
Theoretical validity	H	H	H	H	H	M	L	H
Contextual validity	L	M-H	L	L	L	L	H	M
Flexibility	M	M	L	L	M	M	M	M
Usefulness	M	H	M	L-M	M-H	M-H	H	M-H
Res. efficiency (Training)	M	L	L	M	M	M	M	M
Res. efficiency (Time)	M	L	L	L-M	M	M	L	M
Usability	M	L-M	L	M	M	M	M-H	M
Auditability	M-H	M-H	M	H	M	M-H	H	M-H
Rank	**4**	**3**	**8**	**7**	**5**	**6**	**2**	**1**

Notes: Each rating was converted to a score as follows: L / Low = 1; L-M / Low-Medium = 2; M / Medium = 3; M-H / Medium-High = 4; H / High = 5. These scores were summed to provide total scores, which were used to calculate the rank orders.

The first of the two techniques highlighted for the ATM industry is *(Human) HAZOP*, an established, group-based approach to human hazard identification based on the HAZOP study method, developed in the chemical industry. The HAZOP approach is one that has enjoyed success in a number of industries, and it has become a regulated requirement in some industries to use HAZOP-type approaches as a means of assuring safety, protecting from hazards of all types, including human sources. It therefore seemed logical to try to apply HAZOP in the ATM industry too. However, in industries such as chemical and nuclear power, and in the defense industries, there is not complete reliance on such group-based methods. Rather, there is also often an additional analysis using a more detailed approach in which all possible failure modes are considered step-by-step. Such approaches have their roots in techniques such as FMEA (Failure Modes and Effects Analysis) used for hardware reliability analysis, which led to related HEA derivatives such as SHERPA, involving a thorough analysis usually by a single assessor. It was realized some years ago that whilst these approaches (HAZOP & FMEA) overlapped, the overlap was not complete (Whalley and Kirwan, 1989). Instead, the optimal solution involved application of both approaches, to gain a fully comprehensive range of potential errors. Such an approach amounted to 'defense in depth' in safety assessment terms. Also, whilst HAZOP is powerful, it relies on the availability and quality of experts, and their ability to work together effectively, and neither of these conditions for success can always be guaranteed. Therefore, a new FMEA-style approach was developed, similar to tools such as SHERPA (Embrey, 1986), but adapted to the ATM industry. The result was TRACEr, a comprehensive tool for investigating and predicting errors in ATM, and its derivative *TRACEr-lite*, a relatively new, single analyst-led approach to HEA developed (initially) for ATM. Both TRACEr and TRACEr-lite were developed by Shorrock (Shorrock and Kirwan 1999; Shorrock and Kirwan, 2002). HAZOP and TRACEr-lite have since been applied in test cases in the ATM industry, and have led to safety insights and changes to system designs.

The following sections therefore define HAZOP and TRACEr-lite, outlining the way they work and the type of insights they bring, based on some case studies in ATM. A comparison of HAZOP and TRACEr-lite, via a partial but formal application to three real cases in ATM system design and development is then presented, to further see how these techniques compare. This is followed by a discussion that considers additional future needs in HEA for ATM.

HAZOP

Hazard and Operability (HAZOP) studies provide a formal, systematic and critical examination of the process and engineering intentions of a design. HAZOP assesses the potential for hazard and identifies mal-operation or malfunction of individual items of equipment and the consequences for the whole system. This examination of the design is structured around a set of guidewords, which help to ensure comprehensive coverage of possible problems. HAZOP studies normally involve a team who has experience of the system or design to be studied, including design, engineering and operational personnel, often also including training specialists, Human Factors specialists, and independent safety specialists. There are generally four overall objectives addressed by HAZOP:

- to identify all deviations from the way the design is expected to work; their causes, and all the hazards and operability problems associated with these deviations;
- to decide whether action is required to control the hazard, or the operability problem, and if so to identify the ways in which the problem can be solved;

- to identify cases where a decision cannot be made immediately and to decide on what information or action is required;
- to ensure that actions decided upon are followed through.

The (Human) HAZOP study method can also be applied to the study of a task, human-machine interface or operating procedure. A 'task' can be seen as a set of things including a system goal, resources for accomplishing the system goal, such as information and controls, and a set of constraints on how the goal may be achieved using these resources (Shepherd, 2001). A 'procedure', meanwhile, is defined here as a set of instructions whose aim is to direct an operator to make changes to the state of a system in a safe manner, so that a particular objective is achieved. The first process is to identify the intent of the task or procedure, including the system state at the beginning of the task/procedure, and the required system state at the end, and how it will be achieved. The detailed study is then carried out in a step-by-step manner and for each task or task step the team agrees upon an objective or intent. The team will then view the step as requiring an action at a time in a sequence. If procedural documentation is available, they will also review the wording of the instruction. To assist this process, a set of guidewords is used. For a Human HAZOP study this might include some or all of the guidewords and concepts displayed in Table 15.2.

Table 15.2   HAZOP guidewords and concepts

Basic Guidewords	• More information
• No action	• Less information
• More action	• No information
• Less action	• Wrong information
• Wrong action	*Additional Concepts*
• Part of action	• Purpose
• Extra action	• Clarity
• Other action	• Training
• More time	• Abnormal conditions
• Less time	• Maintenance
• Out of sequence	• Safety

A HAZOP study meeting will follow through a series of steps repeatedly. There are seven stages, which are repeated many times during a HAZOP (see Figure 15.1). Examination sessions are structured with the team leader facilitating the discussion, and a secretary (often a qualified expert and full participant in the HAZOP itself) recording the deliberations on log sheets, detailing tasks, deviation guidewords, causes, consequences, safeguards, risk ranking (if used) and recommendations. The discussions and actions arising are recorded on log sheets.

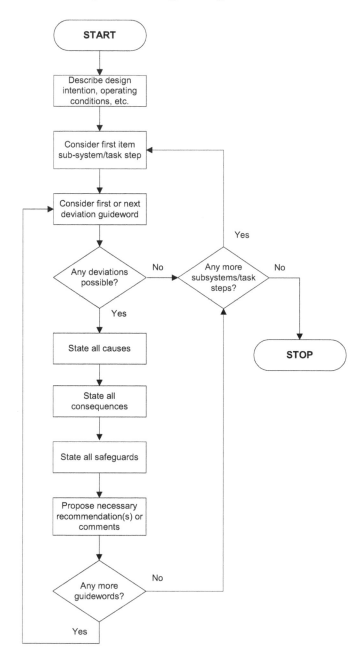

Figure 15.1   Process for conducting HAZOP study

TRACEr and TRACEr-lite

TRACEr (Technique for the Retrospective and Predictive Analysis of Cognitive Errors) (Shorrock and Kirwan, 1999; 2002) was developed from literature findings and operational experience for retrospective and predictive human error analysis in ATM. TRACEr was represented as a set of decision-flow diagrams containing human error modes and mechanisms, intended for use by Human Factors specialists. [25] After approximately two years of research and use of TRACEr, it was further developed to be predictive in focus, and this version of TRACEr was called TRACEr-lite (Shorrock, 2002a; b).

For predictive use of TRACEr-lite, the analyst first scopes the analysis, and then conducts a task analysis, e.g. using Hierarchical Task Analysis (HTA) (see Shepherd, 2001). Using TRACEr-lite and the task analysis, the analyst determines what could go wrong. There are four key components to the TRACEr-lite toolkit (see Figure 15.2). Prior to the TRACEr-lite analysis, during the task analysis process, Performance Shaping Factors (PSFs) are analyzed to set the scene for the analysis. PSFs are those factors, either internal to the controller or pilot, or relating to the task and operational environment, that affect performance positively or negatively, directly or indirectly. The PSFs are used to prepare a general *Context Statement* – a set of questions about the performance conditions under which the controller will be working. PSFs may also be analyzed separately for particular task steps, relating to specific errors, if the analyst so chooses. Each PSF takes the form of a question, eliciting a 'yes'/'no' type response and a statement of justification. These questions occupy a number of categories, such as traffic and airspace, procedures and documentation, training and experience, workspace design, human-machine interface (HMI) and equipment, etc. TRACEr-lite does not predetermine the links between PSFs and error modes/mechanisms because of the many-to-many mapping relationships involved, as well as the uncertainty in making such specific links. However, some general guidance is provided on how particular types of PSF affect cognitive processing.

Once a context statement is prepared the analyst engages in a cycle of activities, applying the TRACEr-lite taxonomies to the detailed task steps within the task analysis. *External Error Modes (EEMs)* are first used as prompts to enable the identification of the observable manifestations of potential errors, based on logical outcomes of erroneous actions, in terms of timing, sequence, selection and quality. Examples include 'Omission', 'Wrong action on right object', 'Mis-ordering', and 'Information not sought/obtained'. EEMs are context-free and independent of cognitive processes (e.g. intention). However, when applied to a task step from a

---

[25] TRACEr was adapted for retrospective use in Europe in the EUROCONTROL 'HERA' project – human error in ATM (see Isaac et al., 2002) – which, in collaboration with the Federal Aviation Administration (FAA), was further developed in a joint project resulting in the HERA-Janus technique.

task analysis, the error mode is converted to a contextual 'external error' (e.g. 'Controller fails to issue instruction to select target').

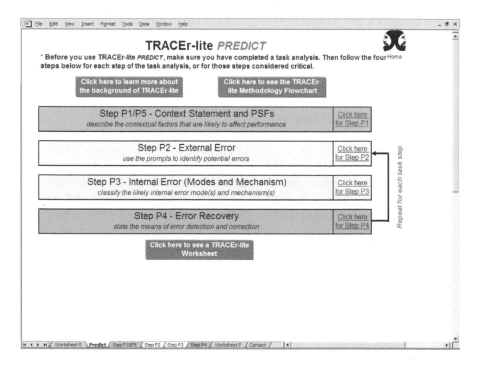

Figure 15.2   TRACEr-lite light prototype interface and predictive task steps

The cognitive aspects of the error are analyzed using a set of Internal Error Modes and Mechanisms. These are structured around four error domains, with an associated question prompting further classification:

- Perception: Does the controller/pilot have to see or hear something during the task step?
- Memory: Does the controller/pilot have to recall information or remember to perform actions in the future during the task step?
- Decision making: Does the controller/pilot have to project required separation, or make a plan or decision during the task step?
- Action: Does the controller/pilot have to perform a manual action or say something during the task step?

Internal Error Modes describe how the controller's/pilot's performance failed to achieve the desired result. For instance, Internal Error Modes within the 'Perception' error domain include 'mishear', 'mis-see', 'no detection (visual)' and 'no detection (auditory)'. One or more Internal Error Mode is used for each error identified in the report. *Internal Error Mechanisms* describe in greater depth the psychological underpinnings of an Internal Error Mode and can better enable the consideration of measures to reduce or mitigate errors. Example error mechanisms within the 'Perception' domain include 'expectation', 'confusion', 'discrimination failure', 'perceptual overload' and 'distraction/preoccupation'. Internal Error Mechanisms would normally be identified only for critical errors (e.g. errors with low Recovery Success Likelihood). The TRACEr-lite error modes and mechanisms are shown in Table 15.3.

Table 15.3   TRACEr-lite Internal Error taxonomy

Internal Error Mode	Internal Error Mechanism
Perception	
Mishear	Expectation
Mis-see	Confusion
No detection (auditory)	Discrimination failure
No detection (visual)	Perceptual overload
	Distraction / Preoccupation
Memory	
Forget action	Confusion
Forget information	Memory overload
Misrecall information	Insufficient learning
	Distraction / Preoccupation
Decision Making	
Misprojection	Misinterpretation
Poor decision or poor plan	Failure to consider side- or long-term effects
Late decision or late plan	Mind set / Assumption
No decision or no plan	Knowledge problem
	Decision overload
Action	
Selection error	Variability
Unclear information	Confusion
Incorrect information	Intrusion
	Distraction / Preoccupation
	Other slip

Following the analysis of Internal Errors, *Initial Consequences* are determined by a process of analysis. Consequences are stated as free text, and are normally restricted to the more immediate and likely effects and consequences. Error likelihood and severity are not rated as part of the standard TRACEr-lite approach, but can be rated by a team of individuals if appropriate, for instance using data (e.g. simulation-derived), or expert judgment.

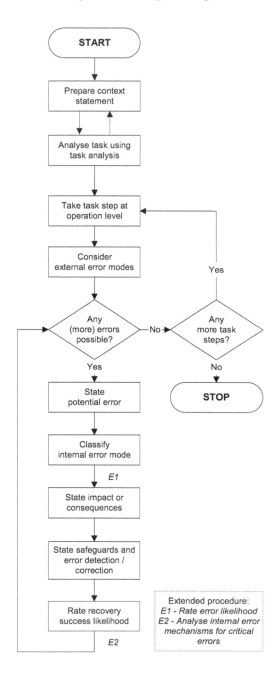

Figure 15.3   Process of using TRACEr-lite

The next step in the TRACEr-lite process involves considering *Recovery* from the error. This may involve stating a future step in the task analysis, or stating the 'detection means', i.e. cues from the work context, e.g. R/T readback, radar monitoring, other controller, etc. The analyst may also at this point rate the 'Recovery Success Likelihood (RSL)'. This is a subjectively rated likelihood of recovering the task successfully without adverse consequences, assisted by the use of an anchored rating scale contained within TRACEr-lite.

On the basis of this analysis, 'Comments' or 'Questions' may be made or 'Recommendations' may be proposed. However, rather than propose recommend-dations in a reactive fashion, these are normally proposed during the synthesis of the data to address common themes within a manageable set of recommendations or requirements. The process above is illustrated in Figure 15.3. An example of TRACEr-lite analysis output is shown in the case study on TRACEr-lite in the evaluation section.

Overview of HAZOP and TRACEr-Lite Applications in ATM

HAZOP Case Study 1 – A Hazard and Operability Study (HAZOP) for an Electronic Flight Strip System

Currently in many ATC centers, en-route controllers use paper 'flight progress strips' (FPS) in conjunction with radar displays to control and monitor aircraft. The generation of new FPS systems, however, shifts the focus away from paper to an electronic medium. A HAZOP analysis was conducted on the proposed HMI of a new flight progress information system being developed by NATS. The study is described more fully in Kennedy et al. (2000).

A HAZOP team assessed the implications of the new interface. The team comprised three designers, one air traffic controller and two Human Factors specialists. A total of 16 hours, spanning three separate HAZOP sessions, were spent interrogating the prototype system. Table 15.5 gives an example of the output of the HAZOP, and Table 15.4 attempts to give an insight into the way HAZOP operates, via a transcript of part of the discussion during the very first HAZOP for this system.

The study identified a number of vulnerabilities in the prototype and opportunities for error that needed to be addressed. A total of 87 recommendations were generated from the three HAZOP sessions, including changes to interface design and menus, improvements in user feedback, training and procedures recommendations, modifications to aircraft status on screen, hardware changes, and further research ideas. The HAZOP group identified what factors needed to be changed in the system and how these changes could be addressed. Since the designers were present and actively involved, any design changes they thought necessary were simultaneously accepted for implementation. The HAZOP therefore had very effective impact on the design process.

Table 15.4   Excerpt from HAZOP discussion

HAZOP Group Member	Comment
HF Specialist	*If these two objects overlap, could the controller operate on the wrong object, i.e. aiming a message for the one on top but actually communicating with the one underneath that is now flying beneath the top one?*
Designer 1	*Well, we expect the controller always to move the objects so that they will not overlap, before transmitting a command.*
Controller	*Well, there might not always be time, but as long as you're operating on the one on top, one coming underneath will not be selected without it being clear will it?*
Designer 1	*Hmm. Well, it is not impossible actually, depending on how long you leave the cursor without entering a command. Presumably that would be no more than a couple of seconds would it?*
Controller 2	*Not necessarily, I mean if I'm in the middle of something and then a higher priority call comes in, I'll leave the cursor there and then come back to it. It could be a while, up to a minute.*
Designer 2	*Right. Okay, we need to take another look at this, and implement some way of highlighting that the original target has been de-selected and must be re-acquired, otherwise the right message could be sent to the wrong aircraft.*
Chairman	*So, are we agreed then that we need an action on the designers here to ....*

Note: This adapted 'excerpt' gives the flavor of the types of discussions that occurred during the HAZOP, and of the utility of a good hybrid group. For commercial sensitivity reasons, the actual technical nature of the item being discussed has been omitted.

In terms of the strength of the HAZOP approach, a number of problems that were identified in the design would have been difficult to detect without a multidisciplinary team present. In particular each member of the team brought a specialism to the group process. All this information was shared effectively, leading to a rich multiple-perspective on the system design and its strengths and weaknesses. Such an interrogation of the system would have been very difficult to achieve in any other way and underlines the potential niche for HAZOP in system design and evaluation.

Table 15.5   Extract from a HCI HAZOP Study (from Kennedy *et al.*, 2000)

Function	Guide Word	Cause	Consequence	Indication	System Defences	Human Recovery	Recommendations
Highlight Object	No	Another item preventing access to target	Difficulty in hooking target aircraft	No highlighting of target	None	Drag blocking object out of way; Strategic management of screen items	Design objects to roll around each other; Height filtering; Flip system to move between object on top and the one beneath; Highlight background
	Other	Clustering results in different aircraft being highlighted instead of target	Instruction may be given to wrong aircraft on the system	As Above	Highlighting is colour coded to indicate direction of travel; Call sign is displayed on all menus	As Above	As Above
Track Pointer	No	Mouse cursor moves off menu	Menu automatically closes when cursor moves away	Loss of menu on screen - controller needs to look at screen and not keyboard	Put gate around object to prevent loss of menus	If noticed, the cursor can be re-positioned over the function and re-clicked for mouse to reappear	As Above
	Other	Mouse cursor knocked-off intended function during keyboard entry	Input made into wrong function due to controller using keyboard whilst not looking at the screen	Different functions have different consequences (e.g. heading and speed); Aircraft not behave as expected; Conflict alert may indicate error		Mouse use results in head-up selection of menus on screen - thus increasing likelihood of identifying an error	Training to encourage mouse rather than key input; Provide head-down display of input function to prevent head-down errors; Lock object so it does not slip out of focus

HAZOP Case Study 2 – Pre-Live Trial HAZOP for Reduced Separation outside Controlled Airspace

As reported in the chapter 'Reducing Separation in the Open Flight Information Region: Insights into a Human Factors Safety Case' by Kirwan et al. in this book, another human error project concerned determining whether separation minima could be reduced between aircraft receiving a radar advisory service (from military controllers) when flying outside controlled airspace. As part of this study, it was decided to conduct a 'live trial' for an initial period of six months in a small number of areas, before deciding whether to go 'nationwide' with the reduced vertical separation (note that this was not connected in any way with the later Reduced Vertical Separation Minimum (RVSM) program in civil airspace). Prior to this live trial, there were concerns about what could happen in safety terms – i.e. would the trial itself be safe, and could there be unforeseen interactions between the military units and aircraft in the participating and non-participating areas? In order to address these concerns, a one-day HAZOP was convened, which considered a number of scenarios, including emergency events and catastrophes, to determine safeguards and 'reversionary procedures' to adopt in case anything went wrong. Additionally, certain monitoring measures were put in place to determine if safety was showing any signs of being compromised.

The one-day HAZOP, including military personnel and controllers, and pilot representatives, fulfilled its function, raising certain issues that had not been thought of, and procedures were amended accordingly. One methodological insight worthy of note was that two of the HAZOP organizers had tried to develop some new guidewords specifically for this HAZOP context. However, in practice, these did not work, and the study reverted to the original and usual guidewords as defined earlier. This suggested at the time that there was little 'technical' adaptation of the technique required, other than deciding how to present ATM concepts and scenarios (in this example this was achieved via use of maps and flipcharts representing the geographical areas involved and potential scenarios).

HAZOP Case Study 3 – Pre-Live Trial HAZOP for Medium Term Conflict Detection

In 2002 a live trial of Medium Term Conflict Detection (MTCD) was being planned. MTCD (like its URET [User Requirement Evaluation Tool] counterpart in the US) detects conflicts much earlier than existing European approaches such as Short Term Conflict Alert (STCA), giving the controller for example eight minutes of warning before a conflict would occur. This allows more efficient resolutions to be applied, and can deliver a bigger margin of safety for the ATM sector it is used in.

Although some formal safety work had been performed on MTCD, it was realized that the live trial would be the first time this version of MTCD was in use with live traffic, based in the new Rome Ciampino Centre. A HAZOP was therefore convened to consider what could go wrong, and to define any reversionary procedures and emergency controls that needed to be developed associated with the trial. Although the HAZOP group was uncharacteristically large (approximately fifteen people), a number of recommendations were made and accepted. The trial then proceeded smoothly in spring 2003. The reversionary procedures were only called upon once throughout the extended trials, but their availability and the knowledge of their existence is likely to have added confidence to those running the simulation. This 'Live trial HAZOP' approach has now been applied for one other Eurocontrol-linked project and is being considered for other future Eurocontrol planned live trials.

The above are some examples of the early application of HAZOP to ATM. Other applications have occurred since this time, concerned with the automation tools being developed for controllers (Springall, 2001).

TRACEr Case Study 1 – HEA for New Scottish Centre

HEA was applied to a series of real-time simulations conducted at NATS (see Shorrock et al., 2001). The overall aim of the simulations was to ensure the safety, efficiency and usability of an ATM system being developed for use within a Scottish en-route environment. More specifically, from a human error perspective, the study aimed to identify any aspects of the Human-Machine Interface that could cause or contribute to controller error. Of particular interest were errors associated with the use of electronic tools developed to assist the controller when re-routing aircraft, predicting conflicts and co-ordinating aircraft.

The intention was to use the HEA findings to feed into the design phase to ensure that the system could be made more resistant or tolerant to errors. A combination of methods was used for the HEA – TRACEr analysis, observation and debrief, and questionnaire. These are described below, as they show that HEA can be done in a fairly straightforward fashion if there is access to real or simulated operations.

TRACEr was first used with an HTA of an automation tool to predict the types of errors, and then to see what actually occurred, and to categorize the causes of the errors. Following this analysis, controllers were observed using the tools during each exercise throughout the simulations (each simulation lasted three weeks and each exercise was 60 to 75 minutes in duration). Observation sheets were developed from the HTA developed prior to the simulations, which also helped the observer to understand in detail how the user should correctly interact with each tool. Written notes were made regarding any errors that the controllers made when interacting with the tools and, where possible, the number of times the task was performed to provide the baseline or denominator figure so that a Human Error Probability (HEP) could be calculated (i.e. the number of errors that occur divided by the number of opportunities for the error to occur). This gave a reasonable basis to estimate the likelihood of some errors (i.e. those with a large corresponding baseline) during operational air traffic control. The study focused specifically on errors with HEPs of 0.05 or above (one or more errors in 20 opportunities).

The types of errors recorded were mainly 'simple' errors in selecting an object on the HMI (e.g. typing errors, menu selections, text field selection and other mouse button presses). 'Cognitive' errors concerning judgment and planning were difficult to observe and verify reliably. Consequently, they were not the focus of the direct observational analysis. A five-minute debrief was conducted for each controller after each observed exercise. This was supported by a list of questions relating to use of the tools and functions of interest, again constructed from the HTA, and used to prompt controllers to recall any errors that the observer may have missed. The debrief also allowed the observer to confirm whether certain observed actions were, in fact, errors (this is not always evident to an observer). Finally, a human error questionnaire was distributed at the end of each simulation. The questions focused on types of errors (e.g. selection, typing, judgment) and associations or causes (e.g. system feedback, distraction, confusion). The observation and questionnaire methods confirmed many of the errors predicted before simulations, producing a 'hit' rate of approximately 92 per cent.

The HEA revealed a variety of errors and their potential causes. The types of errors occurring were associated with typing and mouse selection/manipulation, mistaking/confusing flight level information, increased workload caused by tool interaction, and failure to detect alerts. A number of 'Performance Shaping Factors' associated with the HMI and system performances were identified.

The HEPs were combined with controller knowledge of the potential impact on safe operation. As a result, it was possible to prioritize re-design effort in terms of safety and usability. The combination of methods provided additional information that may otherwise have been omitted, and allowed verification of findings. A number of remedies were suggested to reduce or resolve the potential for error as far as practicable, which influenced the detailed design of the HMI.

The methodology was well accepted by the client, who commissioned a continuation of the data collection for three further simulations.

TRACEr Case Study 2 – Predicting Errors in using a Future Approach Sequencing Tool

Evans, Slamen, and Shorrock (1999) describe the use of HEA to address the potential for human error during use of the NATS Final Approach Spacing Tool (FAST). The analysis was able to conclude that most potential errors were likely to be detected by the controller team involved in using and developing FAST. However, the TRACEr application identified some unconsidered errors, some of which would be important if they did actually occur. In a subsequent simulation, some of these key errors were seen. This led to a more widespread application of TRACEr for two main reasons: first, as a safety assurance tool, to eradicate significant error potential; and second, to get rid of basic 'irritation' errors that might not affect safety, but would or could lead to user population disaffection during user trials and tests. This function is important because if controllers are irritated by small but frequent errors in a new tool, their dissatisfaction, not with the concept, but with a premature prototype, may lead to generalized controller disaffection with the tool concept being developed.

Evans et al. concluded that TRACEr explores the potential for many types of errors and provides a systematic assessment. They further argued that user trials alone are unlikely to provide an extensive understanding of error potential because errors might not occur regularly with sufficient frequency to be observed during the relatively short life-time of a trial or simulation. In addition, the process identified problems that had not been identified by the design team, and provided a further layer of assessment of FAST beyond that achieved by user trials.

TRACEr Case Study 3 – Reduced Separation Outside Controlled Airspace

As with HAZOP, TRACEr was also used in a predictive fashion during the study mentioned above. The specific use of TRACEr was to determine which types of errors could occur caused by reducing separation. TRACEr was also used to classify the causes of the incidence of losses of separation that had been observed during a period of investigation of this type of UK airspace. The results are shown elsewhere in this book, but it was concluded that this summary of what could happen did help to decide whether vertical and/or lateral reductions in separation minima were advisable or not.

The above case studies illustrate a number of ways that these two approaches have been applied. While the detail of results publishable has been limited by commercial confidentiality associated with the systems being evaluated by TRACEr and HAZOP, it can be said that these tools have had impact on a number of designs and concepts. Following these initial successes, it was decided to compare and contrast them more formally via a study at the EUROCONTROL Experimental Centre (EEC) in Brétigny, France. This comparison study is detailed in the next section.

Contrast and Evaluation of HAZOP and TRACEr-lite

The two HEA approaches were compared via independent analyses of three EUROCONTROL research and development concepts: *Co-space, Time-Based Separation* and *CORA 2*. These concepts represent varying degrees of change in the way that ATM is performed, and in the level of computer support involved. The objective of the comparison was to assess the relevance of the techniques, show what they can deliver in terms of safety and design insights, and to show the relative advantages of each for human error analysis purposes. The study did not attempt to provide a full safety assessment, but rather to provide an illustration of how human error could be considered prospectively.

The ATM Concepts

Co-space The Co-space project aims to increase controller availability through a reorganization of tasks between controller and pilot, thereby achieving a more effective task distribution that is beneficial to all parties. It is expected that increased controller availability could lead to improvements in safety, efficiency and/or capacity. Delegation of the aircraft spacing task from the controller to the flight crew is envisaged as a possible option to help achieve this. For aircraft within an arrival stream, the delegation could consist of tasking the flight crew to maintain a given spacing value to a lead aircraft, as defined by the controller.

Within the Co-space concept, spacing tasks are delegated to flight crews upon controller initiative; the controller decides to delegate if appropriate and helpful. The delegation is limited since the controller can only delegate 'low-level' tasks (monitoring and implementation) as opposed to 'high-level' tasks (conflict detection and resolution). The delegation is flexible since the controller has the ability to select for each situation the level of task to be delegated from monitoring up to implementation. The delegation takes advantage of emerging technologies such as Cockpit Display of Traffic Information (CDTI) and Airborne Separation Assistance System (ASAS). This project focuses on near-term applications being considered by certain ATC organizations for both en-route airspace and terminal areas.

In order to define a new task distribution between controllers and flight crews, two key constraints were identified and adopted. The first one is related to human aspects and can be summarized as 'minimize change in current roles and working methods of controllers and flight crews'. The second one is related to technology and can be expressed as 'keep it as simple as possible'.

At the time of this evaluation study, the project had just completed a third cycle of simulations. A prototype set of delegation procedures existed, which was used as the basis for analysis by HAZOP and TRACEr-lite.

Time-Based Separation: The Time-Based Separation (TBS) project aims to define and investigate the relevance of a new concept of operation applied to the arrival phase of flights. This concept involves replacing actual distance-based separations with time intervals. More specifically, the project will investigate the possibilities of preventing loss of runway capacity under strong wind conditions while maintaining required levels of safety performance. The project aims to do the following:

- assess a new concept of separation based on time interval as opposed to Radar or ICAO wake vortex separation criteria;
- investigate the use of lateral separations of less than 3nm (or less than 2.5nm if this is in use);
- explore possibilities of compensation of wind effects by aircraft speed adjustment and required ATC techniques.

This project was in a very early stage of background research and concept development, with no outline procedures defined, at the time of the study.

CORA 2 The Conflict Resolution Assistant (CORA) tool provides computer-based support for air traffic controllers in the detection, identification, prioritization and resolution of predicted conflicts in the en-route flight phase. Conflict identification and resolution is a core ATC task today, carried out by Planning Controllers (longer-range) and Tactical Controllers (shorter-range) by scanning the radar display and paper or electronic strips. Without change, increased traffic will naturally bring more conflicts, leading to higher workload for controllers and more complex conflict resolution problems. CORA aims to improve planning and anticipation processes through earlier conflict notification and resolution decisions with the introduction of new computer-based ATM tools, and the associated evolution of ATC procedures, roles, tasks and working methods. This improvement should also help to smooth peaks in controller activity, redistribute workload between Planner and Tactical Controllers, and improve the level and quality of service for airlines by minimizing deviations from airline optimal trajectories via earlier and more strategic resolutions.

The 'CORA 1' system will identify conflicts for controllers and support the planning and decision-making process by helping to test the impact of tactical clearances on the traffic situation. 'CORA 2' will provide a set of ranked, conflict-free resolution advisories for the controller, who can directly select and implement one of the suggested advisories, or employ a different, self-generated resolution. The CORA 2 project is defining and developing operational requirements and prototype enhanced concepts for conflict resolution. Its equivalent in the US is the PARR (Problem Analysis, Resolution and Ranking System) system.

Study Method

Following initial scoping interviews and high-level task definition for each project, the HAZOP and TRACEr analyses were conducted separately and led by different analysts. The time permitted for each analysis was roughly equivalent – about 10 person-days for each of the three projects. In order to maintain objectivity, the results of each analysis were only pooled when each analysis was complete. The general approach is described below.

HAZOP Approach The HAZOP studies for the three test projects proceeded according to the method described above. For each project, the HAZOP considered one or two high-level tasks taken from a task analysis (Table 15.6). The HAZOP considered both controller- and pilot-related errors, since all were judged relevant and useful to consider. Due to restrictions in the project team's availability, each HAZOP was conducted over one day, and was attended by three project team members (e.g. operational specialists, designers, Human Factors specialists). The sessions were facilitated and recorded by independent safety engineers or Human Factors specialists.

Table 15.6   Tasks represented in the Hierarchical Task Analysis, and indication of tasks further analyzed by HAZOP (H) and TRACEr-lite (T)

Co-space HTA Top Level Tasks	TBS HTA Top Level Tasks	CORA 2 HTA Top Level Tasks
1. Conduct task of Extended TMA controller using Co-space	**1. Control terminal and approach air traffic using TBS**	**1. Resolve conflict situation using CORA 2**
Plan: Do 1.1 at start of shift. Do 1.2, 1.3 and 1.4 as appropriate. For delegation aircraft do 1.5, 1.6 and 1.7 in order. Do 1.8 near sector boundary. Do 1.9 at end of shift.	*Plan: Do 1.1 at start of shift, then do 1.2. Do 1.3 to 1.9 as appropriate depending on controller position. Do 1.10 at end of shift.*	*Plan: Do 1.1 to 1.5 in order. Do 1.6 throughout as appropriate.*
1.1 Take over from off-going controller (T)	1.1 Take over from off-going controller (T)	1.1 Detect conflict situation (T)
1.2 Receive aircraft (T)	1.2 Receive aircraft (T)	1.2 Prioritize conflict situation(s) (T)
1.3 Maintain traffic separation within sector (T)	1.3 Maintain traffic separation within sector (T)	1.3 Analyse focussed situation (TH)
1.4 Form sequence plan / Follow sequence formed by AMAN (T)	1.4 Hold aircraft (T)	1.4 Act on focussed situation (T)
1.5 Conduct identification Phase (TH)	1.5 Sequence aircraft / Follow AMAN sequence (TH)	1.5 Check resolution progress (T)
1.6 Issue delegation instruction (TH)	1.6 Turn aircraft onto base leg (T)	1.6 Monitor situation (T)
1.7 End delegation (T)	1.7 Turn aircraft onto intercept ILS (T)	
1.8 Transfer to next sector (T)	1.8 Establish aircraft on ILS (T)	
1.9 Handover control to relief controller (T)	1.9 Transfer to next sector / tower / controller (T)	
	1.10 Handover control to relief controller (T)	

TRACEr-lite Approach The TRACEr-lite analysis began with the development of a Hierarchical Task Analysis (HTA) for each project. Following initial consultations, various documents were reviewed, including procedures, operational concept documents, human-machine interaction protocols and diagrams, operational scenario documents, and simulation briefing documents. For each project, the task steps were represented in detail in an HTA, to the 'keystroke' level where necessary. The draft HTAs were presented to each project team member, who helped to shape and revise the HTAs until agreed versions were formed.

Comparison of HAZOP and TRACEr-lite Findings

The comparison presented some difficulties, which need to be borne in mind:

- TRACEr-lite analyzed more tasks than HAZOP. For the three projects, TRACEr considered between six and ten of the tasks represented in the HTA (see Table 15.4), while HAZOP considered one or two tasks for each project. This was because HAZOP is more time- and resource-intensive, and only a one-day meeting was possible for each HAZOP.
- HAZOP considered both pilot and controller errors, as well as a limited number of information/equipment problems relating to the HMI for example. TRACEr-lite considered only controller errors. This was because the HTAs concentrated only on controller tasks.
- The HAZOP and TRACEr-lite approaches studied errors at different levels. TRACEr-lite analyzed errors at a more detailed level than HAZOP.

Since HAZOP analyzed fewer tasks, it is only possible to compare HAZOP and TRACEr-lite findings for those tasks analyzed by HAZOP. A comparison was performed for each study of errors and issues identified by HAZOP and TRACEr-lite (jointly and separately). Table 15.7 presents an extract of the errors and issues identified by HAZOP and TRACEr for a section of the Co-space HTA.

In order to facilitate the discussion and comparison of HAZOP and TRACEr-lite as applied to EUROCONTROL projects, a range of criteria for the evaluation of HEA methods is used, as described in the introduction.

Comprehensiveness HAZOP and TRACEr-lite performed differently on the issue of comprehensiveness. Both HAZOP and TRACEr-lite identified critical errors that could occur for all projects. HAZOP identified both controller and pilot errors when considering each task, and also examined potential information problems. However, in the time available, HAZOP could only examine two tasks for Co-space, and one task for TBS and CORA 2. TRACEr-lite examined the whole range of controller tasks for each project, and provided a highly comprehensive and detailed 'register' of potential errors. However, TRACEr-lite examined only controller errors and not pilot errors (because controller tasks were the focus of the HTA).

Table 15.7 **Errors and issues identified by HAZOP and TRACEr-lite for Co-Space Task 1.5 'Conduct Identification Phase'**

Error/Issue	In-scope HAZOP?	In-scope TRACEr-lite?
Issues identified by both TRACEr-lite and HAZOP		
Controller instructs pilot to select wrong target	Yes	Yes
Controller goes straight to delegation without confirmation from pilot (fails to detect / query missing readback or target identification)	Yes	Yes
Controller goes straight to instruction of delegation omitting to identify target (fails to instruct pilot to select target)	Yes	Yes
Controller issues correct target selection instruction to wrong aircraft	Yes	Yes
Controller gives wrong target to correct aircraft	Yes	Yes
Etc.		
Issues identified by HAZOP only		
Controller may give other instruction along with delegation instruction	Yes	Yes
Pilot does not confirm he has heard the target *	Yes	No
Pilot does not read back the target reference *	Yes	No
Pilot selects wrong target *	Yes	No
Pilot identifies correct target and goes straight to next action (anticipates controller's instruction) **	Yes	No
Etc.		
Issues identified by TRACEr-lite only		
Controller fails to detect / query erroneous readback	Yes	Yes
Controller fails to select aircraft on radar	Yes	Yes
Controller selects unintended aircraft (not part of delegation) when selecting aircraft on radar	Yes	Yes
Controller fails to detect / query failure to position	Yes	Yes
Controller fails to query spurious position	Yes	Yes
Etc.		

* Failure(s) of associated ATCO hearback identified

** Failure(s) of associated ATCO monitoring identified

TRACEr-lite tended to analyze errors in a more detailed and systematic fashion than HAZOP. This is largely because TRACEr-lite uses a detailed HTA, while HAZOP is not normally conducted at such a fine level of task modeling, and instead used the high-level tasks in Table 15.6 along with a brief description of what was involved. However, one of the potential dangers of detailed TRACEr-lite analysis is getting 'lost in detail' and failing to identify more fundamental issues. While this did not seem to occur in the current analysis, it is a potential problem to be aware of.

The analysis showed that TRACEr-lite identified approximately 91 per cent of the errors/issues identified by HAZOP that were within the scope of the TRACEr analysis (i.e. controller errors). For issues and errors identified by HAZOP that were outside the scope of the TRACEr-lite analysis (such as pilot errors and general performance conditions), TRACEr-lite was able to predict related errors, such as failures in ATCO responses to pilot errors for approximately 36 per cent of the issues. HAZOP identified approximately 42 per cent of the errors identified by TRACEr-lite for Co-space, TBS and CORA 2. However, as previously stated, HAZOP identified other issues such as pilot errors and controller performance conditions that were not analyzed by TRACEr-lite.

HAZOP and TRACEr-lite identified a similar range of consequences for the three studies, though the TRACEr-lite consequences tended to be limited to the more immediate, short-term consequences while HAZOP sometimes focused only on the 'bottom line' consequences. HAZOP, however, identified significantly more safeguards than TRACEr-lite, reflecting the experience of the HAZOP team members. This may suggest that TRACEr can produce more 'pessimistic' analyses than HAZOP.

The Recovery Success Likelihood (RSL) rating used by TRACEr-lite was useful in filtering the errors predicted, and, in the case of the Co-space analysis, the ratings were moderated by members of the Co-space team to help ensure that they were realistic. Only around 12 per cent of the total number of RSL ratings for all errors predicted was modified in this review, thus suggesting that the initial RSL ratings were, on the whole, realistic. The RSL concept may be useful to consider incorporating into a group-based technique such as HAZOP. It would also have been useful to rate error likelihood within the TRACEr-lite analysis using expert judgment (see Kirwan, 1994).

Lifecycle applicability Both HAZOP and TRACEr-lite demonstrated that they could be used throughout the formative and latter phases of system design lifecycle. TBS was in an early conceptual stage, while CORA 2 was in the mid-design stages, and Co-space was in a later design stage. However, TRACEr-lite's need for a more detailed task analysis would prompt the conclusion that HAZOP is more suited to projects in the conceptual and pre-design stages of development. A useful approach may be to conduct a preliminary HAZOP initially, to focus the analysis, identify the primary sub-tasks of interest, and identify the fundamental errors, and then follow this up with a detailed TRACEr-lite analysis to help ensure that errors are captured at a detailed and comprehensive level.

Theoretical Validity TRACEr-lite is based on a model of human performance, with a theoretically plausible internal structure (Wickens, 1992). While HAZOP does not have this foundation, the guidewords are valid human performance outcomes.

Contextual Validity HAZOP and TRACEr-lite account for context in different ways. HAZOP primarily uses the expertise of the HAZOP team. This is an established method of ensuring that such analyses are contextually relevant. TRACEr-lite uses project personnel and documentation to construct and review the Context Statement and Task Analysis. The project personnel would also then review the TRACEr-lite analysis. This is an important post-analysis step, which would involve face-to-face discussions between the TRACEr analyst and project personnel, perhaps also considering error likelihood and severity, but was not possible during this study due to time and resource constraints[26]. Some of the TRACEr-lite-predicted errors may, therefore, appear somewhat 'naïve'. However, equally, there are positive arguments in favor of this. Project team members who are very 'close' to the project' may consider certain errors 'incredible' in a HAZOP analysis, and choose not to propose or record them. From a TRACEr-lite point of view, on the other hand, such errors would be logical possibilities, and so would need to be considered (and possibly rejected with suitable justification) by the project team. Overall during this study, considering the errors and issues predicted, consequences, safeguards and recommendations, HAZOP outperformed TRACEr-lite on this criterion.

Flexibility Both HAZOP and TRACEr-lite are flexible in that they allow different levels of detail in the analysis. HAZOP has the advantage that early in the concept development/selection process; a preliminary HAZOP can be performed, using the creative brainstorming and knowledge of the group. The full HAZOP technique can then be performed later. TRACEr-lite, however, requires a more developed task analysis, since the individual analyst is not qualified to perform such a 'brainstorming' approach. TRACEr-lite does, however, provide flexibility in the use of Internal Error Modes and Mechanisms – the use of Mechanisms in the analysis can be omitted or used only for critical errors. TRACEr-lite can also be used in a small group-based format, using the Error Domains Perception, Memory, Decision, and Action as prompts. Indeed, this method has previously been incorporated into the Human HAZOP method to better account for cognitive aspects of errors in other HAZOP studies.

Usefulness Both HAZOP and TRACEr-lite helped to produce error reduction or mitigation measures. The HAZOP analysis produced more traditional recommendations while TRACEr-lite tended to produce 'performance requirements', which could be fulfilled in a number of ways. The HAZOP method, however, was clearly much more productive in this respect. This productivity can

---

[26] The TRACEr-lite synthesis was reviewed by the Co-Space project team.

be traced to the contribution of the project teams. The project team could, in future, participate similarly in generating recommendations based on TRACEr-lite findings. Indeed, this was done by in a study reported by Gordon et al. (2004).

The TBS project team found the HAZOP a very useful exercise even at the very early stages of the project, and felt that the HTA helped to identify tasks that would be affected by TBS, and where tasks still needed to be specified or considered.

Resource Efficiency (Training) This criterion has not been formally tested, but training to become a HAZOP facilitator would normally involve three to four days training, and prior experience as a recorder. No training is required to act as a participant but awareness or familiarization training is very useful to speed up the process. HAZOP's training requirement is largely associated with the ability to facilitate a group-based process. The role of the HAZOP leader or TRACEr-lite analyst may not suit every individual, and the HAZOP leader role requires different skills to the TRACEr-lite analyst role. Training in the use of the TRACEr-lite technique would normally involve 1-2 days, plus a further 1-2 days to learn how to use Hierarchical Task Analysis. Hence, the training demand for each technique is quite similar.

Resource Efficiency (Usage) Overall, TRACEr-lite was the more resource efficient technique, analyzing more tasks in a similar amount of time. When HAZOP and TRACEr-lite are compared for an equivalent analysis of all controller high-level tasks in the CORA HTA, it is estimated that HAZOP would require a total of 26.5 person days, while TRACEr-lite would require 16 person days – about 60 per cent of the HAZOP time. These figures cover background familiarization, task modeling/analysis, analysis, data formatting/tidying, reporting, as well as review and revision at each stage.

Usability Both HAZOP and TRACEr-lite are usable techniques. HAZOP has stood the test of time, and does not demand complex analysis from participants. However, the process can prove frustrating, particularly where several unbroken days of analysis are performed. In this respect, it is wise to break up sessions that occur over several days (though care should be taken to ensure that a 'flow' is maintained in the study). TRACEr-lite, similarly, can be frustrating to the analyst due to the repetitive nature of analysis.

Auditability Again, both HAZOP and TRACEr-lite provided a fully auditable process, with worksheets demonstrating the reasoning behind the analysis. HAZOP, additionally, visually projected each worksheet during the session so that all of the participants could verify the findings 'on-line'.

Discussion of HEA in ATM

HAZOP and TRACEr-lite

TRACEr and HAZOP both led to human error insights in the three projects. However, some distinctions were drawn after this study. First, TRACEr-lite can not be applied as early as HAZOP – this is often due to the lack of clear ideas on controller working practices with the proposed tools or concepts. Therefore, for very early concept studies, HAZOP appears more useful. HAZOP's reliance on the presence of good expertise was also noticed, and if for logistical or technical reasons this is not available, then the results will suffer. TRACEr, being less reliant on such factors, had a higher degree of 'stability' than HAZOP.

The feedback from the 'client' projects was positive, and there were requests for further studies using these approaches, and broadening to other projects. Overall, both Human HAZOP and TRACEr-lite proved useful methods to support designing for safety. It may be most useful to use a hybrid approach by performing a preliminary Human HAZOP to identify the core tasks and critical, high-level errors followed by a detailed TRACEr-lite analysis. This preliminary Human HAZOP could also identify the relevant safeguards and consequences for use in the detailed TRACEr analysis, to improve the contextual relevance of the technique. Additionally, HAZOP can be modified to assess Human-Machine Interfaces (e.g. Kennedy et al., 2000) when interface design mock-ups are available.

These two approaches are variants of already successful approaches in other industries. Therefore this chapter shows the adaptability and usefulness of these approaches in another new field, that of Air Traffic Management. The emphasis has been on enriching design, making it safer, rather than independent assessment of design, and it is intended to have these approaches seen as the province of 'safe designers', as well as safety assessors.

Future ATM HEA Needs

An insight can be inferred from the earlier summarized review of techniques – if HAZOP and its TRACEr-equivalent (e.g. SHERPA) were good enough in other industries, there would not have been a proliferation of other types of techniques in those industries. This is true, and probably also applies to ATM – HAZOP and TRACEr may well not be enough to deal with all types of error forms. At present, ATM may be thankfully free from what other industries would call 'violations', but 'errors of commission' have arisen, most notably in the form of the Lake Constance mid-air collision in Germany. This tragic accident on 1st July 2002 involved, from a human error perspective, an error of commission interacting with a safety net (TCAS – Traffic Alert and Collision Avoidance System). As systems become more complex, errors of commission become 'afforded' by the complex system nature, and this applies equally to cockpits (e.g. mode errors) as it does to ATM. There may therefore be the need to develop a more specialized error of commission approach for ATM.

Similarly, there may be a need for more cognitive simulations. Whilst these have not proven so useful yet in other industries, they would be very useful for studying early design concepts which by definition may have no real controller 'expertise' to draw from. Cognitive simulations, or 'fast-time' consideration of error potential could prove useful in developing future automation support for controllers (see also chapter 14 in this book).

A further need in this area relates not to the error identification method, but the medium to which such techniques should be applied. Typically this is a task analysis, but more generally it may be seen as an operational scenario or environment. ATM is very dynamic, and whereas Hierarchical Task Analysis has fared very well as a precursor to HEA in other industries, it may be that other methods are required that capture the full richness of ATM behavior and its cognitive elements.

Another less obvious aspect of consideration relates to the interactions between sub-systems and between concepts. ATM is growing and changing rapidly, and rather than adding one concept onto an existing and established one, several challenging concepts are being developed in parallel, e.g. datalink, free flight, conflict detection and resolution, etc. It is sometimes difficult even to be clear about the intended operational environment when carrying out the safety assessment and design assurance of future systems. There is continually a danger of failing to see an unintended but threatening inter-connection between functionally diverse but nonetheless inter-connected or inter-connectable systems or system components. Perrow (1984), in his landmark book 'Normal Accidents', gives many examples of the tragic consequences of failing to see such inter-connections beforehand. This inter-connectivity is there because the system boundaries are hard to define in ATM – everything is connected and part of the ATM system. This is different from, say, a nuclear power plant, which is a bounded system. Therefore, a new approach, incorporating human error considerations (because often humans themselves enable the links between systems and system elements), needs to be developed to ensure we do not carry out 'compartmentalized' or 'blinkered' assessments. The approach being discussed both in the US and in Europe is HAZOP-based, and has the working title of Cross-Boundary HAZOP. This is an area for urgent development of approaches.

Conclusions

In conclusion, ATM has seen fit to borrow certain approaches from other industries and adapt them for ATM use, and this appears to be working. It remains for these approaches, and parallel ones as no doubt will be developed, to increase in usage, and to become a formalized part of the design and safety assurance process. If this is done, there will be fewer nasty surprises, and the controllers themselves will be better prepared and equipped to do what they do best, controlling the traffic safely.

Acknowledgements

The authors gratefully acknowledge the contribution to the study of the following people: Roisin Johnson (DNV Co-space HAZOP facilitator), Emma Noble and Joanne Ramsay (DNV HAZOP recorders); Isabelle Grimaud, Laurence Rognin and Karim Zeghal (EUROCONTROL Co-space team); Peter Crick and Antoine Vidal (EUROCONTROL TBS team); Mary Flynn, Sophie Dusire and Nicolas Dardenne (EUROCONTROL CORA 2 team). The views expressed in this paper are those of the authors; they are not necessarily those of EUROCONTROL, The University of New South Wales, National Air Traffic Services, or DNV.

References

Baber, C. and Stanton, N. (1991), 'Task analysis for error identification: towards a methodology for identifying human error', in E.J. Lovesey (ed.). *Contemporary Ergonomics*, pp. 67-71, Taylor and Francis, London.

Baber, C. and Stanton, N.A. (1994), 'Task analysis for error identification: a methodology for designing error tolerant consumer products', *Ergonomics*, Vol. 37, pp. 1923-1941.

Billings, C.E. (1988), 'Toward human centred automation', in S.D. Norman and H.W. Orlady (eds.), *Flight Deck Automation: Promises and Realities,* pp. 167-190, NASA-Ames Research Center, Moffet Field, CA.

Embrey, D.E. (1986), 'SHERPA: A systematic human error reduction and prediction approach', paper presented at the *International Topical Meeting on Advances in Human Factors in Nuclear Power Systems*, Knoxville, Tennessee.

Embrey, D.E., Kontogiannis, T., and Green, M. (1994), *Preventing human error in process safety.* Centre for Chemical Process Safety (CCPS), American Institute of Chemical Engineers, CCPS, New York.

Evans, A., Slamen A.M., and Shorrock S.T. (1999), 'Use of human factors guidelines and human error identification in the design lifecycle of NATS future systems', *Proceedings of the Eurocontrol/FAA Interchange Meeting*, Toulouse, France.

Gordon, R., Shorrock, S.T., Pozzi, S., and Bosciero, A. (2004), 'Using human error analysis to help focus safety analysis in ATM simulations: ASAS separation', *Proceedings of the Human Factors and Ergonomics Society of Australia Annual Conference*, Cairns, Australia.

Hollnagel, E. (1998), *Cognitive Reliability and Error Analysis Method: CREAM*, Elsevier, Oxford, UK.

Kennedy, R., Jones, H., Shorrock, S. and Kirwan, B. (2000), 'A HAZOP analysis of a future ATM system', in P.T. McCabe, M.A. Hanson and S.A. Robertson (eds.), *Contemporary Ergonomics 2000*, Taylor and Francis, London.

Kirwan, B. (1992), 'Human error identification in human reliability assessment. Part 2: Detailed comparison of techniques', *Applied Ergonomics*, Vol. 23(6), pp. 371-381.

Kirwan, B. (1994), *A Practical Guide to Human Reliability Assessment*, Taylor and Francis, London.

Kirwan, B. (1998a), 'Human error identification techniques for risk assessment of high risk systems – Part 1: Review and evaluation of techniques', *Applied Ergonomics*, Vol. 29(3), pp. 157-177.

Kirwan, B. (1998b), 'Human error identification techniques for risk assessment of high risk systems – Part 2: Towards a framework approach', *Applied Ergonomics*, Vol. 29(5), pp. 299-318.

Kletz, T. (1974), *HAZOP and HAZAN – Notes on the Identification and Assessment of Hazards*, Institute of Chemical Engineers, Rugby.

Kletz, T.A. (1988a), *HAZOP and HAZAN: Identifying and Assessing Process Industry Hazards*, Institution of Chemical Engineers, Bradford.

Kletz, T.A. (1988b), *What Went Wrong? Case Histories of Process Plant Disasters*, Gulf Publishing, US.

Rasmussen, J. (1981), *Classification system for reporting events involving human malfunction*. Riso-M-2240, DK-4000, Risø National Laboratories, Denmark.

Reason, J.T. (1990), *Human Error*, Cambridge University Press, Cambridge, UK.

Sarter, N. and Woods, D.D. (1995), *Strong, silent, and out-of-the-loop: properties of advanced (cockpit) automation and their impact on human-automation interaction*, Technical Report CSEL 9-TR-01, Cognitive Systems Engineering Laboratory, Ohio State University, Columbus, OH.

Shepherd, A. (2001), *Hierarchical Task Analysis*, Taylor and Francis, London.

Shorrock, S.T. (2002a), 'The two-fold path to human error analysis: TRACEr-lite retrospection and prediction', *Safety Systems, Newsletter of the Safety-Critical Systems Club*, Vol. 11 No. 3.

Shorrock, S.T. (2002b), 'Error classification for safety management: finding the right approach', *Proceedings of the Workshop on the Investigation and Reporting of Incidents and Accidents*, University of Glasgow, UK.

Shorrock, S.T. and Kirwan, B. (1999), 'TRACEr: a technique for the retrospective analysis of cognitive errors in ATM', in D. Harris (ed.) *Engineering Psychology and Cognitive Ergonomics: Volume Three – Transportation Systems, Medical Ergonomics and Training*, pp. 163-171, Ashgate, Aldershot, UK.

Shorrock, S.T. and Kirwan, B. (2002), 'The development and application of a human error identification tool for air traffic control', *Applied Ergonomics*, Vol. 33, pp. 319-336.

Shorrock, S.T., Kirwan, B., MacKendrick, H. and Kennedy, R. (2001), 'Assessing human error in Air Traffic Management systems design: methodological issues', *Le Travail Humain*, Vol. 64(3), pp. 269-289.

Springall, L. (2001), 'Presentation on use of HAZOP in NATS UK', *FAA-Eurocontrol Workshop on Safety*, Bournemouth, Dorset, UK.

Swain, A.D. and Guttmann, H.E. (1983), *A handbook of human reliability analysis with emphasis on nuclear power plant applications*, USNRC-NUREG/CR-1278, Washington, DC.

Whalley, S.P. (1988), 'Minimizing the cause of human error', in G.P. Libberton (ed.) *10th Advances in Reliability Technology Symposium*, Elsevier, London.

Whalley, S.P. and Kirwan, B. (1989), 'An evaluation of five human error identification techniques', *Proceedings of the 6th International Loss Prevention Symposium*, pp. 31/1-31/18, Oslo, Norway.

Wickens, C.D. (1992), *Engineering Psychology and Human Performance*, 2nd Edition, Harper Collins, New York.

Wiener, E.L. (1985), 'Cockpit automation: in need of a philosophy', *Proceedings of the 1985 Behavioural Engineering Conference*, pp. 369-375, Warrendale, PA.

Wiener, E.L. (1988), 'Cockpit automation', in E.L. Wiener and D.C. Nagel (eds.), *Human Factors in Aviation*, Academic Press, San Diego.

Wiener, E.L. and Curry, R.E. (1980), 'Flight-deck automation: promises and problems', *Ergonomics*, Vol. 23, pp. 995-1011.

PART V
HUMAN FACTORS
INTEGRATION PROGRAMS

This section concerns Human Factors programs, i.e., integrated approaches to developing and managing a Human Factors support function in an organization. It is likely that this section will prove useful to anyone running such a program, to non-Human Factors (HF) management personnel who wish to understand how HF programs or sections can work, and to those working in such programs or sections who wish to see the 'larger picture', or a 'top down' perspective. The 'programmatic' approach and perspective is necessary if HF is to achieve sustained, widespread and consistent results. Otherwise, human-system integration will be only partially achieved, and overall system performance may suffer due to inadequacies in other parts of the system. A programmatic approach therefore shows a committed approach to Human Factors by an organization. The question is then one of how to deliver such an integrated and encompassing approach to Human Factors across many elements of the ATM system. This question is answered in the five chapters in this section, showing different approaches to evolving and maintaining an effective Human Factors competence for an ATM organization.

The first chapter by Hewitt et al. gives a comprehensive and compelling account of achieving integration of Human Factors into a sophisticated development and acquisition program of activities. A major flavor of this chapter is the commitment to high Human Factors standards, and assurance that sufficient trained Human Factors expertise exists to carry out the work. The chapter is also useful in its honesty about what does not work, and where some major pitfalls lie.

The chapter by Allendoerfer et al. is not a Human Factors program as such, but is a very large and important study, spanning a number of years and involving considerable effort. It is therefore included to show how a major Human Factors challenge can be tackled in a comprehensive and programmatic way. It is also included because many Human Factors units or programs often arise in the aftermath of successful involvements in a large and high profile study. Human Factors units or sections can therefore sometimes be developed 'on the back of' such efforts. As for the chapter content itself, it is concerned with the detailed HMI development of the FAA Standard Terminal Automation Replacement System (STARS) program, and has good illustrations of the display analysis and design. It also has sound and pragmatic advice in terms of 'lessons learned' from this extensive project.

Josefssons's chapter is a very frank account of the evolutionary path of Human Factors in the Swedish ANSP LFV, including a useful timeline of the major events in recent years. This chapter differs from the others in an important respect, in that much reliance was placed on an outside agency (a Swedish University research group) to develop and apply Human Factors for LFV, particularly in the area of safety culture. LFV also focused on certain key Human Factors areas recognizing that to tackle all would not be feasible within their resources. This chapter also highlights (in common with the others) the importance of communication, of openness, and of dissemination of information to controllers and management.

The last two chapters by Kirwan and by Burrett et al. are connected, in that the first details the initial three-year program of Human Factors at NATS, and the second then describes the following three years. It is therefore possible to see how the Human Factors Unit was initially set up, and then how it became firmly embedded in the company business infrastructure. In fact, these two chapters show two steps in the evolution of a Human Factors presence in an organization. The first effectively winning the trust of the organization and showing that Human Factors can deliver, the second showing how to build on that trust and make Human Factors considerations in ATM a matter of course. The net result is an effective and mature Human Factors Management System for an ATM organization.

Successfully addressing Human Factors in an organizational context presents a particular challenge since it is very often dependent on the cooperation of actors that might be unaware or skeptical towards the benefits of Human Factors. Hence these activities have to be carefully planned, often in the context of limited resources. Experience has shown that progress is much easier to achieve through the involvement of all parties concerned and this involvement depends on the often very subjective sentiment that Human Factors 'can deliver' for them. Addressing the most relevant, but sometimes also the most visible, areas; speaking the organization's language rather than 'HF'; and making Human Factors achievements tangible and visible are critical aspects for the success of Human Factors programs. Trust thus achieved can be built upon to the point where Human Factors becomes 'invisible' since it permeates the organizational culture.

Chapter 16

The Management of Human Factors Programs in ATM Applications: A Case Study

Glen Hewitt, Paul Krois and Dino Piccione

Introduction

Ample material has been written about the definition and execution of Human Factors activities in system acquisition programs and the steps supporting system engineering. However, there is insufficient information about the macro view of how to address a broad Human Factors program across many diverse organizational environments and in applying Human Factors to the air traffic management (ATM) arena. This chapter uses the experience of developing a Human Factors program at the Federal Aviation Administration as a case study of the 'macro' view of applying Human Factors to research and engineering in system acquisitions. This FAA experience demonstrates that three essential vectors are required to achieve a successful Human Factors program: a) relying on Human Factors research and requirements that are relevant and robust, b) developing and managing the Human Factors infrastructure in a way that is flexible and timely, and c) identifying, assessing, and mitigating Human Factors risk in a way that is interactive and iterative.

Relying on Human Factors Research and Requirements

No properly developed Human Factors program can thrive without establishing a strong foundation of Human Factors research and well-defined requirements, especially during the early phases of acquisition programs involving mission analysis, functional analysis, and requirements determination activities. A human-centered approach to mission analysis provides Human Factors research with the basis for establishing concepts of operation, concepts of use, and an ATM architecture that recognizes and incorporates human performance limitations and capabilities. A fundamental challenge for civil aviation research is aligning the forces associated with mission analysis to address the vision of architecture evolution with the energies of researchers pushing technology innovations.

This challenge was articulated in the 1995 National Plan for Civil Aviation Human Factors (FAA, 1995a) that defined human-centered automation as keeping the operator in the loop to facilitate situation awareness of automated system performance. Such an approach to system performance involves balancing operator workload and resolving issues related to the degradation of basic skills should the automation fail. Research on new system capabilities needs to be assessed to ensure a proper allocation of function between the human operator and system functionality to achieve consistency with current and proscribed roles and responsibilities for the human operator as part of the integration of future concepts of use into baseline ATM concepts of operation. This activity goes beyond the conventional allocation of function with its perfunctory assessment of human and system performance strengths and weaknesses, and is necessary to understand the impact of those capabilities on current and evolving responsibilities of the human operator. The dependence upon the early precursors to technology implementation is supported by a comprehensive and robust research base, a flexible and detailed requirements baseline, and a resilient approach to evaluating the technology readiness of research and requirements to move to the next development step.

The Research Base

 Human Factors challenges to achieving ATM objectives at the FAA have been articulated as a 16-point framework (Grossberg, 1998). This framework of Human Factors and human performance issues, which synthesizes considerations from the ATS Concept of Operations for 2005, the NAS Architecture Version 4, and strategies for operation concept validation, has served as a model upon which to build an ATM research program.

Grossberg's approach recognized that properly addressing Human Factors issues early and recursively throughout the development of new operational concepts promotes successful implementation. This approach helps mitigate the potential for degraded performance and supports incorporating tolerance for and recovery from human error in operation of the system. Achieving this goal requires open and insightful collaborations between researchers, users, and system designers to ensure consistent application of Human Factors guidelines, information, principles, and best practices throughout the National Airspace System (NAS).

This 16-point approach ensures that Human Factors considerations are addressed across the different operational domains and services comprising the NAS. This includes flight planning, airport surface, departures and arrivals, En-route or cruise operations, oceanic operations, NAS traffic flow and infrastructure management, and management of personnel and facilities. The 16-points comprising this research and requirements analysis framework are addressed below.

Phased Technology Implementation The phased implementation of enhanced capabilities and advanced technologies poses changes in the roles and responsibilities of human operators. Phased implementation requires assessing how much change is needed to produce tangible effects, but this change should not be so large as to induce obstructions from those who will be instrumental in implementing change.

Research implications include the need for performance baselines tailored to the different operational domains of the NAS; measuring controller and pilot task sequences and timing; and assessing workload, task complexity, sector team communications, human error, and system-level performance indicators such as traffic delays. Research also needs to assess potential relationships between system-level metrics and human performance metrics.

NAS Performance Management Changes to the NAS assessed using fast-time models are typically expressed in macro terms using categories of metrics such as capacity, efficiency/flexibility, predictability, productivity, and safety. Such measures do not consider performance at the controller or sector team level because it is difficult to link measures of human performance to measures of traffic flows and NAS performance. In contrast, human-in-the-loop (HITL) simulation affords measurement at the level of the individual controller. Nevertheless, Human Factors implementation issues must be addressed at the macro level to avoid appearing at the micro level in a manner that constrains benefits intended with enhanced capabilities.

Redistributed Roles and Responsibilities The 2005 Concept of Operations provides for the delegation in separation responsibility from the controller to the flight deck. Implications for Human Factors research point to the need for HITL simulation to assess the effects from re-allocation of responsibilities and tasks among controllers, aircrews, and automation within each of the operational environments. This assessment should provide metrics for calibrating human performance limitations to achieving the intended benefits of enhanced capabilities and new technologies. The timing of communications between controllers and pilots across different traffic situations should also be examined to understand and mitigate the impacts on safety and traffic flow efficiency.

Employing a Human-Centered Implementation Approach The human-centered approach to phased implementation ensures that human capabilities and limitations of controllers and pilots remain a primary consideration in NAS development and implementation. Unfortunately there is no 'gold standard' by which to define a design centered on the human. While system requirements provide a framework for supporting human-system information requirements and interactivity, experience has shown differences among acquisition teams in how designs are held to be human-centered, including how performance constraints should be addressed during NAS implementation.

Distribute Timely and Consistent Information Across the NAS The evolution and increasing complexity of the NAS necessitates an increased exchange and sharing of an ever-widening range of information. New and enhanced sources are generating additional information for use in different applications in ATC, ATM, and maintenance. For example, as part of the FAA's Target System Description (TSD) for 2015, system-wide information management (SWIM) provides for currency of NAS information. The TSD envisions use of digital data link to continuously distribute this information to air and ground systems. Research should examine how to provide this information to the user in a manner that is structured, integrated, and efficiently formatted.

Dynamic Airspace Boundary Adjustments Dynamic changes in sector configurations and airspace boundaries provide for increasingly efficient traffic flows to offset constraints otherwise imposed by weather or congestion, and present new operational challenges to traffic flows, communications, training, and procedures. Implications for Human Factors research involve controller and pilot performance and workload. Transferring responsibility for airspace from one facility to the next poses impacts on controller knowledge and skill relative to certification for working that airspace. Researchers should develop guidance on how human performance limitations should be addressed through design, training, and staffing.

Seamless Communications Transitions to the field of automation systems along with dynamic restructuring of airspace pose increased demands on controller communication and coordination. The need for seamless inter- and intra-facility communications can be supported via conventional contact (i.e., face-to-face or telephonically) as well as via system-assisted coordination using electronic messaging.

Human Factors research can assess operational and functional requirements relative to concepts of use for system-assisted coordination and associated advanced procedures. A human-centered approach intends for communications that are seamless relative to changes in roles and behaviors of controllers, supervisors, and traffic management specialists.

Airspace Flexibility Development of a NAS-wide information system provides for the dynamic assignment of airspace assets between facilities to meet contingencies such as equipment and facility outages. Special consideration would be needed for conducting initial and recurrent training for controllers to confidently assume responsibility over airspace dynamically assigned to them, taking into consideration changes in traffic volume and complexity. Performance would need to be equally proficient to mitigate the potential for human error as well as operational errors.

Collaborative Decision-Making Collaborative decision-making (CDM) in air traffic management facilitates real-time handling of traffic flow problems that result from congestion and adverse weather conditions. Collaboration among airline operations centers and the FAA Air Traffic Control System Command Center (ATCSCC) occurs 24 hours per day, 7 days per week.

Human Factors research has contributed to improved CDM through development and national implementation of the Post-Operations Evaluation Tool (POET), which is addressed in another chapter of this book. Research continues to examine prototype decision aids and study baseline synchronous and asynchronous communications. Research should examine how changes in procedures, new and refined automation tools, and training might further increase efficiency of the NAS.

Fault-Tolerant Systems The monitor and control of NAS systems and equipment ensures a managed balance in the reliability and redundancy of ATC and ATM capabilities. This also includes the provision for procedural backups to minimize the time required to restore or replace failed functions.

Human Factors research should address how airway facilities (AF) specialists and technicians respond when particular capabilities fail and how partial failures are detected and diagnosed. Fielding of ATC decision aid sub-systems poses the potential for fractious monitoring of additional alerting and status information that may impact situation awareness, workload, and communications. Identification of best practices could feed back into recurrent training to leverage past successes such as in how failing conditions were saved.

Automation Aids Automation aids that augment controller decision-making intend to increase controller productivity while maintaining safety. Productivity can be increased as a function of expanded capabilities for handling more aircraft, and accomplished through provision of decision aids, dynamic changes in airspace boundaries and sector design, reduced vertical separation minima, improved air/ground communications, and integrated air/ground capabilities.

Regarding controller use of electronic flight data, research needs to continue to examine variations in usage in different operational environments for maintaining situation awareness and record-keeping. Automation programs should assess the usability of electronic flight data relative to controller workload and performance under the envelope of normal and failure operational conditions.

Timely Implementation of Procedures Procedures dictate how the concept of use for an enhanced capability will be followed. Consequently, changes to procedures should be defined and evaluated to ensure that the operational environment is consistent with the operational concept. Even after field implementation, procedures may be modified at facilities due to changing circumstances, with or without accompanying equipment function or capability changes.

Procedures need to be expansive in terms of going outside the boundaries of particular subsystems and domains in order to address the inter operability of collocated tools as well as air/ground interactions. As an example, data link procedures could be envisioned as transparent across En-route, oceanic, terminal, and tower domains to provide seamless communications between pilots and controllers.

Infrastructure Management Enterprise management of the NAS infrastructure relies on the monitor and control of extensive real-time NAS-wide information systems. AF develops and uses enterprise management tools for centralizing the integration of maintenance operations and planning for NAS infrastructure activities. In particular, AF works with air traffic managers and controllers as their customers who have a stake in setting priorities and scheduling software and hardware maintenance activities.

Research implications for infrastructure management underscore the importance of timely accurate information critical to assessing the relative contribution of systems or subsystems that have failed, are malfunctioning, or scheduled for servicing in terms of their impact to NAS functions. Continuously updated information will be needed so AF system specialists can track the progress of maintenance activities and share information with affected operational elements and customers.

Enhanced Weather Information Increasingly accurate weather data are being made available to enhance NAS safety and efficiency. These data include hazardous weather alerts for convective weather, wind shear, microburst, wake vortex, and areas of precipitation, icing, and low visibility. Weather information needs to be tailored to the role of the user, whether it is the air traffic controller, traffic flow coordinator, airline operations center specialist, or pilot. Requirements for the types of information and its use will consequently influence how it is filtered and displayed.

Operational Supervision Air traffic supervisors serve a critical role in providing the primary management presence in the operational area. They are responsible for such critical functions as the staffing of positions and training of developmental controllers. Research needs to examine better the interactivity of the supervisor position with enhanced capabilities, procedures, and airspace. Research should also examine supervisory best practices to characterize effective operational supervision and help sharpen the focus for controller-in-charge training.

Facility Management Facility managers need to be provided with appropriate tools with which to manage budgets, staff, and operational costs. These tools provide information to address short-term problems and more strategic management of resources. Research issues include the development of performance management databases and the training needed to support their use. In addition, planning the succession of facility supervisors and managers including development and career progression of regional and headquarters managers needs to be addressed.

Technology Readiness Paradigm

Experiences in the FAA Human Factors program demonstrate that in addition to identifying research requirements and conducting the required research properly, transitioning research results itself poses myriad challenges. The paradigm for managing research on new and emerging technologies during Concept Exploration to Concept Development phases and transitioning them to acquisition employs a model comprising Technology Readiness Levels (TRLs). Human Factors can make important contributions to the numerous technical and program (cost and schedule) issues associated with the transfer of research products to acquisition engineering.

The joint FAA and NASA Inter-agency ATM Integrated Product Team (IAIPT) uses the TRL model (FAA, 2002a), shown in Figure 16.1. Concept Exploration involves TRLs 1 through 3. During Concept Exploration, emerging and evolutionary concepts are described and assessed for their feasibility and potential benefit for improving NAS performance. Promising concepts mature rapidly, and early assessments may filter out concepts that do not meet the targeted operational improvement. A preliminary operational concept would be developed describing how operational users would use the capability to deliver intended benefits for airspace users. The nature and magnitude of expected benefits would be modeled. Technical and operational issues would be identified including their impacts on intended benefits.

The Concept Development Phase spans TRLs 4 to 6. During Concept Development the operational concept is revised which includes resolving major design problems such as integration and interoperability, mitigating key acquisition risks associated with technical and operational issues, and refining projections of intended benefits including costs for continued research, development, and implementation. To transition to TRL 6 for prototype development, there must be data-driven concurrence that the expected benefits warrant further expenditure for prototype development.

In TRL 6, the fully functional prototype is developed and demonstrated in a field setting, and acquisition documents are developed supporting transfer of the prototype capability to an acquisition office. This TRL ensures that important design features are assessed, potential development risks identified and mitigated, and integration and interoperability issues resolved. A thorough cost-benefit analysis is completed to validate intended benefits. To exit from TRL 6, a specification is developed to minimize risk to implementation. The pre-production prototype may continue in operational use for several years to continue to provide some level of benefits prior to deployment of the production units.

In Figure 16.1, the lines labeled A and B denote investment decision points. These decisions are made by a council of key agency executives. The council determines the level of investment for the implementing organizations, called Integrated Product Teams (IPTs), and their acquisition activities.

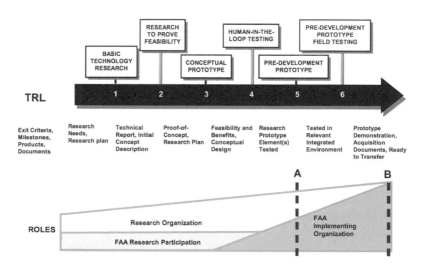

Figure 16.1   Technology Readiness Level (TRL) paradigm

Developing and Managing the Human Factors Infrastructure

Equal to the importance of a research and requirements foundation and mechanisms for assessing and mitigating risk is the role of developing and managing a mature Human Factors infrastructure. Developing and managing the Human Factors infrastructure at the FAA during the past ten years has entailed creating some of the fundamental building blocks of a good human factor program. These building blocks include: a) establishing a management plan to guide the agency in the development of Human Factors support, b) acquiring, organizing, and training the human intellectual capital and expertise required to achieve a viable Human Factors program, and c) creating and institutionalizing policy, processes, and practices within the agency or organization.

Establishing a Management Plan

The first step taken in developing and managing the Human Factors infrastructure entailed creating an Implementation Plan for Human Factors Research and

Engineering (FAA, 2002b). The basic concepts of this plan were identified in a 'National Plan for Aviation Human Factors' (FAA, 1995a) even before the most rudimentary elements were in place. The plan outlined the major objectives upon which several years of Human Factors program development were to be based. This plan established an outline for Human Factors technical and managerial support of acquisition systems/programs, domains, and other ATM acquisition activities and applications. The implementation plan included elements to identify specific annual tasks and milestones; measure, maintain, and document changes and progress in the program; achieve program technical management and community workforce development; and allocate the appropriate expenditure of available funds supporting these endeavors.

Management Plan Concepts Under many well-defined acquisition management policies such as in the FAA, integrated requirements teams (IRTs), investment analysis teams (IATs), and IPTs serve as empowered, cross-functional teams that have the responsibility for participating in the delivery of a product or service that meets the needs of their customer. The integration of Human Factors into these teams, and the application of Human Factors to acquisition programs help ensure that the system design is human-centered, meets program goals and objectives, reduces risk, lowers lifecycle costs, and achieves a higher probability of program success. Human Factors research and support to IRTs, IATs, and IPTs ensure the most effective use of human capabilities and minimize the effects of human limitations and errors on the overall performance of the system. The degree of Human Factors support required varies by program and considers the complexity of the system, the acquisition strategy, the phase of development, interaction and integration with other systems, and the level and type of human involvement.

In accordance with ATM acquisition goals, all systems and domains are to be adequately supported by Human Factors activities and professionals in a coordinated but distributed approach. The concept of coordinated management and decentralized execution of a Human Factors program consists of two primary elements: a) a centralized element of a small number of resources to help direct, manage, coordinate, support, and integrate disparate research and engineering acquisition with Human Factors activities across domains/environments such as shown in Figure 16.2, and b) a decentralized element collocated with the ATM domains and environments to coordinate and integrate Human Factors research and engineering activities within the domains/environments. The Human Factors 'decentralized' role is executed by HF coordinators (HFCs) collocated in appropriate quantities within the business lines and IRTs, IATs, and IPTs to ensure that adequate Human Factors support is rendered. In acquisitions, the HFC's major responsibilities include participating in the development of integrated requirements, supporting investment analysis of alternatives, assisting in request for (vendor) proposal preparation or market surveys, serving on source selection panels, preparing human performance test and evaluation criteria, preparing data collection and analysis plans, and participating in post-contract award activities (such as attending preliminary and critical design reviews). This decentralized but coordinated management concept accommodates considerations of critical mass, the exchange of discipline-unique information, sharing of lessons learned, efficient use of resources, and the relationship between research and applications.

Figure 16.2 FAA Human Factors organization

Management Plan Execution Strategies Once the organizational concept for providing Human Factors support is designed, achieving Human Factors integration and ATM acquisition objectives involves two primary strategies. The first strategy (Research) entails conducting Human Factors research to provide the knowledge base and foundation for the integration of Human Factors into the acquisition of FAA systems and applications. The second strategy (Acquisition Engineering) applies Human Factors policies, processes, and best practices through system engineering activities and reviews to ensure Human Factors issues are identified and mitigated in FAA acquisitions and applications. The first strategy is achieved through special projects and studies completed and accepted by the acquisition and air traffic management sponsors. The second strategy is accomplished by integrating Human Factors in and across systems ensuring that they apply and satisfy Human Factors principles and best practices.

The implementation of management plan performance objectives entails the coordinated conduct of Human Factors research and acquisition engineering and supporting activities and reviews. The plan requires a direct and continuous exchange of information among agency Human Factors specialists and other members of program offices. The approach outlined in the plan provides a means to accomplish the exchange of critical information for Human Factors planning and execution in system research and engineering.

Management Plan and Program Controls In order to monitor and control the Human Factors program adequately, progress and status reporting was established consisting of periodic and regular technical reviews of Human Factors support for acquisition programs and applications; performance plan reporting (including a quarterly rating) on the status of detailed Human Factors tasks/milestones during each fiscal year; and an annual Human Factors review of 100 per cent of systems (including a summary of the status of Human Factors issues in and across systems and applications). Feedback during these monitoring mechanisms provides the basis for an overall FAA Human Factors program review of the institutionalization of Human Factors using established measures.

Acquiring, Organizing, and Training the Required Human Intellectual Capital

One of the most critical elements of developing and managing the Human Factors infrastructure is to ensure that the appropriate level of Human Factors expertise is available. A plan covering staffing, training, and professional development was prepared to manage and document this important objective. The staffing, training, and professional development document is a component of the annually updated, five-year Human Factors management plan which provides goals, strategies, and tasks to meet Human Factors objectives. The staffing, training, and professional development plan is divided into three areas: Human Factors Staffing for the Acquisition Lifecycle; Human Factors Practitioner Qualifications and Functions; and Human Factors Professional Development and Training.

Human Factors Staffing The most critical action for improving or maintaining Human Factors support is to provide the necessary resources (expertise and funds) to conduct Human Factors activities. These resources are provided by hiring FAA employees or FAA support contractors, and by creating other opportunities for obtaining needed Human Factors expertise, such as arranging for Independent Personnel Assignments (IPAs) or developing a 'personnel succession plan' from alternative sources such as National Aeronautics and Space Administration (NASA). In addition to the direct support provided to product and project teams, the FAA Human Factors community includes its internal laboratories (e.g., the Human Factors Laboratory at the William J Hughes Technical Center and laboratories at the Civil Aerospace Medical Institute) and extends to other resources at the FAA's government partners (e.g., NASA, Volpe Transportation System Center), Federally Funded Research and Development Centers (e.g., MITRE), universities and colleges, and other contractor supported organizations.

To provide an estimate of the Human Factors professionals needed to support ATM acquisitions and related projects, a baseline for required Human Factors resources was established. Positions critical to the success of the Human Factors program were identified as 'minimum essential.' Resource requirements for each FAA organizational element were based on the best available estimates using internally generated staffing guidelines and rules of thumb. Acquisition programs are reviewed periodically (and at least annually) in order to prioritize programs to receive Human Factors technical support from available Human Factors expertise and to identify shortfalls. Because the review consists of a 'snapshot' in time, results depend upon the current status of the program/project and the schedule of program activities. Human Factors resource allocations are based on prioritization factors (not listed in order of importance) shown in Table 16.1. Additional requirements and the need for redistribution of resources are identified as new information becomes available. The status of key human capital resources is reflected in Figure 16.3.

Table 16.1   Prioritization of Human Factors engineering resource requirements

PRIORITIZATION CRITERIA

The prioritization for Human Factors resources will be based on the following prioritization factors:

1. Sponsor's assessment of priority and user input
2. Visibility of the program
3. Importance to the NAS
4. Congressionally mandated programs
5. Complexity and level of the operator and/or maintainer interfaces including the 24 study areas from the FAA HF Acquisition Job Aid (FAA, 2003)
6. Program schedule
7. Acquisition strategy (COTS vs. development)
8. Number of systems to be fielded/number of people to be affected by the system
9. Dollar value of the program
10. Opportunity to influence design

Human Factors Practitioner Qualifications and Functions In order to institutionalize an approach to staffing, training and professional development, a viable mechanism to support a qualified workforce needed to be defined. Accordingly, guidelines about the key qualifications for Human Factors professionals were established (see Table 16.2). Additional guidance about qualifications, functions, roles and responsibilities, limits of authority, accountability, lines of communication, and reporting relationships was covered in detail within a report entitled 'Human Factors and Union Involvement in the Lifecycle Acquisition Process Implementation Plan' (FAA, 2000) and is summarized here. Information about the functional role of the Human Factors coordinator/specialist and the general roles and responsibilities of Human Factors professionals was outlined and documented on the FAA Human Factors web site.

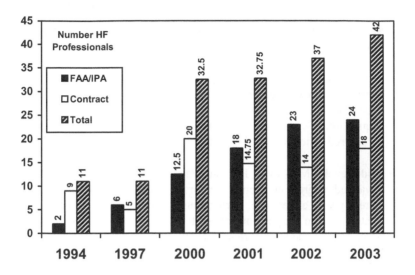

Figure 16.3   Availability of Human Factors expertise

Roles and Responsibilities To achieve Human Factors objectives for systems acquisitions, the leadership of and responsibility for the Human Factors effort must be clearly specified during each stage of the acquisition cycle. The Human Factors coordinator or specialist must be a full member of acquisition work groups and teams, functioning as a fully empowered team member within the Integrated Product Development System (IPDS) framework. When feasible, the Human Factors specialist serves as a core team member rather than an extended team member of IPTs and project teams (PTs) to help ensure that Human Factors receive appropriate priority and resources when IPT/PT commitments are made. The specialist occupies a position in a specific FAA organization with authority to represent it; however, the specialist also maintains a direct communication path to the central Human Factors Research and Engineering Division to support carrying out broader FAA objectives. During each phase, Human Factors specialists from acquisition organizations provide support to IPT/PTs (see Table 16.3).

Table 16.2   Key qualifications for Human Factors personnel

Senior-Level: Senior-level Human Factors personnel shall have an advanced degree in Human Factors or related field, or equivalent. Certification as a Professional Ergonomist or Human Factors Professional, although not required, provides an indication of the appropriate mix of education, experience and competence. Ten years experience performing Human Factors tasks directly related to the acquisition of systems that are similar in type or procedures to the FAA is required. The individual must be able to work without supervision.

Mid-Level: Mid-level Human Factors personnel shall have an advanced degree in Human Factors or related field, or equivalent. Three years experience performing Human Factors tasks directly related to the acquisition of systems that are similar in type or procedures to the FAA is required. The individual must be able to work with minimal supervision.

Junior-Level: Junior-level Human Factors personnel shall have a degree in Human Factors or related field, or equivalent. One year of experience performing Human Factors tasks directly related to the acquisition of systems that are similar in type or procedures to the FAA is desirable. The individual must be able to work with moderate supervision and be able to contribute significantly to a Human Factors program and perform appropriate analyses with minimal supervision.

Developing Human Factors Understanding and Expertise To ensure that best practices are employed, the Human Factors program provides Human Factors development and training support to individuals and teams. This effort recognizes the importance of developing a cohesive and competent community of experts and the emphasis placed upon professional development and training by the FAA, Department of Transportation, General Accounting Office, and other agencies. The necessary activities entail: a) professional community development, b) generic, cross-cutting Human Factors technical information exchanges, and c) individual and community training that varies by program and considers the differences among the systems, the acquisition strategy, the phase of development, interaction and integration with other systems, and the level and type of human involvement. Training associated with the Human Factors integration activities consists of the basic elements of awareness, tailored, and professional training outlined below.

Table 16.3   Human Factors Engineering (HFE) coordinator roles in system acquisition management

Human Factors Coordinators perform, direct, or assist in conducting the following activities:

- Mission analysis and requirements determination (human impacts, constraints)
- Human-system interface considerations in market surveys/investigations/trade studies
- Generation and update of Human Factors engineering plans
- HFE input to solicitation package preparation
- Identification and analysis of critical tasks performed by operators and maintainers
- Generation, refinement, and analysis of operational scenarios, human-system modeling, and human in the loop simulations
- Development, demonstration, and evaluation of Human Computer Interface design requirements, prototypes, design, and development efforts
- Review/analysis of Human Factors Engineering documentation
- Coordination of Human Factors engineering working group activities
- Conduct of task performance analyses and coordination with training and logistics
- Conduct and coordination of Safety and Health Hazard Analyses
- HFE concepts, analyses, and reviews of engineering change proposals, preliminary design, and critical designs
- HFE input to test and evaluation plans, measures, criteria, and data collection efforts

Awareness Training This training consists of instruction that is focused on providing basic information about Human Factors in general. Areas addressed may include such topics as the definition and scope of Human Factors; demonstrations and illustrations of Human Factors concepts, principles, and conventions; generic and common approaches to Human Factors engineering in acquisition and other environments; standard practices and lessons learned; and common problems in Human Factors application.

Tailored Program Training This training consists of instruction that is focused on an IPT, acquisition program office, or acquisition system. It provides information that is specifically relevant to the program or system being acquired and addresses risks or opportunities that may be peculiar to the systems or program.

Professional and Community Developmental and Training Professional acquisition community development and training consists of activities that promote the technical and professional development of the FAA Human Factors workforce.

Closely related to these activities are those that promote the professional community's cohesiveness, consistency, and cooperation. Professional development and training also includes instruction that is focused on technical areas of Human Factors that may be of specific or general interest to the acquisition or FAA population, such as color in ATC displays, designing Computer-Human Interface (CHI) and CHI prototyping techniques, and survey and questionnaire design. The professional development and specialized technical training provide developmental initiatives (such as Human Factors certification support), resources, and information that are relevant to the application of Human Factors in a variety of ATM applications.

Creating and Institutionalizing Policy, Processes, and Best Practices

Institutionalization of an effective ATM Human Factors program is sometimes as dependent upon the organizational culture as it is upon the organizational technical capabilities. Thus, expending some energy in the direction of establishing organizational policy, guidance, and processes sets the stage for formulating and adhering to Human Factors best practices. Accordingly, employing Human Factors best practices entails establishing appropriate policy/guidance; monitoring FAA acquisition processes and proposing revisions to integrate better Human Factors in ATM acquisitions; acquiring the necessary tools, capabilities, and techniques; and applying lessons learned about overcoming the barriers to achieving program goals. During reviews conducted by the Human Factors program, discussions and reporting include information that will assist in meeting performance plan objectives for Human Factors related to research and acquisition policy, processes, and best practices. Consequently, the topics identified in Table 16.4 are of interest.

Policy Few Human Factors programs can generate the necessary momentum (much less sustain it) without legitimizing the integration requirements of Human Factors through established policy. FAA policy at the highest level (signed by the Administrator) directs that Human Factors will be integrated into every function and every organizational element of the agency. This policy identifies the objectives of the Human Factors program and assigns responsibility broadly for all senior leadership positions. The policy provides the basis upon which every other activity of the Human Factors action is founded, and establishes an anchor by which the ship of Human Factors is moored. In addition to this agency-level Human Factors policy, Human Factors policy is inserted and integrated into other agency policies that relate to acquisition procedures, requirements, investment analyses, cost/benefit studies, system development status reporting, testing, and other important research and engineering functions.

Table 16.4   Performance plan Human Factors items of interest

STRATEGIC AREA	ITEMS OF INTEREST TO REPORT ON
Research: Identify and conduct required Human Factors research	Areas where Human Factors research studies need to be conducted to identify or mitigate human-system performance risks
Acquisition Engineering: Institutionalize Human Factors policy, processes, and best practices; support Human Factors conventions, guidelines, and tools; conduct Human Factors training; and apply Human Factors engineering principles	Areas where policy/guidance/practices need to be changed to facilitate Human Factors Resources (number of qualified people and funding) needed vs. available for Human Factors research and engineering activities Tools, techniques, guidelines, and databases needed vs. available Professional development and training required and training available Human Factors risk areas and required/planned mitigation activities

Processes In addition to the policies that establish agency requirements, FAA directions describing how activities are to be carried out include Human Factors processes and guidelines. These 'how to' processes include those related to mission analysis, system engineering procedures, process improvement (such as in capability maturity models), risk assessments, in-service review checklists, statement of work generation, data item descriptions for vendor data calls, and many other research and engineering functions that support the Acquisition Management System (FAA, 1995b).

Technical Tools Support of the Human Factors community requires the procurement of capabilities to perform the technical work of the Human Factors researcher and engineer. Accordingly, Human Factors research organizations and laboratories have been established to execute the technical work that complements the efforts of the contracted vendors. These laboratories and technical organizations must have available to them the modern tools and techniques of the Human Factors trade. In overseeing the Human Factors program broadly, a key role for the central Human Factors organization is in conducting reviews and facilitating the development or acquisition of capabilities and technical tools such as simulation capabilities, diagnostic tools, organizational ATM standards and conventions, process improvement and capability maturity models, and job aids. For example, the FAA Human Factors Acquisition Job Aid provides a how-to guide and depicts key Human Factors tasks and activities to be conducted during the ATM system acquisition phases.

Lessons Learned in Managing the Infrastructure Managing a Human Factors program reaps a high volume of useful information and feedback about all elements of the program. Applying the lessons learned about such topics as communications, facilities, tools, decision-making, organizational culture, and external factors helps overcome the barriers to achieving acquisition program goals. Table 16.5 provides a list of major obstacles that were encountered in creating and executing an ATM Human Factors program (FAA, 1997). A more comprehensive FAA lessons-learned database for system acquisitions is available to the acquisition workforce, replete with key Human Factors findings and conclusions.

Table 16.5   Obstacles to achieving Human Factors objectives

1. Undefined or unrefined requirements or concepts of use
2. Undefined or varying roles, responsibilities, and accountability among management, union, and acquisition workforce members
3. Weak or interrupted lines of communication among organizational elements
4. Differences in culture, strategy, philosophy, and doctrine among organizational elements
5. Failure to apply Human Factors lessons learned
6. Unrealistic schedules
7. Incompetent vendor/ contractor
8. Unrealistic technical solutions or acquisition strategy
9. Lack of technical knowledge in workforce
10. Inadequate Human Factors research, analysis or data
11. Failure to attach appropriate priority to Human Factors
12. Lack of an integrated Human Factors program across acquisitions
13. Lack of an understanding of Human Factors
14. Insufficient oversight of contractor Human Factors
15. Lack of a well-defined acquisition process
16. Unrealistic cost
17. Lack of trust among participants (e.g., labor/management)
18. False assumptions about Commercial Off-The-Shelf acquisitions

Acquisition Program Risk Assessment and Mitigation

One of the important functions of Human Factors in the acquisition of a major system is to assess and mitigate the Human Factors risks that are identified during the various stages of technology readiness. The FAA's experience in ATM acquisition research and engineering demonstrates that risk identification and mitigation is an essential vector required to achieve a successful Human Factors program. The application of Human Factors engineering to the acquisition of ATM systems poses many of the same technical and programmatic challenges and rewards that may be found when dealing with complex systems in any work environment. The formula for success requires that Human Factors professionals with energy, vision, and resources apply the discipline in a fertile environment where the role of the human in system performance is recognized. As in any other applied environment, Human Factors practitioners must gain an in-depth understanding of the users, systems, and operating environment. To be successful, the Human Factors professional also needs a good understanding of their own operating environment including the organizational mechanisms and interpersonal relationships that affect design decisions and the allocation of resources.

Environment of Applied Human Factors Engineering

The applied environment for the Human Factors professional is very different from the research environment, as are the criteria for success. From the Human Factors engineering perspective, success has been achieved when the total system reaches its performance objectives with the human in the loop. The Human Factors engineering practitioner must be able to show others the fruits of one's labor in the design of the system. This is but one of many areas where Human Factors engineering (HFE) differs from Human Factors research. In the applied arena, the referees that grade the professional's success are the system users and those that benefit from the effectiveness of a well-designed system.

The HFE professional objective is to send safe and effective systems to the users. This statement of the desired end state often leads to a series of questions that deal with the acquisition process and the application of Human Factors 'best practices' in the identification and mitigation of potential risks. The foundation for Human Factors in acquisition lies in the analysis of the mission and the development of human performance requirements that are achievable within a tolerable level of program risk. These analyses must identify key functions and critical tasks that can emerge in the requirements documents in the form of testable characteristics that comprise the human-system performance capability and contribute to the operational improvements.

ATM Acquisition Environment

In the air traffic arena there are significant challenges to the application of Human Factors in the acquisition process. Historically, air traffic systems have been added to workstations on a piecemeal basis and the process has been largely driven by the introduction of technology rather than by operational concepts. To enhance the efficiency and capacity of the NAS there has been considerable attention paid to the development of decision support tools that make use of computer processing power to move toward optimizing the management of air traffic. Advances in sensor and computational capability have brought additional capabilities that augment the information available to the controllers, strategic planners, supervisors, and maintenance technicians that operate and maintain the NAS. Each tool or capability introduced has usually been developed independently, and since multiple tools and systems are being developed on concurrent, parallel paths, they cascade onto the workstation and create an interesting array of Human Factors problems as seen in typical tower cab workstations (see Figure 16.4). The temptation to the Human Factors professional is to dive into the situation and participate in the process of fixing the problem. While this is important, the overriding concern should be to participate in the definition of the total workstation design objectives and to clearly define the operational concept, roles and responsibilities of the operators and maintainers, and the desired operational and design outcome. Laying this foundation and moving forward to develop functions and detailed design requirements increases the probability of success.

The peculiar challenge in the air traffic arena is that of dealing with organizations and users that are (rightfully) risk averse and cannot tolerate misadventures during operations. Controllers are held responsible for each incident where aircraft separation is in jeopardy. The operating agency (i.e., FAA) is held responsible whenever the safety of the flying public is compromised. Expansion of mission envelopes is done with caution and reluctance. Even efforts associated with enhancing the capacity of the NAS are approached with deliberation since inadvertent impacts on safety are not tolerable and even potential or perceived adverse impacts on the workforce or bargaining units can result in unintended outcomes. Consequently, the role of Human Factors in acquisition in terms of 'best practices' continues to evolve.

Models for Human Factors Application The model for the application of Human Factors in the development of air traffic systems needs to consider an interactive and iterative role in the definition of the mission and the concept of operations. Traditional Human System Integration models for top-down definition of mission needs, concept of operations and system requirements may work well for military and commercial systems, but the history of air traffic systems indicates that these linear models may not be viable. Therefore, traditional Human Factors models that were meant to operate in a top-down environment will not be effective without modification. Air traffic systems development tends to be driven by technology. Mission analyses and requirements documents have tended to rely heavily on the products of existing research or development programs as tools to define future concepts of operation. Rarely have these analyses exposed needs and concepts that are not being met by an existing research or development program. Requirements tend to be driven by the availability of a solution. This mode of operation is in concert with the image that there are organizations that conceptualize or invent new air traffic systems or tools, and others that implement, operate, and maintain them. The potential problem is that research may take a path that is unsupportable, undesirable, or unsuitable for the users in the operating environment. From another perspective, operating organizations tend to be conservative and rooted in existing technology and procedures, which limits progress toward an enhanced level of service to a reduced pace.

Figure 16.4 Typical tower cab workstation

Future Model The model that is emerging in the air traffic system development process is to engage technology futurists in the research community with operational subject matter experts (SME), Human Factors professionals, air traffic procedures specialists, and air traffic requirements specialists to develop a supportable concept of operations. Human Factors practitioners use structured methods, such as cognitive walk-through, to determine baseline mission functions, critical tasks, decisions required, and information flows between various parties and the source of information (Krois and Rehmann, 2002). This basic set of information acts as the initial set of requirements to construct an integrated workstation that contains all the elements of information to build and sustain situation awareness and perform the specified mission. The first version of the workstation acts as a catalyst to determine if the previously defined mission functions are viable when combined with technology in operational scenarios. The mission definition and functional requirements are tested to determine if the new tools or information elements provided by advances in technology can, or should, expand the mission envelope. This method provides the analytical foundation needed by the Human Factors community to design displays, controls, and integrated workstations that effectively transfer information to the user to enable decisions and actions that conform to the mission functions. The construction of a workstation enables the various air traffic subject matter experts (SMEs) to see and manipulate a workstation that is a working element of the concept of operations. In this iterative model, the concept, workstation design, and technology feed on one another until a state of stasis is found where there is a balance between the mission requirements, operational capability, tolerance of the user, and the ability of technology to deliver a benefit.

Air Traffic Service Consumer To move forward effectively in the evolution of the NAS from the Human Factors perspective, one element is missing from the model described above – the air traffic service consumer. In most cases, this will be the pilot community. The Human Factors discipline prides itself on taking a user-centered approach, but the occupational hazard is to focus on the visible user that manipulates the system of interest – in this case the air traffic controller or maintenance technician. In the past, only a voice radio link and a common lexicon connected the air traffic service provider and the pilot. The availability of data link communications that will enable digital air-ground messages, traffic information, and weather displays in the cockpit may alter the communications process between pilots and controllers. Other technology, such as Global Positioning System (GPS) and Automatic Dependent Surveillance-Broadcast (ADS-B), has the potential to alter the traditional roles and responsibilities of controllers and pilots. The concepts of shared separation responsibility, or self-separation may change the relationship between controllers and pilots if these concepts are a viable means to manage air traffic. The availability of timely and accurate weather products and aircraft intent information may change the types of services provided to the pilot by the air traffic service community. To determine how the NAS could, or should, evolve requires the Human Factors professional to broaden the scope of how the 'user' is defined so that total system performance is addressed and the expectations of all the vested parties are met. Giving the pilot community a voice in the development of air traffic management systems has usually only occurred at the policy level, not at the service level. The Human Factors community must be prepared to use the HFE tool set to accommodate a broader definition of 'user' during the initial phases of the lifecycle.

Communicating Human Factors Requirements Once the definition of the air traffic system moves past the concept stage where the mission capability and approach are specified, the new concept must be communicated to the acquisition community that is charged with fielding a system that meets the need. This is normally done through the Requirements Document that is arguably one of the most important and difficult acquisition documents to generate. It has been proven difficult to specify human performance requirements in terms that flow from the operational concept and mission need: such terms must be testable and make a meaningful contribution to the definition of the capability in clear terms that do not originate from the design of a specific solution. The requirements document should not be oriented toward the process of applying the body of Human Factors knowledge, analytical methods, or design standards. Those processes should be articulated in Human Factors program plans, statements of work and system specifications. Rather, the requirements should be specified in terms of human-in-the-loop functional performance that is needed to satisfy the operational improvement identified in the concept of operations and mission statement. To the extent possible, human-system performance should be in quantitative terms. The arena that has had greatest success with this approach has been that of maintainability. Specifying Mean Time

To Repair for major systems is an example of a human performance requirement that can often be incorporated. A tool such as the Top Down Requirements Analysis (TDRA) provides a means for generating human requirements at the outset of the system development process (Carson and Malone, 2002).

Levels of Specification There was substantial promise in the method for the specification of human performance in system requirements proposed by the Army Human Engineering Laboratory (Kaplan and Crooks, 1980) in that a structured approach to human performance specification was offered. Kaplan and Crooks advocated the identification of system missions, decomposing the mission into functions and then into tasks and task conditions. However, the analysis to generate the task list and then generate the performance attributes of each is time consuming. The analyst must sort the tasks to identify the critical ones and those that are the functional drivers that contribute substantially to mission success. Requirements documents are reasonably brief and need a concise statement of the criteria for success that will be used for system development and testing. A drawback of Kaplan and Crooks' method is that the analyst must perform an extensive system function and task analysis to then backtrack with a few dominant tasks that are worthy of inclusion in the requirements document. An analyst with substantial domain and subject matter expertise may be able to narrow the focus to those functions and tasks that have a high likelihood of contributing to mission success and operational improvement.

Time and accuracy are the key attributes of human performance specifications used by Kaplan and Crooks. These parameters are familiar to the Human Factors practitioner and the inclusion of this type of human performance specification in the requirements document is appealing. Total time of performance was comprised of reaction time and performance time. The measurable specification for accuracy was its reciprocal: errors. The human performance specification was expressed in terms of probability of success, mean time, and errors. In many arenas, such as manufacturing processes and military systems operations, this has been a reasonable and acceptable approach. Weapon system designers deal with probabilities as a normal part of their modeling. Each weapon has a probability of hit (P_h), and probability of kill given a hit (P_k) to characterize its performance. Assigning a probability to human task performance fits into the scheme quite nicely. If the reliability of human task performance degrades overall system performance below the established threshold, additional resources should be dedicated to either improve human performance or redesign the system to eliminate that task from the critical path for mission success.

Minimum Safe Altitude Warning (MSAW) system In the cases where air traffic systems designers try to accommodate human performance in terms such as total time of task performance, we sometimes run into interesting situations. For example, the Minimum Safe Altitude Warning (MSAW) system is an alerting system that provides the controller with an alert to indicate that an aircraft in the sector is on a trajectory that may result in collision with terrain. The broad requirement for this alerting system was to provide the controller with sufficient warning of an impending collision with terrain such that the control instruction will result in aircraft terrain avoidance. There is substantial anecdotal evidence that many past MSAW alerts were nuisance alarms in that there was no impending threat to the aircraft that required intervention on the part of the controller. Note that these are not characterized as false alarms since the conditions specified in the MSAW operating principles were satisfied in each case. These are true alerts (from the standpoint of software designers) that must be acknowledged and reviewed by the controller and either dismissed or acted upon. The MSAW alert had a high enough nuisance factor that it was turned off in some facilities for a period of time. To improve the system performance of the MSAW alert, a number of enhancements were introduced including upgraded resolution of the terrain and accommodation of human performance when setting lead-time for the alert to allow aircraft to avoid terrain. Total human performance time includes the task time for the controller (reaction time from the onset of the alert plus the time to formulate and issue a control instruction) and the task time of the pilot (reaction time from the transmission of the controller's message to the point of making a flight control input). Using reasonable values for human performance time and the existing values for aircraft maneuvering and recovery, the project was successfully concluded.

Integration in the Requirements and Development Teams An important function of the human integration section of the Requirements Document is that it provides an indication to the development team of the scope of activities that must be planned. The HFE practitioner on the team has the responsibility to assure that human performance objectives are met and to coordinate design and testing activities with other members of the technical development and operational community. Cost estimates must be prepared to assure that resources to conduct the necessary activities are available. Responsibilities and interdependencies must be allocated among the members of the procurement staff. Budget planning is initiated several years in advance of actual program execution and the inclusion of human performance criteria in the requirements is the mechanism that will enable the Human Factors practitioner to develop an initial cost estimate, schedule of events, and participate in generating an integrated program plan. Human Factors must be viewed as an integral part of the team with assigned responsibilities that make a contribution to program success. The Human Factors function must be prepared to share the workload borne by the development team and contribute to the effort as a full responsible member that deserves a share of the resources. The orientation of the Human Factors effort should be to help the team achieve success. Acquisition professionals are judged by their success in the areas of schedule, cost, and technical performance. Therefore, assuring that Human Factors technical risks are identified, mitigated, and tracked as part of the systems engineering effort during the early stages of the program will aid in securing resources to perform the necessary program activities during the system development stage. If resource planning is performed and the Human Factors program is executed as part of a coordinated plan, the two crucial areas of cost and schedule will fall in place.

User Acceptance is Not Human Factors Engineering In some cases, development activities in air traffic system acquisitions have served largely to structure the collection of user opinions. This is particularly true for those programs where the emphasis is on marketing the solution to the operational community as a means to manage expectations and to reduce risk associated with user acceptance. If Human Factors is seen as the application of a few principles and guidelines to aid the articulation of user preferences, it will get little attention until the latter stages of the program and will likely be expected to perform a limited function with few available resources using 'just in time' tactics. When user acceptance (rather than user performance) is the primary evaluation criterion, there is no assurance that system performance will achieve the benefits that made the cost of the program justifiable. Because users may gravitate toward the familiar characteristics of the existing system to increase their comfort or, may expect the system to provide attributes that were not part of the targeted operational capability enhancement, this user satisfaction approach to design is programmatically risky. On more than one occasion such an approach has resulted in modernization programs that failed to provide enhanced services, capabilities, capacity, or safety. In any operational arena, particularly air traffic, user acceptance may be an essential ingredient to the program, but it should be based upon demonstrated human-system performance – not merely opinions and preferences.

Government Provided Information In non-ATM sectors of the government (e.g., military system acquisition) a vendor is sought through a contract mechanism that contains a statement of work for the vendor and the system specification. If an aircraft is being procured, there is an expectation that the vendor will design the cockpit as an overall part of the design effort and take ownership of the crew station design. Development of air traffic systems has evolved in a somewhat different direction. Often, government generates the design of the human interface and provides this design as part of the system specification. Vendors are reluctant to take ownership of the human interface and have not done well on the few occasions when they have elected to do so. The FAA Standard Terminal Automation Replacement System (STARS) and Operational and supportability Implementation System (OASIS) programs are notable cases where deficiencies in the vendor's human interface design led to significant program delays with a cost impact. Alternatively, the procurement program for a low-cost ground surveillance radar for the airport movement area called Airport Surface Detection Equipment Model X (ASDE-X) developed the user interface design and published that design as a specification that was part of the request for proposal package when the FAA asked vendors to bid on system production. The IPT responsible for procuring the system engaged the services of the Human Factors practitioners at the FAA

Research and Development Human Factors Laboratory to develop the display and control design with the aid of a user team. This example is indicative of a trend where the air traffic operating agency develops the design of the workstation and provides this design information to vendors that are content to follow this direction.

Systems Approach Regardless of whether the operating agency or a system vendor develops the user interface design for a system, a systems approach must be taken and a program plan in place that maps out the Human Factors activities to support the overall objectives of the acquisition. The program plan provides the over-arching strategy that specifies what activities will be conducted. It should spawn detailed plans for specific activities such as simulations, analyses, design support efforts, and evaluations. The essential elements of a Human Factors program plan are provided in Table 16.6.

The objective of the plan is to participate in the development process in a manner that contributes to program success and reduces risk for management. The Human Factors program and the tools and methods employed can be a useful means for all the stakeholders to view the progress of system design and gain comfort that the design of the system is moving in a direction that will achieve all the program goals.

Table 16.6   Essential Human Factors program elements

- Program objectives.
- Identification of user characteristics.
- Scope of the Human Factors program and integration with the acquisition program.
- Identification of Human Factors issues that need to be addressed.
- Identification of human performance factors and the risks associated with achievement of performance levels.
- Identification of potential sources of human error and the impact on system safety and human performance.
- Roles and responsibilities of the procuring agency, vendors, user groups, supporting organizations and others that will contribute toward meeting program goals.
- Identification of the major Human Factors activities and tasks associated with system development.
- Schedule of major Human Factors activities to support success on the overall program schedule.
- Resource requirements to conduct the program.

Post-Deployment Assessment One area of Human Factors involvement in the lifecycle of air traffic systems that has received scant attention is that of fielding and post-deployment assessment. The in-service management phase of a system results in a change in focus from development to maintenance. The fielding process should be viewed as a large-scale test and evaluation. The challenge is to find a means to collect the appropriate data and address the needs of a new set of decision makers. The earlier concerns about design and testing to meet a specification change to concerns about how a system is used in operations, training, staffing, and assessing effectiveness. Decisions regarding product improvements can be supported by Human Factors data collection by addressing shortcomings in usability or effectiveness. Subjective data regarding operator/maintainer perceptions can be gathered using conventional questionnaire and interview techniques. Gathering objective data on job performance or effectiveness can be more difficult but may increase its value. It is important that Human Factors engineering be conducted in a closed loop environment.

There are several purposes in gathering data during the in-service portion of the system lifecycle. One is to judge the effectiveness of the system in its operational setting from a Human Factors perspective. The practitioner should be continually assessing whether or not the system is performing its intended function. This is important not only from the perspective of system design, but also as a means to assess and improve the Human Factors process that was used to procure the system. As an extension of that concept, program managers need to assess the value of proposed product improvements. Human Factors analyses and data collection efforts aimed at a fielded system can motivate a product improvement, reinforce the need for an improvement offered by a third party, or refute the need for the improvement by determining that the target problem is not serious. Finally, an important purpose of field data collection on an existing system is to feed Human Factors information forward to more advanced systems. It is often useful to understand the user's perceptions and performance history of predecessor systems when designing new systems.

Lessons Learned in ATM Acquisitions

There are several experiences relevant to understanding human-centered integration that serve as 'lessons learned.' One of the most important lessons is that where the technology push is strong, system research and development efforts can easily lose sight of the human-system performance impacts on the end user (operator, maintainer, and support personnel) especially those related to cognitive tasks. This experience was one of several key Human Factors lessons learned from the FAA STARS acquisition program, which showed that these performance impacts must be fully addressed before and during detailed prototyping and design activities.

Failure to adequately address the role of the human operator and understand the allocation of function has encountered difficulties and challenges in many attempts to field ATC capabilities. For example, difficulties with the program called FAST, which was one of the family of controller decision support tools developed as part of NASA's Center TRACON Automation System (CTAS) program, highlighted the need to prescribe changes in the roles between the controller and the system (Cardosi, 2003).

In addition to the complexities of the Human Factors cognitive aspects of the system, Human Factors research and development is affected by how comprehensively the human-system integration efforts are carried out. One lesson learned is that because the scope of the Human Factors effort is often not well understood by all members of the product team, a common set of Human Factors study areas can be used to represent a systematic definition of research and analysis needs. A common set of study areas helps non-practitioners comprehend the components of the Human Factors program and supports consistency in the practitioner's approach across acquisition programs and operational domains. The FAA has developed such a list which is included in the FAA Human Factor Acquisition Job Aid and summarized in Table 16.7.

Table 16.7   Human Factors study areas

1. **Allocation of Function:** Assigning those roles/functions/tasks for which the human or equipment performs better while enabling the human to maintain awareness of the operational situation.
2. **Anthropometrics and Biomechanics:** Accommodating the physical attributes of its user population (e.g., from the 1st through 99th percentile levels).
3. **CHI (Computer-Human Interaction):** Employing effective and consistent user dialogues, interfaces, and procedures across system functions.
4. **Communications and Teamwork:** Applying system design considerations to enhance required user communications and teamwork.
5. **Culture:** Addressing the organizational and sociological environment into which any change, including new technologies and procedures, will be introduced.
6. **Displays and Controls:** Designing and arranging displays and controls to be consistent with the operator's and maintainer's tasks and actions.
7. **Documentation:** Preparing user documentation and technical manuals in a suitable format of information presentation, at the appropriate reading level, and with the required degree of technical sophistication and clarity.
8. **Environment:** Accommodating environmental factors (including extremes) to which the system will be subjected and understanding the associated effects on human-system performance.
9. **Functional Design:** Applying human-centered design for usability and compatibility with operational and maintenance concepts.
10. **Human Error:** Examining design and contextual conditions (including supervisory and organizational influences) as causal factors

contributing to human error, and consideration of objectives for error tolerance, error prevention, and error correction/recovery.

11. **Information Presentation**: Enhancing operator and maintainer performance through the use of effective and consistent labels, symbols, colors, terms, acronyms, abbreviations, formats, and data fields.

12. **Information Requirements:** Ensuring the availability and usability of information needed by the operator and maintainer for a specific task when it is needed, and in a form that is directly usable.

13. **I/O Devices:** Selecting input and output (I/O) methods and devices that allow operators or maintainers to perform tasks, especially critical tasks, quickly and accurately.

14. **KSAs:** Measuring the knowledge, skills, and attitudes (KSAs) required to perform job-related tasks, and determining appropriate selection requirements for users.

15. **Operational Suitability:** Ensuring the system appropriately supports the user in performing intended functions while maintaining interoperability and consistency with other system elements or support systems.

16. **Procedures:** Designing operation and maintenance procedures for simplicity, consistency, and ease of use.

17. **Safety and Health:** Preventing/reducing operator and maintainer exposure to safety and health hazards.

18. **Situation Awareness:** Enabling operators or maintainers to perceive and understand elements of the current situation, and project them to future operational situations.

19. **Special Skills and Tools:** Minimizing the need for special or unique operator or maintainer skills, abilities, tools, or characteristics.

20. **Staffing:** Accommodating constraints and efficiencies for staffing levels and organizational structures.

21. **Training:** Applying methods to enhance operator or maintainer acquisition of the knowledge and skills needed to interface with the system, and designing that system so that these skills are easily learned and retained.

22. **Visual/Auditory Alerts:** Designing visual and auditory alerts (including error messages) to invoke the necessary operator and maintainer response.

23. **Workload:** Assessing the net demands or impacts upon the physical, cognitive, and decision-making resources of an operator or maintainer using objective and subjective performance measures.

24. **Work Space:** Designing adequate work space for personnel and their tools or equipment, and providing sufficient space for the movements and actions that personnel perform during operational and maintenance tasks under normal, adverse, and emergency conditions.

Other lessons in the application of Human Factors to ATM systems are captured in FAA Human Factors program documentation and include those summarized below.

Use of Anecdotal Information Relative to the degree of rigor in studies and analyses, applied research and engineering of complex human-system integration issues usually necessitates a well-controlled environment. However, data sources such as anecdotal information or short videos of real problems can be powerful, for example, in demonstrating compelling evidence of the need of a Human Factors program. Use of such information should be considered for its relative value for different purposes.

Documenting Study Results Research studies and engineering design solutions should be followed-up with appropriate reports and specifications to ensure that similar programs can benefit from past experiences. This documentation can be invaluable in supporting the development of guidelines and standards, and also help in the derivation of lessons learned to avoid repeating systemic problems, especially when formatted succinctly for the design and engineering communities.

Performance Criteria System performance criteria need to consider the human component in a degree of specificity that will enable testing and data collection of technical human-system performance. Human performance constraints can pose limitations in achieving the desired level of system performance and also pose impacts to program cost and schedule. For example, changes to resolve usability issues may require unplanned software development. Programmatically such changes increase the cost of system development.

Risk Analysis Generation of Human Factors risk analyses and human performance data needs to occur during the mission analysis and investment analysis phases of the acquisition lifecycle. These data are important to help estimate and direct the resource allocations and expenditures associated with the human performance aspects of the system.

Collecting Human Performance Data The distinction should be clear in collecting data on human performance in contrast to obtaining personal preferences. The engineering basis for Human Factors demands obtaining structured user performance information to assure that program decisions are based on the scientific performance information (Krois and Rehmann, 2002; Cardosi, 2001). Using personal preferences as the basis for decision-making thwarts objectivity and empiricism.

System-level human performance baselines – often lacking in a systems development – need to be established. Human performance measures and baselines are used to assess program progress, evaluate test results, and analyze the value of proposed future system capability enhancements.

Micro and Macro Human Factors Issues Human Factors and human performance issues should be recognized as ranging along a continuum from micro to macro issues. Addressing this variation necessitates applying a broad range of human resource and human-system performance parameters from ergonomic workspace design to cognitive tasks.

Human Factors Requirements Clear and specific definition of the human performance considerations is required in the operational concepts and requirements documents. Without this clarity and detail, changing or adding requirements downstream related to Human Factors is likely to result, which almost always creates adverse cost and schedule impacts. A commercial off-the-shelf (COTS) acquisition does not eliminate the need for these Human Factors information requirements.

The potential for under- or over-specifying human performance requirements may affect the adequacy of government direction and control of the vendor. Devising appropriate and adequate human-system performance specifications and requirements necessitates explicating the intended role of the controller and the expectations for both performance proficiency and tolerance for human error.

Human Integration Design Changes There is a general lack of understanding of the limitations and implications of suggesting human interface design changes or modifications outside of the official contractual channels. Some suggested changes may have legal or programmatic ramifications when dealing with contractors; others may impact union bargaining options; all have human-system performance implications.

Organizational Roles The roles and responsibilities related to Human Factors must be clearly defined for those individuals and organizations involved in Human Factors support of the system acquisition. Lack of clear, designated functional responsibility (and expertise) for Human Factors often leads to an expensive ambiguity about who is to do what and how much Human Factors is enough.

Conclusions

In the first section of the chapter, Relying on Research and Requirements, a human-centered approach to mission analysis was suggested as the basis for establishing concepts of operation, concepts of use, and an ATM architecture that incorporates human performance limitations and capabilities. The 16-point Human Factors program elaborated on the critical elements to achieve ATM objectives. Technology readiness levels and implementation readiness levels offered a means for managing technology transfers.

In the chapter's second section, Developing and Managing the Human Factors Infrastructure, FAA insight into some of the fundamental building blocks of a good human factor program identified guidelines for organizing and using the expertise, facilities, and tools required to achieve a viable Human Factors program. The value and risks associated with creating and institutionalizing policy, processes, and practices within the organization were identified, including what type of training program needs to be established to achieving program goals.

In the third section of the chapter, Acquisition Program Risk Assessment and Mitigation, a practical view of the Human Factors program addressed the engineering imperatives and lessons learned in the application of Human Factors to ATM and non-ATM systems, and translated those imperatives and lessons into practices that ensure issues, potential impacts, and risks are identified, documented and resolved.

From the discussions in this chapter, one can appreciate how to support and manage a broad Human Factors program to ensure Human Factors methods and best practices are applied to ATM systems throughout their lifecycle. The macro view of the management of a Human Factors program across systems suggested a refined definition of the activities to be performed and clarification of the role of Human Factors and Human Factors practitioners in ATM acquisitions. FAA experiences during numerous acquisition programs revealed several important topics, including: a) the major components of a broad Human Factors program, b) why these components are important to the outcome of Human Factors engineering efforts within product teams, c) objectives and guidelines for addressing the components successfully, and d) what attributes and measures can be used to evaluate the health of the program and its components. Examples from several systems highlighted how the various components of the Human Factors management effort have affected the acquisition products and provide evidence of the impact of such efforts on ATM system design, development, and operations.

Experiences in establishing and managing a broad ATM Human Factors program at the FAA demonstrate the essential elements to ensure Human Factors methods and best practices are applied throughout the system development lifecycle. The activities to be performed and the role of Human Factors and Human Factors practitioners in ATM acquisitions entails a well defined program that recognizes the importance, breadth, and dependence upon Human Factors research and requirements; that matures and nurtures the Human Factors infrastructure; and that establishes viable mechanisms for assessing and mitigating Human Factors risks at each stage and for each component of the Human Factors program. Applying the lessons learned at the FAA in each of these vectors will assist other Human Factors programs to overcome the significant obstacles to achieving Human Factors and acquisition system program goals.

References

Cardosi, K. (2003), *Human Factors integration challenges in the terminal radar approach control (TRACON) environment*, DOT/FAA/AR-02/127, Federal Aviation Administration, Office of Aviation Research, Washington, DC.

Carson, F. P. and Malone, T. B. (2002), 'HSI top down requirements analysis', *Naval Engineers Journal*, Vol. 115(2).

Federal Aviation Administration. (1993), *Human Factors Policy, Order 9550.8*, Washington, DC.

Federal Aviation Administration (1995a), *National Plan for Civil Aviation Human Factors: An Initiative for Research and Application*, Washington, DC.

Federal Aviation Administration. (1995b), *FAA Acquisition Management System*, Washington, DC.

Federal Aviation Administration (1997), *STARS Human Factors Management Process*, STARS Human Factors Process Group, Washington, DC.

Federal Aviation Administration (2000), *Human Factors and Union Involvement in the Lifecycle Acquisition Process Implementation Plan*, Vol. I, DOT/FAA/AR-04/35, Washington, DC.

Federal Aviation Administration. (2002a), *Integrated Plan for Air Traffic Management Research and Technology Development, Version 6.0, Volume 1: Executive Summary/Overall Plan*, Washington, DC.

Federal Aviation Administration. (2002b), *Implementation Plan for Human Factors Research and Engineering*. Washington, DC.

Federal Aviation Administration. (2003), *FAA Human Factors Acquisition Job Aid*, Washington, DC.

Grossberg, M. (1998), *Implications of the Air Traffic Services' Concept of Operations on Human Factors Research, Engineering and Development*, Federal Aviation Administration, Washington, DC.

Kaplan, J.D. and Crooks, W.H. (1980), *A concept for developing human performance specifications*, Technical Memorandum 7-80, U.S. Army Human Engineering Laboratory, Aberdeen Proving Ground, MD.

Krois, P. and Rehmann, J. (2002), 'Integrating Human Factors in air traffic control research and acquisition', *Proceedings of the 21st Digital Avionics Systems Conference*.

Chapter 17

The Development of Effective
STARS User Interfaces

Kenneth R. Allendoerfer, Tanya Yuditsky,
Richard H. Mogford and Joseph J. Galushka

Introduction

Federal Aviation Administration (FAA) terminal air traffic controllers are responsible for sequencing and separating aircraft operating within approximately 40nm of the airport and below 10,000 ft. Terminal controllers work in terminal radar approach control (TRACON) facilities and control towers. Many TRACONs serve multiple airports and several serve multiple major airports and numerous towers, such as the consolidated facilities located in the New York, Potomac, and Southern California areas (FAA, 2002).

Controllers use a variety of information, automation, and communication systems to complete their air traffic control (ATC) tasks. Since the 1960s, the most important of these is the Automated Radar Terminal System (ARTS). In its most advanced versions, known as IIIA and IIIE, ARTS provides the following information to controllers:

- the location and altitude of aircraft, based on surveillance data;
- aircraft information, such as flight number and aircraft type;
- flight planning information, such as the next fix and assigned runway;
- controller jurisdiction indicators that show which controller is responsible for each aircraft;
- special condition codes that result from aircraft transponder settings, such as emergencies, radio failures, and hijacks;
- conflict alerts and minimum safe altitude warnings; and
- the location and intensity of precipitation.

However, ARTS is obsolete. It uses computers and displays that are no longer manufactured and are increasingly difficult to maintain. Some versions of ARTS do not use modern programming languages and do not have sufficient capacity to handle the increased traffic volume anticipated in the next decade. ARTS does not have the capability to incorporate new functionality such as decision support tools.

The FAA has long recognized this problem and has been working to replace ARTS since the 1980s.

The Standard Terminal Automation Replacement System (STARS) is a joint FAA and Department of Defense (DoD) program that replaces the ARTS at 171 FAA TRACONs, 362 FAA towers, and 140 DoD radar approach control (RAPCON) facilities. The STARS program will cost more than $1.3 billion and will take more than 10 years to design, build, test, and deploy. As of 2003, versions of STARS are operational at 11 FAA and 2 DoD facilities.

STARS Equipment

STARS consists of many components and subsystems. For controllers, the most important components are the Terminal Controller Workstations (TCWs) and the Tower Display Workstations (TDWs). Each TCW consists of a 28-inch (72 cm) diagonal cathode ray tube (CRT), commonly referred to as a 'Sony 2K' because of its 2048 pixel by 2048 pixel resolution, and up to three keyboards and trackballs. Each TDW consists of a 20-inch (51 cm) liquid crystal display (LCD) and accompanying keyboards and trackballs. The TCW and TDW contain identical information about aircraft locations, flight data, and weather.

For maintenance technicians, the most important component is the Monitor and Control Workstation (MCW). Each MCW is a computer workstation outfitted with a processor unit, monitor, keyboard, and mouse. The MCW contains information about the health of system components, provides usage and diagnostic reports, and allows technicians to configure the system.

User Interface Development Process

The STARS user interface development activities followed a process that evolved over time. First, a usability assessment was conducted to identify and document Human Factors issues. Second, working groups consisting of system stakeholders were assembled. Third, the working groups proposed ways to address the Human Factors issues and created user interface prototypes of these solutions. Fourth, the proposed solutions were documented and provided to the system vendor. Finally, once the changes were implemented, the solutions were validated using a methodology similar to the initial usability assessment.

Usability Assessment

Usability assessments were structured activities designed to solicit feedback from users regarding the user interface. The methodology employed is known as a 'structured walkthrough' and consisted of five components:

- identification of the most common and most critical functions of the system,

- development of human-in-the-loop simulation scenarios or written scripts that exercise the selected functions in a realistic manner,
- completion of the simulations or scripts by users,
- observation and interviews by Human Factors specialists, and
- group meetings in which issues are consolidated.

Computer-Human Interface Working Group

Once the usability assessments were completed, the FAA assembled multidisciplinary computer-human interface (CHI) working groups. The working groups consisted of representatives from the FAA, Department of Defense (DoD), the unions, and Human Factors. In addition, users from field sites were selected by the unions to serve as working group members and subject-matter experts. In most cases, these users had served as participants in the usability assessment, were already knowledgeable about STARS, and understood the Human Factors issues. Engineers from Raytheon Company, the STARS vendor, attended working group meetings and provided technical guidance about how modifications could be made to the system. The working groups followed a consensus process. That is, all members had to agree to a decision before it was documented and forwarded.

Iterative Rapid Prototyping

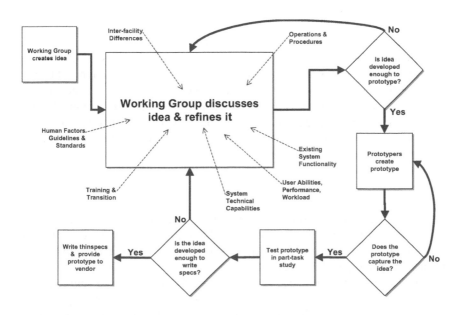

Figure 17.1   Flowchart showing the rapid prototyping process (Allendoerfer and Yuditsky, 2001)

The working groups embarked on a process of iterative rapid prototyping, following the process shown in Figure 17.1. The highlights of this process are the meetings of the group that include gathering and analyzing information from many sources, followed by multiple assessments, validations, and three critical decision points.

Decision 1: Is the idea developed enough to prototype? Prototyping was rapid but not immediate. Not every idea invented during brainstorming could or should have been prototyped. It was not necessary or even desirable to achieve full consensus on an idea before it was prototyped. It was important, however, for the working group to agree that the idea had enough merit to proceed further. The group agreed how each idea would behave and appear. The ideas were then expressed clearly to the prototype developers. The developers asked for clarification when the ideas were unclear or as issues arose during the development.

Decision 2: Does the prototype capture the idea? The engineers who created the prototypes were not present for every working group meeting or decision. This led to misunderstandings about aspects of the idea or how to implement it. Before the merits of an idea could be assessed, the working group needed to agree that the prototype accurately reflected the idea. If the prototype was incorrect or incomplete, the working group clarified the idea to the developers who then revised the prototype.

Decision 3: Is the idea developed enough to write specifications? The group decided when an idea was sufficiently mature or which idea was best among several. Included in this decision was whether the idea, as prototyped, met the operational requirements, was straightforward and simple to use, minimized human error, and allowed users to complete tasks effectively. If a prototype did not achieve these goals, the idea was judged to be incomplete and was returned to the working group for more discussion and development. Passing this decision point typically required several iterations of the previous steps, depending on the complexity of the idea.

'Thinspecs' and Decision Papers

'Thinspecs' is the name the working groups gave to documents that were developed alongside the prototypes. The purpose of the thinspecs was to describe the prototype in a format similar to a traditional specification document and to provide details that the prototype did not capture.

Though the prototype provided good guidance for the Raytheon engineers, the thinspecs were their official user interface design documents. The thinspecs were considered 'thin' because they were written less formally than a traditional specification. This was because the working group members were not professional specification developers and the amount of time available to write the thinspecs was limited. The thinspecs were continually revised to describe the working group's latest decisions.

In some cases, the working groups developed formal decision papers in addition to the thinspecs. Decision papers were created when the decision involved areas other than the user interface. For example, the TDW working group created a decision paper that called upon the STARS Program Office to conduct periodic market surveys of display technology. This decision was not appropriate for a design document and was published as a separate paper instead.

CHI Validation

Once the working group agreed upon the prototype and the accompanying thinspecs, and the FAA formally tasked Raytheon to implement the changes, Raytheon modified STARS to reflect the group's decisions. Because user interface changes were made to both the Full Service Level (FSL) and Emergency Service Level (ESL) components of STARS, two groups of Raytheon engineers in Massachusetts and California worked separately on the implementation. In some cases, Raytheon hosted Early User Involvement Events (EUIEs) at their facilities. These were opportunities for the working groups to see works-in-progress and to ensure that the implementation was proceeding correctly.

Once the changes were completely implemented, the modified versions were brought to the FAA William J. Hughes Technical Center for validation. The goals of the validations were:

- to ensure that the vendor had accurately implemented the designs described in the thinspecs;
- to ensure that the designs, which seemed satisfactory during prototyping, were still satisfactory in a more complex, dynamic, and interactive environment; and
- to determine if any new usability issues had been generated by the changes.

The validations repeated the methodology of the usability assessment. Users completed selected tasks in a realistic environment and provided their feedback to Human Factors professionals. The validations tended to be more structured than the usability assessments because most issues were already defined and understood.

Terminal Controller Workstation (TCW) Development

TCW Usability Assessment

The TCW usability assessment was conducted at the Technical Center ARTS Transition Laboratory. The laboratory contained four TCWs and four ARTS displays. Controllers from 12 TRACONs nationwide used the STARS equipment to complete ATC scenarios simulating Boston Logan International Airport (BOS) airspace. The controllers received training on STARS and the BOS airspace and procedures. Human Factors specialists sat with the controllers as they worked, took notes, asked questions, and administered questionnaires.

After the scenarios were complete, the Human Factors specialists consolidated the many issues that had been generated. The assessment participants reviewed the consolidated issues and refined them further. In the end, the assessment generated 98 usability issues, categorized into the following nine problem areas: data input, workspace ergonomics, on-screen windows, target symbol attributes, data block attributes, other display attributes, cognitive issues, on-screen menus, and system functionality (STARS Human Factors Team, 1997a; 1997b).

Air Traffic CHI Working Group

The Air Traffic (AT) CHI Working Group was formed to develop resolutions to the 98 issues. The group consisted of representatives from the FAA AT Requirements (ARS), AT Procedures (ATP), and AT Acquisitions (ATA) offices; the National Air Traffic Controllers Association (NATCA), the Professional Airway System Specialists (PASS), DoD, and FAA Human Factors. In addition, controllers from four large TRACONs were selected by NATCA to serve as working group members and subject-matter experts. These controllers had served as participants in the usability assessment, were already knowledgeable about STARS, and understood the Human Factors issues. Engineers from Raytheon attended working group meetings and provided technical guidance. During the history of the working group, the membership expanded to include tower controllers and members of the FAA Supervisor Committee (SUPCOM).

TCW Prototyping

The working group began to develop solutions for each of the 98 issues. Their first activity was to categorize the issues into solution areas or, as they became known, threads of development. The threads were:

- display control mechanisms (no opaque windows);
- target presentation and history trails;
- input devices (keyboard and trackball);
- weather presentation;
- consistency between service levels; and
- tower display and console ergonomics.

The development threads were different than the nine categories developed during the usability assessment. The threads were solution-oriented rather than problem-oriented. That is, disparate problems that could be addressed with a single redesign were grouped together in a thread. For example, issues that had been categorized into the data input, windows, display attribute, menu, and cognitive issues problem areas when viewed from a problem standpoint, were categorized as Display Control Mechanisms from a solution-oriented, development thread standpoint, as shown in Figure 17.2. Each thread led to the creation of major user interface changes that addressed usability problems in many problem areas. In the sections that follow, we describe several of the most important and interesting threads.

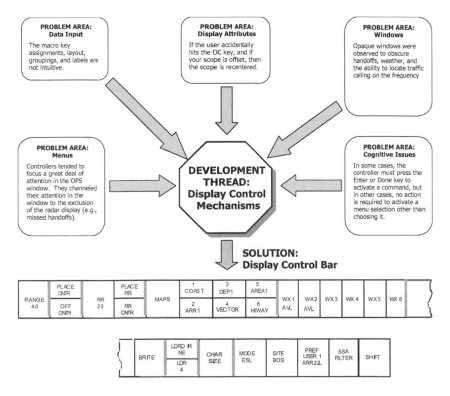

Figure 17.2   Example of how comments in disparate problem areas were grouped into a solution-oriented development thread that led to a single major design change

Before the group attempted to work on the threads, we identified several general design maxims to help structure the process. The maxims were not firm rules or requirements. Rather, they were intended to help the group focus on its priorities and justify changes. For example, one maxim asked that the interface design 'limit memory requirements and minimize controller actions and there should never be more than two levels to any menu hierarchy.' This maxim became the foundation for the changes to the display control mechanisms described in the next section.

Display Control Mechanisms Controllers adjust the original ARTS displays using physical knobs and switches located on the display console. These knobs change parameters such as range, brightness, character size, and weather levels. In the original STARS user interface, the display control functions were provided by on-screen windows and accompanying keyboard shortcuts. The windows overlaid and obscured the information behind them. This was a major criticism made by the controllers. They feared that aircraft would be hidden behind the windows and could go unnoticed until they emerged or caused a loss of separation. The windows needed to be eliminated or, at least, made transparent to aircraft. In addition, the windows had hierarchical menus that required a great deal of attention to navigate. Controllers naturally resist any function that causes them to look away from the radar for an extended period of time. Functions that were readily available, fast, and intuitive on ARTS seemed hidden, slow, and awkward on STARS.

The eventual solution was to create a toolbar, called the Display Control Bar (DCB) along the edge of the radar display, similar to toolbars used in many other graphical user interfaces. The DCB is not a drop down menu; no parts of the toolbar fall into the radar display when lower levels of the menu are accessed. Controllers interact with the DCB by moving the cursor ('slewing') or by pressing a function key on the keyboard that is linked to the graphical buttons on the DCB. Once the cursor is over a button, the function is activated by pressing the trackball Enter key. The layout of the DCB is shown in Figure 17.3.

The original DCB concept was developed at the FAA Research Development and Human Factors Laboratory (RDHFL) on a whiteboard. These rough sketches were presented to the working group where they were evaluated and revised. The working group discussed exactly how the design would operate to determine if it met their operational requirements. The sketches and descriptions of the behavior of each button were noted and provided to prototype developers for a first attempt at the look and feel of the interface. This was a relatively low-fidelity representation of the DCB functionality, created in Macromedia Director. The most important aspect of the low-fidelity prototype of the DCB was that it conveyed the basic visual characteristics of the DCB and was easy to modify quickly.

After multiple iterations of feedback and revisions using the low-fidelity prototype, the focus shifted toward operational realism. That is, the DCB seemed to be acceptable in isolation but how would it work when controllers were attending to targets, data blocks, and weather on the same display? This question could not be answered on the low-fidelity platform. The RDHFL had recently acquired a sophisticated prototyping tool called the ODS Toolbox (Orthogon GmbH, 1999) from German-based ISA-Orthogon (now Barco Orthogon). The ODS Toolbox has been used by the EuroControl Experimental Centre in the development of its user interfaces and provides extensive ATC libraries that developers can use to configure the display (Galushka and Vögele, 2000). An ODS Toolbox-based version was built of the prototype STARS user interface, including the new DCB, and simulated air traffic from the Technical Center Target Generation Facility

(TGF). The ODS Toolbox and the TGF together provided a rich environment for assessing the user interface.

RANGE 40	PLACE CNTR		RR 20	PLACE RR	MAPS	1 COAST	3 DEP1	5 AREA1	WX1 AVL	WX2 AVL	WX3	WX4	WX5	WX6
	OFF CNTR			RR CNTR		2 ARR1	4 VECTOR	6 HIWAY						

BRITE	LDRD IR NE / LDR 4	CHAR SIZE	MODE ESL	SITE BOS	PREF USER 1 ARR 22L	SSA FILTER	SHIFT

Figure 17.3   STARS Display Control Bar (DCB). The buttons are shown in two rows here but form a single row on the actual system

The DCB was probably the most critical thread that decluttered the display and provided the main display control method. By allocating a three-quarter inch band across the TCW display for the DCB, the design provided a clean radar display. In later iterations, the DCB was modified so that it could be positioned either horizontally or vertically along any of the four display edges. In addition, the initial trackball sensitivity parameters were modified to provide the right feel so that controllers could use the trackball without selecting the wrong DCB button.

Target Presentation and History Trails In a number of cases, a thread led to multiple proposed solutions. In particular, the design of the targets and history trails required many iterations and comparisons between proposals. In ARTS, radar targets appear in what controllers call a 'top hat,' with a small rectangle for the primary radar return and a thinner and longer rectangle for the beacon return. The targets orient toward the radar that generates them, as shown in Figure 17.4.

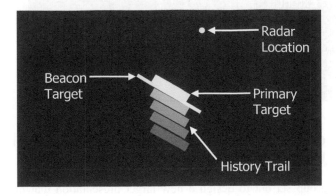

Figure 17.4   ARTS target presentation. The size and brightness of the targets have been exaggerated for the figure

When radar data are updated, new targets are drawn in new positions. For the first few moments after the update, the targets are very bright and easy to see. Then they fade very rapidly. This fade is an artifact of the phosphors used in the analog ARTS displays. By the next radar update, the targets have faded so much that they form a so-called history trail that is visible for as long as one minute. If a controller increases the brightness of the targets, the phosphors take longer to fade and leave behind a longer history trail. Controllers rely on the targets and history trail for the following information:

- current aircraft location (targets);
- direction of flight (overall direction of the history trail);
- aircraft speed (spacing of history trails); and
- turn initiation and rate (displacement of the target from the history trail).

The controllers in the usability assessment reported that the original STARS target and history trail presentation hindered their ability to make these determinations. More importantly, the targets, the basis of ATC, were simply too different from the ARTS targets for the controllers to feel comfortable and accept in a single upgrade. In particular, the original STARS targets did not distinguish between the primary and beacon targets and the history trails did not fade, as shown in Figure 17.5.

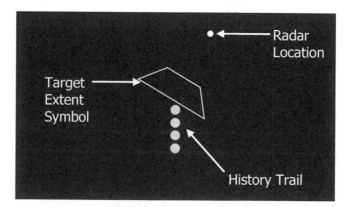

Figure 17.5   Original STARS target presentation. The size of the target has been exaggerated for the figure

The working group set out to design a new target presentation that provided the information needed by controllers in a more familiar format. Raytheon and the STARS Program Office concluded that precisely re-creating the ARTS would negatively affect STARS graphics resources. In particular, emulating the dynamically fading phosphors would require redrawing the targets and history trail many times per second. Was the fading really necessary? What did it provide controllers? The working group determined that the fading histories provided the following benefits over non-fading histories:

- fading histories create less clutter on the display than non-fading;
- fading histories do not look like any other symbols on the display; and
- fading histories make it easy to determine which trails belong to which target.

The working group developed and compared numerous proposed symbols and drawing methods. The best of these options were directly compared in a 'fly off.' The fly off was conducted using the STARS prototype displaying maps, weather, range rings, and so on. Targets were flown over the other display elements and the controllers were asked which design best met their needs.

The working group selected the design shown in Figure 17.6. The top hat target presentation of ARTS was preserved with separate primary and beacon targets. The targets orient to the radar and change in size due to distance from the radar and range setting. The targets and history trail preserve some of the fading of the ARTS but they do not dynamically fade. That is, targets and history trails do not fade *between* radar updates. Rather, when a new radar update is received, the previous target position becomes a history dot and steps down one level in brightness.

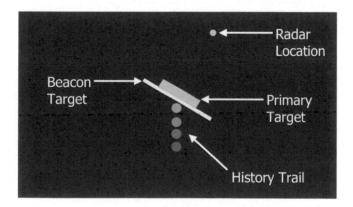

Figure 17.6   STARS target presentation. The size and brightness of the targets have been exaggerated for the figure

From a Human Factors standpoint, the other proposed target and history designs met all the information requirements. However, it was very important to the controllers to focus on solutions that were similar to ARTS. Given STARS was needed in the field as quickly as possible, the working group selected the most familiar, least controversial designs. In addition, this design preserved the positive aspects of the ARTS history trails discussed earlier.

Thinspecs and CHI Validation

After the working group completed the prototyping process, their decisions were documented in thinspecs. The working group reviewed the thinspecs and prototypes until they agreed that they accurately reflected their decisions. When the working group was satisfied with the thinspecs, they provided the thinspecs and prototype to Raytheon as government-furnished information. The prototype served as a 'visual spec' that provided an unambiguous reference for the thinspecs.

Raytheon implemented the changes into the baseline STARS software. Once this was complete, the structured usability assessment was repeated to validate that the original usability problems had been corrected and no new ones had been created. In general, it was found that this had been successful (STARS Working Group, 1998). Several features of the user interface, however, were not completely satisfactory and needed further development. These items were:

- target symbol presentation, including history trails;
- trackball sensitivity;
- highlighting, brightness, and colors in the DCB; and
- display and indication of weather.

The working group took the findings from the CHI validation and continued to work on the designs through the prototyping process until the group judged them to be satisfactory and approved them. We conducted a CHI validation for each major release of STARS. In each case, a smaller number of new functions were added and fewer new issues were identified. The validation for each version occurred approximately 12 months prior to initial operations at the key site using that version. As of 2003, we had conducted the following validations:

- STARS Early Display Configuration (EDC) for ARTS IIIA sites (December 1998);
- STARS EDC for ARTS IIIE sites (March 2000);
- Full STARS Release 1 (December 2000);
- Full STARS Release 2 (June 2001); and
- Full STARS Release 2+ (November 2001).

Tower Display Workstation (TDW) Development

Bringing STARS into the tower domain required its own Human Factors engineering and evaluation process. The TDW contains the same functionality as the TCW and controllers may work in both the tower and TRACON environments.

Therefore, it was a design goal for the TDW to preserve as much of the TCW user interface as possible. The functions, symbols, colors, and so on are nearly identical between the two. This reduces training costs and reduces confusion at 'up-down' facilities (Mogford, Krois, and Allendoerfer, 1999). Although the user interface is essentially the same, the tower environment requires smaller, daylight-readable displays and moveable input devices. Because the new STARS user interface had been designed for the TCW, the smaller display real estate in the tower required some additional modifications to the user interface.

The tower is a very different working environment from the TRACON. Controllers primarily look out of the windows to identify and track aircraft close to or on the runway surface. The radar display is used as a reference to keep track of arrivals, departures, and overflights. The ARTS tower display is the Digital Bright Radar Indicator Tower Equipment (DBRITE). The DBRITE is a 20.5-inch (52 cm) monochrome green CRT that typically hangs from the ceiling of the tower or is sometimes mounted in or on the console. This monitor provides a contrast ratio of 1.7 to 1 under 6000 foot-candles (fc) of incident illumination.

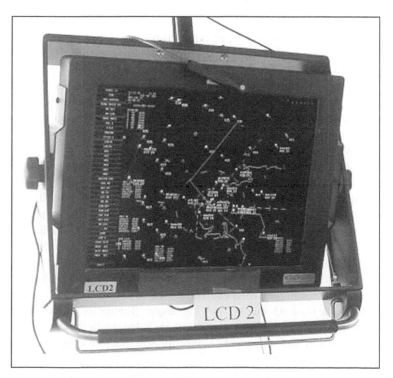

Figure 17.7 Prototype TDW mounted in an articulating arm

The STARS TDW is a 20.1-inch (51 cm) active matrix LCD monitor with anti-reflective filters and a special brighter backlight to increase its usability in the bright ambient conditions of the tower. Earlier laboratory testing showed that this monitor meets the STARS contrast ratio requirement for the colors that STARS uses for primary data (white, green, yellow, red). The TDW is shown in Figure 17.7.

Monitor Legibility and Suitability

The aspect ratio and size of the LCD are different than the DBRITE, so the operational impact of these differences needed to be assessed. First, the LCD has a 1.25 to 1 aspect ratio rather than 1 to 1. This changes the amount of airspace that appears on the display. At the same range setting, more airspace appears on the LCD in the horizontal dimension than the DBRITE. For example, at a 30-mile range setting, the DBRITE presents 30 miles in the vertical and horizontal dimensions. The LCD, however, presents 30 miles in the vertical and 38 miles in the horizontal. Controllers can use this extra room to display tabular lists or toolbars and remove clutter from the main radar area.

Second, the LCD is about 13 per cent shorter and 8 per cent wider than the DBRITE. This changes the scale of the map and reduces the distance between targets and data blocks on the display. For example, at a 30-mile range setting, the DBRITE displays 2.1 miles per inch. At the same range setting, the LCD displays 2.4 miles per inch. In theory, controllers could zoom in proportionally and achieve the same scale. In the example, controllers would need to zoom in to about 26 miles to achieve 2.1 miles per inch. However, the amount of airspace that controllers must monitor is defined procedurally so they may not be allowed to zoom in enough to achieve the same scale as the DBRITE. This caused considerable concern among the controllers on the working group. Raytheon thoroughly reviewed state-of-the-art display technology and confirmed that the proposed monitor was the largest commercially available LCD that that could meet the ambient illumination requirements. As a result, the Human Factors testing included phases to determine if the slightly smaller LCD was acceptable to the controllers and a suitable replacement for the DBRITE.

In the legibility assessment, controllers read text on two LCDs with different anti-reflective filters and the DBRITE. Controllers stood 4, 7, or 10 feet away from the monitors at a 0 or 45 degree viewing angle. Simulated traffic and weather appeared on the displays. Lighting conditions included daylight with shades down, daylight with shades up, and night. Legibility was measured by the number of correct call signs recognized.

At the 4-ft viewing distance, the distance most frequently observed during tower field site visits, controllers were able to accurately identify almost all of the call signs (see Figure 17.8). There were no substantial differences between the scores for the three test monitors. However, the recognition accuracy fell off at the farther viewing distances, with the LCD displays not performing as well as the DBRITE.

Although there were subtle differences in the STARS and DBRITE fonts, as well as better controller familiarity with reading the DBRITE, there did not seem to be any clear explanation for the superior legibility performance of the DBRITE at the greater viewing distances. There were no consistent text legibility differences between the two LCD filter types. The same general trend in the results also applied to the 45-degree angle daylight and both night viewing conditions (for complete results of this work, see Lenorovitz, Krois, Mogford, Masterson, and Kopardekar, 1998).

The working group agreed that the legibility of characters was acceptable for viewing distances less than seven feet. This viewing distance should be achievable at nearly all sites, especially those with an articulating arm or multiple TDWs. For farther viewing distances, however, the working group recommended that controllers select a larger character size. STARS provides six user-selectable character sizes and the working group concluded that this provided enough flexibility.

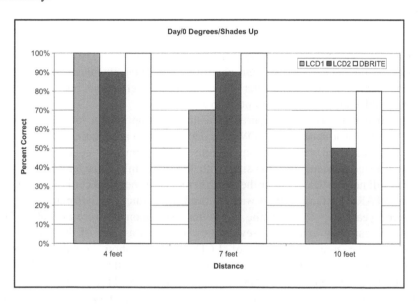

Figure 17.8   Sample legibility results comparing the DBRITE to two prototype TDWs with different anti-reflective filters. These data were collected under daytime conditions with the shades up from a 0-degree viewing angle (i.e., straight on)

The monitor suitability evaluations focused on the display real estate issue. We asked controllers to view each monitor and rate its suitability for ATC on a number of factors. Controller ratings for the LCDs at a 20-mile range setting were about evenly split between 'about the same as DBRITE' and 'worse than DBRITE.' At greater range settings, the ratings were less favorable. The rating and comment data showed that there was substantial concern that the LCD had significant shortcomings compared to the DBRITE, especially in the 60-mile, combined operations configuration. The main areas of concern were the size of the screen available for display of radar data, the perceived screen clutter, and problems with screen reflections and glare. These results, along with the data from the filter questionnaire, were also clear in stating that neither of the LCD filter combinations equaled the DBRITE in controlling or eliminating annoying reflections.

Working Group Agreement

The findings of the various assessments led to a formal working group agreement. The group agreed that the LCD was an acceptable replacement candidate for the DBRITE and the program proceeded to acquire, test, and deploy the LCD given certain conditions were met (STARS TDW Working Group, 1998). Some of the conditions and the steps taken by the working group and the STARS Program Office are listed below.

- A contrast-increasing filter was required. Ultimately, the TDW was deployed with the non-etched, 60 per cent neutral density filter examined in the TDW suitability assessment. The filter still created some reflections but the working group judged it to be the best option available.
- An articulating arm was required as a mounting option for sites where it is appropriate. Ultimately, the TDW was deployed with articulating arm, desktop mount, and fixed ceiling mount options. The original articulating arm was found to be unacceptable for sites with heavy seismic considerations. These sites will need to select either the desktop mount or the fixed ceiling mount.
- The STARS Program Office was required to conduct another market survey several years later to determine if a better TDW monitor were available. The FAA is currently evaluating several new monitors as part of this survey. In general, the new generation of LCD monitors show greatly improved performance in clarity, brightness, and viewing angle than the original TDW LCD. The newer generation monitors, however, are not larger than the original LCD so the size issue remains unresolved. The new generation monitors will be used in STARS deployments in the 2005 timeframe.
- Some changes to the user interface were necessary, primarily to accommodate the different shape of the TDW and the brighter viewing environment. The TDW user interface included these changes in Full STARS.
- Airway Facilities (AF) concerns were required to be addressed. During the TDW formal tests, AF technicians identified the relatively short life of the LCD backlight bulb and the potential for the articulating arm to damage the cables feeding the LCD as important maintainability issues. These issues were resolved by agreement with PASS prior to Full STARS becoming operational at Philadelphia International Airport.

Monitor and Control Workstation (MCW) Development

A critical component of keeping ATC operations running smoothly is the technical health of the system. The hardware and software must be configured, monitored, and maintained. System faults must be addressed quickly. In STARS, maintenance technicians use the MCW to identify, troubleshoot, and fix system problems. They also use the MCW to configure the system for operations. Like the TCW and TDW, the MCW user interface evolved through a series of assessments, design activities, and validations. The specific issues, however, are quite different for the MCW. In systems that include many components, the monitor and control task is complex. The system must provide the user with an understanding of the overall system and its health. At the same time, the system must provide easy access to detailed information about individual components for accurate diagnostics. When there is a change in the system, the user interface must alert the user about which system components are affected and the severity of the fault.

MCW Usability Assessments

Two usability assessments were conducted for the MCW. In April 1997, a team of Human Factors specialists and one Airway Facilities (AF) subject-matter expert conducted the first assessment (Mogford, Rosiles, Koros, and Held, 1997). The team used a script of representative monitor and control tasks to test MCW functions for ESL and FSL. They identified 89 usability issues including inconsistencies between ESL and FSL, difficulties interpreting alerts and alarms, and awkward data entry formats. After nearly six months of revisions to the interface, a team that was more representative of the AF user community conducted a follow-up evaluation (STARS Human Factors Team, 1998). The team included Human Factors specialists, several AF systems specialists representing operational facilities from across the country, representatives from the STARS Program Office, DoD, and PASS. The team followed the same methodology and procedure script as the original evaluation and found that 47 of the original 89 issues had been resolved. They also identified 17 new usability issues. There were remaining problems in the areas of color coding, error reporting, fonts, tabular displays, and consistency between the subsystems. For example, neither subsystem used auditory alarms for catastrophic failures. Because AF technicians in the field are responsible for monitoring several systems simultaneously, a visual indication alone may not be sufficient to draw their attention.

MCW Prototyping

An AF CHI Working Group was assembled, following the model set by the AT CHI Working Group. In the sections that follow, we describe two areas of the MCW user interface design that the group addressed: consistency and organization of information. Many issues identified during the follow-up evaluation were related to these areas. The working group resolved these issues following the prototyping process described earlier. The solutions were initially drawn with paper and pencil. Once refined, they were implemented in an interactive prototype that was built in Microsoft PowerPoint with Visual Basic for Applications. The prototype allowed the working group members to evaluate how all of the solutions worked together. Thinspecs were developed based on the prototype and were provided to Raytheon for implementation. Validation activities were conducted for each major release of STARS to ensure that the issues identified during the usability assessments had been adequately addressed. Any new issues that were identified during the validations were brought back to the working group for resolution.

Consistency STARS is made up of two subsystems, ESL and FSL, which add complexity to the monitor and control task and to the user interface. Each subsystem is monitored through an independent software application and the technicians must keep track of two sets of resources (e.g., processors, interfaces, networking equipment) simultaneously.

As shown in Figure 17.9 and Figure 17.10, the original FSL and ESL MCW user interfaces were very different. There was little consistency in the color coding schemes and the user interaction style between the two subsystems. This was a serious issue for the users because if FSL failed, the technicians would be forced to use the unfamiliar ESL interface precisely when fast, accurate maintenance actions were needed most (STARS Human Factors Team, 1998).

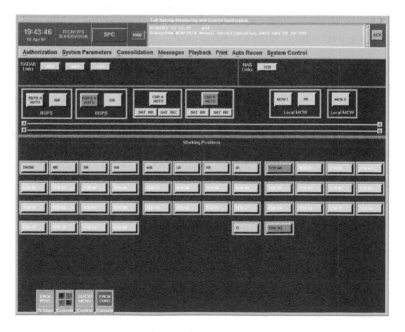

Figure 17.9   The original MCW user interface for FSL

Figure 17.10 The original MCW user interface for ESL

Maintaining consistency between the subsystems was established as a primary goal of the working group. They redesigned the ESL and FSL user interfaces to be as similar and consistent as possible. Pull-down menus were reorganized to provide the users with comparable menu structures. A single set of rules was established for indicating the health and availability of resources. Every change that was made to one subsystem was mirrored, to the extent that it applied, in the other.

Organization of information AF technicians are responsible for monitoring the status of many systems other than STARS. Therefore, the MCW user interface should provide the users with an at-a-glance status of the overall system and, in the case of a problem, an immediate impression of where the problem occurred. To achieve this objective, the information on the display must be organized in a way that is logical and intuitive to the technicians. That is, the organization should match their mental model of the system and the monitor and control tasks. The AF technicians' mental model is largely rooted in how those tasks are accomplished using existing systems. The technicians consult schematic engineering diagrams that represent the flow of information through the system, identify where the flow is disrupted, and then troubleshoot from there. The original ESL and FSL interfaces represented the STARS resources categorically. The ESL resources were divided into groups such as Radar Gateways and TCWs, whereas the FSL resources were in groups such as Working Positions and Radar Links. Neither user interface provided any indication of how the resources are connected.

The working group found that the categorical organization did not match their mental model of the system. Furthermore, without a representation of how information flows between resources, it was difficult to quickly assess how a problem with one resource affected the rest of the system. The original categorical organization was replaced with schematic block diagrams (see Figure 17.11) that represent the basic architecture of the two subsystems.

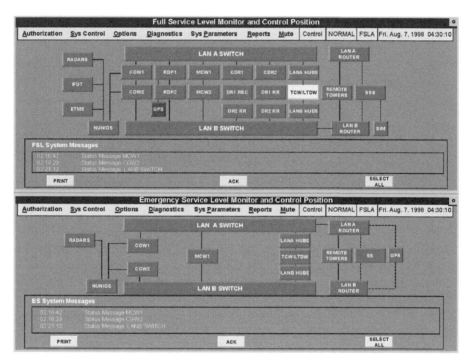

Figure 17.11 New MCW user interface prototype. Both subsystems are monitored on a single screen and have a consistent look and feel

Designing the schematic presented several challenges. The first was deciding the appropriate level of detail to display. Presenting too much information would have resulted in a cluttered display where visual alarms and alerts could be overlooked. Too little information would have resulted in a top-level display that could not be used effectively to monitor and control the system. The users provided valuable input into this process by describing what information would be needed at a minimum to feel comfortable that the system was functioning properly. Once the appropriate level of detail for the top level was established, the working group designed second and third level displays that depicted increasingly greater detail in a hierarchical model. The users move down through the hierarchy by 'drilling down' or double clicking on elements in the display.

The working group found that presenting system information in the context of a schematic diagram was more effective because it was more consistent with how they understood the system and its components. This organization facilitates troubleshooting by showing the user not only which resources have been affected but also how they are connected to each other. It also provides them with the information needed to quickly assess how a fault may contribute to problems downstream so that they can take preventative action.

Lessons Learned and Other Issues

In this section, we provide a few lessons we learned during the STARS user interface assessment, development, and validation process. We hope that future user interface developers can learn from our successes and avoid repeating our missteps.

Prototyping Has Real Benefits

The benefits of prototyping are well established in the human-computer interaction literature (Hardgrave and Wilson, 1994; Wilson and Rosenberg, 1988). Creating and modifying a prototype, even a highly realistic one, is much faster and cheaper than making code changes to a fully functional, complex system like STARS. Operational code is subject to many requirements that prototypes need not meet. Prototypes have only minimal testing and documentation requirements. The size and complexity of prototypes are intentionally restricted; a small team of programmers can build and change one. The prototype is small and simple enough that every programmer on the team can understand and work on all the code. If changes cannot be made quickly, working group members may lose interest and momentum will be lost.

User Interface Requirements Should Be Written at a Detailed Level

The commercial off-the-shelf (COTS) acquisition strategy may be appropriate for specifying many areas of FAA systems but it is probably not appropriate for specifying user interfaces. As we learned in STARS, this is especially true when the intended users are highly specialized, organized, and accustomed to particular ways of operating. Without requirements that directly describe or contain pictures of the desired screens, buttons, menus, symbols, or colors, vendors will very likely create user interfaces that are not acceptable to the users. Success in other countries or domains is not a guarantee of success in the FAA. User interface design in the FAA is a process of balancing system and human performance requirements while also accounting for dynamic political, managerial, budgetary, and labor situations.

There Will Be Many Bad Ideas Along the Path Toward a Good One

In many cases, the working groups developed five or more designs before they found one satisfactory. Many ideas that seemed fine on the whiteboard were simply not usable in practice. The group sometimes had difficulty abandoning bad designs, especially when those designs resulted from hours of intense discussion or had been generated by one of the group leaders. Group members should be encouraged to adopt an attitude that all ideas should be submitted to scrutiny and that genuinely bad ideas, no matter how good they seem at the outset, need to be eliminated as quickly as possible.

Do Not Demonstrate Designs Too Early

'You never get a second chance to make a first impression' quickly became a motto of the working groups. Early in the process, there were several design proposal 'fumbles' that demonstrated premature concepts. The users, although receptive to new ideas, quickly became skeptical of any design that appeared unfinished or raised too many procedural or implementation issues. Potentially good ideas were eliminated because they were introduced without first thoroughly working out the design. To remedy this situation, we arranged dry runs with subject-matter experts from the Technical Center before the working group saw it for the first time. This lesson later became formalized as the first two decision points in the prototyping process described previously, 'Is the idea mature enough to prototype?' and 'Does the prototype capture the idea?' If the prototype developers could not answer both questions positively, the prototype design was not presented to the whole working group.

Design Decisions Have Short Lives

Many smaller design changes were made in nearly real time with users providing input directly to the developers as they tweaked the prototype. When the changes were more extensive, we needed to allow the prototype developers more time to program and test. By the time the prototype was ready to demonstrate, the working group members had sometimes already changed their minds about what they wanted to see. By breaking up the prototyping into threads of development, we were able to finish a round of design proposals, send them to the prototype developers, and stop working on that thread temporarily. This allowed the working group to accomplish things quickly while also giving the developers enough time to get the prototype ready.

Match the Fidelity of the Prototype to the Decision Being Made

Throughout the design process, the prototypes incrementally increased in realism as well as functionality. As the designs matured, so did the need for more realistic demonstrations and assessments. We found that the controllers needed a more robust venue to consider design implications from an operational and safety point of view.

First, the operational task needs to be nearly complete. That is, users must be able to accomplish nearly all aspects of the task to judge the individual pieces. For example, judging the acceptability of the DCB was difficult without the controllers also attending to their other ATC tasks like communicating and separating aircraft.

Second, the environment needs to be realistic. In particular, using the operational lighting levels and sitting or standing positions was critical for controllers to assess the implications of designs.

Third, for STARS functionality that was new or significantly different from ARTS, it was critical to understand the intent of the function and what procedures would be required to use it. For example, STARS multisensor mode provides functionality that the ARTS does not have. As such, accompanying operations concepts and procedures did not exist for its use so controllers had difficulty determining how its user interface should appear and behave.

The progression from whiteboard sketches through interactive prototypes to validations using an actual system served to increase confidence in our approach and validate specific design features.

Consider Technical Capabilities During the Design Process

There was one downside of using a prototype to demonstrate and refine the user interface. The prototype was robust and easily reconfigured but was not subject to all the requirements and constraints of an actual system. A number of the concepts developed by the working group were impossible to implement on the actual system given all the other requirements that STARS had to meet. It is important to have design maxims that constrain the solution set to an existing hardware and software architecture and to include the input of the system vendor in the development of prototypes. Engineers from the vendor can help keep the working group 'inside the box' and help them avoid situations where the users were shown something that they ultimately cannot have (Virzi, Sokolov and Karis, 1996).

Establish and Follow Exit Criteria

The iterative assessments, particularly the validations, intentionally strained the user interface by surrounding it with the other parts of the full controller task. For example, the DCB worked fine during a demonstration in the laboratory, but did it continue to perform well when a controller was attending to multiple aircraft and making radio calls? To the extent possible, we took a performance-based approach to measuring usability and ultimately user acceptance. During the part-task evaluations, we used rating scales, questionnaires, and rankings to help users compare the new design to the ARTS to ensure that STARS did not add undue workload or create new operational problems.

Form Truly Multidisciplinary Working Groups

The composition of a working group is a major factor in its success. The group should have a representative from the most important stakeholder organizations. However, the group cannot be so large that it is no longer agile and able to make decisions quickly. The members also must be empowered by the organizations they represent. Nothing sours a working group faster than having its decisions overturned by outsiders. In addition, it is critical to have engineers from the vendor as members or consultants to the group. This keeps the vendor up to date with group's thinking and helps the group avoid paths that are unacceptably difficult or expensive to implement. Participation in the working group also helps the vendor interpret the user interface requirements correctly when the time comes for them to implement the requirements in the actual system.

Remaining STARS User Interface Issues

Though the assessments, prototyping, and validations resolved most of the original user interface issues, some remain and others have arisen. We recommend that the FAA and industry engage in focused research and development work to resolve these issues. Many are not specific to STARS but reflect gaps in knowledge regarding terminal ATC user interfaces in general.

Color

STARS and other FAA terminal systems use their own color schemes and assign their own meanings. For example, in STARS, yellow data blocks are converging runway 'ghost' targets that are used for visualizing spacing. In the ARTS Color Display (ACD), yellow data blocks are point-outs. Significant research exists examining the maximally discriminable colors for ATC (Cardosi and Hannon, 1999) and readable colors on displays (Krebs, Xing, and Ahumada, 2002), but this work has not yet transformed into a terminal domain color standard. Having no standard color set forces each FAA program to invent its own and assign its own meanings. The lack of standardization makes developing procedures and training more difficult. It also makes it difficult for users who must remember different color sets and associated meanings across systems, which increases the chance for human error.

Target Presentation and Controller-Jurisdiction Indicators

As of 2003, the AT CHI Working Group continues to wrestle with how to best display targets and controller-jurisdiction indicators (CJIs, also called position symbols). Initially, the CJI was difficult to read because it was displayed as small green letters overlaid on green targets. Numerous improvements were made and the

CJI is now readable, but now the targets can be obscured by the CJI at certain distances from the radar and range settings. The root of this problem is the scaling of radar target symbols. When close to a radar or when the controller has zoomed out the display, targets appear very small and can be completely obscured. When far from a radar source or when zoomed in, targets appear large. Various proposals have been made to address this issue, but none so far has proved completely satisfactory. Ultimately, it may be necessary to use a non-scaling target or move the CJI into the data block similar to the Display System Replacement (DSR).

Data Block Offset Algorithms

The second most common data entry that terminal controllers make is to move data blocks. Various algorithms have been tried in ARTS and STARS to automatically move data blocks to prevent overlap. None of these has been widely adopted by FAA controllers. The latest STARS algorithm has been judged to be adequate and a definite improvement over earlier ones, but we remain skeptical that controllers will use it much. A good algorithm must account for readability and also provide ways for controllers to use the position of data blocks to convey meaning, such as placing data blocks of southbound aircraft on the left and northbound on the right. In addition, the algorithm should not distract from operations by making too many or too severe adjustments.

Data Entry Standardization and Simplification

The current STARS data entry syntax is based on ARTS, which evolved over 30 years as functions were added and modified. Many of the entries were arbitrarily chosen because they were the key combinations available when the function was developed. This syntax hinders training and does not follow any modern usability standard. Though the current controller population may resist such an effort, we believe that it is in the FAA's long-term interest to modernize the data entry syntax. A large percentage of the controller workforce will be eligible to retire during the next five years. New controllers, raised on mouse-based interfaces and standardized shortcut keys, may expect data entry methods more familiar to them.

Local Adaptation

Each terminal site has authority to optimize STARS for its own operation using local adaptation parameters. However, there is little standardization or research into how sites use these parameters. Will the sites use STARS in a way that makes best use of its capabilities or are parameters being selected because they mimic how things were done in ARTS? In addition, how can a site learn what other STARS sites have done? There is no good mechanism for sites to communicate their local strategies, configurations, and rationales to other sites.

Technology Refresh

STARS makes extensive use of COTS hardware and software. COTS technology is constantly changing as upgrades and new products are released. Manufacturers routinely phase out and stop supporting old versions when new ones are introduced. Because STARS is a multiyear deployment, models and versions that were new at the beginning of the deployment will be outdated and unavailable by the end. Sites at the end of the deployment schedule will necessarily receive equipment that is somewhat different from sites at the beginning. Once STARS is fully deployed, the hardware and software will require frequent upgrades to keep pace with the marketplace. For example, the TDW monitor selection process described earlier will need to be repeated approximately every four years as newer display technologies become available and the current models fall out of production. Even when every STARS site has a TDW, some will fail every year and the original model will no longer be available for spares. New models and versions will be phased in as the program runs out of the previous ones. The Human Factors considerations of technology refresh include: How well will upgrades integrate into existing equipment? Will modifications to the user interface be needed to make full use of the upgraded hardware or software?

QWERTY Keyboard

Though the ABC keyboard was a major component of the user interface redesigns, the QWERTY keyboard remains part of STARS. Standardized keyboards will be cheaper in the long run for the FAA and will ultimately reduce training needs. The AF technician community insists on the QWERTY keyboard, and it is required for most MCW functions. When new capabilities are added to STARS, such as improved weather information, new surveillance sources, or decision-support tools, the ABC keyboard may not have the functionality or literally enough keys to work. Once STARS has reached a facility and the controllers become comfortable using it, they may come to request the QWERTY keyboard. As the FAA hires new controllers who were raised on a QWERTY layout, the workforce may eventually demand it. In our opinion, the shift toward the QWERTY keyboard will occur eventually but not for several years.

Integration with Flight Data

Currently, all flight plan information is entered separately into the Flight Data Input/Output (FDIO) system and not into STARS. The FDIO has its own input and output devices. Having separate STARS and FDIO systems creates redundancy in equipment and training. More significantly, controllers using STARS do not have direct access to the flight plan information contained in the FDIO. This differs from the En-route domain in which controllers may modify flight plan information in the Host using their DSR workstations. In addition, controller tools in STARS, such as conflict alert, have no access to the flight plan data and are less useful as a result. STARS cannot make full use of flight plan information for the purposes of trajectory prediction or other advanced functions.

Eventually, the FDIO system should become part of STARS. When these functions are brought into STARS, there will need to be significant development to make the new commands work with the existing commands and, if the ABC keyboard is still exists by then, to allow controllers to make flight plan entries using it.

Visualization Tools

ARTS does not provide any tools to help controllers visualize separation or show predicted trajectories. STARS provides a tool called the Predicted Track Line (PTL), which extrapolates the current aircraft heading and speed up to five minutes ahead. The PTL functions similarly to the vector line used in the En-route system. In addition, STARS provides a tool called the Minimum Separation (MinSep). MinSep allows a controller to designate two aircraft and display how close they will come to each other and where this will occur, based on their current speed and heading. Because terminal controllers have never had these sorts of tools, it is not known if or how they will use them. STARS does not provide tools similar to the halo ('J-ring') or the graphical flight plan readout that are available in En-route. We recommend analysis of data from field sites to examine how the PTL and MinSep tools are actually used and to determine if additional tools would be helpful.

The user interfaces for the tools must motivate controllers to use them. We have seen little interest among the controllers to use the tools that STARS already provides. The tools must be designed to be so fast, accurate, and simple so that they spark the controllers' interest and earn their confidence.

The Future of the STARS User Interface

The redesigns of the user interface added major schedule delay and cost to the program. Without these changes, however, the AT and AF users would not have accepted STARS and the program likely would have failed. The experience led to important changes in the program and to the FAA acquisition process in general. Most importantly, STARS helped the FAA recognize problems with pure COTS acquisition strategies. The STARS program now follows a spiral development, 'build a little, test a little, deploy a little' philosophy. This is characterized by the development of prototypes, multiple small releases, and risk assessments at each phase. The program now has user teams with experienced and stable membership. The working groups have the knowledge, processes, and tools in place such that they are able to quickly address interface issues when they arise.

This trend is reflected in FAA acquisition programs in general. Programs since STARS, such as the Airport Surface Detection Equipment (ASDE)-X and the

Automated Flight Service Station Voice Switch (AFSSVS), have involved Human Factors engineers and user representatives during the requirements exploration phase. These programs have developed and evaluated user interface prototypes and provided them to the system vendors. We believe that this significantly increases the likelihood that these systems will have acceptable interfaces and reduces the risk of cost overruns and delays due to user interface redesigns during implementation.

User Interface Development Process for STARS Enhancements

Since the initial user interface development, new work has begun on enhancements such as the addition of advanced weather and surveillance data to STARS. Having learned lessons from the development of the baseline STARS interface, the program now involves Human Factors and user representatives in the requirements exploration, system design, implementation, and testing phases. The program now supports the development of prototypes and the conduct of simulation-based research before requirements are written. The overall process that will be followed in the user interface development work is shown in Figure 17.12. This process is marked by the following tasks.

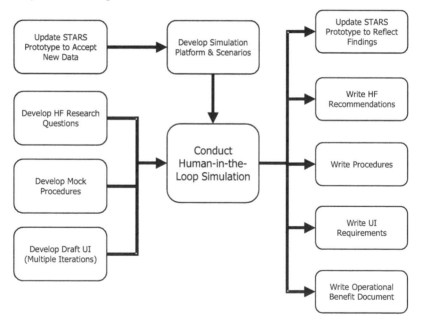

Figure 17.12 A general process for developing user interface and procedural requirements for STARS

Update STARS User Interface Prototype to Accept New Data As new capabilities are proposed in the terminal domain, we will modify our STARS user interface prototype to incorporate them. The research questions will determine the amount of integration necessary between the STARS prototype and the new external system. In some cases, incorporation of actual data from the external system will be necessary. In other cases, the prototype needs only to appear and operate as if it had incorporated the new system. For example, we have recently modified the prototype to accept advanced weather data from external weather processors. These data are extremely complex and cannot be realistically presented in the prototype without an interface between the prototype and the external processor. On the other hand, a new capability that adds information to the data block, such as an Area Navigation (R-NAV) indicator, is simple to emulate in the prototype without a link to an external processor.

Develop Simulation Platform and Scenarios We will develop a simulation platform and scenarios to examine the new capability. In particular, we will create scenarios where the capability would be most applicable. For example, to test a new weather capability, we would create scenarios that include the applicable weather patterns and associated traffic flows. Where appropriate, we will create scenarios that explore the complexities and practicalities of new operational concepts including the transition from the current operation to the new. For example, when a new surveillance capability becomes available, not every aircraft will be equipped immediately. We would develop scenarios with varying levels of equipage to explore the implications of the transition period.

Develop Human Factors Research Questions Each new capability to be incorporated into STARS will carry its own set of questions that should be addressed by Human Factors research. Many of these questions will focus on topics such as the following:

- How will controllers or technicians use this information?
- How can this information be best displayed and presented?
- How will the users interact with this information?
- How will the new information affect the information already on the display?
- What procedures will controllers or technicians follow?
- Does providing this information improve the ATC operation in a measurable way?

Develop Mock Procedures This activity will create mock procedures for users to follow during the simulation. The procedures should include any changes to responsibilities, airspace, phraseology, and so on. In the case of AF technicians, these would include mock written maintenance procedures and instruction manuals.

Develop Draft User Interface This activity creates a reasonably mature user interface for the simulation. All issues need not be resolved by the time of the simulation. In fact, the simulation may be the forum for resolving some issues. The interface should be mature enough that it does not significantly interfere with the controller's ability to accomplish the task.

Conduct Human-in-the-loop Simulation Conducting a simulation is critical for the addition of major new functionality. Many functions that seem satisfactory during laboratory demonstrations turn out not to be when the controllers are simultaneously doing all the other tasks they do in the operational environment. In particular, keeping track of aircraft and making radio calls significantly distract controllers from interacting with the automation system. There can be no truly thorough assessment of the user interface without asking evaluators to do these actions. The most important part of the simulation will be to demonstrate the operational benefits and utility of the new information or capability. Does the information improve ATC in a measurable way? Areas where operational benefit and utility can be tested include:

- safety;
- capacity;
- efficiency;
- workload;
- situation awareness;
- quality of service provision; and
- radio frequency congestion.

Update STARS Prototype to Reflect Findings The simulation, like the usability assessments and CHI validations, may reveal problems with the design of the user interface. Solutions to these problems should be generated following the prototyping process discussed earlier.

Some examples of user interface problems that may be identified are

- inconsistencies between the existing STARS functionality and the new capability;
- inconsistencies between the new information displayed on STARS and the same information on a standalone system;
- necessary information that is missing, hidden, or hard to interpret;
- functions and commands that are slow, cumbersome, or hard to remember;
- displays that are cluttered or confusing; and
- messages that are hard to understand or misleading.

Write Other Human Factors Recommendations In addition to changes to the user interface or the mock procedures, the simulation may generate necessary changes to other parts of ATC. For example, suppose the simulation reveals that the new capability requires too much additional effort for one controller to handle. The Human Factors recommendation might be that a second controller be assigned to the position to work with the new capability. These recommendations might be in the areas of staffing, training, working environment, schedule, and so on.

Write Procedures The findings of the simulation should be provided to ATP so that they may write the legal policies and national procedures. The national and local standard operating procedures will be a matter of negotiation between the FAA, the regions, the facilities, and the unions. As such, they may differ from the mock procedures used in the simulation. Once draft procedures are prepared, they should be compared to the validated mock procedures to determine if any differences merit additional testing.

Write User Interface Requirements Once the simulation has shown that the interface design works well in an operational environment, formal user interface requirements can be written by ARS or the Terminal Business Service (ATB). In our opinion, user interface requirements need to be more detailed and specific than most other areas of requirements in a COTS acquisition.

Write Operational Benefit Document Besides helping write procedures and user interface requirements, the simulation should help justify the operational benefit of incorporating the new capability into STARS. Some available capabilities may provide no operational benefit and should not be pursued further. Some capabilities may provide benefits that can only be measured qualitatively. Other capabilities may provide benefit as a standalone system but no additional benefit when incorporated into STARS. Some capabilities may provide benefit in one area but be detrimental in another. The ideal is for a new capability to provide measurable benefit without negatively affecting the existing functionality.

Conclusions

STARS is a significant step toward modernizing the equipment in the FAA terminal domain. The experience of acquiring, evaluating, and improving the components of STARS offered many lessons in working with the Human Factors issues of a large, complex system. Procedures and techniques were developed that eventually proved to be very effective in determining what needed to be done and carrying the changes through to completion. The crucial elements of this process were not only identifying and developing effective techniques and tools, but also creating and maintaining productive relationships with the users and developers. Human Factors engineering is both a technical and social endeavour. We present our experiences for the consideration of others who may become involved in similar projects.

Numerous Human Factors issues remain in STARS that should continue to be researched. The development and improvement of STARS may never be truly complete but a successful process is now in place to identify and deal with future Human Factors issues. We hope that the lessons learned from the STARS program will inspire more user-centered design methods throughout the FAA and will encourage programs to consider the user's needs as an essential component of the success of the system.

Acknowledgements

The authors thank all past and present members of the Standard Terminal Automation Replacement System Air Traffic and Airway Facilities Computer-Human Interface Working Groups for their participation in the activities described here. In particular, we thank the FAA Terminal Business Service, the National Air Traffic Controllers Association, the Professional Airway Systems Specialists, and Raytheon Company for the many contributions of their employees and members. We would also like to thank the numerous Human Factors specialists, simulation support personnel, and engineers who helped us plan and execute the assessments and prototyping activities described here. In particular, we would like to acknowledge the contributions of Paul Krois, Mark McMillen, and Robert Oliver for their contributions to usability assessment and development process.

References

Allendoerfer, K.R. and Yuditsky, T. (2001), 'Further lessons learned in the design of computer-human interfaces for terminal air traffic control systems', *Proceedings of the Human Factors and Ergonomics Society Annual Meeting*, Vol. 45, pp. 195-199.

Cardosi, K. and Hannon, D. (1999), *Guidelines for the user of color in ATC displays*, DOT/FAA/AR-99/52, Federal Aviation Administration, Office of Aviation Research Washington, DC.

Federal Aviation Administration (2002), *Blueprint for NAS Modernization 2002 Update*, Washington, DC.

Galushka, J.J. and Vögele, U. (2000), 'ATC system development – the role of visual specifications', *Air Traffic Technology International 2000*, Dorking, UK.

Hardgrave, B.C. and Wilson, R.L. (1994), 'An investigation of guidelines for selecting a prototyping strategy', *Journal of Systems Management*, Vol. 45, pp. 28-35.

Krebs, W.K., Xing, J., and Ahumada, A.J. (2002), 'A simple tool for predicting readability on a monitor', *Proceedings of the 46th Annual Meeting of the Human Factors and Ergonomics Society*, Vol. 46.

Lenorovitz, D.R., Krois, P.A., Mogford, R.H., Masterson, S., and Kopardekar, P.H. (1998), *STARS TDW Monitor Suitability Evaluation Results Report*, Internal report, Federal Aviation Administration, Atlantic City International Airport, NJ.

Mogford, R.H., Krois, P.A., and Allendoerfer, K.R. (1999), 'Human Factors lessons learned in terminal air traffic control system procurement', *Proceedings of the Human Factors and Ergonomics Society Annual Meeting*, Vol. 43, pp. 1022-1025.

Mogford, R.H., Rosiles, A., Koros, A.S., and Held, J.E. (1997), *Computer-human interface evaluation of the Standard Terminal Automation Replacement System Monitor and Control Workstation*, Internal report, Federal Aviation Administration, Atlantic City International Airport, NJ.

Orthogon GmbH (1999), *ODS Toolbox*, Version 4.06, Bremen, Germany.

STARS Human Factors Team (1997a), *Standard Terminal Automation Replacement System Human Factors review, Volume I*, Internal report, Federal Aviation Administration, Atlantic City International Airport, NJ.

STARS Human Factors Team (1997b), *Standard Terminal Automation Replacement System Human Factors review, Volume II*, Internal report, Federal Aviation Administration, Atlantic City International Airport, NJ.

STARS Human Factors Team (1998), *Report of the computer-human interface re-evaluation of the Standard Terminal Automation Replacement System Monitor and Control Workstation*, Internal report, Federal Aviation Administration, Atlantic City International Airport, NJ.

STARS Working Group (1998), *Standard Terminal Automation Replacement System Early Display Capability validation study: Issue results report from the working group meeting, December 17, 1998*, Internal report, Federal Aviation Administration, Atlantic City International Airport, NJ.

STARS TDW Working Group (1998), *STARS CHI Working Group TDW Hardware Recommendations*, Internal report, Federal Aviation Administration, Washington, DC.

Virzi, R.A. Sokolov, J.L. and Karis, D. (1996), Usability problem identification using both low- and high-fidelity prototypes, *Proceedings of the Association for Computing Machinery Conference on Human Factors in Computing Systems*, pp. 236-243.

Wilson, J. and Rosenberg, D. (1988), 'Rapid prototyping for user interface design', in M. Helander (ed.) *Handbook of Human-Computer Interaction*, pp. 859-875, Elsevier, Amsterdam.

Chapter 18

Human Factors and the Management of Change – a Personal Perspective

Billy Josefsson

Introduction

Human Factors (HF) in air traffic control is an area that can be very challenging but also very rewarding – if successfully introduced, managed and applied to daily operations. The characteristics of the service provided by controllers working at air traffic control centers are dynamic, the workload is sometimes high and regulations regarding airspace are frequently altered. Further, traffic is also on the increase in an already very dense airspace and capacity at the airport is at its maximum level. On top of this there has been an inadequate level of staffing over a long period. The equipment is sometimes a mixture of old and new, and can contribute to operational inconsistencies. It is obvious that management must be made aware of this in order to be able to react accordingly. Can safe and efficient air traffic service (ATS) be provided in such an environment? This chapter shares some thoughts and experiences which are based on the Swedish Civil Aviation Administration's (LFV) status concerning the Human Factors in air traffic management change process. The chapter provides an initial overview of how Human Factors has been integrated, and then focuses on two particular areas (safety culture and human-machine interface design) in more detail.

Recent Changes in the Swedish ATM System

LFV ATM System Change Number One

In 1994 LFV started the process of modernizing its national ATC system (see Figure 18.1 and Figure 18.2). Knowledge about Human Factors and system-design was in its early stages for LFV. Thanks to insight from high-level management, two motivated air traffic controllers (ATCOs) were encouraged to carry out studies at Universities focusing on areas believed to be critical for the ATM system change process at that particular moment. The areas studied were computer science, cognitive psychology, human-computer interaction, applied economics and ergonomics. The whole organization soon realized that Human Factors had to be included and evolving in order to keep pace with the surrounding activities. This in turn put demands on the LFV organization and management to create the prerequisites for such activities.

In the early stages, Human Factors (HF) was constantly confused with Human Resources (HR) and vice versa. There was very little knowledge and awareness of HF in the organization, whereas Human Resources was already established. In order to improve this situation Human Factors was included in management training and also explained to and discussed with all air traffic controllers at their yearly summit.

In the ATM system change process, a big step for LFV is the implementation of the transition to and modernization of its two major air traffic control centers during 2004. These units are located in Stockholm and Malmö. Together they comprise approximately 400 members of staff, who are in turn subject to significant changes. The 2004 Operational Concept includes changes towards a strip-less system, based on direct interaction with the flight label and online data entry.

Figure 18.1   Pre-modernization ATCC operational equipment

The starting point from a system perspective was a traditional national mainframe-based ATM system in operational use since the mid-seventies. There were four Area Control Centers (ATCCs) operating the airspace. The recent reduction from four to three caused a lively debate on several levels in the organization. In 2004 the third ATCC will be closed. This will happen despite of the fact that this particular ATCC is the most efficient ATC unit in Europe according to the European Performance Review Committee (PRC). Even though the closure is required to pave the way for future safe and cost-efficient air traffic services, there were and still are different views within the ATCO groups concerning the closure.

Figure 18.2 Post-modernization ATCC operational equipment

LFV ATM System Change Number Two

From 2004 onwards only two ATCCs will be controlling the LFV airspace and the plans for the future involve the development of one common Nordic Upper Area Control (NUAC) centre. The NUAC cooperation is believed to start with the formation of a joint company owned by Sweden, Denmark, Norway, and Finland. A harmonized work culture and climate in the countries involved must be ensured since the mutual understanding of the different actors' preferences forms the basis for further migration towards a common and defined safety culture. The relevance of Human Factors in this process is obvious to management, who support cross-border activities within this field. The management of cross-organization and cross-border issues is catered for within the Human Factors in Air Navigation Service (ANS) project (known as 'HUFA'). The HUFA project maps and identifies the ATM safety culture within the major control centers and ANS Headquarters. The outcomes from different activities are then fed back to the different local groups. The rationale for this is to help the control centers identify significant elements in the ATM culture, i.e. safety culture elements, and hence be able to use this material to communicate within and between the control centers and ANS headquarters. This method will be applied to the NUAC transition project. Suffice it to say at this stage that safety culture is seen by management as a key ingredient to success in LFV's ATM evolution, and that Human Factors is seen as the discipline supporting safety culture measurement and assurance.

ATM System Change and Follow-up

In general the process of modernizing the ATC system led to a focus on Human Factors issues and the creation of usable guidelines in order to address them. There was also an emphasis on a valid, efficient and interactive reporting and follow-up system to monitor the progress of the various items issues during the change process. The information and subsequent analysis should then enable proper follow-up actions to be identified based on the actual status. This mechanism should give feedback to the ATCO as well as providing some initial data analysis for the management. The presence of this type of mechanism is essential in order to be able to focus and identify the necessary actions. This means that as a general principle, the management of change and effects on HF issues was not only dealt with at a management level – the controller was kept in the loop as well. This 'inclusiveness' has become a hallmark of the LFV approach and a key ingredient of its success.

Fundamentals

In order to establish user trust for the HF issues it was necessary to start with something pertinent and visible without being vague. Confidence in the HF area was gained through HMI design, and also by highlighting HF aspects in the context of upgrading the present reporting system. In a wider aspect HF could be seen as a vehicle to keep the users in the loop, create commitment and hence be able to develop the ATC area which in turn comprises both 'soft' and 'hard' issues.

The incorporation and promotion of HF in ATM systems requires a commitment from management to invest the required resources. Further, it appears wise to establish a dialogue with management to ensure that all actors really understand the importance and potential leverage that can be obtained by adopting a true and pragmatic Human Factors approach to daily service whilst renewing and modernizing the ATC system.

The main objective of Human Factors in this process was to modernize the ATC system whilst at the same time ensuring the involvement by the end-user. At LFV management was aware of and committed to this and the project team saw the potential of pushing Human Factors further. As a starting point six areas were identified which were believed to be relevant from a HF point of view: marketing, and recruitment (MRU); Team Resource Management (TRM); incident report analysis and follow up; Critical Incident Stress Management (CISM); Ergonomics; and Human-Machine Interface design (HMI). These activities are elaborated briefly below, discussing some contextual examples.

Human Factors Areas in ATM

The overall strategy was therefore to select certain key areas for HF to focus on, rather than trying to tackle all possible areas. Figure 18.3 presents the six areas that were selected by LFV.

Incident Report Analysis and Follow- Up	Critical Incident and Stress Management (CISM)	Ergonomics
Human-Machine Interface (HMI) Design	Marketing and Recruitment (MRU)	Team Resource Management (TRM)

Figure 18.3 HF areas of focus for LFV

Marketing and Recruitment It is necessary to market the job of the air traffic controller, and then to select those who are able to successfully complete training within the limited time available. MRU (Brehmer, 1996) is a systematic way to deal with the cycle from marketing to controller validation, also incorporating the classification of the instructors. At LFV the success rate, including validation, is 79 per cent and the trend is stable (see also chapter 8).

Team Resource Management (TRM) training is introduced during ab-initio training. The TRM course raises awareness for team-related Human Factors issues and creates a common ground and terminology that help improve communication and safety within the ATM organization. TRM is therefore seen as valuable by LFV controllers and management alike (see also chapter 11).

Incident Report Analysis and Follow-up There are several reporting formats that contain valuable information. However, the majority are the incident reports that are filed in conjunction with any abnormality such as equipment failure, failure of systems to meet operational demands, loss of separation or organizational mishaps. The feedback to those concerned is through an investigation/analysis report which includes HF aspects relevant for the incident.

CISM (Critical Incident Stress Management) Even though incidents are unwanted they do in fact occur. CISM aims at alleviating the psychic consequences for air traffic controllers who are witnessing or, in some cases, contributing to an incident. Depending on the severity of the consequences debriefings are carried out by trained 'peers' and, failing a sustainable recovery, a psychologist is brought in to continue the treatment (see also chapter 10).

LFV learned that is also very important to share information concerning incidents as soon as possible in order to keep the workforce informed. It is felt that by sharing the experiences it is easier to avoid rumors of 'who and what and when'.

Ergonomics The layout of the controller working position and the location of equipment are often labeled as ergonomics in ATC. By using HF knowledge we hope to avoid the unfavorable working conditions often caused by the introduction of commercialized off-the-shelf (COTS) products such as the need to operate several keyboards, the operation of different input devices, and the need to correlate data from different displays. The short and long-term effect of using COTS products on operations will be closely monitored.

Human-Machine Interface Design is a very rewarding area for the application of Human Factors knowledge and it was indeed here that LFV first gained confidence for applied Human Factors. HF might be seen as constraining the design of the HMI but it does also contribute to a dramatic increase in safety, efficiency and usability. Design of the HMI and the Human-Computer Interaction (HCI) principles used are critical to the usability and acceptance of the system. One pitfall is that often the HMI design is pushed too far before the underlying system architecture and functionality are clearly defined. A lesson learned at LFV is that system usability testing should precede the definition and creation of the final HMI. The internal LFV Behavior and Style-guide (see below) for modern ATM system document is the main tool for the update and new design of any operational LFV ATM system.

Enlarging ATCO Competencies From an operational point of view it is obvious that the ATCO should be in control, should be 'in the loop' and able to stay there when carrying out the ATC task in a safe and efficient way. From a system development point of view it has also become obvious that the controllers must be 'in the loop' from the specification stage, capture of requirements, development, validation, training, operational acceptance and follow-up of the new system. LFV also learned that ATCOs familiar with Human Factors performed much more efficiently in the system development process. This means that there is a positive effect due to adding a competence on top of the ATCO profile. Additional competencies such as knowledge and experience in areas of safety case work, incident investigation, computer science, ergonomics, and Human Factors are becoming a sought-after profile for controllers.

Cooperation between Industry and Regulator LFV also has invested significant effort in closing the loop with industry in order to establish the transfer of operational HF requirements to industry. Further a continuous and good communication with the regulatory function overseeing ANSPs activities is very desirable. Often, operational HF issues are a a starting point for these discussions.

The current regulatory view is that industry has started to take note of HF aspects in a more beneficial way. Inclusion of Human Factors aspects in their specifications and prototype processes enables better HF practices to be integrated into the products. However, there is some variation in the degree of commitment from industry. One example is the definition and HMI design for the nationwide Voice Communication System (VCS) applicable to LFV operational sites (see

Figure 18.4). The original contractor was unable to implement the desired functionality and HMI. The only way forward was to terminate the relationship and seek a solution elsewhere. Thanks to an ongoing user involvement and the use of generic HMI/HCI principles from both the new contractor and LFV staff, it was possible to reach approved site acceptance test for the first centre within seven months. LFV also found that a successful system development team within ATM needs at least operational, technical and HF knowledge in order to be successful.

Figure 18.4   Voice Communication Interface and equipment to the left

Summary of Overall Integration of HF into LFV

The starting point for promoting Human Factors is defined by the existing cultural and organizational situation. If certain HF key areas have already been defined in the organization it might be wise to map ongoing activities to these areas thus making HF as clear as possible. Furthermore, it is important to establish functional and relevant co-ordination in these areas, thereby making HF activities more visible and allowing the involvement of both management and ATCOs.

Presently the HF activities are well-integrated in the reporting and follow-up system and the LFV will continue refining this reporting and follow-up system. Concerning HMI/HCI and ergonomics these are now widely integrated into daily operations and the change process. The selection and training department has established quality functions to monitor its progress. The CISM responsibility is delegated to the local ATS manager but there is central support available. TRM is fully integrated in the operational organization. Thus, integration has been effective.

Having given an overview of HF integration into LFV, the following two sections focus in more detail on two particular aspects of the work, namely safety culture (and safety climate) and HMI design. Afterwards, the 'milestones' in the integration of HF from 1985 until the present are summarized, before making some final comments about the overall program.

Safety Culture and Safety Climate

As noted in the introduction to this chapter, management at LFV has recognized the importance of safety culture, of attitudes and ways of working safely, during a period of extensive change. Safety depends on appropriate attitudes and working habits, as they exist in their various operational and management 'sub-cultures' in an organization, and also to an extent on the organizational climate and 'style' that allows safety culture to flourish during such change periods. The work outlined below therefore addressed two sections separately, namely safety culture and safety climate, the first dealing with more specific safety attitudes, and the latter with more diffuse organizational climate aspects that may nevertheless affect safety, positively or negatively. Both of these attributes have been measured using largely subjective measures (since they are subjective phenomena), with a view to understanding the trends in these areas for LFV. Since both of these measurements are relatively recent, only the presentation of results, and not the managements' nor controllers' reactions to those results, can be presented.

Safety Culture Measurement

The LFV has established, over a long period of time, joint activities between the university and the ATM division. However, the exchange and mutual benefits have been intensified since the ATCC in Malmö Center started a very focused and pragmatic work program. The subject of that program was to understand the underlying factors which have contributed to an increase in the amount of incidents reported. After about 15 different Human Factors studies had been conducted it was recognized that there was a difference in the ATM culture between LFV's two ATC centers. Differences in culture do not necessarily indicate any problems. However the conclusion was that this was an area that needed to be investigated and, if ATM culture could be observed, disseminated and understood, then a mechanism could be created with the objective of monitoring the ATM culture. The objectives of that particular culture study (Olsson and Åkerberg, 1998) were to discover and then describe the culture and sub-cultures and then identify the difference in reporting styles, participation, willingness to accept changes etc. During the dissemination of this study the first ideas for a future project were discussed together with the two ATCCs, management, user federation and Human Resources. The outcome was the HUFA project that focuses on safety culture and operational leadership for the two ATCCs and ANS Headquarters.

The first attempt to create a culture group in 1996 failed but the idea was reinforced when the Olsson and Åkerberg (1998) study was presented and disseminated. During 1999 and 2000 LFV/ANS were involved in a study (Andersen, Jensen and Madsen, 2001) that carried out a comparison between the Swedish and Danish ATC reporting cultures. The result was indeed surprising, noting not only the differences but also the large number of similarities. This culminated in the definition of the HUFA project, which comprises work in the area of defining the actual safety culture, making it visible and measurable as well as minute-by-minute operational leadership at the different ATC units. The definition of the HUFA activity took about a year to perform and it included dialogues with management, HR staff and the user federation. The result was a four-year activity funded by LFV/ANS. Staffing of this project comprises three PhD students and university professors. The particular expertise in the project includes organizational psychology, safety culture and analysis expertise working with the disturbance-effect-barrier analysis (DEB) approach (Ternov and Akselsson, 2004).

The first set of feedback from the HUFA project is currently under evaluation. The feedback process is twofold, first to the management itself in terms of how they perceive the safety culture and how it relates to the organization; secondly, how the staff perceives their management. Especially the comparison between the two centers and Headquarters is interesting: in general there is a harmonized view but differences exist in areas such as trust, and how to handle difficult issues. The positive attitude to safety was higher at management level than reported by the operational ATCOs. All agree that there is a blame-free culture; however there is room for improvement in terms of how to give feedback to the staff not performing as requested. A general objective with the HUFA project is to foster prerequisites for a learning organization that is able to deal with individuals and groups within the organization who deviate from the safety culture baseline.

Figure 18.5 presents the mean score for each safety culture (SC) dimension at the three study locations (scale range: 1-5). Generally, the study yielded positive results for all nine SC dimensions, with similarity across the three study locations. A strong and positive dimension was risk perception, which received the highest mean values. Communication received the lowest (ANS division). Generally, working situation, learning, reporting, attitudes towards safety and safety-related behaviors received high scores. Justness, flexibility and communication received somewhat lower scores (1). The results showed that at both ATCCs (Stockholm and Malmö) the top management group reported a better perception of attitudes towards safety compared to non-managers. At the En-route center the comparisons showed that individuals in a leadership position compared to non-leaders perceived significantly better the following: communication in daily work; better reporting; better learning; and more positive views on attitudes towards safety and safety-related behaviors. No differences in SC perceptions were found between leaders/non-leaders at the arrival and departure center. At the ANS division, individuals in leading positions compared to other groups had higher mean scores on seven of the nine safety culture dimensions (see Table 18.1). The existing differences in safety culture between different organizational levels could be due to different prerequisites at each level that will affect the answers even if the culture is the same.

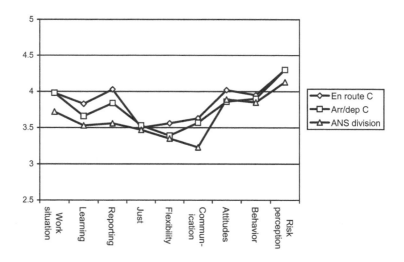

Figure 18.5   Mean score for each safety culture dimension

Table 18.1   Group differences in perceived safety culture dimensions (t-test)

	Safety culture dimension								
Group	Work situation	Learning	Reporting	Justness	Flexibility	Communication	Attitudes towards safety	Safety related behaviors	Risk perception
Leadership/ non-leadership									
En-route		*Lsp +	**Lsp +			*Lsp +	**Lsp +	*Lsp +	
Arr/dep									
ANS division		**Lsp +	**Lsp +	**Lsp +	*Lsp +	**Lsp +	*Lsp +		*Lsp +
ANS/ATCCs	***A NS-	**AN S-	***A NS-		*ANS -		***A NS-	**AN S-	

*p < .05, **p < .01, ***p < .001, 2-tailed; Lsp = Leadership
+/- = the group had higher/lower mean score on the dimension relative to the comparison group

Safety Culture – Detailed Results

ANS Division versus the Two ATCCs As compared to the personnel at the two ATC centers, ANS personnel reported a more negative view of their working situation, less flexibility in their work, poorer communication in daily work, poorer reporting, lesser learning and a more negative view of the safety related behaviors (see Table 18.1).

Communication in Normal Work A parallel result was found for the two control centers concerning the dimension communication. Respondents said they would receive the information they needed and at the proper time in order to perform their work in a safe way. A negative result was that large groups at the control centers and at the ANS division thought that they had not received enough training in emergency communication. This was reported by 32 per cent of the respondents at the En-route centre, 53 per cent at the arrival and departure centre and 49 per cent at the ANS office.

Justness The result concerning justness contained both positive and negative components. Positive components were that the majority of the respondents experienced that the organization to a very low degree wanted to find a scapegoat when something went wrong at work. Furthermore, the majority did not hesitate in taking initiatives in their work, because of anxiety of what would happen if it turned out wrong. However, approximately one third of the respondents at the respective control centers were of the opinion that those who did not perform their work in a safe way were seldom made aware of this fact. When the question was asked from an alternative point of view, i.e. whether those who performed their work in a safe way were acknowledged for this, a more negative result was found. About half of the respondents were of the opinion that they very seldom received such acknowledgement.

Safety Climate Measurement

Safety Climate Ekvall et al. (1983) define the organizational climate as a conglomerate of the attitudes, feelings and behaviors which characterize life in an organization. This definition of organizational climate is just one of many definitions in the literature. Even if there are some disagreements of the exact meaning of the term, most authors seem to assume that the organizational climate is rather stable over time with respect to attitudes and that it affects people's behavior. The organizational climate is important because it seems to affect different organizational and psychological processes. Communication, problem-solving, decision-making, learning and motivation can all be affected by the organizational climate. This in turn might affect the effectiveness and the productivity of the organization as well as the working environment and well-being at the work place (Ekvall, 1985), and it might also affect safety standards. The approach described and the supporting material is labeled GEFA (Göran Ekvall FA council).

Within the HUFA project different dimensions of climate according to the GEFA approach were studied at ANS headquarters and the two ATCCs using questionnaires and interviews.

The comparisons between the obtained GEFA mean scores and available reference data indicate the overall climate to be somewhat above average for the three units. With the exception of the score in the dimension 'Debate' all scores at the En-route centre are placed somewhere in between the reference scores. At the arrival and departure centre, the two dimensions 'Trust' and 'Conflicts' shows a somewhat more extreme score than the reference material. At the ANS division the dimensions 'Freedom', 'Support for ideas', 'Trust', 'Debate' and 'Conflicts' are close to but below the reference data from the ten innovative reference organizations. The rest of the dimensions are placed somewhere in between the reference scores, see Figure 18.6 (filled symbols indicate statistically significant differences).

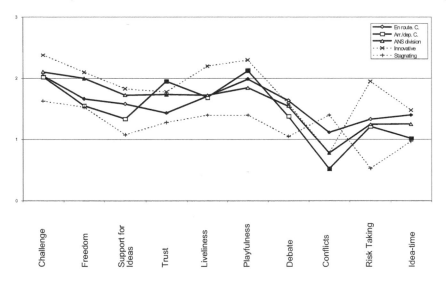

Figure 18.6   GEFA mean scores for each dimension and for each unit in comparison with reference data

Analysis of Variance

For two of the ten GEFA dimensions, 'Challenge' and 'Liveliness', no statistically significant effects were noted. For the other eight dimensions, significant main effects will be reported in Table 18.2 and Table 18.3.

The Arrival and Departure Centre versus the En-route Centre Significant main effects were found between the two ATCCs in four of the GEFA dimensions. At the arrival and departure centre, 'Trust', 'Playfulness/Humor' and 'Conflicts' were rated more positively than at the En-route ATCC independently of the air traffic controllers' work and position. However, 'Idea time' was rated more positive at the En-route centre compared to the arrival and departure centre.

Table 18.2   Significant main effects between ATC-unit, management position and work

GEFA-dimension	Challenge	Freedom	Support for ideas	Trust	Liveliness	Playfulness	Debate	Conflicts	Risk taking	Idea time
En-route C. – Arr./dep. C.				***A./d.+		*A./d.+		***A./d.+		**E.r.+
Administrative-Operative Managers-non managers		*Adm.+	*Adm.+				*Adm.+			

*p < .05. **p < .01. ***p < .001.

Administrative versus Operational Personnel A significant main effect was found for the GEFA dimension 'Debate', indicating that administrative personnel experience the organizational climate as characterized by more different views, ideas and experiences than the operational personnel.

Management versus Non-management No main effects were found for the management variable in the comparison between the two air traffic control centers.

Table 18.3   Significant main effects between ATM unit and management

GEFA-dimension	Challenge	Freedom	Support for ideas	Trust	Liveliness	Playfulness	Debate	Conflicts	Risk taking	Idea time
ANS – En-route C.		*ANS+		*ANS+				**ANS+		
ANS – Arr./dep. C.				*A./d.+				**A./d.+		
En-route C. – Arr./dep. C. Manager – Non manager				*A./d.+				**A./d.+		

*p < .05. **p < .01. ***p < .001.

ATCCs versus ANS division Significant main effects were found between the two ATCCs and the ANS division in three of the GEFA dimensions. At the ANS division the dimensions 'Freedom', 'Trust' and 'Conflicts' were rated as being significantly more positive than at the En-route center, independent of working position. At the arrival and departure centre, on the other hand, the dimensions 'Trust' and 'Conflicts' were rated more positive than at the ANS division.

Management versus Non-management No main effects were found for the management variable in the comparison between the two air traffic control centers and the ANS division.

Collaborative Decision-Making (CDM)

LFV fully acknowledge and understand that Human Factors aspects must be included during the ATM system change process in a way that comprises all actors. The ATM system change process is continuous and operational concepts are improved with methods, procedures and technology such as ADS-B, TCAS, and FMS development. Introduction of Collaborative Decision-Making (CDM) procedures between airlines, airport operators and ATM also put demands on the HF areas. LFV strives to carefully balance the technical part of the ATM system change process with proper amount of HF knowledge.

LFV Behavior & Style Guide for Modern ATM System

Below, an extract from the LFV 'Behavior & Style Guide for Modern ATM System' is presented. First, some high-level strategic leads and 'guiding stars' that LFV found very usable for prototyping and development activities are shown. Then, a sample of the HMI for a Departure List (Figure 18.7) is presented.

HF applied at HMI Level It is necessary to agree upon some proven principles, i.e. HF Guidelines that are favorable to the human operators' performance characteristics:

- minimum information principle (Hicks' law);
- minimize mouse travel (Fitts' law);
- minimize mouse clicks (prevent repetitive strain injuries);
- standards for the use of color (Merwin and Wickens, 1993; Reising and Calhoun, 1982; Martin, 1984);
- avoid needless tasks/load (task focus); and
- use of and adherence to display design guidelines (Wickens, 1992).

Figure 18.7 Sample departure list from LFV behavior & style guide

Adopting these principles in the HMI design process will contribute to usable and implicitly safe and efficient HMIs with the following characteristics:

- clean;
- able to focus on relevant information;
- few and subdued colors;
- direct manipulation;
- integrated actions;
- grey-shades and color coding;
- intuitive and object-oriented at the user level;
- flexible; and
- consistent and easy to maintain task focus.

Examples of three of these characteristics in terms of style guidance are as follows:

- *Clean* It will be possible to use the LFV HMI without having numerous windows and icons on the screen. Fixed HMI components will be accessed from the roll-down bar and from the roll-out forms and will only occupy a fraction of the total screen area when not activated.
- *Focused* The situation display will always be presented in the root window of the main display and can not be closed or 'iconified'. The radar information is put in the centre by using colors where it is relevant: in the target. The surrounding information uses different grey scales which put the target information in focus.
- *Context Sensitive* It will be possible to activate the LFV Components that represent most of the operators work from context sensitive pop-ups in direct connection to the different objects presented in the situation display.

LFV Operational Concept Document OCD 2007+

Below is an excerpt from the LFV 'Operational Concept Document 2007+'

Human Factors knowledge is considered to be well integrated in ATM activities. In particular, the following areas are identified as relevant in the OCD context: human-machine Interface design, human-computer interaction issues, system design, system follow up, reporting and monitoring activities, ergonomics, team resource management, critical incident stress management, recruitment, training and career development.

Most of the issues in the bullet points below have already been addressed in the design and realization of Eurocat2000E, I-ACS and the A-SMGCS. However, it is important to understand that these issues must be revisited when new ATM concepts and methods are being planned, designed and implemented. Whereas Eurocat2000E will form the platform on which the future ATM concept in Sweden will be built, the new concept includes entirely new components that will need very careful analyses and design. Even though many of these issues will be addressed in the EUROCONTROL EATMP work programme (with LFV participation), the integration of new concepts into the Swedish ATM system is a national matter.

- Human Factors principles and methods have to be built-in from the start to increase the human commitment to required ATM changes. Introduce systems thinking which integrates human activity into system design. The use of mixed teams where operational, technical and Human Factors expertise work together using an iterative development approach with a human centered perspective.
- The human and technical aspects of new concepts must be carefully balanced by asking what is operationally safe and useful, rather than what is technically feasible. Non-technical options are often ignored even when they offer greater chance of success or greater cost-effectiveness.
- Involvement of operational ATM staff in HMI requirements definition, the design of operator workstations and new operating methods as well as in their implementation is essential to achieve real usability and acceptance of the new technology.
- New systems are expected to reduce current forms of ATM-related errors, but at the same time will introduce new forms of errors. These new errors could relate to loss of operational awareness, over-reliance, mistrust and skill degradation, and should be mitigated by careful error analysis and the design of error-tolerant and fault-resistant concepts.
- User acceptance of the new concepts is essential for their success. Operational staff, through adequate training programs, must be made aware of the overall benefits offered by the CNS/ATM system and its translation into concrete concepts that will influence their daily work. Likewise, operators should be made aware of evolving technologies that will have an impact on the design of ATM concepts.

Guidelines such as those above have been accepted by the company and are therefore in use in design, development and upgrade activities. Such guidance in the form of internal company 'style guides' help to reinforce the adoption of good Human Factors practice, in a way that has HF well-integrated into design engineering processes.

Human Factors Milestones for LFV/ ANS

Table 18.4 below describes significant steps and action within the field of HF for LFV/ANS. It should be noted that without the open discussion and the sharing of information between Human Factors specialists at Eurocontrol, FAA and other Research & Development sites and Universities the substance in the table below would have been significantly shorter and less valid.

Table 18.4   Major milestones in the integration of HF into LFV

Year	Issue /event	Approach	Result/ lessons learned
1985	First ATCC closure	Top Down decision	Hard feelings
1990-	Change to a new marketing, recruitment and selection model (MRU)	Left the pen and pencil tests for task dependent tests and situational interviews.	Over a 10 year period an increase from approximately 50 per cent to 79 per cent throughput for ATCO training.
1994	New ATC system (S2000) definition phase	HF included in the system development process, mainly HMI and ergonomics.	High usability ambition not supported fully by system functionality
1995	Inclusion of HF issues for incident investigation	Include HF in the report and analysis and attempt to communicate and present findings in terms of HF. This was possible thanks to close collaboration with and attending courses at the Human Factors in Aviation (HFA) centre of expertise. Further this work was given to Lund University and the change@work centre of expertise	No reaction but training for HQ staff was successful and a yearly workshop is established since 1997 which comprises the ATCO union, pilots, scientist and HQ staff.
1995	First draft of 'LFV/ATM Behavior & Style guide for modern ATM systems'	Based on proven HCI principles and prototyping with trained users.	Significantly contributed and still contributes to harmonization of all the operational ATM systems as well as methods.
1996-	General promotion of HF in the organization	Information to ANS staff about HF using the intranet to host facts, studies and information. A HF discussion forum is available but the usage is very low. HF is a mandatory topic in the management training.	Broadened interest and general acceptance of HF as an integrated item. This means that management now is the one who often raises the HF issue in the first place.

1996	First attempt to create a 'culture group' with delegates from incident investigation with very good insight into operational issues, operational managers from the major air traffic control centers and HF expertise from ANS headquarters.	Harmonize the two major ATCC operational issues. The follow-up system used at that time indicated differences between the two sites in terms of willingness to report and what they reported.	Was turned down by high-level management because it was considered to be established already.
1996	SweDen96 real-time simulation at the EUROCONTROL Experimental Centre.	Pragmatic evaluation of several aspects including HMI and ergonomics.	Confirmation and lead for further development. It is all connected, chair, console, HMI, lightning, procedures etc.
1996-99	Initiation of Denmark Sweden Interface project (DSI) followed by the DSI TWR project.	Definition of ATC functions and HMI with informed users. Team consisted of HF, ATCO and technical staff.	Confirmation of the composite team approach. Result used in specification and development of real systems at various sites.
1996	Promotion and establishment of mechanisms to cater for usability, user acceptance.	Acceptance from organization to include HF in all aspects of system / organization development i.e. human centered approach.	HF knowledge merged with management, operational and technical skills.
1997-	The creation of a HF 'smorgasbord' in order to provide focus for academia actors.	Inventory of problems and anticipated problems by talking to staff and management.	A very thorough coverage of the issues.
1997	Creation of an ANS HF strategy.	A quite broad description of HF activities relevant to LFV.	The 6 HF areas: Marketing & Recruitment, Incident report analysis & follow up, HMI / HCI, CISM, Ergonomics and TRM was found useful.
1998	First study addressing culture differences between the ATCCs.	Questionnaires and interviews and dissemination feedback with key staff.	Awareness of differences and management was given a rough 'tool' to deal with the differences in climate and culture.
1998	SweDen98 real-time simulation.	Observation of HF aspects. Methods, procedures linked to safety.	The operational concept was working and confirmed. Further it was a useful case and preview for method development.

1998	LFV became member of Eurocontrol Human Factors Sub-Group (HFSG).	Interactive membership, involvement of operational sites and staff in Sweden.	Increased awareness of HF in LFV organization and understanding of how and where HF should be applied.
Since 1998	Workshop with focus on Flight Safety and HF ATCO user federation – HQ – Airspace Users.	Focus on safety highlighting common operational issues including educational aspects.	Very appreciated and is building network and increasing awareness about HF within the air transport sector relevant to LFV.
1999-2000	Definition of the HUFA (Human Factors in ANS) project.	Addressing the two ATCCs and ANS HQ with the same approach concerning safety culture issues.	Awareness of different focus but also highlighting common views on safety culture. HUFA acknowledged as containing the fundamentals for harmonizing operations.
2000	Client in the HERA project (see chapter 4).	Introduced a new and fresh way of thinking for the incident investigation staff.	
2001	Inclusion of HF in the national Operational Concept Document (OCD).	Pragmatic approach that defines what, when and how for the actual HF item.	Appreciated and enabled a possibility to map HF activities development of other items.
2002-	Update of the HF strategy.	A cooperative activity within ANS HQ, ATCCs, and the union.	
2002	First interactive debriefing of the HUFA phase 1 to the stakeholders.	Distribution of report followed by two half day seminars with key staff.	It is now up to management how they progress with further dissemination to their staff. Feedback to the project is mandatory.
2002-2003	HF considered well-integrated into the LFV ANS organization.	Handbooks and manuals reflect applied HF within ATM	LFV have an urge to share such experiences.
2003	Preliminary definition of HUFA II started.	Will comprise ATCC, TWRs and ANS HQ. Cross-border activities will be performed.	Not yet known.

Conclusions

Trying to assess the amount of work spent in the HF area in the period between 1994 and 2004 for LFV is very difficult but rough figures can be retrieved. An effort of approximately 100 person-years was invested in the HMI/HCI and ergonomics activities. Incident reporting and follow-up amounts to approximately 20 person-years. MRU, CISM and TRM activities account for approximately 30 person-years. The HUFA project has been running for 4 years and a continuation is foreseen. There are some pertinent HF activities going on such as the HUFA project, but the paradox is obvious. The overall goal is to integrate HF activities within the ATM lifecycle and when that is happening no one will label a certain activity as HF, it will just be integrated. One might need to be sensitive to and detect the consequences of the slow but never-ending ATM system change process, accepting that new areas that must be subject for HF activities will emerge. The approach when addressing new HF issues should nevertheless remain co-operative, as this has been a key ingredient to the successful integration of HF into LFV and its management of change process.

What are the Residual Challenges for HF in LFV?

Five key Human Factors challenges remain for LFV in the future;

- maintaining a Human Factors friendly culture in the ATM community;
- monitoring the ATM Safety Culture development in the light of mergers and takeovers (Single European Sky deployment) and reacting accordingly;
- assessing the HF role for TWR /TRACON operations;
- creating ownership of HF activities by communicating and remaining proactive; and
- defining a real usable HUFA II project.

References

Andersen, R.J. and Madsen, D. (2001), *Air Traffic Controllers' Perceptions of the Reporting of Human Errors*, Technical Report R-1249, Risø National Laboratory, Roskilde, Denmark.

Brehmer, B. (1996), *Issues in the Selection of Air Traffic Controller Candidates*, Report No 24, MRU project, LFV, Sweden.

Ek, Å. (2002), *Human Factors in ANS*, LFV internal progress report, Lund University change@work, Sweden.

Ek, Å., Akselsson R., Arvidsson M., Johansson C.R. (2003), 'Safety Culture and Organizational Climate in Air Traffic Control', *Proceedings of the XVth Triennial Congress of the International Ergonomics Association,* Vol. 5, Safety I.

Ek Å., Arvidsson M., Akselsson R., Johansson C.R., Josefsson B. (2003), 'Safety Culture in Air Traffic Management: Air Traffic Control', *Proceedings of the 5th USA/Europe ATM R&D Seminar,* Budapest, Hungary.

Ekvall, G. et al. (1983), *Creative Organizational Climate: Constructions and Validation of Measuring Instrument,* FA-rådet – The Swedish Council for Management and Organizational Behavior, Report 2, Stockholm, Sweden.

Josefsson, B. (1999), 'Integrating Human Factors in the Life cycle of ATM systems', *4th ICAO Global Flight Safety and Human Factors Symposium.*

Olsson, J. and Åkerberg, M. (1998), *Individual and Organizational Factors related to Air Traffic Control Incidents,* Undergraduate thesis at Lund University, Sweden.

Ternov, S. and Akselsson, R. (in press), A method, DEB analysis, for proactive risk analysis applied to air traffic control, *Safety Science.*

Chapter 19

Review of a Three-Year Air Traffic Management Human Factors Program

Barry Kirwan

Introduction

This chapter reviews a program of activities that took place between September 1996 and March 2000 in the Human Factors Unit for National Air Traffic Services (NATS), the main ATM service provider in the UK. The origins of the Human Factors Unit and the program are briefly summarized, and then ten brief case studies are outlined showing the types of impacts the program achieved. There is then a brief summary of the impacts, their cost-effectiveness, and the sustainability of the Human Factors initiative.

Disclaimer: The author was formerly head of Human Factors, Air Traffic Management Development Centre, National Air Traffic Services, Bournemouth, Dorset, UK. The opinions in this chapter are those of the author alone, and do not necessarily reflect those of his employers (past or present).

Events Leading to the Development of the Program

For some years prior to the program described in this chapter, Human Factors work had been carried out both by NATS itself and by its then 'parent' organization, the Civil Aviation Authority (CAA: e.g. see Hopkin, 1995). There had also been a Human Factors presence at the ATMDC (Air Traffic Management Development Centre), in particular because the ATMDC was the place where many large-scale real-time simulations were carried out, with up to 30 controllers for three weeks per simulation (e.g. simulating new sector designs, new interfaces, etc.). Such simulations required a large amount of Human Factors support, in terms of experimental design, and development and analysis of controller measures (mainly subjective measures such as questionnaires and workload assessment). Additionally, some Human Factors work was carried out by contractors on a range of projects, including some support for the developing New En-Route Centre (NERC) at Swanwick, which although scheduled for 1998/9, was finally to go operational in 2002. NATS itself was at the time (and remains today) the major service provider for the UK. Its controllers were based at a number of UK airports (e.g. London Heathrow, Gatwick, Birmingham, Manchester, etc.), and principally were in two main centers in London and Prestwick, handling all the En-route traffic in the UK region. In Prestwick, the controllers also handled the Northern and mid-European oceanic traffic to and from the United States, up to a point mid-way in the Atlantic (15 degrees west).

In the mid-1990s, NATS underwent a Business Efficiency Review. It was identified as part of that review that Human Factors was a growth area and likely to be important in future Air Traffic Management (ATM). Although there had been a significant Human Factors presence in NATS for some time, the Human Factors resource within NATS had been mainly focusing on the real-time simulations carried out at the ATMDC, with some limited Human Factors support to certain other projects. Human Factors work that had occurred had often been carried out by external agencies, or related ones such as the Civil Aviation Authority, or the Institute of Aviation Medicine at Farnborough, for example. The Review recommended instead that a more significant Human Factors 'presence' be developed. It was therefore decided to set up a NATS Human Factors Unit (HFU), and to base this within the Research and Development arm of the organization. The ATMDC was the most obvious location, since there were already five Human Factors staff there, and it had two additional advantages of being co-located with the College for ATC Training (CATC), and it was the 'home' of real-time simulations. It was also geographically relatively close both to the main existing ATM centre at London, and the proposed new centre site at Swanwick. It was decided to select someone from outside of ATM to run the new Unit, to bring a fresh perspective into the area, and to gain experience from other safety and HF-critical industries. Therefore in September 1996, a Head of Human Factors with a track record in the nuclear power industry was recruited externally, and the Human Factors Unit formally came into existence. The new position reported directly to the Head of ATMDC, who reported to the NATS Director for R&D.

Although the Human Factors Unit was 'new', it had an ongoing work program and could thus not start with a 'clean sheet'. Therefore from the outset, the main work was in support of real-time simulations. These are simulations of proposed new airspace procedures or 'sectorizations', or else simulations testing out new equipment or interfaces. The Human Factors personnel assisted in designing such experiments to test certain factors and proposed controller working organizations and also to measure mental workload and a host of other measures to determine the impact of the proposed changes on human performance and sector efficiency. The HF personnel worked closely with other scientific analysts and controllers in the preparation, running and analysis of the simulations. There was also prospective workload analysis for future tools using a tool called PUMA (see Kilner et al., 1998) – this work was led by particular Human Factors Unit members and carried out in conjunction with other scientific staff and controllers. The support to simulations remained a major part of the HF Unit's work program through the next three years. This is partly because it was a major Human Factors responsibility to

NATS' developing airspace, but also because it gave the HF team members such good access to real controllers facing new airspace procedures and interfaces – the HF personnel therefore could keep good contact with operational reality.

There was also some additional Human Factors work ongoing outside the HFU at the time of its official creation, with external contractors assisting with the design of future controller tools, and a small team assisting specifically with the development of the next generation Air Traffic Control Centre at Swanwick, the New En-Route Centre (NERC). These respective teams were not under the auspices of the new Human Factors Unit.

However, the task of the new unit was to develop a strategy for addressing the critical Human Factors issues in existing and future ATM. Therefore some objectives were needed, to develop a strategic program, which could then be implemented and evaluated. Three years was specified by the Director of R&D as a reasonable timeframe in which to develop, implement, and see measurable results from a Human Factors program.

A Three Years Human Factors Program at NATS

Program Objectives

The first task was to broaden the Human Factor Unit's (HFU) remit, whilst still maintaining the support its members gave to the real-time simulations. It was decided early on that the Unit, with relatively limited resources and a fairly large standing commitment to real-time simulation support, could not address all Human Factors issues. A strategy was therefore developed (Kirwan, 1997b; 1997c; Evans et al., 1997) which focused the activities of the HFU on certain key areas. Three of these key areas were:

- human-machine interface (HMI) development;
- training;
- safety (human error assessment and reduction).

Some work had already started in the training area, and there had been a number of informal requests for HMI design support. Human error was seen as a possible future problem for ATM. This was because of the changes that were going to happen to ATM generally in Europe over the next two decades, in terms of a predicted doubling of traffic in the next fifteen years, and a related shift towards more support for the controller in terms of automated tools. Three major objectives of the program were therefore to assure the safety of the move towards this higher capacity and automation-enriched environment, to enhance NATS' operational success, and to give Human Factors support throughout NATS wherever it was needed. The Human Factors Unit was based in a Research and Development (R&D) environment, and therefore could focus on safety, design and operability issues. There were three main stages in the resulting three-year Human Factors program launched in 1997:

- understanding the controller – it was argued that certain research first had to be undertaken to ensure there was an adequate understanding of the controller's cognitive activities, to underpin the other two stages;
- measuring controller performance – the dominant measure at the time was workload, but new areas of situation awareness, and measures such as eye-movement tracking, were available and relevant to air traffic management;
- application of Human Factors – applying the understanding and measures gained in the above stages, and the knowledge and technique base of Human Factors as a whole, to a range of design and operational projects.

Since this paper is concerned with impact and cost-effectiveness, this chapter focuses on the third objective of the program (application). Nevertheless, the other two have been mentioned because they were part of NATS' investment in building specialized expertise within the Human Factors Unit. However, it is worth stating that a fundamental aim of the people in the Unit when it was first formed and as it grew, was to achieve more impact on actual operational systems, and to be seen as supporting operations and design, and not 'just more research'. As will be seen below, this personal goal of the team members was achieved.

Resources Available: Background, Development, Structure and Management

At the outset, the HFU had 5 members plus two contractors – this was expanded within a year to 7 staff plus 5 contractors, and stayed relatively stable at this level during the period this paper refers to (1998-2000). All personnel were qualified in Human Factors, most to a Masters Degree level. It was a relatively 'young team', with the age range from 21 to 36.

All personnel had real-time simulation commitments, but also wanted to work in other areas of Human Factors. This was initially difficult to realize, as the Unit had to form and develop new areas of expertise and new customers, within a relatively short period of time. Therefore, within a year, a 'matrix' organization style was developed for the activities of the Unit. This meant that most members participated in several activities, including real-time simulations (all participated in this core activity), HMI work, safety (human error), training, and PUMA (predictive workload assessment) work. Each of these areas was called a Human Factors 'cell', and each one had a 'cell manager' (see Figure 19.1). Concerning reporting up the management ladder, the Head of HFU mainly interacted with the General Manager of ATMDC and the Director of R&D, in the early phases. Later on there was more interaction laterally at the Senior Management level (one above Head of HF and one below Director) particularly with Operational Departments running NATS ATC Centers and certain Safety Managers, and at the end interactions with Safety and Operations at Director level (members of the Executive Board).

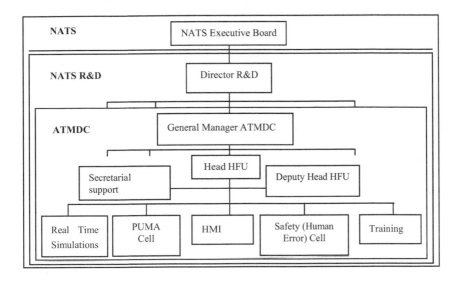

Figure 19.1 Structure and organizational location of Human Factors Unit (1998-2000)

Some of the areas could build on existing personnel experience and skills, but some required development of staff. Furthermore, for the matrix structure to work with a relatively small team and a large number of projects, certain common core competencies were needed for all team members, to facilitate communication and co-ordination of activities. Therefore, to facilitate competency in this range of application areas, several courses and 'development projects' were developed and implemented. This development of personnel included external and internal training in task analysis, human error analysis, PUMA applications, HMI evaluation, situation awareness measurement, and eye-movement tracking analysis. Real-time simulation methods (experimental design; questionnaire design and analysis; workload measurement; debriefing; etc., all tailored to ATM simulations) were learnt on-the-job by initially supporting at least two real-time simulations before being able to support one alone.

There were also certain 'development' projects for the staff (and contractors), with each staff member carrying out a Hierarchical Task Analysis (Shepherd, 2001) for a part of the Air Traffic management 'gate-to-gate' process. These task analyses included the ATM areas of tower operations, area control, oceanic control, terminal maneuvering area (TMA) control, approach and departure operations, (military) distress and diversion emergency control, etc. The task analyses were carried out via interviews and observation of live ATM operations. This meant that each member of the team gained detailed insight into at least one specialized area of ATM, and also a task analysis of the whole ATM process was derived. Additionally, most staff were involved in a project called cognitive 'picture' interviews, which involved interviewing controllers about the nature of the mental picture they used to control traffic (MacKendrick et al., 1998; Kirwan et al.; 1998). As with the HTA, this gave the Human Factors Unit members insights into the cognitive aspects of the task, including 'losing the picture'. Lastly of note, an external 'referee' was appointed to give periodic feedback to the Unit on their work and its adequacy from an academic Human Factors standpoint, throughout the three-year period of the NATS Human Factors program.

This training and development program enriched the competencies of the staff and contractors working in the HFU, and added a degree of homogeneity to their expertise, at least in core areas.

Lastly, in terms of resource management, as already noted, there were five cells of activities. In practice one manager could have up to two cells. There was therefore a 'core' team of three managers (one of whom was deputy Head of HF) and the Head of HFU, managing all the activities of the Unit. This core team met weekly, and the whole Unit met monthly for progress meetings and presentations on one or two of the ongoing projects. The Unit was also allocated one full-time secretary. The Head or the Deputy Head also attended the ATMDC General Manager's meeting once per month to liaise with the other Heads of Units at ATMDC.

Development of the Strategy

The strategy (Kirwan, 1997a, 1997b), as noted earlier, had three main and connected strands of R&D. The first was to better understand the controller. This strand encompassed work such as the task analysis and picture studies mentioned above, the consideration of the future role of the controller in a more automated environment, and understanding how controllers make mistakes, and how they detect and correct them.

The second strand, concerning measurement of controller performance, focused particularly on the areas of mental workload measurement, situation awareness, and eye-movement tracking. The former was a well-established HF measure, but it was considered timely to review it to see if existing measures could be improved. Situation awareness (SA) and eye tracking were seen as relatively new and unproven techniques in ATM, but potentially useful given that automation tools could affect SA, and that new interfaces were being developed that could affect controllers' visual search patterns.

The third strand was less specific in its detail, but more challenging in its 'mission', since it simply stated that there must be applications of HF methods and data, and changes resulting in operational and future ATM systems. The strategy document itself had 15 specific targets for delivery by 2000. Each of these is listed below, with a brief commentary on its outcome:

Continued Quality Support to Real-time Simulations This did occur, although there were occasionally conflicts between simulation support and other new projects – however, simulations remained the priority, as they represented the base load of the HFU's activities, and were a high priority for NATS.

Consolidation on Simulation Measurement Methods (workload; situation awareness; other) A review of methods for workload assessment was carried out (Kirwan et al., 1998). This review, plus trials of various tools, led to the continued usage of the ISA (Instantaneous Self-Assessment tool – developed formerly at NATS) approach and the use of NASA TLX in real-time simulations, and PUMA for predicting workload with future systems (see Kilner et al., 1997).

Physiological methods were seen as being of interest, but also as rather specialized and beyond the capabilities of the ATMDC at that time. A situation awareness method (Endsley, 1990) was tested at the ATMDC in a small number of simulations, and a variant was applied to several simulations including a military one (see Donohoe et al., 1998). Eye-movement tracking measurement was utilized in a small number of simulations for interface design and evaluation purposes. Therefore mental workload measurement remained the dominant measure used across all simulations, with eye tracking and SA reserved for more specialized simulations.

Development of an Incident Analysis and Feedback Approach A system for analysing human error aspects and causes of incidents was developed, called TRACEr (Technique for the Retrospective Analysis of Cognitive Errors in ATM: Shorrock, 1998; Shorrock and Kirwan, 1999; Shorrock and Kirwan, 2002). This approach was applied to the analysis of a number of incidents to determine the Human Factors/Safety priorities.

Task Analysis of Key Controller Tasks, including Cognitive Aspects As already mentioned, a large number of task analyses were carried out, creating a baseline understanding of how NATS controllers worked in the UK.

Research on Human-Centered Automation To help develop concepts for the future role of the controller a project was carried out in the area of Human-Centered Automation (HCA). This project (Kirwan and Cox, 1999; Kirwan, 2002) distilled a number of HCA design principles focusing on a number of areas including the role of the controller. Several R&D projects looking at the design of new systems and tools took up a number of these principles.

Provision of HF Guidelines for Support to Design Projects A review was carried out to develop an internal 'reference guide' for Human Factors for design projects (MacKendrick, 1998; Shorrock et al. 2001). The review included well-known HF texts, as well as standards from the nuclear power industry (e.g. Kincade and

Anderson, 1984; NUREG 0700 version 2; etc.), and more ATM-specific works such as Cardosi and Murphy (1995). The resulting database was then computerized and used by all members of the HFU, particularly for the HMI Ops Cell. Others in NATS also had access to the document and the database.

Development of a Profile Description of the Controller (in terms of skills and aptitudes requirements) This project turned out to be more difficult than anticipated, but a description was in fact developed, though its connection to other projects remained limited. This exposed a general weakness in the area of relating 'selection' type projects to more general HF work.

Development of Predictive Human Error Analysis Approaches Two approaches were developed: one based on TRACEr (Shorrock, 2002) and one based on the chemical industry's Hazard and Operability Study approach (Kletz, 1974), altered to focus on the human error aspects. These techniques were applied successfully on several projects aimed at developing future automation support tools for controllers (e.g. see Evans et al., 1999; Shorrock et al., 2002; Springall, 2001; Kennedy and Kirwan, 2001).

Formalization of Experimental Control Protocols in Real-time Simulations This was carried out to ensure homogeneity and efficiency of Human Factors work in real-time simulations.

Visits to FAA and Eurocontrol These were carried out to try to see what Human Factors issues were being pursued in the rest of Europe and the US.

Annual Review by a Visiting Professor Professor John Wilson of Nottingham University spent several days per year reviewing key projects, and giving independent feedback to the team members.

One New Project in a New NATS Area In fact there were more than a dozen new projects, some large, some small, that occurred during the three-year period, mainly in the area of interface design support.

One European Project The HFU became involved with the RHEA (Role of the Human in European ATM: Nijhuis, 1999) project, and also won a Eurocontrol contract to develop a human error incident analysis system (HERA: Isaac et al., 1999).

One Non-ATC Project This did not occur due to the rapid growth of internal work. The Unit was initially asked to do external work, but NATS' management then decided that there was too much internal work, so this strategic objective was rescinded.

Publication Targets (10 conference papers, 2 journal papers) This was achieved (see references section later).

The methodological aspects (the second strand) to an extent unified the whole strategy, as these approaches helped both to enable an understanding of the controller and controller methods, and also to develop ways to improve real and proposed systems, i.e. applications. Evans et al. (1997) summarized the strategy from a methodological point of view, as shown in Figure 19.2. This figure contains the main methodological approaches, cast in a Venn diagram to show their relationships in terms of applications to developing better tasks, better interfaces, and having the right controllers for the job in the first place ('right task', 'right kit', and 'right stuff', respectively).

Effectively, the measures worked well, although within the resources constraints during the program the relationships between team resource management and target audience description and the rest of the work were not developed.

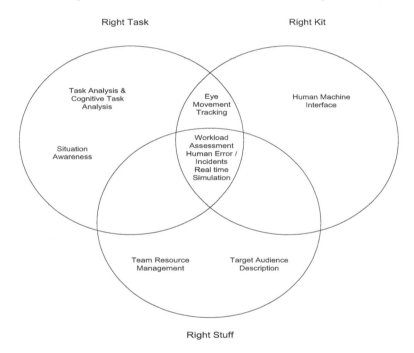

Figure 19.2   Areas of Research in the Human Factors Strategy

Applications

Having outlined the origins of the Unit, and the basic strategy including 'enabling' projects such as those associated with human performance measurement (SA, mental workload, eye tracking), HF Guidelines, error assessment, etc., the remainder of this chapter focuses on the third 'deliverable' of the program, namely the applications. The main 'clients' of the Human Factors work were the operational departments, i.e. the existing control centers and their controllers and management, and also certain future systems design projects, and the College of Air Traffic Control (CATC) training centre. Later on in the program the various safety functions and departments within the company became an additional client area.

Therefore, having set the scene with some background information, the following sub-sections outline some of the key projects that had impact on NATS' systems and future projects. More detailed references are given where possible. The following ten summaries of studies in NATS are some of those that led to changes in real systems or system designs. They are grouped into three of the areas already mentioned, namely HMI design, training and safety. There were impacts in the areas of real-time simulation and prospective workload assessment also, but these are more difficult to summarize briefly, as they are more immersed in the context of the simulations and the airspace procedures being tested. The following examples are therefore more 'accessible' within the confines of a chapter.

HMI Impacts

Mode S Evaluation

Currently most aircraft transponders are known as 'Mode-C', and they transmit certain basic information from the cockpit to the ground (the controller), i.e. they can be interrogated by ground radar for basic information. It has been the intention in aviation for some time to enhance the system by transmitting more information from the aircraft to the ground, via a new system called Mode S. The Human Factors study investigated the prototype interface information for controllers relating to Mode S (see Evans et al., 1998).

A series of interviews and walk-throughs with the proposed system led to a realization that there was much more functionality on the VDU interface than the controller actually needed. This was causing problems for the controllers in terms of visual and cognitive workload in trying to find the function they needed, and in terms of wasted time using functions that were not found to be relevant to the task. A progressive series of small-scale trials and a detailed questionnaire was developed to determine what information was needed and useful, and therefore what could be removed from the system interface. This information was gained systematically from a large number of controllers, and a large amount of additional functionality (approximately a third) was removed from the system interface. The controllers were far happier with the resulting interface, and more inclined to use it.

Distress & Diversion (D&D) Cell – Control Room Workspace Layout Design

A Distress & Diversion (D&D) Cell is aimed at helping military aircraft in distress, getting them to the nearest safe destination, giving general assistance to the aircraft, preparing a clear way through for the aircraft to the aerodrome, and notifying the aerodrome for emergency services to be prepared etc. The D&D cell for NERC (the New En-Route Centre) was being re-designed, and a member of the Human Factors Unit assisted with the design of the new layout. His analysis was based on a task analysis of the existing system at LATCC (London Area and Terminal Control Centre), including observation of actual handling of distress calls, and interviews with the controllers (Lamoureux, 1998). This was a short study, but due to its timing its results could be incorporated directly into the design of the new D&D control room.

Controller Workstation Design

A detailed anthropometric[27] study for the layout of the new Scottish controller workstations occurred, dealing with aspects such as VDU placement, desk and chair design, cable management, lighting, and communications optimization. As with other studies, there was significant use of trials and mock-ups, with real controllers and simulated tasks. This increased the ownership of the recommendations by the future users of the system. A large number of recommendations were adopted for future use by the future centre.

If there had been more such studies, it would have been useful to utilize manikins and HF-based software tools for determining anthropometrically optimized workplace layouts. In practice at the time, the different projects that could benefit from workplace layout study were not using the same design software tools (some of this work was contracted out to different design contractors), and so it was not cost-effective (then) to use advanced HF tools in this area. However, rapid prototyping tools have advanced quickly, and so such approaches can now be envisaged, even for varied design software approaches.

---

[27] Anthropometry or anthropometrics deals with ensuring that the workplace will fit the majority of its users, in terms of visual and physical access, reach envelopes, 'handed-ness' of controls, etc. Systems that are not anthropometrically optimized will lead to physical or reading errors, for example. Anthropometry relies on extensive analyses of body sizes for personnel populations and their statistical variability.

NERC System Control Centre Alarms Management

Although generally the Unit had little formal involvement with the development of NERC's interface[28], one area of involvement was in the form of assistance with the handling of 'alarm flooding' in the System Control Centre (SCC) (see Shorrock and Scaife, 2001). The SCC does not control air traffic, but supports the systems that the controllers use – it is therefore critical in keeping the system running smoothly, and a failure of the SCC could be a very serious event. Some potential problems were identified whereby the SCC operators could receive a very large number of alarms simultaneously (called 'alarm flooding'), making it difficult to determine the problem nature and hence the appropriate and timely corrective action. Some NERC managers knew that Human Factors in other fields (e.g. nuclear power) had experience with this problem, which was relatively new to ATM. An extensive study ensued, with many Human Factors analyses occurring during night shifts (when the system was most accessible), and detailed recommendations were made and implemented (on alarm prioritization and display) to render the system more manageable during such 'turbulent' conditions. The alarm flooding threat was therefore eliminated from the system.

With the exception of the last case study above, these studies represented basic HF approaches to workstation and interface design, albeit backed up by a well-developed HF Design database. There was no 'rocket science' involved, yet such studies were seen as highly useful by the customer, giving the HFU street credibility, and leading to increased requests for work whether of a HMI nature or not. Paradoxically for the HF practitioner perhaps, the HF studies that in some sense are easiest to perform, are also those that are highly valued by the customer, more so than tackling more HF difficult areas. This reinforces the 'soft systems' approach of Human Factors going for quick and relatively easy 'wins' in order to gain a stable foothold in the corporate landscape.

Impacts on Training

Enhancing Training Success Rates

Controllers go through extensive training – first an 18-month period of training at a college such as the College of Air Traffic Control (CATC), co-located with the ATMDC, and then approximately a further 18 months at an operational centre. Being a controller is a difficult and sometimes stressful job, and quite simply, not everyone is cut out for it. There are selection criteria and tests, but there is also generally a recognized high fall-off rate during initial training, as students either

---

[28] Almost all of the HMI support work for NERC had been contracted out before the setting up of the Unit, or else was being handled internally by the NERC engineering team. The HFU therefore had little ability to be involved with NERC.

realize it is not for them, or else fail to keep up with the pace of the training. When the HFU was formally set up, there was a perceived problem within CATC of a higher than desirable failure rate of students. CATC therefore commissioned a study from the HFU (see also chapter 8 in this book).

It was recognized that a short study was unlikely to unearth the causes, as there are many differences between students and even between instructors. A longitudinal study therefore occurred over more than a two-year period, tracking the progress of a series of student courses as they progressed through the training system. Interviews with most of the students who failed the course (and some that passed), and interviews with instructors, were held during the study period. This generated a large amount of sensitive material, and necessitated a number of feedback sessions to try to derive practicable means of improving student success rates. In particular, the timings of exams throughout the early training cycles were seen to be too soon, i.e. before the theory had necessarily been fully assimilated and integrated by the students. Delaying the exams by a matter of a few weeks would not cause large problems with the training schedule, but could allow the students more time to integrate what they had learned. Additionally, students needed more training media to carry out their own off-line study, on the areas they were not certain of, or to polish skills.

This approach led to the identification of ways of changing the training system to improve the student success rate (Donohoe et al., 1999), and these were implemented, resulting in a 25 per cent increase in the training success rate. This single result was probably the most significant one during the work program, as it directly impinged on an identified Key Performance Indicator for the company, and was a measurable and quantified effect. Work is in fact continuing on developing ways of further improving training and supporting training for future developments, via Computer-Based Training (CBT) systems, as will be mentioned in the next chapter (Burrett, Weston and Foley, this volume).

A key lesson here, in contrast to that of the HMI area above, is that patience can pay off. It was decided early on that no quick fixes existed, and so a longitudinal study must take place. Another key aspect of this study was the handling of sensitive and confidential data, and dealing with people on the precipitous subject of 'failure'. There was therefore a need for sensitivity and proper interview and data-handling protocols throughout this study, in order to gain and maintain the trust necessary to yield the insights required to resolve the problem.

Impacts on Human Error and Safety

Development of an Incident Error Classification System

A new error classification system (Technique for the Retrospective Analysis of Cognitive Errors; TRACEr – Shorrock, 1997; Shorrock and Kirwan, 1998; 2002) was developed by extensively adapting and extending tools from other areas, tailoring the resultant tool for ATM applications. This tool aims to classify errors in their operational context, but also to unravel the complex causes that led to the error, so that learning can take place, and future errors can be prevented or recovered. The tool was validated and used as a basis for a number of studies (two of which are mentioned shortly). However, it is worth mentioning that the TRACEr technique was the basis for the European (Eurocontrol) initiative called HERA (Human Error in ATM: Isaac et al., 2001; 2002) which is attempting to implement a common human error classification system for ATM incidents in Europe. Furthermore, HERA has more recently merged with a US system called HFACS (Shappell, 1999), to form a common error classification system (called JANUS) between Europe and the United States of America (Isaac and Pounds, 2001). TRACEr therefore became the basis for a substantial initiative in classifying and learning from incidents. The HERA system itself is currently being implemented in a number of European countries.

TRACEr was applied predictively to a future tool designed to help Approach controllers with the spacing of aircraft approaching airports such as London Heathrow and Gatwick, and was able to identify key failure modes. The tool was also applied to the proposed Human-Machine Interface (HMI) for a new centre (New Scottish Centre) at the design stage by a single analyst, [29] by developing a task analysis for the tool and then carrying out human error analysis using TRACEr (see Evans et al., 1999; Shorrock and Kirwan, 2002). A number of errors were identified, and together with designers the analyst was able to determine their potential consequences and associated error reduction measures. Shortly after the evaluation, an independent real-time simulation was carried out using the tool with a number of controllers. Ninety-four per cent of the errors that occurred in the simulation had been previously predicted by TRACEr, including a small number of significant errors, and the design was altered to avoid these errors in the future. The six per cent of errors that occurred and were not predicted by TRACEr related to tasks that had not been entered into the task analysis (and so could not be predicted by TRACEr). TRACEr also predicted errors that were not seen in the simulation, but the status of these is unclear, since it can be argued that if the simulation continued long enough, they may well have occurred. In any case, the 'hit rate' for TRACEr was good, and led to changes in the design proposals.

Human HAZOP of the Electronic Strip Interface

A parallel study to that cited above using TRACER occurred using a Hazard and Operability study format (Kennedy et al., 2000; Kirwan and Kennedy, 2001) adapted to analyzing the interface for an electronic strip system. [30] At the time, a

---

[29] Steven Shorrock, the principal developer of the TRACEr technique.

[30] Currently most controllers use paper strips, one for each aircraft, which contain the essential information about that aircraft for the controller. These are being replaced in the

controller would use paper 'flight progress strips' (FPS) in conjunction with a radar screen to control and monitor aircraft through their sector. The generation of new FPS systems, however, shifts the focus away from one form of information presentation and usage (paper) to a completely different presentation medium (computer screen). A HAZOP analysis was therefore conducted on the proposed HMI (Human Machine Interface) of the new flight progress information system. The study is described in Kennedy et al. (2000), and is described in more detail in the chapter by Shorrock, Kirwan and Smith (chapter 15 in this book).

The HAZOP team that assessed the implications of the new interface was made up of three designers, one air traffic controller and two Human Factors specialists. A total of 16 hours, spanning three separate HAZOP sessions, was spent interrogating the prototype system. The study identified a number of 'vulnerabilities' in the prototype system and 'opportunities' for error that needed to be designed out or worked around (e.g. via procedures and training). A total of 87 recommendations were generated from the three HAZOP sessions and these were classified as follows:

- changes to interface design and menus (34 per cent);
- improvements in user feedback on actions / inputs (25 per cent);
- training / procedures recommendations (16 per cent);
- modifications to aircraft status on screen (13 per cent);
- hardware / equipment changes (9 per cent); and
- further study / future research ideas (3 per cent).

The HAZOP group identified what factors needed to be changed in the system and how these changes could be addressed. Since the designers were present and actively involved, any design changes they thought necessary were simultaneously accepted for implementation. The HAZOP therefore had very effective impact on the design process.

In terms of the strength of the HAZOP approach, a number of problems that were identified in the design would have been difficult to detect without a multidisciplinary team present. In particular each member of the team brought a 'specialism' to the group process. All this information was shared effectively, leading to a rich multiple-perspective on the system design and its strengths and weaknesses. Such an interrogation of the system would have been very difficult to achieve in any other way and underlines the niche for HAZOP in system design and evaluation. The success of this project led to the formal application of the technique on other future tool concepts (e.g. see Springall et al., 2001).

UK as in many other countries by electronic 'strips' on a computer screen, that contain similar information.

Reducing Vertical Separation Outside Controlled Airspace

A long-term study (see Kirwan et al., chapter 6 in this book) commenced in 1998 concerning the required aircraft separation minimum distances in areas outside normal controlled airspace. In such airspace, a service called Radar Advisory Separation (RAS) is often available from military controllers under which aircraft will be given advice and controlling instructions if they desire it. RAS advice is based on separation minima of 5nm laterally between two aircraft, or 5000 feet vertically. These criteria were determined several decades ago when transponder and radar accuracy were inferior by today's standards. Therefore, in order to improve controller workload and service to aircraft, a safety case was needed to determine whether the criteria could be reduced to 3nm laterally and 3000 feet vertically. Human Factors expertise was requested at the outset by the client.

The Human Factors part of the study used a number of approaches. Civil and military incident records were analyzed; and two simulations were carried out which looked at errors, mental workload and situation awareness using the current and proposed separation standards. Controllers were interviewed at three military control centers; the human performance literature was analyzed to look at contextually-relevant reaction times; and human error analysis occurred using the TRACEr technique (see Scaife et al., 2000). The net result was that the vertical reduction appeared feasible with negligible decrement in safety terms, but the lateral reduction to 3nm was inadvisable due to turn rates and reaction time considerations.

A six-month trial was then designed, following a HAZOP of the trial process itself, and the new vertical standard of 3000 feet was adopted by a number of military air traffic service units in the Vale of York area of the UK, one of the busier areas for such traffic. During this period any errors or events were recorded and analyzed and evaluated to see if the 3000 feet reduction was having a negative impact. Questionnaires were also received each week from each air traffic service unit taking part in the study. The results after six months were very positive. The trial was then extended for a further six-month period, and then the safety case was documented. The 3000 feet vertical separation standard was then introduced nationally in December 2000 and is now the national standard for RAS usage outside controlled airspace in the UK.

This was a significant and measurable impact in two ways. First, it gave a concrete answer to a difficult question that had been seen as being rather subjective, and a difficult one for which to construct a safety case. Secondly, the year-long trial was a significant period in which to detect problems, or at least the threat of potential problems, none of which were observed.

Position Hand-Over Review

During a period of some months, it had been noticed that most reportable incidents (themselves of small risk) had occurred in the first ten minutes after one controller had handed over his or her operational position to another controller. This is called position hand-over, which allows the outgoing controller to have a rest break. The Human Factors Unit was asked to investigate this trend to see if it was a real problem and if so, determine how to stop it.

Observation of live control situations and hand-overs occurred over a large number of shifts at LATCC, and an analysis revealed a great diversity of hand-over practices, i.e. in terms of what the outgoing controller would tell the incoming controller. It was agreed with the controllers themselves that a special procedure would itself be unwieldy as there is little time for reading anything, when both the controllers need to be focused on what is on the screen. Additionally, sometimes there is simply little to be said, depending of course on how busy the situation is. A mnemonic was therefore produced which the outgoing controllers could use to quickly think about what needed to be transmitted to the incoming controller. The mnemonic developed was PRAWNS (standing for Pressure, Runways, Airports, Weather, Non-standard events, and Strips to display – NATS, 2001). All the controllers were trained on this technique for achieving 'best practice' on position hand-over, and it is now in active use (see chapter 2 in this book).

A major lesson learned in this area was not to lose people in jargon and excessive detail, as can sometimes happen in the field of human error and human reliability assessment. Considerable efforts were taken to make TRACEr understandable by controllers, and to run the HAZOP sessions in a way so that participants (usually controllers) could focus on the problem and its ATM context, rather than having to focus on the technique itself. Additionally, the fundamental aim of these studies in improving safety in practicable ways was always at the forefront. This continual emphasis on improving safety, rather than on polishing human error techniques, was appreciated by the participants. The participants in any case often had better suggestions for safety improvement than were arising from the techniques in some cases. The techniques were therefore seen as guiding and structuring safety improvements, rather than replacing the controllers' expertise in this respect.

Ad Hoc Study Areas

The above was the official strategy for the three years discussed in this paper. Of course, other areas arose during the three-year period, including new areas of research. Two of these are worthy of mention here, namely the study of complexity, and of recovery from system (automation) failure, both of which were aimed at understanding, modeling, and resolving issues in these areas. Both of these areas were addressed as 'fast-track' research, aimed at developing rapid insights into certain key developing HF concerns.

The first area, complexity, was of interest because a number of controllers and researchers alike were using this term, and talking more of problems of complexity than of workload, for example. Thus a controller when asked how many aircraft could be handled in a sector, would respond in terms of complexity rather in absolute numbers: e.g. 'one difficult aircraft situation can lead to the workload normally associated with 10 aircraft'. There was therefore a desire to try to understand what makes the task complex for controllers, and how airspace design itself can help ease complexity, so as to maintain safety and potentially raise capacity.

The approach was very human-centered, using a mixture of complexity factor elicitation techniques to derive complexity-inducing factors, and to rank them in their importance (see Kirwan et al., 2001a; 2001b). Approximately 40 controllers from various sectors in the UK took part in the study. Examples of the derived complexity factors are shown in Table 19.1. These led to a suggested list of 'do's and 'don'ts' (Table 19.2) for airspace designers, though this resultant list has not necessarily been used by airspace and sector designers. The study did however raise the need to go deeper into this area, and showed that complexity could be 'deconstructed' meaningfully.

The second area of study was recovery from system failure. The idea was simple: as automation support increased, controller dependence on the automation would increase, especially since the human-automation partnership would enable higher traffic capacity in normal conditions. However, when the automation failed, this would mean that controllers would suddenly have to use 'old' and unpracticed skills in a high traffic density/complexity situation. This does not necessarily mean that the controllers would be 'out-of-the-loop', as the automation concepts were intended to support awareness rather than replace it, but it could put a significant strain on controller performance.

Therefore, a study took place to investigate this area, using a small range of failures concerned with existing automation (so that the controllers were fully used to it, and hence any dependence relationships with it would have matured). The environment used was a fully functional training centre at an operational ATC centre (London Terminal Control), and the methods used are reported in Low and Donohoe (2001). Several failures were tested, including complete radar failure, and a less 'revealed' failure wherein the callsigns of two aircraft are transposed on the screen (callsign conversion coding error). A mixture of measures was used, including situation awareness and eye-movement tracking, and debriefing whilst watching a replay of the simulation.

The results were firstly that in most cases the controllers were able to recover within a matter of minutes (e.g. for radar failure, achieving procedural control of all aircraft on frequency within three minutes), although in one scenario an aircraft was temporarily 'lost' from situation awareness. The eye-movement tracking proved invaluable in determining what information sources were being used. It also led to some interesting revelations, such as some controllers continuing to glance up at the radar screen after it had gone blank, to 'see' or remember where the aircraft had been prior to radar picture loss. Significantly, the controllers exhibited very effective flexible team structures to assure safety during these failure scenarios. This made the evaluation team realize there was a need to be able to measure teamworking in such situations, so as to develop best practices and training for such emergency scenarios.

Table 19.1   Derivation of top complexity factors rank ordering

Factor
1. *Volume/ flow/ growth rate of traffic* – *including the effect of 'bunching' of traffic at peak periods.*
2. *Airspace design* – *including sector shape, number of levels in the sector, route structure and number of crossing points.*
3. *Shared understanding* – *e.g. between adjacent sectors; between different ATM functions; etc.*
4. *Communications & co-ordinations* – *relating to time pressure on communications, both with radio-telephony (controller –pilot) and other calls on 'landlines'.*
5. *No soft option* – *lack of sufficient non-busy times to think and be proactive – related to so-called 'over-loads'.*
6. *Procedures* – *some procedures seen as overly complex or even clumsy, or with a high rate of change of sector design.*
7. *Presentations to sector* – *handing over aircraft from one controller/sector to another (e.g. too early or too late, etc.).*
8. *Human resources* – *relating to staffing issues.*
9. *Non-standard flights* – *concerned with unusual traffic such as survey aircraft and training aircraft, for example.*
10. *Aircraft performance* – *some areas have a large range of aircraft performance capabilities, from 'lows-and-slows' to fast-climbers.*
11. *Military* – *largely relating to the unavailability of military danger areas for civil usage given the desire for increased civil capacity.* [31]
12. *Weather*[32] – *e.g. turbulence, wind shear, thunderstorms, and weather interference for radio-telephony.*

---

[31] Military airspace is obviously not available for civil usage when military exercises/ preparations are occurring. However, flexible usage of such airspace by civil users, when no such exercises are underway, could improve capacity and lower complexity by reducing current airspace constraints, some of the time.

[32] If weather is extreme, it can dominate complexity.

These two ad hoc studies also showed a shift from the original method of working, i.e. of having more time available to understand the problem area before modeling it and before carrying out an application. The shift was towards more fast-track research that could yield results in a shorter time-frame – it was therefore a shift of focus from research to development. This can be seen partly in the light of the prevailing economics of the industry and in particular NATS' move towards privatization at the time, as these factors both were pushing for more 'results-oriented' work. It also reflected a change in attitude in the company towards Human Factors – there was a realization that HF could not merely research and discuss and write learned papers, but that it could also deliver meaningful solutions. Clients were no longer just demanding study, but also answers. This had been the ultimate aim of the personnel in the HFU, to 'make a difference' to operational systems, but it also meant that expectations in the company became higher.

Table 19.2   Sample of prototype sector design 'complexity checklist'

Sample of checklist items	
Have climb-throughs been minimized?	*Have conflicting departure patterns from two adjacent airports been avoided?*
Have number, location, and orientation of holds been optimized?	*Are sector splits even?*
Has traffic been standardized as far as possible?	*Have funneling and choke points been avoided/minimized?*
Is the sector long enough given the route profiles (e.g. long enough for climbs and descents)	*Have inbounds been based on reasonable and realistic a/c performance profiles, rather than optimal ones?*
Can a 'dual carriageway' concept be implemented where possible?	*Have too many reporting points been avoided?*

A Review of the Impact of the HF Program

The three-year program largely delivered the desired impacts, and led to numerous improvements to actual systems and future design projects. There were some failures, notably a lack of allowed involvement in NERC, and there was little penetration of the Airports sector of work, but generally positive impact was achieved. The Human Factors Unit gained a reputation for being able to deliver practicable results. This was to some extent a change of orientation from R&D, but one that all seemed happy with. This also led to a small number of the HFU staff being seconded into other parts of the company, rather than being purely located in R&D. In particular, during the three-year program, key Human Factors personnel moved into one of the larger new (Scottish) ATM Centre development projects, into the Safety department, and a specialized Human Factors cell was set up within the future systems/tools design group. This organizational proliferation and diversification of Human Factors within the company helped to strengthen its application and support base, and ensure a focus on the 'real' issues facing the company.

Cost-effectiveness of the HF Program The individual costs of the projects above have not been cited for two main reasons. First, such information currently has commercial value and the author is not at liberty to disclose it. Second, it is misleading to cite these resources and ignore those for the under-pinning research that took place in first two supporting strands of the program. For example, a HMI evaluation might take a month, but the development of the guidelines underpinning such evaluations might take a year, so what figure should be quoted? Perhaps from this latter standpoint the true costs, those borne by the industry, were those for the full program, therefore of the order of around thirty person years over a three-year period. Against these costs must be measured the full impact of the program, but unfortunately it has only been possible to cite a proportion (e.g. around a third) of the more tangible impacts from the program, within the confines of this paper.

Sustainable Impact One of the objectives of the R&D Director when the HFU was set up, was to achieve sustainable impact. A residual question therefore, is whether the impacts such as those seen above were followed by further impacts, i.e. whether the general Human Factors impact was sustained after the initial three-year program, or was perhaps a passing 'fashionable item'. This is primarily the objective of the next chapter in this volume, but it briefly merits discussion here.

Since the ending of the three-year program, and the transition to a new strategy and a new Head of HF, the size of the group has not only been sustained but has grown, and the HFU has successfully survived the privatization of the company. There has been in the intervening three-year period a rising demand for more practical applications of Human Factors in operational centres, as will be seen in chapter 20 in this book. This has been in large part due to the Human Factors Unit providing concrete solutions for a number of problems. It therefore seems that the Human Factors work has achieved a high degree of credibility and perceived usefulness inside the company. Therefore, the Human Factors program was judged to be a success not only in terms of achieving impact and change on real projects, but also in terms of aligning HF work more closely with the operational culture within the company, moving from the status of a purely research function to an engineering support function. Given the 'applied' mission of the Unit, and more generally Human Factors itself, this was a significant achievement and impact.

Acknowledgements

HFU staff and main contractors and students during the three year period: Andy Kilner, Laura Donohoe (now Voller), Alyson Evans, Tab Lamoureux, Heather MacKendrick,. Adrian Gorst, Toby Atkinson, Richard Kennedy, Irene Low, Richard Scaife, Steve Shorrock, Paul Fearnside, Abigail Phillips (now Fowler), Susie Foley, Martin Cox, Paul Nicholson, Charlotte Cumming, Mike Hook, Ted Megaw, Avril Williams, and Maggie James.

Also thanks to Peter Brooker, Paul Thomas, Al Lewis, Steve Garner, Jim Downing, Fergus Cusden, Mike Strong, Graham Vernon, Lloyd Brown, John Wilson, Ted Megaw, Mica Endsley, Chris Wickens, Gurpreet Basra, Damien Forrest, Huw Gibson, Sarah Harris, Helen Jones, and Nicky Heath.

References

Chapters in this book
Chapter 2: Voller, L., Glasgow, L., Heath, N., Kennedy, R., and Mason, R., 'Development and Implementation of a Position Hand-Over Checklist and Best Practice Process for Air Traffic Controllers'.
Chapter 6: Kirwan, B., Shorrock, S., Scaife, R., and Fearnside, P., 'Reducing Separation in the Open Flight Information Region: Insights into a Human Factors Safety Case'.
Chapter 8: Voller, L. and Fowler, A., 'Human Factors Longitudinal Study to Support the Improvement of Air Traffic Controller Training'.
Chapter 20: Burrett, G., Weston, J. and Foley, S., 'Integrating Human Factors into Company Policy and Working Practice'.

General cited references
Cardosi, K. and Murphy, E.D. (1995), 'Human factors checklist for the design and evaluation of air traffic control systems', *Volpe Centre ATC Human Factors Program*, US Department of Transportation, Research and Special Programs Administration. Federal Aviation Administration.
Hopkin, V.D. (1995), *Human Factors in Air Traffic Control*, Taylor and Francis, London.
Isaac, A., Kirwan, B., Kennedy, R., Andersen, H., and Bove, T. (2000), 'Learning from the past to protect the future – the HERA approach', *Proceedings of the European Applied Aviation Psychology (EAAP) conference*, Crieff, Scotland.
Isaac, A., Kirwan, B., and Shorrock, S. (2002), 'Human Error in European Air Traffic Management: the HERA Project', *Reliability Engineering and System Safety*, Vol. 75(2), pp. 257-272.
Kincade, R.G. and Anderson, J. (1984), *Human factors guide for nuclear power plant control room development*, NP-3659, Electric Power Research Institute (EPRI), Palo Alto, California.
NRC (1996), *NUREG 0700 Vols 1 and 2, Human system interface design review guideline – Process and guidelines,* Final Report, Nuclear Regulatory Commission, Washington DC.
Shepherd, A. (2001), *Hierarchical Task Analysis,* Taylor and Francis, London.
Springall, L. (2001), 'A Human HAZOP of the FACTS toolset', *FAA-Eurocontrol Seminar on Safety,* NATS ATMDC, Bournemouth, UK.

References on the HFU Strategy

Evans, A., Donohoe, L., Kilner, A., Lamoureux, T., Atknison, T., MacKendrick, H., and Kirwan, B. (1997), 'ATC, automation and Human Factors: Research Challenges', in H.M. Soekha (ed.), *Aviation Safety, Proceedings of the International Aviation Safety Conference*, pp. 131-151, Utrecht, The Netherlands.

Kirwan, B. (1997a), *Human Factors Unit Review and Outline Plan*, NATS R&D Report 9644, ATMDC, Bournemouth, UK.

Kirwan, B. (1997b), 'Human Factors in ATC: strategic research issues', *IEA '97*, Vol, 7, pp. 243-245. Helsinki, Finland.

Kirwan, B. (1997c), 'Human Factors in the ATM System Design Life Cycle', *Proceedings of the Europe/FAA ATM R&D Seminar*, Eurocontrol, Paris.

Kirwan, B. (1998), 'Human Factors techniques in the NATS ATM System Development Process', *Proceedings of the the Eurocontrol Workshop on the Integration of Human Factors into the System Design Life Cycle*, Luxembourg.

Understanding the controller

Gibson, H., Megaw, E.D., and Donohoe, L. (2001), 'Failures in pilot-controller communications and their implications for datalink', in D. Harris (ed.) *Engineering psychology and cognitive ergonomics* Vol. 5, pp. 325-334, Ashgate, Aldershot, UK.

Kirwan, B., Donohoe, L., Atkinson, T., Lamoureux, T., MacKendrick, H., and Phillips, A. (1998), 'Getting the Picture: – Investigating The Mental Picture Of The Air Traffic Controller', *Ergonomics Society Annual Conference*, Cirencester, UK.

Kirwan, B. (2001), 'The Role of the Controller in the Accelerating Industry of Air Traffic management', *Safety Science*, Vol. 37, pp. 151-185.

Lamoureux, T. (1997), 'The influence of aircraft proximity data on the subjective mental workload of controllers in the air traffic control task', *Proceedings of the Conference on Cognitive Science Approaches to Process Control* (CSAPC '97), Baveno, Italy.

MacKendrick, H., Kirwan, B. and Atkinson, T. (1998), 'Understanding the Controller's Picture in ATM', *Proceedings of the 2nd Conference on Engineering Psychology and Cognitive Ergonomics*. Oxford, UK.

Methodological

Cox, M. and Kirwan, B. (1999), 'The future role of the air traffic controller: design principles for human-centred automation', in Hanson, M.A. (ed.), *Contemporary Ergonomics*, Taylor and Francis, London.

Donohoe, L., Shorrock, S., and Kirwan, B. (1998), *Feasibility of situation awareness measurement in ATM simulations*, NATS R&D Report 9857, ATMDC, Bournemouth, UK.

Forrest, D. and Lamoureux, T.M. (2000), 'Future system state by novice and expert air traffic controllers', *Proceedings of the Annual Ergonomics Conference*, Taylor and Francis, London.

Harris, S.L. and Lamoureux, T.M. (2000), 'The future implementation of datalink technology: the controller-pilot perspective', *Proceedings of the Annual Ergonomics Conference*, Taylor and Francis, London.

Low, I. and Donohoe, L. (2001), 'Methods for assessing ATC controllers' recovery from automation failure', in D. Harris (ed.) *Engineering psychology and cognitive ergonomics,* Vol. 5, pp. 161-170, Ashgate, Aldershot, UK.

Kilner, A., Hook, M., Fearnside, P., and Nicholson, P. (1998), 'Developing a predictive model of controller workload in air traffic management', in M.A. Hanson (ed.), *Contemporary Ergonomics*, pp. 409 - 413, Taylor and Francis, London.

Kirwan, B., Kilner, A.R., and Megaw, E.D. (1998), *Mental workload measurement techniques: a review*, NATS R&D Report 8RD/14/16/2040, ATMDC, Bournemouth, UK.

Kirwan, B. (1999), 'Cognitive Error Analysis Of Future Automation Options In Air Traffic Management', in M.A. Hanson (ed.), *Contemporary Ergonomics*, Taylor and Francis, London.

Kirwan, B., Shorrock, S., and Isaac, A. (1999), 'Human Error in European Air Traffic Management: the HERA Project', *Proceedings of the Conference on Human Error, Safety, and System Development, (HESSD '99)*, Liege, Belgium.

Kirwan, B., Scaife, R., and Kennedy, R. (2001a), 'Investigating complexity factors in UK air traffic management', *Proceedings of the Conference on Engineering Psychology and Cognitive Ergonomics (EPCE)*, Edinburgh, UK.

Kirwan, B., Scaife, R., and Kennedy, R. (2001b), 'Investigating complexity factors in UK air traffic management', *Human Factors and Aerospace Safety*, Vol. 1(2), pp. 125-144.

Kirwan, B. (2002), 'Developing Human-Informed Automation in Air Traffic Management', *Human Factors and Aerospace Safety*, Vol 2(2), pp. 105-146.

Lamoureux, T., Cox, M., and Kirwan, B. (1999), 'Cognitive task analysis in training system re-design', in M.A. Hanson, E.J. Lovesey, and S.A. Robertson (eds.), *Contemporary Ergonomics*, pp. 17-21, Taylor and Francis, London.

Lamoureux, T. (1999), 'The influence of Aircraft Proximity data on the subjective mental workload in the air traffic control task', *Ergonomics*.

Low, I., Timmer, P., and Kilner, A. (1999), 'Are inconsistencies of outcome between predictive and descriptive mental workload techniques systematic?', in M.A. Hanson, E.J. Lovesey, and S.A. Robertson (eds.), *Contemporary Ergonomics*, pp. 7-11, Taylor and Francis, London.

MacKendrick, H. (1998), *Development of a human machine interface guidelines database for air traffic control centres*, NATS R&D Report 9822, ATMDC, Bournemouth, UK.

Shorrock, S. and Kirwan, B. (1998), 'The development of TRACEr: a technique for the retrospective analysis of cognitive errors in ATM', *Proceedings of the 2nd Conference on Engineering Psychology and Cognitive Ergonomics*, Oxford, UK.

Shorrock, S.T., MacKendrick, H., Hook, M., Cumming, C., and Lamoureux, T. (2001), 'The development and application of human factors guidelines with automation support', *People in Control Conference*, Manchester, UK.

Shorrock, S., Kirwan, B., MacKendrick, H., and Kennedy, R. (2001), 'Assessing human error in Air Traffic Management systems design: Methodological issues', *Le Travail Humain*, Vol. 64, pp. 269-289.

Applications

Atkinson, T., Donohoe, L., Evans, A., Gorst, A., Kilner, A., Kirwan, B., Lamoureux, T., and MacKendrick, H. (1997), *Human Factors Guidelines for Air Traffic Management Systems*, NATS R&D Report 9739, ATMDC, Bournemouth, UK.

Donohoe, L., Lamoureux, T., Atkinson, T., Kirwan, B., Phillips, A., and Brown, L. (1999), 'Human Factors support to training for future air traffic controllers', *Proceedings of the International Conference on Training*, CAMI, Oklahoma City.

Kennedy, R., Jones, H., Shorrock, S., and Kirwan, B. (2000), 'A HAZOP analysis of a future ATM system', *Ergonomics Society*.

Kirwan, B., and Kennedy, R. (2001), 'Assessing Safety and Usability of Air Traffic Management (ATM) Systems Using a HAZOP Approach', *Human Error and Safety System Development (HESSD 2001)*, Linkoping, Sweden.

Lamoureux, T. (2001), 'Development of datalink systems for air traffic management', in D. Harris (ed.), *Engineering psychology and cognitive ergonomics,* Vol. 5, pp. 211-220, Ashgate, Aldershot.

NATS (2001), 'Best practice in the hand-over and take-over of operational positions', *Safety Matters,* NATS, UK.

Scaife, R., Fearnside, P., Shorrock, S.T., and Kirwan, B. (2000), 'Reduction of separation minima outside controlled airspace', *Aviation Safety Management Conference,* London, UK.

Shorrock, S.T., Scaife, R., Foley, S., MacKendrick, H., and Kirwan, B. (2001), 'The practical application of error analysis in UK air traffic management', *Proceedinsg of the 2nd People in Control Conference,* Manchester, UK.

Shorrock, S.T. and Scaife, R. (2001), 'Evaluation of an alarm management system for an ATC centre', in D. Harris (ed.), *Engineering psychology and cognitive ergonomics,* Vol. 5, pp. 221-230, Ashgate, Aldershot.

Shorrock, S.T. and Kirwan, B. (2002), 'Development and Application of a Human Error Identification Tool for Air Traffic Control', *Applied Ergonomics,* Vol. 33, pp. 319-336.

Awards

The Human Factors Unit won the GATCO (Guild of Air Traffic Controllers) Prize (1998), for the following piece of work: Atkinson, T., Donohoe, L., Evans, A., Gorst, A., Kilner, A., Kirwan, B., Lamoureux, T., and MacKendrick, H. (1997), *Human Factors Guidelines for Air Traffic Management Systems,* NATS R&D Report 9739, ATMDC, Bournemouth, UK.

Chapter 20

Integrating Human Factors into Company Policy and Working Practice

Gretchen Burrett, Jennifer Weston and Susie Foley

Introduction

Human Factors principles, methods and tools are beginning to gain widespread acceptance throughout most industries, and there are many examples of how this has provided significant benefits. All too often, however, this success is dependent upon the talent and goodwill of individual people. We have not yet reached the point where Human Factors is addressed as a matter of course and with optimal cost-benefit.

This chapter discusses why the integration of Human Factors into company policy and working practice is important, and gives guidance on how to achieve this. These concepts are then illustrated with some examples from the experience of the Human Factors Group at National Air Traffic Services (NATS).

Business Benefits of Human Factors

The benefits of Human Factors will be well known to the readers of this chapter, but links to operational and business benefits can be overlooked. Some examples of these are provided below.

- Safety: Human error, often a function of poor system design, is a causal factor in over 90 per cent of Air Traffic Management (ATM) incidents (FAA, 1990; Kinney et al., 1977). Although it will never be possible to eliminate human error completely, Human Factors can help in managing both the causes of error and the impact of any errors that do occur.
- Performance: People play a vital role in both the safety and effectiveness of complex systems. With projected increases in air traffic and the transition to new technology, the human elements of the ATM system will be under increasing pressure in the future. Human Factors can help to ensure that the controller makes an effective contribution to the overall system performance.
- Cost management: Human Factors can have a significant impact on both the development cost and the cost of owning and operating the system (lifecycle

costs). In terms of development cost, Human Factors helps to ensure the 'fitness for purpose' of the system from the outset, reducing the need for costly re-design at a later stage. Human Factors can reduce the lifecycle costs by improving usability (thereby reducing training costs), optimizing workload (to obtain the correct manning levels) and increasing user satisfaction (thereby increasing staff retention).

- Legal considerations: Health and Safety legislation increasingly imposes a responsibility for companies to provide a safe working environment for their employees and protection from hazardous accidents for the community at large. Human Factors methods can help improve health and safety and demonstrate compliance to the legislation.

The benefits described above will be achieved when Human Factors is properly applied, but in order to maximize the benefit, Human Factors Integration is necessary.

The Integrated Approach

So, what is Human Factors Integration, and what benefits does it bring? The goal of Human Factors Integration is to move away from a reactive state, where Human Factors is used to address specific issues, to one where Human Factors principles are applied proactively in a timely and considered manner. This means that Human Factors is not solely understood by a specialist group of people, but is established throughout the company and is routinely considered alongside other business activities. The core concepts surrounding Human Factors Integration are given below:

- Total system focus: There is a focus on the system as a whole, which includes a combination of the human and the equipment working together to achieve the operational and business objectives. As a result, the human element of the system becomes an implicit part of strategies, procedures and working practice, alongside other elements. In addition, the human element is included in the risk management process and cost benefit analyses, in order to achieve the best overall outcome. If this is not done, one of the largest contributors to risk and cost is being overlooked.
- Organization: Roles and responsibilities for Human Factors are defined and there is support for this at all levels of the organization. This allows multidisciplinary teams (e.g engineers, safety specialists and operational staff) to work together effectively to address Human Factors issues, enabling trade-offs to be made collaboratively; for example, enabling a trade-off to be made between the cost of building a more intuitive design and the cost of increased training.
- Process: Human Factors activities are part of procedures and processes and there is a common language across the company to refer to these activities.

This ensures timely input of Human Factors to key decisions and processes, which in turn leads to increased consistency of results and reduced development risks.

• Knowledge: There is a company-wide knowledge base and re-use of knowledge and lessons learnt between projects. This includes human error rates and trends, operational task definitions and training needs analyses. The re-use of knowledge between projects makes the development process more cost effective and facilitates continuous improvement based on past experience. Building a reliable set of Human Factors data improves the ability to make objective decisions, to predict future performance and to validate these decisions over time.

The benefits described above will be realized if Human Factors is integrated into any company, but are particularly important for an industry such as ATM, which is characterized by the fact that it is safety critical and involves highly complex and expensive systems. Not only are the systems themselves expensive, but the costs and time involved in maintenance and development are high. It is therefore vital that mistakes are not made in the system design – the cost of getting it wrong could be extremely high in both human and financial terms.

Why Isn't It Done Already?

Although the potential benefits of Human Factors are increasingly gaining support, there is still a long way to go before it is fully integrated into company policy and working practice. So, why is this?

Human Factors focuses on the interaction between the people, the equipment, the environment and the procedures. As such, it is easy to run into problems with clarity over roles and responsibilities for Human Factors issues, which can lead to duplication of effort or failure to complete key tasks. Of course, this situation can be solved, but it may require changes in the company culture, organization and working practice, all of which take time.

Since Human Factors Integration requires commitment throughout the company, it means that people outside of the Human Factors community need to believe in the value of Human Factors. When there are tragic accidents, such as Three Mile Island (nuclear power industry), Piper Alpha (offshore oil and gas industry) or Ladbroke Grove (rail transport), the cost of failing to apply Human Factors is all too obvious. However, the positive result of an intuitive user interface or an effective training program, is harder to see. By its very nature, Human Factors attracts the label of being 'just common sense'. However, as illustrated in a study by Weston and Haslam, 1992, this is not necessarily the case. They demonstrated that on two simple tasks (one design task and one evaluation task), people who had been trained in Human Factors performed better than people who had not. Another problem associated with the very nature of Human Factors is that a well designed and easy to use system is unlikely to attract comment (Geesink, 1990).

Human Factors Integration depends on having the right tools and methods in place. Whilst many of these have been available for some time, there are still a lot of individuals throughout most organizations who have not used them as part of their projects or working practice. Naturally, this requires additional time and effort for familiarization. The state-of-the-art in Human Factors continues to evolve and, as this happens, one of the challenges is to keep relevant parties informed about how developments in Human Factors practice may impact their work.

In addition to the right tools and methods, Human Factors Integration depends on good quality information upon which to base decisions about Human Factors issues. It is necessary to have answers to questions such as: what are the targets for operational effectiveness and cost?; what is the expected human contribution to these targets?; and what human performance is currently achieved? Efficient consideration of Human Factors issues requires basic information to be collected for each project, with methods for allowing re-use where appropriate and refinement over time as the amount of data increases. Although the long-term effect will be a cost reduction, this takes an initial investment of time, money and effort.

Building Human Factors Integration

Before describing how to approach the integration of Human Factors, it is important to stress that this is a long-term and iterative process – there are no quick and easy answers. It is something that will happen over years rather than months, and will always be evolving. The approach used by the NATS Human Factors Group is based on the five steps illustrated in Figure 20.1 below.

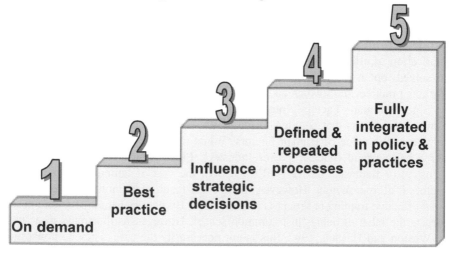

Figure 20.1   Steps towards Human Factors Integration

It should be noted that these steps are only meant as a guide towards Human Factors Integration, and are not intended to be used as any sort of measurement or audit tool. It should also be noted that different parts of the organization may be at different steps. The following table describes each of the five steps. For each step there is a description of its characteristics and what is needed in terms of people, tools and processes.

Table 20.1   Five Steps towards Human Factors Integration

Step 1 – On Demand	
Human Factors is used, but only to solve specific problems or as an 'after thought' near the end of the project. The company is gaining some benefit from Human Factors, but the process and results are not optimal.	
People:	There are some Human Factors experts in the company.
Tools:	The use of advanced Human Factors tools and methods may be limited by the time available.
Process:	The process is limited to problem solving and providing practical solutions to meet customer requests.

Step 2 – Best Practice	
There are examples of good Human Factors involvement throughout certain projects. There is demonstrable evidence of sound Human Factors results, but with limited impact on company decisions. The involvement of Human Factors in projects is more frequent, and tends to start earlier, however, it depends on individual motivation, rather than being part of working practices.	
People:	There is an established team of Human Factors experts.
Tools:	'Best practice' tools and methods are used. Appropriate tools are identified and there are initiatives to develop or modify tools to suit the requirements of the company.
Process:	There are established Human Factors processes, but these are not necessarily integrated with other company processes. Customer feedback is collected after each project.

Step 3 – Influence Strategic Decisions

There is evidence that Human Factors influences key strategic and tactical decisions. Human Factors issues are considered alongside other issues such as financial or system performance targets. However, the level of influence depends upon the skill and reputation of individuals, rather than on company working practice.

People:	In addition to the established team of Human Factors experts, other key people in the organization have an appreciation of Human Factors.
Tools:	Mechanisms exist for ensuring that objective information is available in time to influence key decisions regarding Human Factors issues, and consider Human Factors risks in relation to other risks. The data required to influence key decisions is understood and developed.
Process:	Processes where Human Factors influence is required are identified and Human Factors information is integrated within these processes in order to influence key decisions.

Step 4 – Defined and Repeated Processes

There is a long-term strategy which maximizes reuse of information and lessons learnt, and reduces the cost to individual projects. There is an agreed process that enables various parts of the organization to work together and contribute effectively to the resolution of Human Factors issues. This means that there is also a clear definition of roles, interdependencies, resourcing, timescales and deliverables. This requires commitment from all levels of management.

People:	Human Factors knowledge extends to other teams. There are methods for developing the relevant level of Human Factors competency across the organization. Key people are trained in relevant aspects of Human Factors, such as project managers, management teams, engineers, incident investigators and end users.

Tools:	There are mechanisms that enable the re-use of data within and between projects (e.g., taxonomies and databases) and to ensure that data is progressively enhanced and validated. Tools are developed that can be used by non-Human Factors specialists to help them apply Human Factors principles.
Process:	There are defined and repeated processes that are reliably applied to all projects. Human Factors is a part of key company processes, such as project planning, requirement definition, design and development, training and incident investigation. Human Factors is part of the checklist at key project milestones.

Step 5 – Fully integrated in policy and practices

Human Factors is a fundamental part of the way the company works. Disciplines and strategies are aligned, e.g., Engineering, Human Factors and Training. The company is a learning organization, that monitors and continuously improves its approach to proactively addressing Human Factors.

People:	Human Factors principles are appreciated throughout the organization. There is a balance in who does Human Factors work, with optimal use of Human Factors experts.
Tools:	Processes and tools continue to be optimized to yield the greatest risk reduction at lowest cost.
Process:	Human Factors is fully integrated into company policy and working practices, with the capability to monitor and improve integration.

Some Practical Considerations

As mentioned previously, Human Factors Integration will not happen overnight. The following paragraphs give some thoughts from the authors' experience about making the transition smoothly.

Human Factors will never be integrated within a company simply by writing it into procedures. It is necessary that people understand and believe in the benefits of Human Factors. The Human Factors community therefore needs to communicate effectively and consistently, share information, be pragmatic in their approach and provide advice that demonstrably helps in achieving short and long-term benefits to the company.

Human Factors integration requires a change in attitude of people throughout the company, including Human Factors experts. It is important to work towards a state where all appropriate parties have an equal voice and compromises are made based on all requirements and an objective process. It is important to understand that the end result may not be the *best* solution from a Human Factors perspective, but it will be the *optimal* one from the company's perspective.

It is necessary for the different parts of the organization to understand each other's priorities, concerns and constraints. Therefore, it is not only important that other parts of the organization understand Human Factors, but that Human Factors specialists understand other parts of the organization. For example, to effectively influence HMI design decisions, it is necessary to understand engineering and operational processes. This common understanding can be facilitated by shared workshops, meetings and specific training.

Integrating Human Factors throughout the company means that the most appropriate people should take on the responsibility for Human Factors activities. This is not necessarily a Human Factors expert, it could be an expert in another field who is trained in the use of specific Human Factors techniques. For example, Health and Safety officers are often trained in workstation ergonomics. In order to facilitate this sharing of the Human Factors role, it is useful to have tools (such as guidelines and checklists) that do not necessarily require a Human Factors expert to apply them. The overall aim should be to achieve a balance between helping people to address Human Factors issues for themselves and ensuring that they have the right expertise available when appropriate.

Examples of Human Factors Integration within NATS

Within NATS we are working towards Human Factors Integration. This section will focus on examples of approaches that NATS has adopted to integrate Human Factors into its policies, strategy and working practice.

The examples given below illustrate work the Human Factors Group has done to help change the nature of the company's approach in key areas such as requirements development, safety risk management and ATCO training. The aim has been to enhance consideration of Human Factors issues, to ensure the results of Human Factors activities influence key decisions; and to apply Human Factors at the right time, following defined and repeatable processes, in order to achieve the maximum cost-benefit.

Example One: Total Systems Approach to Requirements (TSAR)

A fundamental problem in the procurement of many systems, particularly complex systems, is that requirements for equipment performance are included in specifications and contracts, but requirements for total system performance (human + equipment + procedures + training) are not. This does not make sense from either a business or risk management point of view because experience shows that people play a vital role in both the safety and effectiveness of complex systems. Therefore, failure to explicitly consider the impact of the design, training and procedures on expected human performance can lead to a gross misjudgment of the likely total system safety, performance, operational effectiveness and personnel-related lifecycle costs. To help address this, NATS is using the TSAR approach to change the focus of requirement definition from an equipment-based view to one based on operational and business need (Burrett and Gorst, 2002).

This change of focus is fundamental to Human Factors Integration, as it can be used to change the perspective of the entire project team. As such, project activities, decisions and acceptance testing are geared towards ensuring that the total system (human + equipment + procedures + training) can achieve its operational and business goals. In addition, this approach provides a common language for use by an integrated product team, where hardware, software, and human solutions can be viewed within the context of the total system requirements. This also tends to promote the iterative development of the procedures and training earlier in the project, alongside development of the equipment.

Total system requirements are expressed independently of equipment solutions. They are expressed in terms of the operations or functions that need to be performed, the level of performance required, and the conditions under which the performance needs to be achieved. These requirements will vary depending upon the type of system being developed. For example, total system requirements for an Air Traffic Control (ATC) system may include requirements such as minimum separation of aircraft, accurate and timely delivery of instructions to aircraft, and the ability to handle a defined traffic density or complexity. For a military aircraft system, they may include ability to evade the enemy, probability of kill and ability to navigate effectively to a precise location at a precise time.

Table 20.2   Traditional and Total System Requirements

Traditional Requirements	Total System Requirements
Specification of the type of equipment and the characteristics of the equipment, e.g.: • A radio, with a given frequency spectrum, • A headset with volume control, • A control panel for selecting frequencies with the option for a certain number of frequencies at a time.	Specification of operational needs, e.g.: • Ability to communicate effectively with a set of aircraft in order to issue instructions, • Accurate and timely delivery of instructions, • Need to transmit a defined set of information, • Ability to hear transmissions accurately in defined ambient noise levels. • Ability to confirm accuracy of the transmission.

The table above illustrates the difference between traditional requirements and those developed using TSAR. The example in this case is requirements for issuing instructions from ATC to aircraft.

A key feature of this approach is that it increases the flexibility about the type of equipment used to achieve a given goal and this highlights the range of options associated with the allocation of function between the human and machine. For example, solutions to the TSAR requirement could range from radios to datalink and might include human or automatic accuracy checking. The important point is that these options are considered in light of how well they meet the business and operational need.

It is recognized that at some point, in order to actually build anything, equipment architecture and specifications must be developed. The advantage of starting the process with total system requirements is that it allows traceability between human and equipment performance to the operational and business goals. For example, the inability of a user (or equipment) to perform an action effectively, can be traced to the potential impact on operational effectiveness and safety. This facilitates trade-offs and prioritization during the development process. It also provides a valuable mechanism for reducing the subjectivity associated with assessing the usability of systems.

Another feature of total system requirements is an emphasis on understanding why something needs to be done, as well as a better definition of what needs to be done (as opposed to how it should be done). In particular, key decisions, actions and information requirements can be defined for each function. These can also be developed in the early stages of projects, independent of allocation of function to human or machine. This can be invaluable in helping to establish comprehensive control and display specifications and sub-contracts. It is also extremely useful to support procedures development, error analysis, and training needs analysis for those activities that ultimately get allocated to the human as tasks during design. For example, better definition of the information required for a function can feed into display specifications to ensure that the right information is available to support effective decision making. If this information is not available, it might impact the likelihood of human error and/or the amount of training required.

One of the challenges of the TSAR approach is verification of the total system specification. Decisions about whether or not a system meets its specification will be based upon a perception of whether or not the design, training and procedures support human operators and maintainers to achieve their tasks safely and effectively. TSAR provides an approach for explicitly and systematically specifying, contracting and testing how well the total system meets its requirements. In particular, this includes:

1. setting total system performance targets, which can be used as measures of effectiveness;

2. identifying a set of conditions and scenarios in which the performance needs to be achieved for use in iterative evaluation of proposed solutions against the requirements;
3. defining functional requirements, independent of allocation to human or machine and understanding the relative priority of functions in terms of operations and safety;
4. defining procedures and training in parallel with the design to the maximum extent practicable (this enables assessment of the total system);
5. specifying assumptions about representative skill and experience levels of personnel;
6. ensuring contractual payments, milestones, and deliverables consider the total system;
7. increasing partnership between customers and suppliers to address total system capability.

One of the factors that can make verification of this sort of requirement difficult is that human performance can vary significantly depending upon the operational situation, the individual or even in a particular individual if they are tired or stressed. Therefore, an important element of this process is step 2, which starts with identification of those operational conditions and events that conspire to compromise the ability of the human operator to maintain the required levels of performance. For example, weather conditions, time of day, unplanned events and equipment failures are all conditions that play a part in the performance of the human. Establishing these conditions is an important prerequisite of deciding on scenarios under which the system must be capable of demonstrating its ability to meet performance and safety targets. It is also helpful to agree a set of baseline assumptions about representative levels of user experience, training and skill. This reduces the likelihood of later disputes about findings associated with human-in-the-loop assessments.

Clearly, it is not practical or cost-effective to develop and/or test all possible combinations of conditions that could affect human performance. The aim should be to define a set of scenarios that characterise the range of functionality and conditions (particularly high priority functions and 'risky' combinations of conditions) in order to achieve sufficient risk reduction in subsequent analyses and tests. These scenarios help to define the scope of subsequent tests of total system performance against the defined performance targets.

Scenarios therefore need to be developed to ensure that the following elements are identified and tested:

- the full range of functions that must be performed successfully;
- the impact of functions that must be performed concurrently;
- the integration of functions that must be performed consecutively;
- the range of events that provide the greatest test of human intervention in operational events (i.e. safety-related and/or non-standard tasks);

- conditions under which human operators are expected to mitigate system failures; and
- team roles, co-ordination and cross-checking procedures and routines.

One final point about the Total System Approach to Requirements is that it can also increase the integration, sharing and re-use of data between projects. This is because the fundamental operational and business objectives associated with ATM are fairly constant. The level of performance required and the solutions to achieving that performance evolve over time. By relating everything back to the key operational and business imperatives, TSAR provides a consistent framework from which to view and assess our evolving ATM system. As such, assessments of the impact of new technologies and procedures can be more efficiently integrated into the context of the overall system.

Example Two: Integration of HF into Safety Management Systems

Another key area for Human Factors integration is company Safety Management Systems (SMS). These systems set the policies for managing safety risks as part of development and operations. In NATS, the SMS provides policy for topics such as incident investigation, safety cases and safety audits. The SMS has recently been updated to include further information on methods for integrating consideration of Human Factors into safety management activities.

Evidence of the overwhelmingly disproportionate contribution human error makes to total system risk (e.g., Kinney, Spahn, and Amato, 1977; Reason, 1990; and Federal Aviation Administration, 1990) indicates that traditional safety risk management programs have struck the wrong balance between consideration of the risks posed by equipment failure versus those posed by human error. In order to redress this imbalance, it is important to ensure that Human Factors best practice, and methods that explicitly consider the human are integrated into the safety risk management process.

This section will give examples of Human Factors integration in two key safety management activities: incident investigation and safety cases.

Considering Human Factors as part of Incident Investigation At the forefront of the identification of system risk is a robust incident investigation process that identifies risks posed by equipment, procedural and human failures evident from incidents or accidents. Once methods are in place for capturing risks, it is vital to ensure there are mechanisms to truly understand the nature of the risks posed. A robust incident investigation process needs to determine how errors and incidents are occurring, what is causing them, which are critical, the conditions under which they are more likely to occur and the effectiveness of mitigations for human and equipment failure modes.

The incident investigation process needs to be structured in such a way as to go beyond describing what happened and determine why it happened. Establishing the reasons why an operator made an error is vital to understanding which interventions might prevent or reduce the likelihood of reoccurrence.

In order to obtain a level of error detail that actually enables error problems to be clearly identified and understood, a robust error classification technique is essential. Whichever specific technique is used, it is vital that it enables the consistent collection of key error detail using a standardized set of classifications in order to obtain an accurate picture of error risk.

The Technique for the Retrospective (and predictive) Analysis of Cognitive Error in air traffic management (TRACEr) is one such technique developed by NATS (Shorrock, 1997). TRACEr is based on models of human information-processing where human error is viewed as a failure of human information-processing. Human error can arise from a breakdown at any stage of information-processing. For example:

- perception: to mis-perceive or fail to perceive information correctly;
- decision: an error of judgement, planning or decision-making can occur;
- memory: information can be forgotten or mis-recalled; and
- action: all of the information-processing can be successfully carried out (correctly perceived, memory accessed correctly and the right decision made), but an error can occur in carrying out the action.

By considering human error in these explicit terms and collecting information that describes which aspect of human information-processing failed, how it failed and why, the process of identifying human error interventions that may prevent or mitigate specific errors becomes more informed. It is important to know how and why the information-processing failed because it is only with this level of detail that effective interventions to specific error risks can be identified.

To take an example, suppose a controller cleared an aircraft through the level of another aircraft. This could be caused by a perceptual error (e.g. misreading the flight progress strip), a decision error (e.g. expecting the aircraft to achieve a rate of climb outside its normal performance), a memory error (e.g. forgetting that another aircraft was already at that level) or an action error (e.g., saying climb to flight level 110 instead of flight level 100). Clearly, the solution to preventing similar errors in the future would be significantly different depending on the actual cause.

Additionally, data on the conditions under which errors and incidents occur should be systematically collected. Errors do not occur in a vacuum; rather their likelihood is influenced by the conditions under which people operate. From many accident and incident reports in a variety of industries, it is apparent that accidents result when active failures (like human error) combine with existing latent conditions lying dormant within the system. It is essential that the incident investigation process rigorously collects data on the conditions that increase safety risk in addition to the key causal factors implicated in incidents. Through developing an understanding of these error-provoking conditions, an additional opportunity to further reduce the risk posed by human error is presented.

TRACEr uses a standardized series of pick lists and decision trees to enable consistent classification of the type of error that occurred, as well as the conditions present at the time. This facilitates evaluation of trends and management decisions about the extent to which risk mitigation is required. The following table provides an example of the standard list of conditions or Performance Shaping Factors (PSFs) for use in categorizing the conditions associated with the incident.

Table 20.3   Performance Shaping Factors (PSF)

Performance Shaping Factors (PSFs)	
Social and Team Factors	
Allocation of function and responsibility	Team pressure
Team relations	Sector manning
Personal Factors	
Alertness/fatigue	Confidence
Emotional or occupational stress	Job Satisfaction
Procedures	
Number	Comprehensiveness or completeness
Complexity	Duration in use or stability
Training and Experience	
Task familiarity	Mentoring
Level of experience	Recency (time on sector)
Workplace Design, HMI and Equipment Factors	
Console (or flight deck) ergonomics	Radar
Electronic tools	Equipment
Ambient Environment	
Noise	Temperature
Lighting	Air quality
Pilot-Controller Communications	
RT workload	Controller RT standards
Pilot language or accent	Pilot RT standards
Traffic and Airspace	
Traffic load	Traffic complexity
Weather	Sector design

Explicit consideration of Human Factors in Safety Cases A number of HF techniques now exist for analysing the potential impact of human error on overall system safety (Scaife, 2001). Active, systematic and explicit consideration of Human Factors as part of safety cases is an important avenue for ensuring that appropriate risk management occurs during system development. The safety case should include evidence that total system requirements have been met and that human error risks have been mitigated to an acceptable level. The type of evidence required to address human error risks for most projects, together with the key techniques for providing that evidence, are summarized in the table below.

Table 20.4   Performance Shaping Factors (PSF)

Required Evidence	Methods[33]											
	Hazard Logs	FMECA	HRA	Fault Trees	HAZOP	Task Analysis	Human Perf. Models	Workload Analysis	Role Definition	Simulation	TNA	Incident Investigation
Analysis that shows that all potential human errors have been identified.	➤➤	➤➤	➤➤	➤➤	➤➤							➤➤
Analysis to identify conditions/ events that increase the probability or impact of human error.						➤➤	➤➤	➤➤				➤➤
Prioritization of the safety significance of human errors, including potential for detection and recovery of those errors.	➤➤	➤➤	➤➤	➤➤	➤➤							➤➤
When the human is to be used to mitigate known design deficiencies and/or system failure modes, evidence that safety levels will not be compromised.	➤➤	➤➤	➤➤	➤➤	➤➤	➤➤	➤➤	➤➤	➤➤	➤➤	➤➤	➤➤

---

[33] FMECA = Failure Modes Effects and Criticality Analysis, HRA = Human Reliability Analysis, HAZOP = Assessment of Hazardous Operations, TNA = Training Needs Analysis.

Required Evidence	Methods[33]											
	Hazard Logs	FMECA	HRA	Fault Trees	HAZOP	Task Analysis	Human Perf. Models	Workload Analysis	Role Definition	Simulation	TNA	Incident Investigation
Evidence that training needs have been identified and that expected training will support necessary levels of human performance.						▶▶	▶▶	▶▶	▶▶	▶▶	▶▶	▶▶
Definition of operational procedures and evidence that human performance in defined operational scenarios using those procedures will be acceptable.			▶▶	▶▶	▶▶	▶▶	▶▶	▶▶	▶▶	▶▶	▶▶	▶▶
Identification of operator and maintainer roles within teams, including supervision, monitoring and cross-checking of safety critical actions.		▶▶				▶▶	▶▶	▶▶	▶▶	▶▶		▶▶
Human-in-the-loop assessments of selected scenarios to verify results of the analyses above.										▶▶		▶▶
Audit trail showing that Human Factors issues have been identified, integrated into relevant project activities and included in the design of equipment, procedures and training to mitigate safety risks.	▶▶											

Example Three: Closing-the-loop between Safety Risk Management Activities

As with equipment reliability, human reliability targets and predictions made during the development of complex systems can be informed by data collected about actual performance in operation. In addition, it is important to ensure that claims made as part of safety cases remain valid once a system becomes operational. To date, some success has been achieved in 'closing-the-loop' between predicted equipment safety levels and safety achieved in operations. This process is equally, or possibly more important for the human element of the system, where data on human error in ATC is still in the early stages of development. The same situation existed for the safety risk management of hardware in the early days of reliability modeling, but this has improved over time.

In addition, methods for addressing human error risk management during development and operations gain additional value when a common language and structure is used to integrate traditionally independent risk management processes. This facilitates the sharing, comparison and validation of human error data across all stages of the system lifecycle. It also increases our ability to deliver more accurate and cost-effective human error management.

This section summarizes the key elements of a safety risk management process and how Human Factors can be explicitly considered at each stage and proposes a framework linking the activities in each stage.

Human Factors in the Safety Risk Management Process

The process described in this sub-section is an iterative loop. For the purposes of this chapter, we begin our discussion of the process during operation. The first steps are to understand the current system risks and then to identify strategies for reducing those risks. Next, risk reduction targets need to be set so that there are clear goals for where system risk should be in the future. There needs to be a proactive attempt to identify the opportunities for error (and their mitigations) that may arise during the development of system improvements. Throughout the design of any improvement to the system, the development of its safety case and its testing and evaluation, its anticipated performance needs to be compared against the risk reduction targets. Once in operation, the actual performance of the system needs to be measured against the targets set. This needs to be a continuous process of risk reduction. Figure 20.2 below illustrates the process.

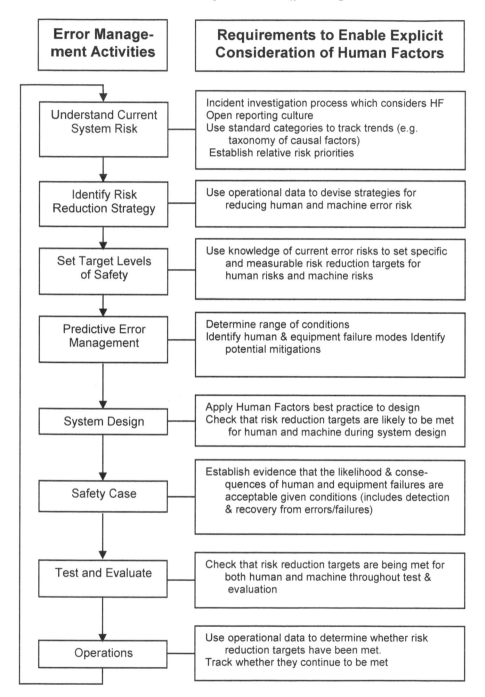

Error Management Activities	Requirements to Enable Explicit Consideration of Human Factors
Understand Current System Risk	Incident investigation process which considers HF Open reporting culture Use standard categories to track trends (e.g. taxonomy of causal factors) Establish relative risk priorities
Identify Risk Reduction Strategy	Use operational data to devise strategies for reducing human and machine error risk
Set Target Levels of Safety	Use knowledge of current error risks to set specific and measurable risk reduction targets for human risks and machine risks
Predictive Error Management	Determine range of conditions Identify human & equipment failure modes Identify potential mitigations
System Design	Apply Human Factors best practice to design Check that risk reduction targets are likely to be met for human and machine during system design
Safety Case	Establish evidence that the likelihood & consequences of human and equipment failures are acceptable given conditions (includes detection & recovery from errors/failures)
Test and Evaluate	Check that risk reduction targets are being met for both human and machine throughout test & evaluation
Operations	Use operational data to determine whether risk reduction targets have been met. Track whether they continue to be met

Figure 20.2   Human Factors in the safety risk management process

Framework to Link Safety Risk Management Activities

This sub-section addresses the goal of closing-the-loop between all stages in the safety risk management process (as presented in Figure 20.2). To achieve this, a framework must be established to ensure a common language and structure. The primary framework elements are listed below, and illustrated in Figure 20.3.

- Goals: The safety, operational and business aims of the system. For Air Traffic Control (ATC), one fundamental safety goal is maintaining safe separation of aircraft.
- Functions: The actions that need to be performed (by either the human or machine component of the system) in order to achieve the system goal. In ATC, this would include detection and resolution of potential conflicts between pairs of aircraft. Performance targets can be assigned to both functions and goals to define safety targets/requirements.
- Failure Modes: The human or equipment failures that could result in the goal *not* being achieved. An example from ATC would be issuing the wrong clearance to an aircraft. Both the type and frequency of failure modes are important to safety risk management. A measure of the criticality of a given failure can be determined by the extent to which it relates to *not* achieving key safety goals.
- Causal Factors: A set of factors that describe how the functions can result in error (e.g., lack of training, poor design, fatigue).
- Conditions: The range of conditions in which the goals must be achieved (e.g., 24 hours a day, in poor weather, in a defined working environment).
- Mitigations: System defenses in place to prevent the incidence/reduce the consequences of failure modes (e.g., procedures, error checking, supervision). In this framework, mitigations can be performed by many elements of the system (e.g., human, machine, training, and/or environment), against either human or equipment failure modes. Tracking the predicted and actual performance of mitigations can be central to ensuring that safety risks are managed effectively.

These factors are generic in the sense that they could be defined in a language that remains constant throughout all stages of the lifecycle. From this it is possible to use related concepts and data throughout the system lifecycle. For example, targets set to reduce a particular causal factor by a specific amount can be made during the requirements phase and throughout the design of the new system. If the same term is used to collect safety data during operation of the system, then comparisons can be made to determine the extent to which the target has been achieved. The assumptions made about the conditions under which the system functions would be carried out can be compared with the conditions under which system errors are occurring.

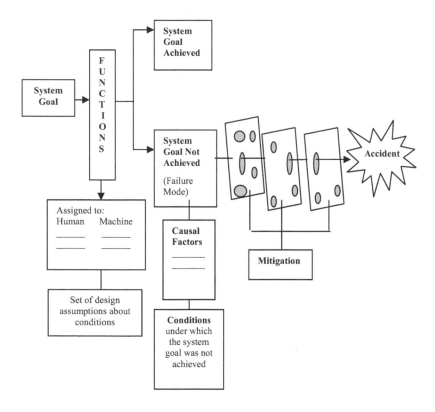

Figure 20.3   Framework to link safety risk management activities

The framework should be continually updated as a result of the information generated by its use throughout the various risk management processes. Thus the framework could provide a mechanism for sharing the output and lesson learning between all of the previously disparate processes as well as a means for increasing efficiency by preventing duplication of effort. Figure 20.3, derived in part from Reason (1990), illustrates the proposed framework. This type of framework can be used to address the following objectives:

- consider human and machine contributions to system success and risk explicitly, and alongside one another, to improve understanding of the balance of risk and prioritization of risk reduction activities;
- represent the complex nature of human error and its many causes;

- set and measure adherence to safety targets at each step of the lifecycle;
- improve our ability to set targets through better operational performance data, as well as validating predictions against operational reality;
- include new or modified sub-systems (which may be added at different times) into safety risk management;
- facilitate trade-offs between risk reduction options; and
- assess the effectiveness of risk reduction measures.

Example Four: Informing Strategic Decisions about Training

Another example of the integration of Human Factors into company strategy and working practice is in strategies for the continuous improvement of training. This also includes use of Human Factors to aid the management of changes to training as a result of new equipment or procedures.

NATS has recently conducted a project called Building Blocks, to inform company training strategy and action plans. This study employed training needs analysis techniques and involved a team of experts representing our operational units, the College of Air Traffic Control, Human Factors, Trade Unions and Change Management. Key elements of the project included:

a. *Definition of the skills necessary for a valid ATCO* We used a combination of existing requirements such as the Civil Aviation Authority's Review Group of ATC Training (RGAT), Eurocontrol's Common Core Competencies, plus expert judgement about the task, cognitive and physical requirements of the controller's job. Particular attention was paid to cognitive skills such as problem solving, teamworking, multi-tasking and situation awareness. The result of this step in the project was a detailed description of the requirements of the ATCO job, plus the skills, knowledge and attitudes necessary to meet those requirements. This gave an agreed and integrated baseline of our training objective.
b. *Establishing consensus on the relative priorities for each skill* This was achieved using a difficulty, importance and frequency rating.
c. *Agreeing measurement criteria for each skill* We identified observable behaviors that could be used to determine whether or not the skill had been acquired to an acceptable level.
d. *Examining how we currently train these skills at each stage of training* This was an important step because it highlighted best practice and differences in our approach at various units.
e. Looking for gaps in what we train, ineffective training methods, the wrong sequence in training, a lack of consolidation time, etc. This forms the basis for the recommendations on how to improve training. Research about how people learn, how to teach cognitive skills and state-of-the-art training methods were used to inform this activity.

f.   *Establishing a consensus on the best way to improve our training system and where it would be most effective to apply resources.*

g.   *Use of this baseline data to enable us to adapt more effectively to change in the future (e.g., airspace re-sectorizations).* The baseline training needs analysis provides an excellent source of information to increase the cost-benefit of activities aimed at identifying new skill requirements and training needs. This also enables trade-offs between increased development costs (e.g., for more usable systems) and the likely downstream lifecycle costs (e.g., from increased training).

h.   *Ensuring that all of the above is done with full participation of all relevant parties and as objectively as possible.*

Overall, the Building Blocks project has provided an invaluable set of objective data from which the entire training community within NATS can judge how best to improve our training in future. Human Factors has been integrated from the very start of the project and will be a key part of the implementation of training improvements to support training departments and instructors throughout the company.

Conclusions

This chapter has outlined the benefits of integrating Human Factors into company strategy and working practice. A five-step process has been proposed to act as a route map to achieving successful integration. This process is aimed at moving from a reactive state, where Human Factors is provided on demand to customers in the company, to a proactive state where:

- Human Factors influences key decisions.
- Human Factors is established in defined and repeated processes.
- the body of company-wide knowledge and data about Human Factors is being built up over time.
- processes and methods are optimized to deliver Human Factors solutions and risk reduction with maximum cost-benefit.

This chapter has also given some examples of where NATS is making progress, by implementing approaches that improve the integration of Human Factors into areas such as Requirements Development, Safety Management Policy and Working Practice, and Training Strategy and Development.

The Human Factors profession has made great progress over the past decades in becoming an established and credible discipline. The next step in the evolution of Human Factors is Human Factors Integration. This progression is essential if we are to optimize the extent to which our systems enable efficient, safe, and cost-effective human performance.

References

Burrett, G. and Gorst, A. (2001), Specifying, Contracting, and Accepting Human Factors Aspects of Systems, *Proceedings of the Human Issues in the Aviation System Conference (HIAS)*, Toulouse, France.

Burrett, G. and Foley, S. (2003), Integrating Human Error Management Strategies Throughout the System Lifecycle, in F. Redmill and T. Anderson (eds.), *Current Issues in Safety-critical Systems: Proceedings of the 11th Safety-critical Systems Symposium*, Springer-Verlag, London.

Federal Aviation Administration (1990), *Profile of Operational Errors in the National Airspace System: Calendar Year 1988*, Washington, DC.

Geesink, A, (1990), Marketing and sports: promotion and professionalism, *Ergonomics*, Vol. 33, pp. 251-252.

Kinney, G.C., Spahn, J., and Amato, R.A. (1977), *The Human Element in Air Traffic Control: Observations and Analyses of the Performance of Controllers and Supervisors in Providing ATC Separation Services*, Report Number MTR-7655, MITRE Corporation, McLean, VA.

Reason, J. (1990), *Human Error*, Cambridge University Press, UK.

Scaife, R. (2001), *Guidance on the Use of Specific Human Factors Techniques for Safety Management*, NATS Internal Document ATMDC/0730/TN2.1, National Air Traffic Services Limited, London.

Shorrock, S.T. (1997), *The Development and Evaluation of TRACEr: A Technique for the Retrospective Analysis of Cognitive Errors in Air Traffic Control*, MSc Thesis, the University of Birmingham, UK.

Shorrock, S.T., Kirwan, B., MacKendrick, H. and Kennedy, R. (2001), Assessing human error in air traffic management systems design: methodological issues, *Le Travail Humain*, Vol. 64(3), pp. 269-289.

Weston, J.D. and Haslam, R.A. (1992), How successfully can non-ergonomists recognise and apply ergonomic principles?, *Contemporary Ergonomics*, pp. 334-339 Taylor & Francis, London.

PART VI
DISCUSSION

Chapter 21

Discussion

In the Introduction to this book, several related questions were raised:

- What does ATM actually need from Human Factors?
- What does Human Factors have to offer?
- What can Human Factors actually deliver?
- Where does Human Factors need to improve?

The book as a whole has largely focused on the third question, aiming to provide a set of case studies as 'evidence' that Human Factors can indeed deliver useful results and outputs that lead to change in real ATM systems. This chapter nevertheless attempts to address all of the above questions. To do this, it draws not only from the other chapters in the book, but also from an appreciation of what the requirements in ATM are now, and will likely be in the future.

This chapter is therefore divided into two principal sections. The first deals with the current situation, focusing on what Human Factors has been able to deliver against the apparent needs of ATM. This shows the strengths of Human Factors, but also highlights some of the existing 'gaps' in Human Factors capabilities.

The second part of this chapter is more 'strategic' in orientation. It considers the major challenges that lie ahead for future ATM, and therefore the type of planning that needs to occur if Human Factors is to meet these demands. These challenges are primarily set by the projected increasing air traffic capacity demand to be placed on the ATM system, and the ATM system designers' proposed solutions to achieve required system performance (e.g. via increased automation and airspace design changes). Human Factors therefore needs to identify and conduct the corresponding Human Factors activities and methodological advancements required to support future ATM.

The second section in this chapter reinforces the need for a more coherent and systemic approach to Human Factors in ATM, building on the type of 'programmatic' approaches that have been illustrated in the last section of case studies. Until now, Human Factors has often been able to work 'piecemeal', with the Human Factors specialist able to tackle individual problems based on the 'tools' in his or her 'toolkit'. However, the projected changes and challenges in ATM are fundamental, and the human will remain the critical factor, so Human Factors as a discipline must work in the future coherently across all aspects of human system performance in ATM. This will require not only new tools in the 'toolbox', but a far more structured approach encompassing research and application closely linked to other ATM system domains such as design and development, human resources, and safety. As with air traffic management itself, Human Factors must become more strategic in nature, dealing with key issues systematically, and anticipating problems well in advance so that the general flow of air traffic system development and implementation is supported without hindrance.

Discussion Part One: The Current Situation

Human Factors in Current ATM – Is It Delivering?

This book has shown a broad cross-section of Human Factors case studies and approaches from the US and Europe, dealing with a wide range of Human Factors and ATM issues. It has attempted to show 'Human Factors in action' in the real world of ATM. These individual chapters in themselves already signify to the reader that Human Factors is indeed delivering useful benefits to the ATM community. However, to make the overall picture clearer, and particularly for the reader who does not have the time to read all the case studies, Table 21.1 below summarizes the foregoing case studies according to several aspects:

- The 'driver' for the work – the ATM requirement or problem that triggered the need for the application of Human Factors methods *[why the study happened]*;
- The lifecycle stage during which Human Factors was applied *[when it happened]*;
- The ATM environment (airport; TMA/TRACON; Approach; En-route; Oceanic) *[what part of ATM it concerned]*;
- The Human Factors approach *[how it was done, with what methods]*;
- The impact achieved *[what difference it made]*;
- Human Factors application areas *[which functional areas of Human Factors were used in the study – confirming which areas of HF are of practical value. The numbers in the last column in Table 21.1 refer to the HF Application Areas listed in Table 21.2]*.

These elements in particular answer the key questions of 'why' the work was done, 'when' it was done, 'what' aspect of the ATM system it affected, and 'how' it was done in terms of Human Factors approaches. These threads of inquiry are each discussed below, and some insights are then drawn on Human Factors adequacy for ATM overall. However, it must be remembered that these are simply a representative set of case studies, and there are certainly other case studies that could be documented that might highlight different insights or conclusions. Nevertheless, it is worth drawing from the collection of evidence represented by the case studies in this book, in order to see where current Human Factors may need to improve.

Table 21.1   Approaches and impacts achieved per case study chapter

Chapter	Authors	Title	'Driver' for the work	Life Cycle Stage	ATM Function	HF Approach	Impact Achieved	Human Factors Application Areas
PART II HUMAN FACTORS IN OPERATIONS								
2	Voller, Glasgow, Heath, Kennedy & Mason	Development and Implementation of a Position Hand-Over Checklist and Best Practice Process for Air Traffic Controllers	High rate of incidents following position hand-over	Operations	TMA	Interviews; observation	Position hand-over procedure derived and widely implemented	3, 10, 12, 16, 17
3	Cardosi	Runway Safety	Need to reduce number of incidents in US airports	Operations	Airport	Analysis of incidents	Critical areas of airport operations identified; development of intervention strategies	3, 10, 11, 12, 15, 16, 17, 18
4	Isaac, Engelen, Polman	Human Error in European Air Traffic Management: from Theory to Practice	Need to look 'deeper' into a particular incident	Operations	All	Human error modeling and derived incident classification scheme	Enhanced tool for incident investigation developed; more detailed insights for incident reduction gained	3, 5, 8, 10, 17, 18, 23
5	Pounds & Ferrante	FAA Strategies for Reducing Operational Error Causal Factors	Need to reduce contribution of human error to operational errors	Operations	All	Incident recreation and operational error classification; investigation decision aid	Method and tool for incident investigation implemented and validated	3, 5, 8, 10, 17, 18, 23
6	Kirwan, Shorrock, Scaife & Fearnside	Reducing Separation in the Open Flight Information Region: Insights into a Human Factors Safety Case	Desire to reduce separation minima in UK OFIR	Operations	Military airspace (OFIR)	Human reliability approaches & simulations	Vertical separation minima reduction from 5000' to 3000'	10, 11, 12, 16, 17,18, 23

Table 21.1   Approaches and impacts achieved per case study chapter (continued)

				Design, Operations	Traffic Flow Mgmt	Application of		
7	Smith, Klopfenstein, Jezzernac & Spencer	Distributed Work in the National Airspace System: Providing Feedback Loops using the Post-Operations Evaluation Tool (POET)	Desire to provide NAS users with feedback to reduce traffic flow delays and increase capacity			Application of design guidelines; Analysis of communication techniques in collaborative decision making	National deployment of Post-Operations Evaluation Tool (POET)	9, 10, 11, 12, 18
PART III HUMAN FACTORS AND HUMAN RESOURCES								
8	Voller & Fowler	Human Factors Longitudinal Study to Support the Improvement of Air Traffic Controller Training	High ATCO student fail rate	Operations support (training)	All	Longitudinal study; interviews	Initial improvement in pass rate by 25%; sustained improvement	14, 20, 21
9	Broach	A Singular Success: Air Traffic Control Specialist Selection 1981 – 1992	US strike led to sudden need to select large number of controllers	Operations Selection & Training	Air traffic control centers (including towers)	Selection tests	US selected an effective and sustainable controller workforce	14, 20, 21
10	Leonhardt	Implementation of Critical Incident Stress Management at the German Air Navigation Services	Need to support health and performance after critical incidents	Operations	Air traffic control centers (including towers)	Critical Incident Stress Management (CISM)	CISM approach developed, implemented, and used following incidents and accidents	17
11	Woldring, Van Damme, Patterson & Henriques	Team Resource Management in European Air Traffic Control: Results of a Seven-Year Development and Implementation Program	Desire to improve teamwork in ATM operations	Operations & Training	All	Team Resource Management (TRM)	TRM taken up by 20 European countries	3, 14, 18, 21

Table 21.1 Approaches and impacts achieved per case study chapter (continued)

	Author	Chapter title		Operations	Air traffic control centers			Ref.
12	Della Rocco & Nesthus	Shiftwork and Air Traffic Control: Transitioning Research Results to the Workforce	Health and safety considerations related to shiftwork	Operations	Air traffic control centers	A range of Human Factors, psychological, and psycho-physiological tests	Shiftwork effects on performance and well-being better understood; workforce provided with educational material on mitigating fatigue and improving performance	8, 20
PART IV HUMAN FACTORS METHODOLOGIES								
13	Manning & Stein	Measuring Air Traffic Controller Performance in the 21st Century	Need to measure user performance to support system design & development	Design and development (simulations)	All	Range of measures including workload and situation awareness	Successful 'batteries' of measures available for a range of issues – proven ability to resolve HF issues scientifically	All
14	Corker	Computational Human Performance Models and Air Traffic Management	Need to assess performance in design & development work	Design & development	All	Principally computer simulation and modeling approaches	Use of models to refine airspace design for efficiency	1, 3, 10, 23
15	Shorrock, Kirwan & Smith	Performance Prediction in Air Traffic Management: Applying Human Error Analysis Approaches to New Concepts	Safety assessment and the desire to protect future ATM from human error	Design, Live Trials and Operations	En Route, Approach, Military airspace	Human error identification approaches (TRACER & HAZOP)	Improvements to interface and safety assurance for live trials, warning of future risk areas	3, 4, 11, 12, 15, 17, 18

Table 21.1 Approaches and impacts achieved per case study chapter (continued)

PART V HUMAN FACTORS INTEGRATION PROGRAMMES							
16	Hewitt, Krois & Piccione	The Management of Human Factors Programs in ATM Applications: A Case Study	Need to integrate HF into the acquisition process	Design through to Acquisition	All	Human Factors study areas tied to Technology Readiness Levels (TRLs); integrated team approaches; HF staffing requirements	Establishment of a strong set of procedures, competencies and processes for HF integration into projects	All
17	Allendoerfer, Yuditsky, Mogford & Galushka	The Development of Effective STARS User Interfaces	Need to upgrade the controller radar terminal	Design (HMI)	TRACON facilities and control towers	Usability, prototyping and simulation; integrated teams	Significant improvement of STARS interface design	2, 4, 9, 11, 12, 13, 22
18	Josefsson	Human Factors and the Management of Change – a Personal Perspective	Need to upgrade the operations room technology	Design and development (including the interface) & safety culture	En route and TMA	HMI design, TRM, HERA, safety climate measures	Improvement in the interface (e.g. stripless systems) and understanding of human error in ATM incidents	2, 4, 5, 9, 11, 12, 22
19	Kirwan	Review of a Three-Year Air Traffic Management Human Factors Program	Company see HF as a growth area – development of a HF team and program	Design and operations	All	3-year HF program; integration strategies	Establishment of a strong HF team in the company – many impacts	All
20	Burrett, Weston & Foley	Integrating Human Factors into Company Policy and Working Practice	Integration of Human Factors into company processes during a period of privatization	Design and Operations	All	Total systems approach; integration into business activities; risk management approaches	Consolidated integration of HF into company processes; risk reduction and training improvement	All

Drivers for Human Factors studies – The Apparent Needs of ATM, and What ATM Entrusts to Human Factors

In terms of 'drivers' for the work, all of the case studies can be seen as falling into one (or more) of the following categories:

- problem or event driven – fixing a problem, reacting to a major event in ATM (e.g. incident rate trend; training success rate decrement; post-strike recovery; etc.);
- opportunity driven – seeing an opportunity to improve system performance (e.g. TRM; Human Factors programs);
- safety assurance driven - e.g. safety case needs (e.g. reduced separation minima);
- health and psychological well-being driven (e.g. CISM; shift work; etc.)
- system design and development driven (e.g. simulation modeling; performance modeling; human error modeling);
- system implementation driven (e.g. the STARS story, etc.).

A first and positive insight from the analysis represented in Table 21.1 is that a number of Human Factors studies described in this book have been initiated in response to actual needs, and have provided positive and sustained impacts (e.g. controller training improvement; handover checklist, etc.).

A second observation is that two studies mentioned here are explicitly concerned with the controllers' mental health and well-being and their effect on user performance. Apart from that a number of studies have been initiated in the context of system development and implementation programs. A potential conclusion to draw from this is that Human Factors specialists are increasingly consulted in a pro-active manner rather than drawing them in once something has obviously gone wrong.

This leads to a third observation that there is already a visible shift in evidence from simple 'tactical' usage of Human Factors to solve (human-related) problems that arise, to more programmatic and strategic embedding of Human Factors into large-scale projects, with Human Factors seen as an important enabler of overall system performance. In particular, whilst some chapters have shown an opportunity to (successfully) test Human Factors in a new area (e.g. safety study support; training improvement), others show that Human Factors is now being trusted and empowered to take on more widespread ('strategic') support activities that are more central or 'core' to ATM (e.g. the STARS story; the UK NATS Human Factors program of work). This indicates that Human Factors has indeed been successful in delivering useful inputs to ATM development, and is being trusted to do more in the future. It is also perhaps recognition of the importance of human performance management in future ATM, so that human performance and therefore Human Factors itself may become one of the drivers of future ATM.

As a fourth observation, the above categorization of 'drivers' for Human Factors in ATM would mean that the typical 'clients' for Human Factors work would generally encompass operational system managers, safety managers, Human Resource departments, controller unions, system designers, and system implementers (the latter two may be the same, though this is not always the case). This client range is indeed broad-based, suggesting that Human Factors is already well-diversified, and not, for example, merely seen as an interesting research area. However, given this client range, an interesting question to pose to Human Factors specialists is how such clients would know what Human Factors can do, and how to contact Human Factors professionals. This question is raised because traditionally Human Factors is seen as something of an 'academic' discipline, publishing in scientific journals and conferences, rather than more general 'trade journals' or industry conferences. Therefore, this is an area perhaps where already a need can be seen for some improvement on the 'marketing' of Human Factors in ATM.

As a final remark on 'why' Human Factors studies are commissioned, based albeit on the limited data set in this volume, with the exception of one chapter presented here (reduced separation minima) the majority of these needs refer to the human element of the ATM system. This seems to suggest that the advantages of Human Factors studies may be exploited to a lesser degree where the overall system is concerned. To put it in a more pointed manner: Human Factors may not be the obvious first thought that comes to the mind when there is a problem or need with an integrated system based largely on technology. This means that whilst Human Factors may be seen as a key enabler, it is perhaps seen as being limited to the domain of human performance – essentially enabling humans to perform given the basic technical and operational ATM concept. Human Factors is not yet it seems at the stage of being asked, for example, 'How would Human Factors re-design ATM (e.g. ATC procedures & airspace design & technical functions) to optimize performance?' This suggests that Human Factors has not yet found its logical limit in being able to enhance ATM.

Lifecycle Stage – When Can Human Factors Usefully Start in the Project Lifecycle?

In terms of lifecycle stage of the various case studies, these are roughly split equally between operations and design/development stages. However, whilst positive impacts are particularly evident for studies in the operational context, the 'visibility' of the impacts of studies conducted in earlier stages of the lifecycle suffers from a lack of direct comparison. Whereas the effects of changes in the operational system can often be quantified through a comparison pre- versus post-change, such a comparison is largely hypothetical for Human Factors-influenced design decisions taken during the system development.

The balance of the various case studies between the operations and design/development phases of the system lifecycle suggests that at a high level, Human Factors is capable of satisfying ATM needs during the full development and operational lifecycle of ATM systems. However, within the design component of this sample of case studies, most studies are relatively later in the design stage rather than early on. This is not surprising, since Human Factors is addressed differently in concept exploration and concept development work than it is when a design is relatively advanced. Consequently, it may be difficult to work on details in terms of what the controller must do, with what equipment and interface, etc. since design details have not yet been defined. Nevertheless, there are certain areas wherein key decisions will be made early on (e.g. allocation of function between human and machine) that will have major implications later on, and may indeed limit the 'degrees of freedom' that Human Factors can work with. Such studies may concern specific design solutions, often concerning the system architecture and task allocation as a whole, but also basic human performance characteristics which highlight the potential need for Human Factors tools in the early design stages. This is why this book contains a short section on developing methodologies (e.g. using simulations and predictive human error analysis) that can bring out Human Factors issues at earlier stages in the design process.

This does suggest however that the greater 'leverage' of Human Factors studies in earlier phases of the design process is not being exploited to its full extent. There still perhaps exists a pre-conception amongst ATM system developers that Human Factors is something that 'comes later' in the design lifecycle, simply because Human Factors may be equated with concrete and detailed aspects such as HMIs, workplace design, etc., rather than fundamental issues and more seemingly 'abstract' issues such as allocation of function. In larger, more 'programmatic' approaches to Human Factors, this is probably less of a problem, since the Human Factors personnel will be given the opportunity to make the case and will have more leeway to influence earlier stages of design and development. Nevertheless, as a general comment, allocation of function would seem to be an area where Human Factors needs to make more effort to show its added value in early stages of design.

ATM 'functions' Served by the Case Studies – What Parts of ATM is Human Factors Helping?

The major focus of attention in this set of case studies has been En-route and TMA/TRACON, with a few chapters focusing on Airport operations and Military airspace operations. Oceanic air traffic management has not been an explicit focus in the chapters, though some of the chapters in the 'integration' section of the book did include Oceanic Control Centre operations.

There is a sense that Human Factors in airports has been less of a priority, and that this comment concerns the airport operations as a whole, including all the sometimes complex interfaces at an airport rather than just the control tower. Cardosi's chapter in particular highlights the fact that even very basic Human Factors interventions could have a major impact on efficiency and safety of some airport operations (e.g. improved airport markings). In that regard, it is important to see the airport as a system and not merely focus on the human-machine interface in the tower (though control tower operations are of course critical). Therefore, it would be beneficial to achieve a more even spread of Human Factors intervention across the whole 'gate-to-gate' operations that encompass ATM. This in turn requires Human Factors not only working within each part of ATM (tower; TRACON; En-route; Oceanic; etc.), but where required across these parts. In this way Human Factors can support and enhance the human-based qualities that facilitate good total ATM system performance (e.g. effective communications; shared understanding; anticipation and flexibility; etc.).

Human Factors Approach – How has Human Factors Achieved its Goals?

A range of tools has been used in the various case studies, from basic tools (e.g. interviews and observation), computer-based toolkits (e.g. POET) and techniques (e.g. TRACER, HERA, CISM, TRM), methodologies and approaches (e.g. simulations, human error analysis), to complete applied sub-domains within Human Factors (e.g. shiftwork studies; performance measurement, HMI design). These approaches have shown their worth by delivering results that have led to changes in real systems. Three chapters in particular have shown the developments occurring in the areas of human performance measurement, simulations, and human error prediction. Additionally, following on from earlier discussion points, it appears that perhaps more methods dealing with the very early stages of design could be developed, (e.g. dealing with issues such as allocation of function), as well as approaches to explore more 'core' aspects concerning relationships between ATM performance and human performance (e.g. airspace design and human performance). Such latter study areas may build on existing methods (e.g. workload and situation awareness measurement) or may require new method development (e.g. regarding aspects of controller-experienced complexity depending on airspace design and traffic characteristics).

The last section of case studies showed in particular how such tools can be utilized together to achieve more widespread and arguably more sustainable impact, by being part of an integrated program. This trend towards more 'systemic' approaches of Human Factors is seen as a healthy progression. In this respect, for the 'Integration' of Human Factors, what is probably most needed is simply more experience in developing and running large Human Factors programs that are fully integrated with other major stakeholders' programs, e.g. large-scale system implementation programs, new concept development programs, etc. Such experience should enable organizational learning to take place in terms of how best to utilize and organize Human Factors to enhance ATM system performance.

Human Factors Application Areas – A More Generalized Picture

In summary, the case study chapters have shown at least partly what Human Factors can offer and deliver, according to the needs of ATM. The second part of the discussion will consider the future needs of ATM, and hence the challenges facing applied Human Factors over the next two decades. However, before considering where Human Factors needs to develop for the future, it is important first of all to establish a more generalized and comprehensive picture of Human Factors now, since the above discussion so far has only related to Human Factors as demonstrated in the case study and methodology chapters in this book. So far this treatment of Human Factors has been 'episodic' in nature and vision. Before leaving this section, it is therefore useful to consider Human Factors more generally, as a 'domain', to see all the areas it tries to address in ATM.

A comprehensive range of Human Factors activities as seen by the FAA has been captured in the 24 Study Areas contained in Table 21.2 (Krois, Mogford & Rehman, 2003). Table 21.1 has also identified the study areas associated with each chapter, all of which have been addressed by the full complement of case studies in this book. This listing of areas shows the breadth of Human Factors approaches and areas available to help ATM system performance. Similarly, Figure 21.1 (Schaefer, 2003) shows equivalent study areas but in a more 'dynamic' pictorial representation known as the 'gearbox model'. This has been used in EUROCONTROL training on Human Factors, and aims to highlight the inter-connectivity between Human Factors issues at a 'system' level (e.g. 'ensure the Human-Machine Interface is adequate') and at a 'human performance' level (e.g. 'ensure the workload afforded by the interface aspects are adequate; ensure the controller has adequate situation awareness when using the new interface;' etc.). This model has been found helpful for people to discriminate between high-level issues and supporting analyses, techniques and research. Both these perspectives indicate what Human Factors can offer 'now' to ATM. The second half of the Discussion next moves to consider the challenges for Human Factors in the future.

Table 21.2   Human Factors Application Areas

1. *Allocation of Function.* System design reflects assignment of operational roles, functions, and tasks to humans or equipment while maintaining the human's awareness of the operational situation.
2. *Anthropometrics and Biomechanics.* Accommodating the physical attributes of its user population (e.g. from the 1st to the 99th percentile levels).
3. *Communications and Teamwork.* Applying system design considerations to enhance required user communications and teamwork.
4. *Computer-Human Interface.* Standardization of CHI to access and use common functions employing similar and effective user dialogues, interfaces, and procedures.

5. *Culture.* Addressing the organizational and sociological environment into which any change, including new technologies and procedures, will be introduced.
6. *Displays and Controls.* Designing and arranging displays and controls to be consistent with the operator's and maintainer's tasks and actions.
7. *Documentation.* Preparing user documentation and technical manuals in a suitable format of information presentation, at the appropriate reading level, and with the required degree of technical sophistication and clarity.
8. *Environment.* Accommodating environmental factors (including extremes) to which the system will be subjected and understanding the associated effects on human-system performance.
9. *Functional Design.* Applying human-centered design for usability and compatibility with operational and maintenance concepts.
10. *Human Error.* Examining design and contextual conditions (including supervisory and organizational influences) as causal factors contributing to human error, and consideration of objectives for error tolerance, error prevention, and error correction/recovery.
11. *Information Presentation.* Enhancing operator and maintainer performance through the use of effective and consistent labels, symbols, colors, terms, acronyms, abbreviations, formats, and data fields.
12. *Information Requirements.* Availability of information needed by the operator and maintainer for a specific task when it is needed and in the appropriate sequence.
13. *I/O Devices.* Selecting input and output (I/O) methods and devices that allow operators or maintainer to perform tasks, especially critical tasks, quickly and accurately.
14. *KSAs.* Measuring the knowledge, skills, and attitudes (KSAs) required to perform job-related tasks, and determining appropriate selection requirements for users.
15. *Operational Suitability.* Ensuring that the system appropriately supports the user in performing intended functions while maintaining interoperability and consistency with other system elements or support systems.
16. *Procedures.* Designing operation and maintenance procedures for simplicity, consistency, and ease of use.
17. *Safety and Health.* Preventing/reducing operator and maintainer exposure to safety and health hazards; ensuring the controller can maintain safety of air traffic.
18. *Situation Awareness.* Enabling operators or maintainers to perceive and understand elements of the current situation and project them to future operational situations.
19. *Special Skills and Tools.* Minimizing the need for special or unique operator or maintainer skills, abilities, tools, or characteristics.
20. *Staffing.* Accommodating constraints and efficiencies for staffing levels and organizational structures.

21. *Training.* Applying methods to enhance operator or maintainer acquisition of the knowledge and skills needed to interface with the system, and designing that system so that these skills are easily learned and retained.
22. *Visual/Auditory Alerts.* Designing visual and auditory alerts (including error messages) to invoke the necessary operator and maintainer response.
23. *Workload.* Assessing the net demands or impacts upon the physical, cognitive, and decision-making resources of an operator or maintainer using objective and subjective performance measures.
24. *Workspace.* Designing adequate work space for personnel and their tools or equipment and providing sufficient space for the movements and actions that personnel perform during operational and maintenance tasks under normal, adverse, and emergency conditions.

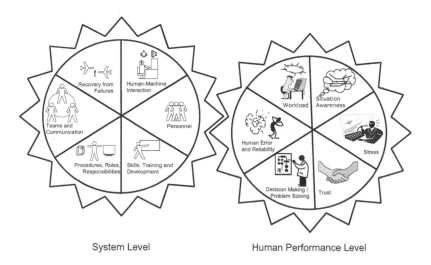

System Level   Human Performance Level

Figure 21.1   The 'Gearbox Approach'

Taking Table 21.2 and Figure 21.1 together, most of the functions are capable of being satisfactorily addressed by Human Factors. Three areas where further methodological research would be useful, however, are allocation of functions; recovery from system failure; and stress (including underload).

Discussion Part Two: Human Factors Challenges for the Future

The Introduction at the beginning of this book briefly traced the history of Human Factors in ATM, including where its future lay, in particular noting ATM's trend towards increasing capacity and complexity, and likely increases in automation for the controller. This section looks in more detail at these probable changes in ATM, and how Human Factors needs to adapt to meet the challenges that they pose.

Two visions are presented. The first is a medium-term vision of ATM system changes that may affect Europe in the next decade. In particular a selection of controller tools suggests the need for specific Human Factors support in key areas. The second vision is of Human Factors for the future US National Airspace System, including near and far-term elements of this vision. Both these visions, and especially the latter one, imply the need for more co-ordination of activities, both within Human Factors as a discipline applied to ATM, and also between Human Factors and related areas of ATM system performance assurance. Applied Human Factors must be coherently applied across a range of ATM projects and issues, and must be an effective partner in the overall ATM development process. Such collaboration will be necessary in order to achieve ATM's capacity and performance goals in the future.

Human Factors in Future ATM – A European Perspective

In the European region, numerous future changes are anticipated, of which the following are notable from the Human Factors standpoint:

- continuing capacity increases (projected doubling of traffic over approximately fifteen years);
- more automation to support the controller;
- the Single European Sky and Functional Airspace Blocks – moving away from traditional national air traffic boundaries;
- limited delegation of aircraft separation control to the cockpit.

Capacity & Automation

The first two aspects are linked. With traffic levels projected to increase significantly over the next ten to fifteen years and beyond, it will become harder for controllers, unaided, to manage traffic safely and efficiently in the future. Additionally, as traffic levels rise, the 'throughput' at airports becomes a critical limiting, or enabling factor. It is essential to be efficient at airports, maximizing runway usage, handling taxiing traffic efficiently and safely, and managing and matching the inflow from arrivals and the outflow from departures.

At a more strategic level, it becomes more important to look ahead further, in order to optimize traffic flow and efficiency. Problems that can occur in a later airspace sector, leading to major deviations for aircraft, and excess workload for controllers, could often be avoided completely by minor changes in an earlier sector, or even a slight change to takeoff time. In effect, for example in Europe, there is currently something of a gap between the Central Flow Management Unit's look-ahead time of 2 hours, and the more usual Air Traffic Control Center's operational planning window of 15 minutes. Ideally there would be tools able to help controllers see into this gap, and make use of such information. This ability to see ahead and predict, and then to collaborate effectively to capitalize on such predictions, is perhaps one of the major keys to future safe and expeditious traffic.

The above reasons have led to the development of a strategic approach to support the role of the controller in handling extra traffic safely. This approach is one of providing extra automation via a set of tools for the controllers in the different 'Gate-to-Gate' phases of air traffic management (i.e. from one airport gate, to departure, to En-route, to approach, to landing and to the arrival gate at another airport). Some of these tools are intended to be used directly by the controller, whereas others will work 'underneath' the interface, although the controller will see their output. However, none of the tools are intended to replace the controller, rather they are there to support the controller, and indeed still require significant controller input, judgment and decision-making.

Near-term Automation & System Performance Objectives

In Europe, a general picture of the types of developments that are likely to happen can be gained from the Eurocontrol ATM 2000+ Strategy (Eurocontrol, 2000). Within this strategy, a number of tools and airspace concepts are proposed, of which some are already developed and being introduced into certain operational air traffic control centers (ATCCs). Some guiding principles for such new automated support tools are as follows:

Safety Safety should not be degraded, and preferably should be improved by the tools. This principle applies obviously to the tools aimed at supporting conflict detection and reduction, but also to the other tools aimed at promoting efficiency. Even these latter tools will be the subject of a safety case to ensure that they are not in any direct or indirect way adding risk to the air traffic system. Human Factors and human error need to be key aspects of such safety cases.

Optimization The tools together should be able to lead to increased system performance in terms of more flights handled per controller, fewer delays, fewer extra 'track miles', etc. The key aspect here is that the tools must work together, via the controllers in different roles and different sectors. A key element here will be training (knowledge and skills) for one's own 'tools' and some (to be determined) level of understanding of the needs of adjacent controllers and their tools. This also requires an element of *Collaborative Decision-Making* (CDM).

Gate-to-Gate Support Following on from the principle of optimization and CDM, the tools should be supporting controllers from departure planning to arrival management, in all types of civil airspace. Since a single controller may have several 'tools' on his or her interface, the integration of such tools into a single usable interface that does not detract from situation awareness nor adversely increase workload will need substantial Human Factors effort.

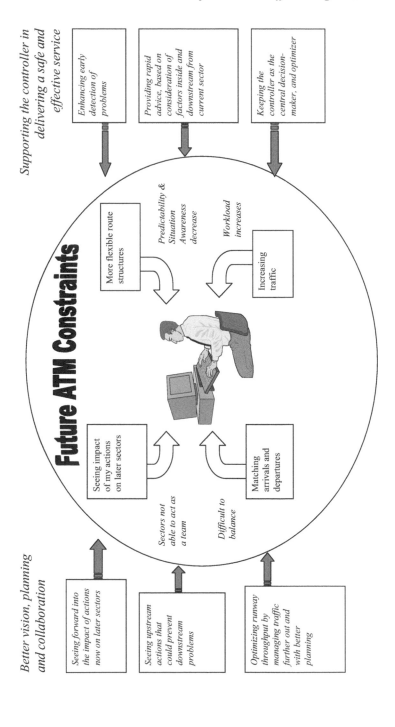

Figure 21.2 Problems in future ATM, and potential 'keys' to solve them

Layered Planning There should be more advanced planning, to anticipate and avoid problems, or at least to deal with the problems at an earlier stage to reduce their impact on the optimization aim. This means more consideration of dynamic traffic flows and airport through-put, and more reliable prediction of potential problems and their resolution. In practice it may mean some new roles or functions for controllers, e.g. a traffic manager role, wherein a controller is able to stand back and predict traffic bunching and determine how to prevent it. Here is an area for new 'allocation of function' and job design work to determine such new roles and their integration with 'normal' air traffic control procedures and teams.

Increased Look-Ahead Time It is essential for optimization and better planning that the look-ahead 'horizon' is increased. This gives more time to act in ways that will lead to real optimization for the whole gate-to-gate cycle, rather than short-term solutions that might favor one sector but penalize a later one. The tools aim to give this increased look-ahead time, via better predictions and thorough but fast analysis of options. In particular the latter will save the controller time. The controller must also be able to see the impact of his or her actions on later downstream sectors, to ensure that problems are not simply being passed 'down the line'. Increased look-ahead time therefore refers both to upstream and downstream vision. There are implications here for the traditional tasks of tactical and planner controller, since potentially more emphasis may move to the planning part of air traffic control. Alternatively, as for layered planning (above), there may be new roles created, along with new displays.

Reducing Complexity and Uncertainty Being able to look ahead is no good unless the picture is clear – otherwise, it may be better to do nothing rather than acting needlessly or actually complicating matters. The tools therefore need to help to reduce the complexity of the situation for the controller, or at least enable the controller to manage the complexity more effectively. Such tools must provide not only more information, but also the fast interpretation of that information and presentation to the controller in useful formats, to allow the controller to make good decisions. The tools also need to predict accurately, and to make the controllers aware of any uncertainty involved, so that they can make a judgment about what to do, and so that they can to a degree place their trust in these tools.

Human Decision-Making The aim is to keep the controller as the decision-maker in the system. The various tools that are being developed will therefore advise the controller, but the controller will decide which option to evaluate, or may implement his or her own solution. It is also important that the controller does not become complacent about the advice of the tools, i.e. always accepting such advice without evaluating it. The goal is to enhance the controller's performance, and indeed the performance of the system. This requires the controller and the tools working together as a team, with the controller clearly leading that 'team'.

Enabling Better Workload Management The aim is to support the controller, particularly when workload increases. The tools will work very fast, so that the advice will be available both early and quickly. The controller will then be able to evaluate rapidly the advice and implement it if it is agreeable, or consider an alternative. Additionally, during the later development phases for these tools, considerable efforts will be undertaken to consider the optimum working arrangements between tactical and planner controllers, for example, as well as potentially new roles for traffic management, or new functions in the tower. This focus on making workload more manageable should enable the controller team to remain proactive, and to forestall overload situations. It should also enable controllers to keep their own skills 'fresh', by not having to rely all the time on the tool set (e.g. for conflict detection and resolution). Implicit here also is that the controller can trust the tools – avoiding the pitfalls of either over-trust (complacency) or under-trust – the former can result in errors, the latter failure to use the tools to their intended advantage.

Maintaining Situation Awareness Because the tools are aimed at supporting certain controller functions, rather than replacing them, the controller should remain aware of the traffic characteristics. The planner will probably have a broader situation awareness than at present, i.e. further upstream and downstream, and certainly a traffic manager will have a very broad situation awareness, if not particularly deep. There should be a net gain in terms of situation awareness with the envisaged tool set, because there will be broader shared awareness across sectors and centers, and also tools for conflict detection will support the early identification of problems. Harnessing this gain in situation awareness to ensure safety and optimization, will however rely on adequate workload management and collaboration between the various controllers and other parties in the ATM system.

Enhancing Collaboration Across Airspace Sectors & Centers One aim for future European ATM is to optimize traffic more widely, both within and across sectors. If optimization in one sector causes significant workload for the next sector, then clearly the net result for improved traffic flows and capacity will be limited. It is better if different sectors and centers can work together as a 'team'. This is only possible if planning can occur across sectors and centers. This is one of the aims of the tool set, allowing multi-sector planning and enhanced arrival management, for example. This aim results effectively in more Collaborative Decision-Making (CDM), and will require and lead to a better 'shared understanding' by controllers of adjacent and up/down stream sectors and centers, and their respective needs and constraints. Another aspect of CDM relates to airport operations, wherein the tools will also lead to more communication and collaboration between controllers and airlines. It is here in particular that optimization relies on the understanding of factors that are highly dynamic, somewhat local, and fluctuate on a daily basis. The tools being developed will enable airlines to make their needs clearer, and have some impact on aircraft arrival and departure sequencing, whilst at the same time making clearer to airlines the operational ATC constraints.

An Automation Toolset for the Controller

The above are the main principles underlying the philosophy for future tools. Some examples of potential tools are as follows:

- DMAN – Departure Manager – optimizes takeoff sequences and runway efficiency, and integrates departures into the overall traffic scheme.

- AMAN – Arrival Manager – organizes the spacing and sequencing of inbound aircraft, optimizing runway efficiency and achieving the most efficient flight path for inbound aircraft, also leading to more accurate Estimated Times of Arrival (ETAs). AMAN will also identify the optimal landing runway for aircraft at multiple runway airports.

- MTCD – Medium Term Conflict Detection – predicts where conflicts will occur, and gives the controller an accurate look-ahead, enabling earlier detection and resolution of conflicts. More generally it aims to reduce the controller workload associated with routine monitoring and conflict detection, and to serve as an enabler for the application of free routes.

- CORA – Conflict Resolution Assistant suggests conflict resolutions, providing en-route Air Traffic Controllers with ranked conflict-free resolution advice. CORA will address both conflicts between aircraft and conflicts between aircraft and restricted airspace, and will give required clearances direct to the tactical (or 'executive') controller.

There are many possible toolset configurations possible, as well as supporting functions such as Data Link. However, the exact timing of the introduction of the tools, and their sequence of implementation, has yet to be determined. Nevertheless, as this book goes to press, for example, some industrialized versions of AMAN and MTCD are already operating in some European airspace sectors, and limited data link is now operating in certain upper sectors.

The future tool set is aimed at supporting controllers throughout the entire gate-to-gate flight process, from preparations for takeoff, to arrival and disembarkation of passengers. It shows that the amount of change anticipated for the controller is significant, though the tools are clearly aimed at supporting rather than replacing the controller.

Single Sky – Airspace Changes

Two additional concepts which enrich further such a future vision are Single Sky and Delegation of Separation Assurance. The Single Sky concept will entail significant collaboration across traditional boundaries, and raises the significant question of how to merge and harmonize working practices in different and adjacent countries in Europe. A major Human Factors challenge is therefore already to carry out some form of task analyses for different centers and nations to have a baseline from which to consider harmonization of working practices. This also raises additional questions about merging safety cultures, and human error and recovery impacts of such airspace changes, as well as more basic (though no less easily answered) questions about workload, task-sharing, teamwork, and situation awareness. However, in this area, the dominant challenge of Single Sky for Human Factors is likely to be the cross-cultural dimension.

Delegation to the Cockpit

Delegation of separation assurance to the cockpit is not a new idea (it is embodied in Free Flight and Direct Route concepts), but it remains an issue because it cuts to the very core of what it means to be a controller. Some Human Factors-related work at the moment in this area is focusing on interface design issues, and running simulations to further explore the concept. It is likely in the future however that more social issues will need to be explored, such as the impact of the concept on controller motivation, situation awareness, and the general ability of controller and pilots to work effectively as a team with respect to delegation, etc. There is also a challenge to ensure that overall system situation awareness is not lost or 'compartmentalized' between the ground and air components in such scenarios, which could result in confusion about 'who is in charge'. Such questions highlight the need for Human Factors and Safety to work closely together when considering delegation to the cockpit.

Human Factors Needs for Future European ATM Concepts

The European ATM challenges have led to the need being identified for certain new approaches in Human Factors in the near term. For example, Kirwan and Rothaug (2001) identified the need for better methods and guidance in seven areas, and a preliminary toolkit (called SHAPE) to address these areas has now been developed (Sträter et al., 2003; 2004):

- situation awareness: maintaining adequate situation awareness for controllers and pilots;
- mental workload: keeping workload within a good performance envelope in all scenarios;
- trust: avoiding over or under-trust in automation and other ATM team members (including pilots);
- new error forms with automation: predicting new error forms that may arise with new technology before they occur as incidents or accidents;
- skill-set changes: learning new skills; not losing key skills;

- recovery from automation failure or degradation: being able to detect false information and recover from system failure, and avoiding the ATM system becoming 'brittle' (prone to sudden catastrophic failure);
- teams and automation: maintaining and enhancing good team-working practices and cultures, including controllers in other parts of the gate-to-gate ATM 'cycle', controller assistants, pilots and technical support.

To this list can be added the need for better understanding of cross-cultural Human Factors, and the impact of core job changes on performance, and a more general need to bring Human Factors and Safety closer together. This latter aspect is happening via an approach recently developed by Eurocontrol (Woldring, 2003) called the Human Factors Case. The Human Factors Case is similar to a Safety Case, and is a detailed investigation of a new concept or development according to Human Factors criteria and considerations. It results in an overall assessment of the adequacy of the system from a Human Factors viewpoint, and detailed recommendations of 'Human Factors Requirements' that need to be embodied by the developing system.

Perhaps the most fundamental challenge in European ATM Human Factors, however, will be to establish adequate means of collaboration that will support a more programmatic approach at a Pan-European level. As European ATM becomes more integrated, this would be advisable to ensure total system coherence and inter-operability. Such a challenge is of course not merely a technical one, but one that demands of Human Factors section leaders and practitioners alike, skills as negotiators, managers and persuaders.

Human Factors in Future ATM – a US Perspective

The above has raised just some of the issues facing Human Factors in its aim to support future European ATM. At this point, it is instructive to review a US vision of the future of Human Factors in ATM. The US operates effectively as one very large, complex and important Air Navigation Service Provider. Given the comparative 'cohesion' of the US National Airspace System (NAS) and its underlying technical and organizational infrastructure (due to the US being one country with one legislation, rather than many that must be aligned), compared to multi-State Europe, the US may take a longer and perhaps deeper view of the future, as outlined below.

Many system changes are proposed or are in development for the next decade in the US National Airspace System. These systems have temporal relationships that present 'Windows of Transformation' (Figure 21.3) that are open relative to information available to operate and maintain a safe, effective NAS (e.g. KSAs of the workforce, procedures, airspace design). The ability to capitalize on the convergence of these windows of opportunity requires less effort across fewer Human Factors application areas the more similar the integration requirements are to the 'legacy' systems. The breadth of study areas that must be addressed and the depth of the activities across those study areas are largely determined by the extent of the change in the system. Therefore, the study areas presented earlier not only represent the study areas necessary to ensure human system integration for the NAS system envisioned for 2015, they also represent the study areas that will be required to develop new concepts associated with the 'System After Next' (SAN).

While these systems allow for a fairly good definition of the Human Factors research and engineering activities necessary to support their integration into the NAS, for systems associated with the longer term, that is, the SAN, it is much more difficult to understand the specific research requirements that support development. That is, the SAN is defined not only by the information available to implement new capabilities, the timing of the opportunities for change, and the infrastructure of the previous system's technology, personnel, and facilities, but also the extent to which changes in the concept of operation are undertaken. If there are changes in the concept of operation that affect roles and responsibilities of the participants in the system, then the research and engineering requirements increase considerably. Given the lack of specificity of the requirements that future concepts may present (e.g. technologies that will be used, training required, selection criteria, and required procedures), it is much more difficult to evaluate the human performance considerations of various concept alternatives since the research must be based on assumptions of the performance characteristics of system elements and their integration.

The Human Factors prisms in Figures 21.4 and 21.6 are an abstraction of underlying policy, process, and best practices that overtly highlights the connectedness of how enablers and drivers of ATM system modernization, as vectors of transformation, are systematically assessed using the full range of HF methods (i.e., the Human Factors Prism) to identify and mitigate human performance issues. Certainly HF research is just one of several research areas associated with these transformations; other examples include design requirements, communications, training, and transition. The vectors of transformation represent new concepts and technologies improving effectiveness and efficiency in the delivery of ATC services. As a starting point for HF engagement, human-centered design intends to integrate humans with changes in machines and procedures by taking advantage of the strengths and accommodating the weaknesses of the various system elements. This engagement consists of applications of scientific HF research methods that continue through the implementation process. These methods are selected, scaled, and used in an integrated manner to ensure HF solutions for issues that might otherwise constrain benefits intended from near- and long-term transitions of new concepts and technologies.

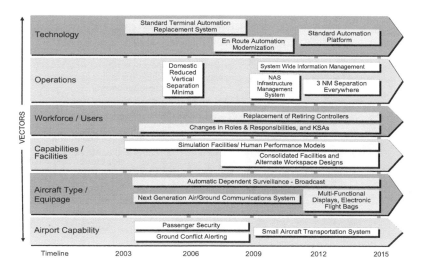

Figure 21.3   Near-term window of transformation in the NAS

Probable Trends

It is always difficult to develop a crystal ball view of things to come, and given the diversity of participants, competing objectives, and the complexity of ATM operations, the future concepts of operation and their associated system designs will be no less difficult to predict. However, both the United States and Europe recognize the need to progress the ATM system to meet the needs of the future and to ensure harmonization of system operations where possible and inter-operability of system operations at a minimum. The FAA report, *FAA Aerospace Forecasts Fiscal Years 2004-2015*, provides the following view of activity over the 2004-2015 period:

> Activity levels at FAA En-route traffic control centers remained basically flat at 43.7 million in 2003, with a 0.8 per cent increase in commercial activity counteracting a 2.0 per cent decline in noncommercial activity. The number of aircraft handled at En-route centers is forecast to increase by 3.2 per cent in 2004 and 3.6 per cent in 2005, largely the result of increases of 4.7 and 4.5 per cent in commercial activity. En-route activity increases by 2.2 per cent annually over the rest of the forecast period, reaching a total of 58.4 million aircraft handled in 2015. Activity at FAA En-route centers is expected to recover to pre-September 11th levels in 2005.

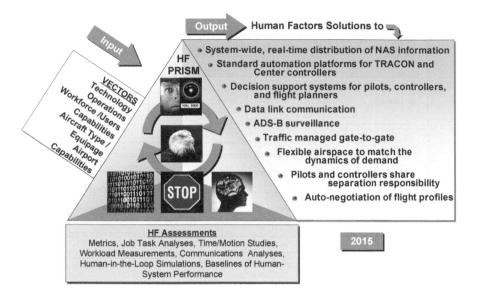

Figure 21.4   Human Factors and the Near-Term NAS

Figure 21.5 presents the forecast increase in US activity at FAA En-route traffic control centers. Demand projections of between 2 and 3 times the current traffic load have been made for the future. These data suggest that continued division of airspace into smaller and smaller units will no longer provide a means to address future demand. Alternative concepts must be developed, and there must be study of the Human Factors considerations and research requirements necessary to understand and address how these concepts can become part of the operational ATM system. For instance, under appropriate conditions it might be possible to allow pilots to provide separation services between limited sets of aircraft (RTCA, 1995). The conditions that allow for this type of transfer of responsibility and the mechanisms through which separation assurance by the service provider is ensured must be fully understood in order for progress to be made in this area. Non-normal operations, failure modes, and a mixed equipage environment present significant challenges to the Human Factors research community relative to defining a concept of operation that assures human system integration to a level that the concept can be safely implemented. Airspace design concepts, facility co-ordination/ communication concepts, air/ground communication integration/inter-operability, avionics/ATM decision support system inter-operability/integration, job redesign and procedures all raise Human Factors considerations that must be addressed to ensure viable system concepts that allow future demand projections to be met.

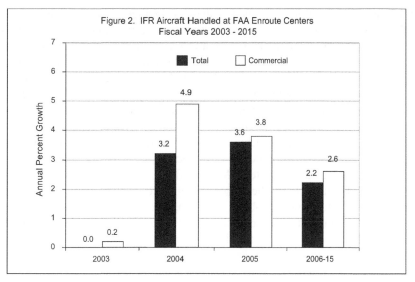

Figure 21.5 **Annual growth rate of IFR aircraft handled at FAA Enroute Centers 2003-2015**

Given that the future will no doubt present a more complex and dynamic ATM system than we have today, it is likely that more automation and decision support systems will be needed to accommodate advanced system concepts enabling these operations.

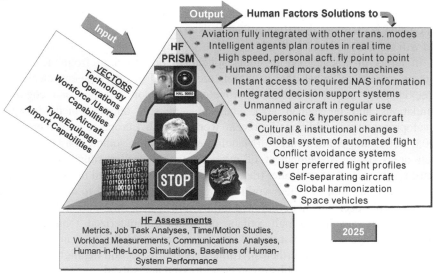

Figure 21.6 **Human Factors and the Long-Term NAS**

Many of these advanced systems require the development of new technologies or the refinement of existing technologies. Without the application of Human Factors science to assist in the identification of minimum performance standards for new technologies, it will be difficult to implement substantive cost-effective changes to the ATM system. Human Factors researchers and engineering practitioners can be of great assistance in establishing the minimum required performance of new technologies to ensure they support human system integration rather than requiring human accommodation of shortcomings in technology. Not often enough is it the case that we end up with higher technology performance (and the associated higher cost) than is needed to support the specific application for which a given technology is being implemented. More often shortcomings in technology performance will either limit the implementation of a new capability, require additional human effort (e.g. task-load, training), or end as failed implementations. Together with an integrated approach toward Human Factors engineering in our product development activities with the upfront work that supports technology definition, Human Factors will assist in defining, developing, and deploying the ATM systems of the future.

Concluding Comments

This book has produced a number of chapters showing firm evidence that Human Factors can help solve and anticipate real problems in ATM. Furthermore, the chapters themselves cover a range of application areas and lifecycle stages, and originate in a wide diversity of air navigation service providers. The overall conclusion is therefore that Human Factors is critical to the advancement of ATM, and will continue to be so in the future.

In terms of whether certain tools appear to be 'missing' from the Human Factors' repertoire, or whether new tools are in need of development, clearly the three chapters on methodologies, particularly those on fast-time and cognitive simulations and on human error analysis, are still ongoing methodological development areas. Additionally, recent work in both Europe (e.g. Sträter et al., 2003; 2004) and North America (Manning & Stein, this volume) has focused on such areas as the development of more robust controller situation awareness measures, measures of controller trust in automation support, and measures of team performance, as well as better workload measures, and a need to address human recovery from decision support system failure and system information degradation, as controllers are increasingly reliant on more computerized tools and data processing. Psycho-social issues such as stress and controller trust in automation also deserve attention. Nevertheless, Table 21.1 in this section highlights that even with today's tools, many problems can be addressed and solved.

However, if Human Factors is to take its proper place as a partner in developing future ATM, the main area for development is in the programmatic area, i.e.

organizing a coherent approach for a range of inter-connected Human Factors activities. Human Factors needs to be more strategic in its approach. The two outlined visions of the future of Human Factors in ATM, one based in Europe and one in the US, both underscore the need for more collaboration. This enhanced collaboration must take place firstly amongst Human Factors practitioners, researchers, and managers themselves, to ensure a strong, relevant and achievable vision of Human Factors within an organization, and secondly between these Human Factors people and other key ATM system development and performance assurance managers, to assure the integration of this vision into ATM system development and operational management. Human Factors in ATM, having proved its worth in the studies cited here and many others around the globe, must take up the challenge to be a main player in ATM development. As with all challenges in the real world, this will fall on the Human Factors people themselves working in research, application and management of Human Factors in ATM.

In conclusion therefore, the application of Human Factors is crucial to solving real systems problems, and developing and implementing successful, safe and operable ATM systems. It could clearly do more, via broader application to ATM system functions (e.g. airports), increased application in different countries, and earlier application in developing better and safer future systems. After all, ATM is a truly global system, one designed and run by humans, and mainly carrying humans themselves as its cargo. A problem or weakness in one area of ATM, or one location, is actually a problem for the whole of ATM. Therefore ATM, for the foreseeable future, needs Human Factors.

In a decade's time, it is hoped that Human Factors will be even more widespread and embedded into ATM system development and operations, and will be seen no longer merely as an adjunct or support function, or purely a research area, or as being there for specific localized issues, but rather as a key and pervasive member of the system development and operational team. The chapters in this book have shown that Human Factors does indeed have this potential. It is hoped that this book, in some small way, may help Human Factors, ATM System Developers and Service Providers, to realize this important role and, in so doing, will help lead to safer, more efficient, and more pleasant air transport for all.

References

Federal Aviation Administration (2004), *FAA Aerospace Forecasts Fiscal Years 2004 – 2015*. Washington, DC: U.S. Department of Transportation.

Joint Planning and Development Office (2004), *An Integrated Plan for the Next Generation Air Transportation System*, First Edition (Draft), Washington, DC.

Kirwan, B. (2002), Developing Human-Informed Automation in Air Traffic Management. *Human Factors & Aerospace Safety*, 2, 2, pp. 105-146.

Kirwan, B. and Rothaug, J. (2001), Finding ways to fit the automation to the air traffic controller, *Ergonomics Annual Conference*, Cirencester, April.

Krois P., Mogford R., & Rehmann, J. (2003), Integrating Human Factors Issues in Technology Readiness Levels for Air Traffic Control Research, In *Proceedings of the Human Factors and Ergonomics Society 47th Annual Meeting*, Santa Monica, CA: Human Factors and Ergonomics Society.

RTCA, (1995), *Final Report of RTCA Task Force 3: Free Flight Implementation*, RTCA Inc., Washington, DC, USA.

Schäfer, D., Woldring, M. & Kirwan, B. (2003), *Human Factors for Project Leaders*, HF Training Syllabus, EUROCONTROL.

Sträter, O., Voller, L., Low, I. & Shorrock, S. (2003), Solutions for Human Automation Partnership in European Air Traffic Management. *International Ergonomics Association 2003*, Korea.

Sträter, O., Voller, L. & Low, I. (2004), Towards an integrated Human Automation Management, *Human Computer Interaction - HCI Aero*, Toulouse, France.

Index